ELEPHANTS AND ETHICS

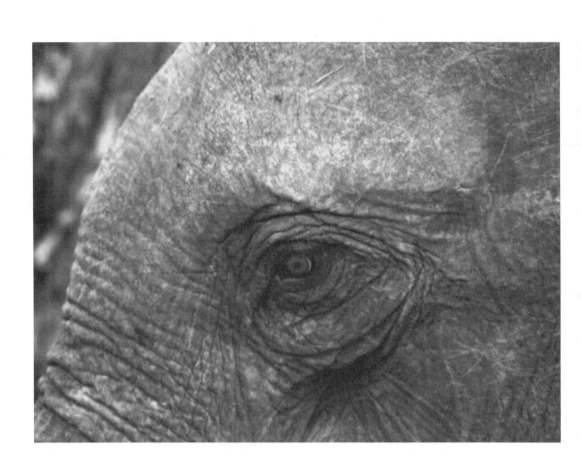

Edited by
CHRISTEN WEMMER AND CATHERINE A. CHRISTEN

ELEPHANTS
AND ETHICS
TOWARD A MORALITY OF COEXISTENCE
Foreword by John Seidensticker

THE JOHNS HOPKINS UNIVERSITY PRESS
BALTIMORE

© 2008 The Johns Hopkins University Press
All rights reserved. Published 2008
Printed in the United States of America on
acid-free paper
1 2 3 4 5 6 7 8 9

The Johns Hopkins University Press
2715 North Charles Street
Baltimore, Maryland 21218-4363
www.press.jhu.edu

Library of Congress
Cataloging-in-Publication Data
Elephants and ethics: toward a morality of
coexistence /
edited by Christen Wemmer and Catherine A.
Christen ; foreword by John Seidensticker.
p. cm.
Includes bibliographical references and index.
ISBN-13: 978-0-8018-8818-2 (hardcover :
alk. paper)
ISBN-10: 0-8018-8818-2 (hardcover : alk. paper)
1. Elephants—Effect of human beings on—Moral
and ethical aspects. 2. Animal welfare—Moral
and ethical aspects. I. Wemmer, Christen M.
II. Christen, Catherine A. (Catherine Ann)
HV4747.N48 2008
179'.3—dc22 2007036045

A catalog record for this book is available from the
British Library.

Frontispiece: Prithiviraj Fernando
Photograph on page v: Christen Wemmer

Chapter Opening Art Credits
1: Jordana Meyer
2: Smithsonian Institution Archives,
SIA2008-0924
3: Donna Nissani
4: Petter Granli / ElephantVoices.org
5: Courtesy John Edwards, London
6: Jessie Cohen, Smithsonian's National Zoo
7: Janine L. Brown
8: Disney's Animal Kingdom
9: Mehgan Murphy, Smithsonian's National Zoo
10: Smithsonian Institution Archives,
SIA2008-0922
11: Feld Entertainment, Inc.
12: Smithsonian Institution Archives,
SIA2008-0926
13: Disney's Animal Kingdom
14: Jessie Cohen, Smithsonian's National Zoo
15: Disney's Animal Kingdom
16: Prithiviraj Fernando
17: Petter Granli / ElephantVoices.org
18: Hank Hammatt
19: Petter Granli / ElephantVoices.org
20: Jordana Meyer
21: Stephen Blake, Wildlife Conservation Society
22: Joyce Poole / ElephantVoices.org

*Special discounts are available for bulk purchases of
this book. For more information, please contact
Special Sales at 410-516-6936 or specialsales@
press.jhu.edu.*

The Johns Hopkins University Press uses environmentally friendly book materials, including recycled text paper that is composed of at least 30
percent post-consumer waste, whenever possible.
All of our book papers are acid-free, and our jack-
ets and covers are printed on paper with recycled
content.

V. Krishnamurthy
(1927–2002)

This volume is dedicated to the
memory of Dr. V. Krishnamurthy,
a remarkable Forest Veterinary Surgeon
of India's Tamil Nadu Forest Department.
Doc devoted his humble life to the care
and welfare of Asian elephants.
He enriched his colleagues by sharing
his peaceful wisdom and his penetrating
understanding of elephants.
His unwavering ethical convictions
are a legacy of hope for those
who carry the torch.

CONTENTS

FOREWORD

We have been defining our relationships with the elephants for as long as we have been people. When discussing the ethics of human-elephant relationships, we should keep in mind a historical reality: In any confrontation, elephants almost always lose. For example, in *The Retreat of the Elephants*, environmental historian Mark Elvin (2004) describes the loss of Asian elephants from China. The history of south China is a tale of the replacement of indigenous peoples engaged in hunting and gathering and shifting cultivation—land uses that favor elephants—by Han Chinese engaged in sedentary wet-rice farming. Four thousand years ago, there were elephants in China's northeast, in what was to become Beijing. A cooling climate made the northeast unsuitable for elephants. For the rest of China south of the Yellow River and east of the escarpments of the Tibetan Plateau, Elvin writes, "the pattern of withdrawal [of elephants] in time and space was, so to speak, the reverse image of the expansion and intensification of Chinese settlement. Chinese farmers and elephants do not mix" (9). Elvin describes what he calls "The Three Thousand Year War" (9) between humans and elephants, in which elephants are killed as crop thieves, captured for work and ceremony, and hunted for ivory and for their trunks, the last considered a gourmet food item. The steady replacement of low-elevation forest with farms has pushed China's elephants right up to the border between Yunnan and Myanmar, now the only place they are found in China.

This tale of elephants and Chinese rice farmers—the former retreating as the latter expanded their dominance over the landscape—recounts just one round in the wars between humans and elephants. Todd Surovell, Nicole Waguespack, and P. Jeffrey Brantingham, in a 2005 paper in the *Proceedings of the National Academy of Sciences* on the global archaeological evidence for proboscidean overkill, conclude: "Spanning ~1.8 million years, the archaeological record of human subsistence exploitation of proboscideans is preferentially located on the edges of the human geographic range. . . . In the present and the past, proboscideans have survived in refugia that are largely inaccessible to human populations" (6231).

Today, there are no more edges of the human geographic range. Satellite images inform us of the heavy and extensive human footprint (Sanderson et al. 2002) on Earth. The landscapes where African and Asian elephants live today are human dominated. Almost no human-free refugia remain, although we like to imagine there must be such places. Given that about one-third of all Asian elephants are now maintained in captivity, it appears that elephants are sliding into domesticity, like camels, which may be the only way they will survive among humans in Asia. In Africa, which lacks a widespread tradition of keeping trained elephants, elephants disappear when human densities and land transformation patterns reach the breaking point in the different ecosystems, as described by Richard Hoare and Johan du Toit (1999).

In March 2003, elephant experts from various disciplines met at the Smithsonian's National Zoological Park Conservation and Research Center in the shadow of this long human versus elephant war to begin the dialogue that has resulted in this volume. There was great unease throughout as we struggled to come to grips with the past and move to a new future in elephant-human relationships. Elephant ecologists, elephant ethologists, zoo-based conservation biologists, elephant conservation practitioners, veterinarians, elephant reproduction scientists, philosophers, elephant rights activists, elephant welfare proponents, and elephant historians all struggled to see the whole elephant while being restrained by having our hands on the proverbial different parts of the elephant.

Philosophers have discovered the remarkable complexity of the lives of elephants and discuss their ability to have conscious experiences. Some see the elephant's demise, as people continue to kill them—as they always have—for their meat and ivory and to save biodiversity because elephants' presence changes ecosystems as their populations grow in restricted areas. If conditions do not improve in most areas of Asia and Africa, however, we know that wild elephants are in the winter of their existence, the human footprint taking its final toll. In addition, we fear that we are nearing the end game for elephants living and working in zoos and circuses.

Elephants have been good for the bottom line; they have been an engine that powers attendance at zoos and circuses. However, some zoos have opted out; they no longer maintain and exhibit elephants. They see the continued exhibition of elephants as a Faustian equation and are wary that keeping elephants in the future is too financially, technically, and ethically challenging. Other zoos see elephants as an opportunity and a means to promote a new dialogue and to build a constituency that can lead to some recovery of wild elephants on their home ground.

The eminent zoo planner Jon Coe has noted that zoos and circuses have had an operational policy based on animal containment and coercion, which is what critics rile against. Many chapters in this volume seek to overcome this history by describing improved concepts in elephant keeping and training based on a new understanding of how wild elephants live and what they need to be healthy and happy. Coe believes a new zoo model (Coe and Mendez 2005) is needed, "the unzoo," which is based on the principles of animal attraction and reward. Elephant sanctuary proponents think they offer the unzoo setting where elephants can live in dignified freedom. However, for elephants living in all North American facilities, the demographic reality for zookeepers is that we are in the hospice stage of the elephant end game.

Because elephants have always lost in their confrontations with people, we have to ask: What is different now that enables us to redefine our future relationship with elephants? I am hopeful because we are learning to sit together and explain our views on elephants to one another. Perhaps, we will begin to see and value the whole elephant and find ways for people to fit into this world with our fellow species intact. This volume and the symposium it derives from represent some first steps in that direction.

John Seidensticker
Smithsonian's National Zoological Park
Washington, DC

References

Coe, J. C., and Mendez, R. A. 2005. The Unzoo alternative. *Proceedings of the 2005 Australasian Regional Association of Zoological Parks and Aquariums (ARAZPA) Conference.* CD format. Sydney, Australia: ARAZPA.

Elvin, M. 2004. *The retreat of the elephants: An environmental history of China.* New Haven, CT: Yale University Press.

Hoare, R., and du Toit, J. T. 1999. Coexistence between people and elephants in African savannas. *Conservation Biology* 13: 633–639.

Sanderson, E. W., Jaiteh, M., Levy, M. A., Redford, K. H., Wannebo, A., and Woolmer, G. 2002. The human footprint and the last of the wild. *BioScience* 52: 891–904.

Surovell, T., Waguespack, N., and Brantingham, P. J. 2005. Global archaeological evidence for proboscidean overkill. *Proceedings of the National Academy of Sciences* 102: 6231–6236.

PREFACE

For several thousand years, human life and human history have been intertwined with elephants. People have mostly benefited from the world's largest land mammals—as religious symbols, as icons of sagacity and objects of entertainment, as beasts of burden during war and peace. However, interactions between humans and elephants have proved largely disadvantageous to the latter, both as individuals and as species. Especially given the modern pressures of expanding human development, wild elephant populations in Asia and Africa have been declining, and, if this trend continues, elephants could disappear from Earth forever. How must human-elephant relations be adjusted to benefit both? How can an ethical balance—by definition fair to both sides—be achieved?

Human-elephant relationships are complex, even more so as they are enacted in so many circumstances and settings. Wide disparities mark people's viewpoints on how these relationships might change to the best mutual advantage. This collection of essays identifies and elucidates contemporary issues in human-elephant associations, both in the wild and in captive settings, and offers an initial attempt to delineate practicable ethical standards for treatment of wild and captive elephants. No matter how controversial the discussion sometimes becomes, some common ground is apparent from the outset. First, all parties to this conversation want to prevent elephant extinction and suffering, and, second, developing and implementing these practicable ethical standards will be essential to that cause.

These chapters concentrate on the central issues and ethical concerns associated with local, institutional, national, and international policies and practices affecting populations of the three identified elephant species, the Asian elephant, *Elephas maximus*; the African savanna elephant, *Loxodonta africana*; and the recently identified African forest elephant, *Loxodonta cyclotis*. In 2003, we assembled a symposium entitled "Never Forgetting: Elephants and Ethics," held at the Conservation and Research Center in Front Royal, Virginia, the research facility of the Smithsonian's National Zoological Park, to address these issues. Our invited paper presenters included an assortment of specialists from Asia, Africa, Europe, and the United States; their diversity reflects the complex and intertwined nature of people-elephant relationships. Their areas of expertise advisedly encompass many disciplines and pursuits in the social and natural sciences, including philosophy, history, anthropology, political science, conservation and wildlife biology, wildlife management, veterinary science, and animal welfare activism. Most of the essays in this book were first presented at the 2003 symposium, but in a few cases, our original invited speakers were unable to participate; happily, we were successful in recruiting highly qualified specialists to cover needed topics in this volume.

Depending on the requisites of their topics, the authors in this volume have variously examined the ethical dimensions of people-elephant interactions by regarding elephants not only as a species but also as individual beings. We have considered elephant issues and associated ethical questions in diverse arenas, from the welfare of elephants in zoos and circuses, to their management in protected areas, to their exploitation for market-driven commodities such as ivory.

Though some strong points of agreement emerge from these chapters as they did during the symposium's workshop discussions—which were, indeed, deliberately aimed at developing areas of agreement and synthesis—strongly conflicting views entangle our collective understanding of people-elephant relationships and will continue to do so for some time to come. The purpose of this volume is not to obscure these controversies because it is through reflection and discussion that ethical standards evolve in human societies. Rather, this book allows the open-minded and yet critical reader to contemplate the ethical issues affecting both people and elephants from vastly different points of view. We hope that the opportunity these essays provide for comprehensive contemplation of this sort will help build a wider tolerance for considering the strong points of many different approaches to saving these species. In the end, this salvation is assuredly the ultimate goal of all the participants in this project.

Many individuals and organizations contributed to the success of the 2003 "Never Forgetting: Elephants and Ethics" symposium and to the preparation of this volume. We thank them for their generosity and commend them for their openness to unfettered discussion and the airing of disparate views. The symposium was part of a larger Association of Zoos and Aquariums (AZA) initiative to stimulate awareness and dialogue about ethical dimensions of human-wildlife relations. Michael Hutchins was instrumental in that initiative and in this symposium's realization, as were Beth Stevens and John Lehnhardt of Disney's Animal Kingdom (DAK). We thank each of them for their support. The symposium was hosted by the Smithsonian's National Zoological Park at the Conservation and Research Center in Front Royal, Virginia, in affiliation with AZA and DAK. Both the symposium and volume preparation were generously funded by Disney's Animal Kingdom. We were fortunate to have the services of skilled mediators and facilitators Lorig Charkoudian and Lisa Bleich to guide the discussions during the symposium's workshop sessions. Their professionalism received unstinting praise from all participants.

Many Smithsonian staff, students, and interns helped greatly in symposium preparations and during the three-day event. We are very grateful to intern Alison Laborderie, who hit the ground running and never stopped, before, during, and for some months after the symposium. Our thanks too to Jocelyn Ziemian, whose services as symposium scribe helped us capture the symposium discussion, and to Elizabeth Freeman for her dedicated volunteer work. Laura Walker, as always, proved invaluable. We also thank accountants Becky Myers and Claudia Luce.

Thanks to all who donated or otherwise helped us acquire chapter opener photographs, including Ellen Alers, Joseph Barber, Tamara Bettinger, Jed Bird, Steve Blake, Janine Brown, Jessie Cohen, Sue Dubois, Pruthu Fernando, Elizabeth Freeman, Petter Granli, Hank Hammatt, Pamela Henson, Mike Kreger, John Lehnhardt, Khyne U Mar, Amy McWethy, Jill Mellen, Jordana Meyer, Mehgan Murphy, Donna Nissani, Moti Nissani, Jenny Pastorini, Tammy Peters, Joyce Poole, Nigel Rothfels, Micaela Szykman Gunther, Dennis Schmitt, and Gary Varner.

We greatly appreciate the efforts of all the peer reviewers who commented on individual chapters and the two who reviewed the entire manuscript. Our thanks go to Vincent Burke, acquisitions editor at the Johns Hopkins University Press, for his steadfast interest in this project. Senior Production Editor Andre Barnett, Acquisitions Assistant Bethany Ross, and Art Director Martha Sewall guided us through the final editing process with efficiency and good cheer. Peter Leimgruber, of the National Zoological Park, has generously shared his knowledge of elephants, insights about wildlife conservation, and editing skills throughout the trajectory from symposium to volume, and we owe him a great debt.

ELEPHANTS AND ETHICS

1 INTRODUCTION

NEVER FORGETTING THE IMPORTANCE OF ETHICAL TREATMENT OF ELEPHANTS

CHRISTEN WEMMER
AND CATHERINE A. CHRISTEN

And so these men of Indostan
Disputed loud and long,
Each in his own opinion
Exceeding stiff and strong,
Though each was partly in the right,
And all were in the wrong!

MORAL.

So oft in theologic wars,
The disputants, I ween,
Rail on in utter ignorance
Of what each other mean,
And prate about an Elephant
Not one of them has seen!

(Saxe, *"The Blind Men and the Elephant,"* stanzas 8–9)

American poet John Godfrey Saxe's oft-quoted retelling of the ancient parable of the blind men and the elephant reminded his nineteenth-century readers, as it reminds us today, how easily our sensory perceptions and our preconceptions can lead to misinterpretations about issues spiritual and worldly. In nine stanzas, the fable's six blind men touch different parts of the same elephant, each authoritatively declaring it to be very like six very different things—a wall, a spear, a snake, a tree, a fan, and a rope. Fittingly, this tale reflects today's differing perspectives and concerns on ethical issues regarding elephants themselves. Though the lives of people and elephants have been intertwined for millennia, there were always places where people and elephants existed in their separate spheres. Now, the situation is different. The fate of elephants anywhere on the planet depends on human decisions and human actions.

People feel strongly about elephants. Their moral views are often motivated by the characteristics of this animal, a "being of the first distinction" (Buffon, quoted in Rothfels, Chapter 5, 107). Most people want elephants to have a natural place to live on Earth, free of human disturbance. This was once possible, but no longer. Human activity and development are almost everywhere, and any concept of undisturbed nature is illusory. Ethical management is fundamental to the welfare of both wild and captive elephants.

Thousands of elephants do not live in freedom; they are obligatory human dependents living in zoos, circuses, and elephant camps. To some people, captivity is a horrible fate for elephants and categorically unethical. Nevertheless, many believe that the suitability and the morality of captivity depend on the circumstances and conditions. Some hold that zoo and circus elephants serve as high-profile ambassadors, educating captivated audiences about elephant conservation, and reminding them of their interest in preserving the species in the wild. There is also the issue of traditional use of domesticated Asian elephants as religious and working animals in some countries, where up to 16,000 serve as temple and work animals (Sukumar 2006). Many would argue this people-elephant culture is worthy of preservation and moral consideration, but those who find redeeming value in some circumstances of elephant captivity usually scorn the living conditions of many other captive elephants. Few find the range of current captive settings acceptable because the conditions often do not provide basic requirements for the well-being of individual elephants nor for the long-term survival of populations.

The participants in this 2003 symposium and subsequent volume have agreed emphatically that elephants have intrinsic value in and of themselves and as members of natural communities. Yet, we have disagreed about many of the moral and managerial issues associated with human actions affecting elephants. We have disputed the elephant's sentient features. We have debated the uniqueness of elephants' social and mental powers and argued whether any of their qualities entitle them to special ethical considerations. Not surprisingly, even when considering the same scientific research findings, we have at times reached different conclusions (compare, for example, the nuanced differences between Varner's comments (Chapter 3) and those of Poole and Moss (Chapter 4) on research findings about theory of mind in elephants). We are like the blind men of the poem, caught up in our own microcosms, "though each [is] partly in the right."

The moral issues regarding elephants in the twenty-first century are complex. They concern elephants as individuals, as populations, as members of natural communities, and as species. Like many wildlife species, elephants exist in a range of settings and circumstances. Individual animals, populations, species, and ecosystems represent different levels of complexity that raise differing ethical concerns. Virtually all elephant populations are influenced by human behavior, or are at least subject to human management, that is, human decisions that bear directly on their welfare and survival. The moral conundrums relating to elephants are deeply embedded in social, economic, and environmental issues with different national, regional, and local contexts. If the problems regarding ethical treatment of elephants were restricted to one region or country, solutions might prove easier to find. Indeed, these problems often appear intransigent, in part owing to the hard stances of protagonists who operate mainly within their own spheres of influence and pursue solutions that would give one side all the winnings. The people who manage elephants in different settings tend to come to their posts already ascribing to different moral philosophies. For example, one does not generally find advocates of animal rights managing protected areas where elephant control is a job requirement.

A book of essays cannot solve all these problems, but it can promote mutual understanding among people holding widely differing views. By making available to a larger audience all the information and perspectives we delved into at the 2003 symposium, we trust we are taking an important step toward finding solutions to elephant problems in all contexts. In this chapter, we have drawn on the symposium exchanges and the collected writings of this volume to generate our own reflections about personal contexts and different ethical perspectives on elephants. We also examine the different circumstances in which elephants live in light of predominating moral considerations. The final section of this introductory chapter is the result of the symposium participants' formulation of major points of consensus during the March 2003 meeting.

Relationships Define Perspectives

Ethicists, philosophers, and biologists have used several criteria to rationalize humane care and to condemn killing or adverse treatment of elephants. These criteria include the ability to feel pain and suffer, and the mental resemblance of elephants to humans, principally in terms of intelligence, sense of self, and social complexity. Those concerned with the ethical treatment of elephants do not share a single view regarding these characteristics. Understanding their differences helps to explain and potentially to mitigate areas of conflict. For us, this symposium brought into strong relief three predominating value systems.

People who have personal relationships with pets or wildlife acquire intimate knowledge of individual animals. Owners and pets develop mutual bonds or at least interdependencies. Field biologists may also develop attachments to their animal subjects. Both groups develop special insights into the psychology and *Umwelt*, or personal universe, of their pets and subjects. Their "humanitarian" perspective is embodied in deep concern for the welfare of individual animals and of animals as individuals. They understand that discomfort, pain, suffering, and death are natural experiences, but they are vehemently opposed to situations in which they believe humans are wrongfully inflicting suffering and death on animals. All of this can apply to elephants.

Land and natural resource managers often have a different perspective.

Their work is concerned with elephants and other species, including flora, as components of ecosystems. One of their priorities is to ensure that the biotic community, usually a protected area such as a national park, maintains its ecological composition within some defined limit. Ecologists accept that natural environments are not static but constantly change and that the same is true for natural populations of many species, including elephants. Nevertheless, natural resource managers often attempt to manage landscapes as though they are unchanging and as though there is an ideal set of conditions that characterizes the area of concern. This worldview can be seen as a predominantly scientific or environmental perspective, about the maintenance of ecosystem functions. The fundamental unit for management is, then, not the individual species (e.g., the elephant) but the entire ecosystem with all its populations of different animals and plants and their interactions. The individual welfare of animals is not unimportant, but individuals are relatively less important than populations and communities. Their notion of elephant management does not forbid killing, which under certain conditions is justified as necessary for the welfare of elephant populations and the integrity of the ecosystem.

Rural people living in close association with elephants often have yet another perspective. Their subsistence lifestyles lead to competition with elephants for vital natural resources. More affluent people can afford to consider punishing and killing elephants to be anathema, believing more humane solutions than shooting can be found. The rural villager may also know about these alternative methods but usually lacks recourse to anything but the most basic and expeditious solution. All people take drastic measures when their own welfare and survival are threatened. Therefore, usually after many attempts to dissuade elephants from their fields, villagers may determine to abandon ethical considerations and kill the elephants.

Moral Tolerance and Adaptability

Personal experience shapes moral beliefs. Unwillingness to understand or accept different ethical systems has often hindered progress towards solving problems associated with elephant welfare. Yet, in practice, it is impossible for anyone to function in isolation from these other views, and most people are limited in their ability to control the different circumstances that affect the fates of elephants as individuals, populations, or species. However, even ardent advocates will sometimes make allowances when confronted with unfamiliar contexts in which their normal expectations and predictions are unrealistic. For example, few humanitarians would find it ethical to thrust blazing torches into the faces of zoo elephants, but in the context of crop-raiding elephants, even the staunchest among them would find the use of torches preferable to shooting (see Kiiru, Chapter 19, and Seneviratne and Rossel, Chapter 17). Thus, people sometimes adjust their values when it becomes clear that their initial position is inconsequential in a given situation. We are less apt to do this if the situations are theoretical and have never been experienced, but becoming informed about unfamiliar circumstances can dramatically shift our perspective. A heightened awareness of the possibilities of moral tolerance, or moral relativity, gives hope for the development of a holistic ethics responsive to the range of issues that threaten the welfare and long-term survival of elephant populations.

Who merits consideration? The uniqueness of elephants is an issue central to determining what should be deemed "ethical treatment of elephants." People set elephants apart from other mammals because of their size, anatomy, intelligence, memory, apparent ability to solve problems, social organization, family life, and certain maternal care practices. The extent to which these attributes are developed is remarkable and lends them a degree of "personhood," but the special features of elephants are not qualitatively different from those of other mammals (see Eisenberg 1981). Cetaceans also demonstrate unique adaptations, memory, long-distance communication of remarkable complexity, and so on (see, for example, McCowan and Reiss 1997). Conversely, the folk wisdom of many cultures attributes intelligence to many species whose abilities to reason are patently exaggerated. In these cases, the perception of wisdom is probably based on an appreciation of the animal's survival skills, a necessity that unites all organisms.

Humans inevitably create circumstances that cause pain and suffering in other creatures, including members of our own species. If we justify ethical treatment of elephants on the basis of a small number of perceived similarities to humans, we exclude a very large number of species from similar ethical consideration. This assemblage of species has a much larger number of other features in common with elephants and for that matter with people. When weighed in the balance of anthropocentric ethics, we exclude these species from eligibility for ethical consideration. This inconsistency is in keeping with current attitudes and practices regarding animal welfare. People accord ethical consideration to pets, but not to livestock, and even less often to pests. Yet dogs, cattle, and rats are similar in their capacity to perceive fear and pain. There are, we should note, some new developments on this front, such as the work of the Chicago Project on Animal Treatment Principles, concerned with activating market forces and legal approaches to promote animal welfare for livestock and best-practices standards and disclosure guidelines for treatment of animals in various industries (Schuler 2004).

In theory, it is possible to classify all mammals with regard to their similarity to humans, by scientifically measuring their cognitive, mental, perceptive, and psychological capacities. On this basis, we could design an ethics of animal welfare that includes all species worthy of ethical consideration. However, we currently lack all the scientific tools to do this. Alternatively, we could focus on a fundamental feature of perception shared by many organisms, for example, the ability to feel pain. This can be inferred from characteristics of the nervous system. We can assert fairly confidently that vertebrates, and some invertebrates, including cephalopods, are able to feel pain, though we still lack a complete scientific understanding of pain (Bateson 1992; see also Rachels 2004; Brown, Wielebnowski, and Cheeran, Chapter 6). An ethical criterion for a creature's meriting humane treatment could be based on its capacity to feel pain. Such an ethic maintains that human activities should attempt to minimize pain to all creatures.

When assessing discomfort, suffering, and stress in animals, people generally rely on subjective rather than scientific criteria. People tend to assess the suitability of living conditions in similar ways. When zoo visitors see an elephant weaving back and forth in a small confined space they usually interpret the situation as abnormal and unacceptable. They are usually pleased to see a mother with a calf at heel, even though the available space is extremely limited compared with natural conditions. It is recognized that pain is an unavoidable and adaptive fact of life that contributes to survival. When it is within their power, organisms avoid pain and suffering. If we adopted a utilitarian ethic, we would assess the degree to which our actions maximize the aggregate well-being and weigh that against the consequences of taking no action. Varner and Alward discuss utilitarianism and related moral theories in Chapters 3 and 10, respectively.

This approach leads to diverging conclusions about management in two cases from this volume: the fates of rogue elephants that become work elephants (see Lehnhardt and Galloway, Chapter 8) and of displaced elephants consigned to internment centers, as reported by Mikota, Hammatt, and Fahrimal (Chapter 18). In the case of Gajendra, the Indian rogue-turned-work elephant profiled by Lehnhardt and Galloway, most would agree that even with the suffering that was part of his adjustment to captivity, this elephant's life as a captive proved better than the "full" freedom in which he suffered greatly as a hunted and reviled crop-raider. Using similar logic, Mikota et al. concluded that the rescue of wild elephants in Sumatra only contributed to the stressful premature death of many and to suffering of the captured survivors.

Living Conditions and Moral Considerations

It is difficult to control nature, but people can much more easily modify the physical features of captivity. They can regulate nutrition and provide health care and behavioral enrichment. Capabilities approach, discussed by Alward (Chapter 10), has much to commend as a way of appraising the adequacy and morality of different *ex situ* settings for animals (see also Nussbaum 2004). The large body of information about the biology of wild elephants provides a basis for determining the capabilities needed for *ex situ* life, though more information is still needed. At least three chapters show that the process of improving *ex situ* environments has started (Mellen, Barber, and Miller, Chapter 15; Hutchins, Smith, and Keele, Chapter 14; and Schmitt, Chapter 11), though perhaps not quickly enough to please all, as we learn from Hancocks (Chapter 13) and Garrison (Chapter 12). Kreger's essay (Chapter 9) sheds light on the intertwined histories of zoos, circuses, and elephants in the Western Hemisphere.

In nature, elephants present a different ethical challenge. They are "keystone species," which means they have a major influence on ecosystems and their processes. The environment responds in dramatic ways to their removal or their abundance. People outside the range countries (where elephants historically have lived in nature) sometimes believe there are no pending ethical issues in places where elephants live naturally and, presumably, in remote pristine wilderness, but this volume teaches us otherwise. Most elephant ranges are not pristine, and elephants may only survive in the future in protected areas, where they must coexist with populations of other endangered plants and animals. Protected areas represent a concerted attempt to ensure the survival of wilderness for many reasons, including safeguarding the spiritual, educational, and experiential enlightenment of people. Protected wildlife areas are laudable from any number of ethical perspectives. The alternative is usually death and suffering of biota doomed by landscape conversion.

This latter scenario, landscape conversion, is a common course of events that often results in conflict between invading people and resident wildlife. Such conflicts are also found in protected areas. Management within reserves often leads to elephant overabundance and environmental degradation that negatively affects many other species. Solutions include benign interventions designed to restore interdependence between elephants and local communities as in Sri Lanka (Seneviratne and Rossel, Chapter 17), and "playing elephant God," that is, using the most appropriate combinations of translocation, contraception, and culling to manage elephant populations, as practiced in southern Africa (Whyte and Fayrer-Hosken, Chapter 20).

These solutions are based on different priorities and have different consequences for elephants. The Sri Lankan solution is management by prophylaxis based on negative conditioning and avoidance. It may be disruptive to the lives of elephants, but it does not deny elephants the functional capabilities discussed by Alward. Therefore, it may be considered ethical within the capabilities framework. Contraception intrudes on reproductive capabilities but precludes death by culling. Translocation undoubtedly causes disorientation and anxiety for elephants, but by maintaining their family integrity, it gives ethical consideration to the needs of elephants for social affiliation and kinship. If the net welfare of the animals is improved by translocation, the practice can be considered ethical. This assumes that translocated elephants survive and adjust to the new environment and that the new environment addresses their capabilities as well as or better than their former habitat.

The culling of elephants, like wildlife hunting in general, is the most ethically problematic intervention. Wildlife and natural resource managers usually justify culling with a logic akin to utilitarian consequentialism, that the moral right of an action is affirmed by the fact that its consequences are more favorable than unfavorable (see Leopold 1949). The consequences of elephant culling are ecologically inclusive in that the effects also act on the community or ecosystem.

Killing is the divide that separates the ethical camps into those who give consideration to animals exclusively as individuals, versus those who give consideration to animals inclusively as interacting parts of ecosystems. The latter view does not condone inhumane treatment to individual animals, but it does not consider killing to be inhumane if done in a way that minimizes pain and suffering. As Duffy (Chapter 22, 451) notes, the "different philosophical positions lead to complex ethical debates about who has the right to use elephants, how they can be used, and for what purpose." Global governance of the wildlife trade through CITES (the Convention on International Trade in Endangered Species of Wild Fauna and Flora) has impinged on the ability of some nations to manage their elephants through culling and to benefit from international trade. The ivory ban, for example, is an attempt at a unitary solution that accords moral consideration exclusively to elephants. An alternative ethical view is that the greater good of people and elephants is served when elephants are sustainably used. This inclusive consequentialist ethic is based on the premise that elephant management will succeed if elephants can be used for profit, with the benefits accruing to the people who share land with those elephants.

The ivory ban can be viewed as an experiment in global governance based on the belief that it is the only acceptable way of saving elephants from overexploitation. Its supporters include those morally opposed to killing elephants and those who believe the ban is the only effective way to shut down the illegal ivory trade and make compliance escape proof. Its detractors include countries where policies of sustainable utilization have demonstrated effective management of elephants. The ivory ban has an ethical dimension that cannot be separated from politics. If the political winds do not change, the experiment may be carried out to its conclusion, and the ethical consequences may one day be analyzed.

We can anticipate at least two conclusions to the ivory ban experiment.

The ban could prove an effective solution to the decline of elephants. For this to happen, tourism, rather than ivory and other products, must become the primary benefit that elephants return to human society (see also Hardin, Chapter 21), and economic solutions must be found for conflicts between elephants and people in large parts of the African and Asian elephants' ranges. If an ivory ban is to be successful, African countries that have managed elephants as a material commodity must adopt effective economic practices and embrace an "elephant ethic" of commodity nonuse. If elephant populations continue to decline as a result of land conversion and elephant-people conflicts, the ban, even if strictly enforced, will only partially succeed in the long run.

Let us consider the "non-ban scenario." What would happen if CITES promoted elephant conservation through free trade of registered elephant products? Elephant range countries could pursue their own means of managing elephants and attempt to use elephants sustainably as a harvestable resource. Within a conducive policy framework, rural stakeholders might organize themselves to police and protect their elephant herds effectively. On the other hand, if past patterns of behavior continue, countries that reject the utilization model may not succeed in controlling poaching and illegal trade. In that case, they will witness the continuing decline of their elephant populations as poachers reap the benefits of the ivory trade. Governments might respond in two ways. They could revert to the ivory ban nationally and lobby CITES to reenact the ban globally, or they could adopt policies of sustainable utilization and imitate their southern neighbors. Certain ethical safeguards could prove instrumental to successful enactment of this scenario, such as implementing chain-of-custody certification that would assure consumers that products were produced in accordance with morally informed regulations.

The essays in this volume offer a glimpse at the complexity of elephant management and the moral issues involved. Perhaps not surprisingly, most of the moral questions discussed are based on predominantly western concerns. Many other questions are equally pressing. Should elephants be used for religious ceremonies? How can the lives of solitary temple elephants be adjusted to address their social needs? Should outsiders even consider interfering with cultural self-determination in Asian countries to the extent of voicing views on these practices? Kurt, Mar, and Garaï (Chapter 16) and others briefly discuss some of these issues, but much more remains to be puzzled out.

Should captive elephants in Asia continue to be used for logging? Using elephant power certainly minimizes some environmental impacts of logging, but there are often risks of injury and other costs to the well-being of the elephants. The loss of mahout culture because of logging bans and unemployment has led to a serious decline in the quality of elephant management and care practices in the largest remaining captive Asian elephant populations. However, some elements of mahout culture, illustrated in Lahiri Choudhury's essay (Chapter 7), may need remodeling according to certain modernizing ethical standards that are starting to have currency in the East as well as the West.

With about 16,000 captive elephants in Asia, is it even rational to question whether any elephants should be in captivity? Pragmatically, we can easily assert that there is nothing to question; captivity is a fact of life for Asian elephants. These animals cannot easily be released back into the wild to fend for themselves. If they stay captive, we have a responsibility to manage them well, which takes us back to the question of what that means in moral terms, given existing economic and cultural backdrops.

Human population growth and burgeoning demands for natural resources continue to cause the loss of elephant habitat. Western consumption patterns are driving much of the loss of elephant habitat in Indonesia and Malaysia, where tremendous natural areas have been replaced with essentially barren oil palm plantations (see Mikota et al., Chapter 18). The harvest feeds into the western consumerscape, for example, in ice cream, commercially produced French fries, cosmetics, and many other products. Questions about our moral values regarding wild elephant populations circle right back to questions of how western economies and polities address fundamental problems of population growth, poverty, and large-scale demands for commodities.

Meantime, mitigation strategies regarding human-elephant conflicts may appear very different from outsider and range-country perspectives. Many humans die every year from elephant encounters in Asian and Africa. Elephants are clearly losing in the bigger picture, but at the same time, when people lose crops, property, and their lives, what kinds of compensation and what means of recourse are appropriate? Who has to pay? Kiiru (Chapter 19) and others in this volume touch on these issues, but the dialogue is only just under way.

No single pathway holds much promise of stemming the decline of elephants and their habitats. Viewed from any moral perspective successful and unsuccessful outcomes have equally troublesome aspects. If elephant populations stabilize at but a small fraction of their present range, the net loss will translate to a great deal of suffering and death of elephants, the disappearance of many other species, and the conversion of natural communities to agricultural landscapes. When the goal of a moral doctrine is to preserve a species under conditions favorable to its continued role within the ecosystem, it can be measured as a success only if the species continues to thrive. As Duffy discusses in her essay (Chapter 22) on CITES and the ivory ban, the politics of global conservation has opted for a unitary solution, which overrides the initiative of individual countries to address the issue within their own moral, social, and political frames of reference.

Toward Consensus: Points of Agreement

We used a consensus-building process during the symposium to identify general areas of agreement. During this workshop, the participants divided into three panels, corresponding to the three major segments of this volume, "Philosophy and Science," "Elephants in the Service of People," and "Elephants and People in Nature." With the aid of professional facilitators, the panels identified the following concerns and points of agreement. Only items on which we were able to agree unanimously—as a reconvened plenary—were included in these points of agreement. The consensus-building process embodied one of our central goals: It allowed people of many disciplines and intellectual persuasions to communicate with one another and seek mutual comprehension.

In the realm of "Philosophy and Science," we concluded that elephants deserve special moral consideration because of their unique combination of physical, cognitive, and social attributes, their ecological significance, and their cultural importance to humans. We shared the conviction that the survival of elephants as a biological species is a moral obligation, but we acknowledged that a unified ethical framework does not guide their management and care in the wild or captivity. Debate is healthy, but it can have dire results when urgent actions are required. Division will hamper progress in finding ethical solutions as long as proponents of different ethical perspectives reject other perspectives. Eventually, partisans of limited outlooks and uncompromising stances may find their own perspectives best served by giving way to or even helping to craft more inclusive solutions.

We discussed cultural differences and ethical relativity in the realm of "Elephants in the Service of People." There will be elephants in captivity in the foreseeable future. Captive situations for elephants differ greatly, and each must be evaluated with respect to local context and culture. We agreed however that standards of management and criteria for assessing the welfare of captive elephants should be based on current knowledge of the behavior and biology of free-ranging elephants and in particular with reference to their special physical, cognitive, and social attributes.

Many zoos do not have appropriate facilities to keep elephants. Those zoos, we agreed, must accept full responsibility for providing optimal facilities and care, or forgo keeping elephants. In Asia, conditions of elephants in service, whether wild or captive born, should also be improved where needed. Zoo associations and regulatory agencies must enact and enforce rigorous regulations to ensure that the welfare of elephants is adequately met. We also agreed that all institutions holding elephants have a moral obligation to support research and *in situ* conservation financially and should serve as conduits of education. Research and evaluation is needed to understand the effectiveness of human education in elephant-holding facilities. However, zoo education should not promote captivity as a substitute for protection and conservation of elephant populations in their natural habitat. Working elephants in Asia can and should contribute to the management and conservation of wild elephants.

We concluded that safety of people is an issue whenever elephants and people share space, but the ways people ensure safety and respond to human injury and death differ greatly between East and West. Captive-elephant caretakers, namely, handlers, owners, and veterinarians, need specialized training and adequate support to deliver high standards of care.

In the realm of "Elephants and People in Nature," we agreed that the governments of elephant range states should make every effort to maintain viable populations of wild elephants as a part of their environmental heritage. The political will to retain natural landscapes already exists in some of these countries, but many obstacles stand in the way of achieving success, including the growth of human populations, poverty, and the resulting need for cultivated land. Human-elephant conflict is the consequence. It is widespread and a major threat to elephant survival across the elephant range states.

Human-elephant conflict will increase on a regional scale, and conflict resolution will always raise ethical questions. The solutions in current use differ in Africa and Asia, and even within regions. We felt that more tools for mitigating and managing such conflict, combined with research on the effectiveness of these tools, are needed to increase options for managing this conflict. More research and additional implementation strategies are needed to ameliorate these conflicts, and local involvement, rather than uniform reliance on regional government is essential and should be sought in finding ethical solutions. We agreed that ecologically sensitive land-use planning can prevent or help mitigate human-elephant conflict and elephant population fragmentation. To balance conservation and human welfare needs effectively, range states must pursue large-scale, multidisciplinary, multilateral approaches to local and regional land-use planning.

Managing wild elephant populations is a necessity to minimize the effects of overpopulations of elephants on the environment. Various methods, from contraception to culling, are currently in use, but all raise ethical issues, from concern for individual elephants to concern for habitats and for wildlife communities. The efficacy of the methods should be evaluated, and additional methods should be developed.

Attitudes toward animals change, and so do the moral yardsticks people use to justify their treatment of animals. The concern with ethical treatment of animals has increased in the past century. Most of western society did not view the killing of wildlife or elephants as unethical one hundred years ago, as the Rothfels (Chapter 5) and Hardin (Chapter 21) essays discuss. It is safe to say that a larger segment of that society today deems killing elephants as unethical under many circumstances. However, an ethic that disallows the utilization of wildlife has not found acceptance in all rural communities that live with elephants or among trophy hunters or some other interest groups. In the end, conserving elephants in the most ethical way will remain a pressing challenge. Their enormity and uniqueness, however, give us hope. They are just too big to ignore.

References

Bateson, P. 1992. Do animals feel pain? *New Scientist* 134: 30–33.

Eisenberg, J. F. 1981. *The mammalian radiations.* Chicago: University of Chicago Press.

Leopold, A. 1949. *A Sand County almanac and sketches here and there.* London: Oxford University Press.

McCowan, B., and Reiss, D. 1997. Vocal learning in captive bottlenose dolphins: A comparison with humans and nonhuman animals. In C. T. Snowdon and M. Hausberger (eds.), *Social influences on vocal development* (pp. 178–207). Cambridge: Cambridge University Press.

Nussbaum, M. C. Beyond "compassion and humanity": Justice for nonhuman animals. In C. R. Sunstein and M. C. Nussbaum (eds.), *Animal rights: Current debates and new directions* (pp. 299–320). New York: Oxford University Press.

Rachels, J. 2004. Drawing lines. In C. R. Sunstein, and M. C. Nussbaum (eds.), *Animal rights: Current debates and new directions* (pp. 162–174). New York: Oxford University Press.

Saxe, J. G. 1878. The blind men and the elephant. In W. J. Linton (ed.), *Poetry of America: Selections from one hundred American poets from 1776 to 1876* (pp. 150–152). London: George Bell & Sons.

Schuler, P. 2004. Conference at Law School will launch discussion of animal treatment, welfare. *University of Chicago Chronicle*, October 21.

Sukumar, R. 2006. A brief review of the status, distribution, and biology of wild Asian elephants *Elephas maximus. International Zoo Yearbook* 40: 1–8.

PART I

OVERVIEW OF ELEPHANT PHILOSOPHY AND SCIENCE

AMERICAN MASTODON   WOOLLY MAMMOTH   FOVR-TVSKED MASTODON

2 ELEPHANTS IN TIME AND SPACE
EVOLUTION AND ECOLOGY
RAMAN SUKUMAR

The Asian elephant (*Elephas maximus*), and the African savanna and forest elephants (*Loxodonta africana* and *L. cyclotis*), are the only living representatives of the elephants, or proboscideans, a group whose evolutionary history goes back about 60 million years, to the Paleocene epoch (Gheerbrant, Sundre, and Cappetta 1996; Shoshani and Tassy 1996). During this period, changes in the earth's geology, climate, and vegetation brought about a spectacular radiation of the proboscideans, which occupied all continents except Australia and Antarctica (Sukumar 2003). They used a diverse range of habitats from swamps to desert, tundra to tropical forest, and riverine lowlands to alpine mountains. Whereas large body size was one of the distinct characteristics of elephant evolution, aberrant dwarf forms occurred on islands. During the Pleistocene epoch (1.8 million to 10,000 years ago), several elephant groups, including the woolly mammoth (*Mammuthus primigenius*) and the American mastodon (*Mammut americanum*) disappeared rather suddenly (Martin 1984; Martin and Klein 1984). These extinctions probably resulted from climate change or overkill by humans and may provide important insights for elephant conservation today.

The amazing evolutionary history of elephants, as well as many characteristics of their biology and ecology, has made this group a favorite in the perceptions and imaginations of people. Elephant ancestors, such as the mammoth and the mastodon, feature prominently in comics, books, and animated movies. Their bones and skeletons are a mainstay in many natural history museums across the world. Their sheer body size, large ears, and characteristic trunk and tusks are among the animal features most often eliciting curiosity and best remembered after a zoo visit, for children and adults alike. In this essay, I briefly review elephant evolution and explain the function and importance of some of the most characteristic elephant features. Although this essay does not directly illuminate ethical issues about the management and treatment of elephants, it provides crucial background information and some insights about why people care about elephants.

Origin and Evolution
Northern Africa was the cradle of the proboscideans, where the early evolution of true elephants was nurtured. The warm and shallow waters of the

ancient Tethys Sea provided the setting for the aquatic origin of the family nearly 60 million years ago, a period when the earth was much warmer than today. For almost a century after its fossil remains were first discovered in Eocene epoch (54.8 to 33.7 mya) deposits of Egypt's Fayum Basin, it was believed that *Moeritherium,* a pig-sized and hippo-like creature, was the ancestral elephant. More recently, however, several fossils older than *Moeritherium* have been discovered, also in northern Africa. Among them, *Numidotherium,* with a tapir-like proboscis and teeth specialized for stripping leaves from branches at head height (Mahboubi et al. 1984), and the closely related *Phosphatherium,* pushed the ancestry of the proboscideans a few million years farther back, into the Paleocene epoch (65 to 54.8 mya) (Gheerbrant, Sundre, and Cappetta 1996). The anthracobunids (represented by five genera) from Eocene rocks of the Indian subcontinent (West 1980) were also contenders for being considered the ancestral proboscideans, but the northern African fossils suggest this to be unlikely.

The habitats of these Paleocene and Eocene proboscideans were shallow, muddy waters along fluctuating shorelines. The dentition of these primitive forms exhibits varying degrees of specialization, from a full complement of teeth to loss of canines, enlargement of incisors, and increased lophs, or ridges, in the molars. An increase in body size can also be seen, with one late Eocene proboscidean, *Barytherium,* having attained 4–5 tons, the size of a modern African elephant. The cooling and consequent decrease in precipitation after the mid-Eocene can perhaps explain the need for an herbivore to adapt, through increased body size, to consuming larger quantities of lower-quality vegetation than had its ancestors. The larger body size effectively increases the surface-to-volume ratio, reducing loss of heat. The cooling and drying trend continued into the Oligocene epoch (33.7 to 23.8 mya) when two types of more advanced proboscideans, known as palaeomastodontids, flourished. These were *Palaeomastodon* and *Phiomia,* each with tusks in both the upper and lower jaws and a distinct gap between the tusks and the molars. However, in strong contrast to their Eocene forebears, the palaeomastodontids were land dwellers.

The Oligocene seems to have been a period of relative evolutionary quiescence for elephant species, but the Miocene epoch (23.8 to 5.3 mya) witnessed possibly their greatest diversification. A new warming trend set in, but this time it was accompanied by a drier climate. Several elephant groups appeared in the Miocene, among them the enigmatic *Deinotherium,* one of the largest of all proboscideans, with tusks curving inward from the lower jaw. Another well-known and successful group of elephants, the mammoths, or mammutids (family Mammutidae), also made its stage debut in the Miocene. The best-known mammoth may be the American mastodon, which survived until the end of the Pleistocene. However, the deinotheres and the mammutids, doomed to extinction, were only sideshows in the great proboscidean drama.

For a long time the gomphotheres—the stem group for true elephants—were a poorly defined group into which paleontologists dumped all proboscideans of uncertain affinity. Now we know, from cladistics (phylogenetic systematics) analysis, that the gomphotheres are distinct from the mammutids in being "long-jawed proboscideans" (Shoshani 1996). Two gomphothere genera that have attracted much attention are the "shovel-tuskers," *Platybelodon* from Eurasia and *Amebelodon* from North America. Both had long, flattened lower tusks, perhaps used to dig up roots and vegetation. As large as a modern Asian elephant, the genus *Gomphotherium* was a conservative form, with a short but thick trunk, tusks in both the upper and lower jaws, and low-crowned molars with three cusps, thick enamel, and a sequence of vertical (erupting from below the existing molar) and horizontal (erupting from the rear of the jaw and gradually moving forward) tooth replacement. During the late Miocene, an advanced gomphothere is believed to have given rise to the ancestral elephant.

By the late Miocene (10 mya), the climate began cooling again, accompanied by further drying of continental interiors. By about 7 million years ago, the reduced atmospheric carbon dioxide—a "reverse greenhouse"—accentuated this cooling and dryness and, more important, allowed the expansion of highly productive grassland on a global scale (Cerling et al. 1997). Herbivores responded to these global changes by adapting to a grazing diet, taking advantage of the abundant grasses. Corresponding dental changes included their development of high-crowned teeth (hypsodonty) bearing complex surface patterns—ridges of enamel with clefts filled in by a special tissue, coronal cementum—providing excellent grinding surfaces. Such teeth allowed them to break apart tough plant tissues to extract whatever nutrients were available. The late Miocene proboscideans show such dental adaptations.

True elephants probably evolved in Africa from one of two genera, *Stegotetrabelodon* and *Stegodibelodon,* of the subfamily Stegotetrabelodontinae. These genera were probably true elephantids, or at least intermediate between gomphotheres and elephantids (Maglio 1973; Todd and Roth 1996). About 6 million years ago in Africa, still during the late Miocene, one of these forms gave rise to *Primelephas,* considered to be the progenitor of the genera *Mammuthus, Elephas,* and *Loxodonta* of the family Elephantidae.

When the Miocene came to a close around 5.3 million years ago, there were alive twenty-two of approximately thirty-eight presently recognized proboscidean genera and a very large number of proboscidean species, including the primitive deinotheres, the mammutids, the gomphotheres, the stegodontids, and the elephantids—an amazing diversity that coexisted around the globe, with the exception of Australia and Antarctica. The radiation of the true elephants began during the late Miocene, proceeded through the Pliocene epoch (5.3 to 1.8 mya), and continued into the Pleistocene. During the Pliocene, the earth's climate continued to cool and remained dry, but Pleistocene climates alternated between twenty or more glacial and interglacial phases. Elephant evolution during these periods included many new adaptations, especially in skull architecture and dentition. The combined grinding-and-shearing design of the gomphotheres' teeth now changed into a design mainly good for shearing (Maglio 1973). Other adaptations in dentition included almost exclusively horizontal replacement of the molar teeth, increase in the number of plates (lamellae) per molar, and an increase in crown height accompanied by thinning of the enamel. These changes enabled the elephants to chew coarse plant material, including abrasive grasses and bark. While the rates of change in *Mammuthus, Elephas,* and *Loxodonta* did not differ greatly during the first 2 million years of their evolution, over the next 4 million years *Mammuthus* and *Elephas* evolved more rapidly in their dental characteristics, as compared with *Loxodonta* (Maglio 1973). The woolly mammoth (*M. primigenius*) had some of the most advanced dental adaptations.

Mammuthus showed not only the most rapid rate of evolution, but among the elephants it also migrated the farthest from its site of origin. The earliest form of mammoth known is *M. subplanifrons* from eastern and southern Africa, followed by *M. africanavus* from northern Africa. The latter is believed to have migrated during the late Pliocene across Gibraltar or the Middle East into Europe, where it evolved into *M. meridionalis.* Mid-Pleistocene cooling caused the mammoth's woodland habitat to retreat in Europe and Siberia, giving way to grasslands. The steppe mammoth *M. trogontherii* can be clearly recognized in Eurasia during the late Pleistocene, finally culminating in the woolly mammoth (*M. primigenius*) about 200,000 years ago. Earlier, at the outset of the Pleistocene, about 1.8 million years ago, one of these Eurasian mammoths migrated during a period of low sea levels across Beringia into North America and evolved into *M. columbi.* The woolly mammoth migrated into North America about 100,000 years ago.

Elephas ekorensis is the first recognizable species of the genus *Elephas,* coexisting with the earliest mammoth in eastern Africa. There, *E. ekorensis* gave rise to *E. recki,* the dominant species of elephant at this time and one that went through several clear stages of dental evolution. It was, however, *E. planifrons,* the Asian derivative of *E. ekorensis,* or a closely related form, that eventually gave rise to the modern Asian elephant (*E. maximus*) through *E. hysudricus,* whose fossils are found in early Pleistocene deposits of the Siwalik hills of the Indian subcontinent.

Loxodonta adaurora, the earliest recognizable species of the loxodont lineage, to which today's African elephant belongs, seems to have changed little through the Pliocene. In fact, *E. recki* may have displaced it as the dominant elephant by the late Pliocene. It is only by the middle Pleistocene, when *Elephas* disappears from Africa, that *Loxodonta* reemerges, in the form of *L. atlantica*, as the common elephant of Africa. The modern African elephants *L. africana* and *L. cyclotis*, however, seem to be derived from an advanced form of *L. adaurora*.

One of the most interesting aspects of proboscidean evolution during the Pleistocene is the rapid dwarfing of certain elephant species on islands isolated by rising sea levels (Lister 1996). The phenomenon is not limited to elephants, but the reversal in body size of these proboscideans is usually explained as an adaptation to limited resource availability on these islands. The best known of the dwarf elephants comes from the islands of the Mediterranean. *Elephas falconeri* stood a mere 1 m tall as an adult. Another example is seen on the Channel Islands off the coast of California. There the Columbian mammoth, *M. columbi*, reduced in size to 1.2–1.8 m and became the dwarf species *M. exilis*. After its extinction in mainland Eurasia and North America by about 11,000 years ago, the woolly mammoth (*M. primigenius*) survived and dwarfed rapidly on Wrangel Island, off the northern coast of Siberia, during the early Holocene epoch (the past 10,000 years). It eventually died out by about 4,000 years ago (Vartanyan, Garutt, and Sher 1993).

During the late Pleistocene (40,000 to 10,000 years ago), a wave of extinctions swept through Australia, Eurasia, North America, and South America, with a devastating impact on larger mammals. The extinctions of the mammoths of Eurasia and North America, the American mastodon, and the gomphotheres of South America occurred toward the end of this period. The causes of this rapid extinction wave are still being debated. Two major classes of explanation have been suggested, one centered on climate change and the other on overkill, defined as "the human destruction of native fauna either by gradual attrition over many thousands of years, or suddenly in as little as a few hundred years or less" (Martin 1984).

The transition from the last Pleistocene ice age to the present climate was perhaps the most severe of any of the Pleistocene's many glacial-to-interglacial alternations. Not only did the globe warm considerably but the seasonal climate patterns changed. Resultant changes in flora and fauna may have led to the collapse of large-bodied mammal species through changes in available nutrition or disruption of reproductive cycles (Guthrie 1984; Kiltie 1984; Webb 1984). Opposing hypotheses invoke humans as the perpetrators of the late Pleistocene extinctions. The overkill idea first appeared when anthropologists proposed that Paleoindians, who had migrated across Beringia into North America about 12,000 years ago, went on a "Blitzkrieg" and hunted the mammoth and mastodon to extinction within a few hundred years, using their characteristic "Clovis" spear points (Mosimann and Martin 1975; Martin 1984). Support for this theory comes from the coincident dates of human arrival and megafaunal extinctions on several continents, together with evidence at a few kill sites that demonstrates how Paleoindians may have killed these giants. Another human-caused extinction hypothesis postulates that humans carried a "hyperdisease" that infected a variety of mammalian species (MacPhee and Marx 1997).

Most likely a suite of factors contributed to the decline of so many proboscidean species. The more synthetic theories incorporate both climate change and human expansion in their reasoning (Haynes 1991). Whatever the true cause of the extinctions, they may hold lessons for the conservation of today's elephants. We may be witnessing the onset of another major phase of climate change, this time driven by human-induced increase of greenhouse gases in the atmosphere. At the same time, populations of both African and Asian elephants are being hunted extensively for ivory and other products, while their habitats are rapidly transforming into human-dominated landscapes. Conditions today may be ominously similar to those prevailing during the late Pleistocene.

By the end of the last ice age, the world had lost much of its megafauna. This period also marks the dawn of agriculture, the expansion of human societies, and, by the mid-Holocene, the rise of the early major civilizations in the Nile, Tigris-Euphrates, and Indus river basins. Concurrently, we find evidence that several human societies independently developed close relationships with elephants. The earliest evidence for elephant taming comes from the animal motifs on steatite seals, about 4,000 years old, of the Indus (or Harappan) civilization; though the elephant may well have been tamed earlier by people of the Indian subcontinent (Lahiri Choudhury 1991). In Asia this association deepened into a multifaceted relationship spanning the entire spectrum of the social, political, economic, and religious life of people. In Africa an early culture of capture of elephants in the north of the continent died out quickly.

Social Behavior

Elephants are among the most sexually dimorphic (different in form) of all mammals. Polygyny is a common mating system in dimorphic species. Polygynous males compete with one another to maximize their success in mating, but only a portion of the males succeed. As a result, variance in reproductive success is much greater in males than in females. The core of elephant society is the matriarchal family (Moss 1988), composed of an adult cow elephant and her immature offspring of both sexes. However, the number of elephants observed in a family or in larger social groups varies across the geographic ranges of African and Asian elephants. I have proposed the term *joint family* be used if families consist of more than one adult cow and their offspring, as one can sometimes observe (Sukumar 1994). Higher levels of social organization such as bond/kin groups and clans have also been suggested for some African elephant populations in eastern Africa (Douglas-Hamilton and Douglas-Hamilton 1975; Laws, Parker, and Johnstone 1975; Moss 1988) and Asian elephant populations in southern India (Sukumar 1989; Baskaran et al. 1995). Indeed, Moss (1988) proposes that social organization in elephants radiates from the basic family unit through "a multi-tiered network of relationships" that eventually encompasses the entire population in a region. However, in central Africa (Turkalo and Fay 1995) and Sri Lanka (McKay 1973; Kurt 1974; Fernando and Lande 2000) observations indicate simpler levels of organization.

One of the primary functions of the elephant family group may be the protection of juveniles from large predators, such as lions in Africa and tigers in Asia (Douglas-Hamilton and Douglas-Hamilton 1975; Gadgil and Nair 1984). The largest elephant groups are indeed found in open habitats. Large carnivores are most abundant in savanna woodlands in Africa and deciduous forests and their associated grasslands in Asia. Elephant group sizes are probably determined by factors such as forage abundance, seasonality, animal density and numbers, human disturbance, natural predation pressure, and possibly even genetic relatedness. Among these factors, food and resource availability probably plays the most important role in determining family or group size. Open savanna or forest habitats tend to have a high abundance of elephant foods. Because these habitats also show much seasonal variation in food abundance, one would expect group sizes to vary likewise. However, larger elephant groups may be seen either during the wet season when resources are abundant (Western and Lindsay 1984; Poole and Moss 1989) or during the dry season when elephants congregate around scarce water sources and their chances of social contact and association increase (Sukumar 1985). In contrast, it is in the rain forests of central Africa where biologists have observed the simplest social groups and the smallest group sizes of any elephant population (Turkalo and Fay 1995). Forage plants are widely dispersed in these forests and large carnivores are absent. Several observers have also noted that elephants may come together in larger groups in regions where elephants are hunted or where there is a history of poaching elephants for ivory (Abe 1994). Here the predators are humans, and the response of elephants is similar to that shown to other natural predators.

In most regions, elephant social units, whether they are clans or family-bond groups, usually show rather fixed patterns of seasonal movement over the year. There is, however, enormous variation in home range size of elephant groups, both in Africa (Douglas-Hamilton 1972; Leuthold 1977; Lindeque and Lindeque 1991; Thouless 1996) and in Asia (Baskaran et al. 1995; Fernando and Lande 2000). Although the home range sizes of most elephant populations are between 500 and 2,000 km², they range from only 50–100 km² in Lake Manyara, Tanzania, and southeastern Sri Lanka, to as much as 10,000 km² in Namibia. In regions of low to medium rainfall, home range size is clearly inversely related to rainfall and productivity. However, in high rainfall areas such as rain forests, home range size seems to be larger because of the greater dispersion of food plants. Bull elephants show greater variation in their movement patterns and range sizes when compared with female elephants.

Social organization into family groups or clans requires well-developed communication among group members. Not surprisingly, different modes of communication seem well developed among elephants, and scientists have focused their research on this particularly interesting topic. Communication based on smell and sound is developed to a high degree of sophistication. Elephants can broadcast an array of chemical signals through urine, feces, breath, temporal glands, and sweat glands between the toes (Rasmussen 1998). The nose of an elephant, its trunk, is highly developed for receiving and assessing these signals. In this function, two specialized organs in the roof of the mouth, the vomeronasal organ and the palatal pits, play important roles.

Elephants also use infrasound, sound waves with frequencies as low as about 10 hertz to communicate over long distances (Payne, Langbauer, and Thomas 1986; Poole et al. 1988; Langbauer et al. 1991; Garstang et al. 1995). Both female elephants in estrus and male elephants in musth broadcast unique infrasonic calls to attract mates (Poole et al. 1988). The ability of a matriarch to discriminate between calls of related and unrelated individuals may eventually be crucial for the survival of her family (McComb et al. 2001).

A female elephant in estrus advertises her condition through characteristic behavior (Moss 1983), an infrasonic call (Poole et al. 1988), and chemical signals (Rasmussen 1998). The specific compound released in urine by an estrous Asian elephant has been identified as (Z)-7–dodecenyl acetate, the same compound used by many female insects to attract mates—a remarkable example of convergent evolution (Rasmussen et al. 1996). An adult bull usually detects estrus through a behavior termed *flehmen*. Flehmen involves touching the urogenital orifice of the cow or urine on the ground with the tip of the trunk and transferring a small quantity to the roof of the mouth where a pair of vomeronasal organs assays the substances (Rasmussen et al. 1982).

An important determinant of reproductive success in bull elephants is musth, a physiological and behavioral phenomenon that has been compared to the rut in ungulates (Eisenberg, McKay, and Jainudeen 1971; Poole

1987). Musth was well-known in the Asian elephant and accurately described in ancient Indian literature (Edgerton 1931; Lahiri Choudhury 1992) but was first recognized in the African elephant only in recent times (Poole and Moss 1981). Usually an annual occurrence in adult bulls, musth is manifested through secretion of a fluid from the temporal glands, a sharp rise in the levels of blood androgens (such as testosterone), and behavioral changes such as increased aggression (Jainudeen, Katongole, and Short 1972; Rasmussen et al. 1984; Poole 1987; Niemuller and Liptrap 1991). The nature of the chemical secretion changes with age; younger bulls secrete sweet-smelling compounds such as esters and alcohols while older bulls secrete foul-smelling compounds such as frontalins and nonanone (Rasmussen, Riddle, and Krishnamurthy 2002). Musth bulls broadcast a ketone (cyclohexanone) that seems to serve as a specific male-to-female signal (Rasmussen, Haight, and Hess 1990).

The intensity of musth usually increases with age and also correlates with good body condition. A bull in musth is dominant over non-musth bulls, even older or larger ones, and has much better chances of mating with a cow in estrus than a non-musth bull (Poole 1989a, 1989b). Bulls within a population seem to show distinct spacing in the time of the year they come into musth. The typical pattern seems to be that the older, higher-ranking individuals come into musth during the most favorable period of the year (wet season), while younger bulls come into musth at other times.

High levels of testosterone weaken the immune system of an individual, increasing its vulnerability to parasitism and disease (Folstad and Karter 1992). Thus, musth can be considered within the framework of the handicap principle, as enunciated by Zahavi (1975). A bull surging with testosterone is advertising its genetic quality to female elephants. A female's choice of a bull in musth over one not in musth would thus be adaptive. Similarly, we can also postulate a sexual selection role for tusk size in elephants (Sukumar 2003). An enormous pair of tusks in a bull elephant (or any proboscideans such as the stegodon or the mammoth) can be a significant handicap; female choice for tusk size thus may serve the same adaptive purpose as female selection of a bull in musth (Watve and Sukumar 1997). Indeed, tuskless bulls and tusked bulls could be following alternative evolutionarily stable strategies, defined in game theory as strategies that, over evolutionary time, are able to withstand the invention of new strategies. The tuskless bulls could be diverting resources to body condition and more intense expression of musth, and the tusked bulls could be benefiting from female choice for tusk size (Sukumar 2003).

Foraging Ecology of Elephants

Elephants are generalist herbivores. Their sheer body size and resulting food requirements means that elephants feed on a variety of plant species and plant parts. Most studies of elephant feeding ecology list well over a hundred species of plants consumed even within one region (Sukumar 2003). Elephants eat leaves, bark, twigs, roots, stem pith, fruits, and flowers. Yet, they select distinctly different plant parts depending on the availability of seasonally changing plant forms, and they show a preference for particular plant families (Sukumar 1989). In quantitative terms, elephants usually prefer grasses, reeds and sedges (Poaceae, Cyperaceae), palms, legumes, many species of the order Malvales, and trees and shrubs of the families Combretaceae, Euphorbiaceae, Rhamnaceae, and Moraceae (Sukumar 2003). In drier habitats such as savanna woodland or deciduous forest, elephant dietary habits usually alternate between predominantly grazing (i.e., feeding on grasses and sedges) during the wet season and browsing (feeding on bamboo, shrubs, and trees) during the dry season (Field and Ross 1976; Sukumar 1989). In moister habitats, such as rain forest, their diet includes several kinds of fruits (Short 1981; White, Tutin, and Fernandez 1993) in addition to the leaves of lianas, vines, shrubs, young trees, and monocots such as palms (Olivier 1978).

Scientists have used the stable carbon isotopic technique to determine the ratio between frequencies of occurrence of different carbon isotopes in elephant bone collagen. These measurements help them determine the dietary ratio of browse or dicotyledonous species to grass or monocotyledonous species (van der Merwe, Lee-Thorp, and Bell 1988; Sukumar and Ramesh 1992) and demonstrate that the contribution of browse plants to elephants' protein synthesis is significantly higher than what could be inferred from the proportion of time they actually spend feeding on browse as compared with grasses (Sukumar and Ramesh 1992). There is a clear positive relationship between rainfall and the extent of browse feeding (Sukumar 2003). Further, elephant populations may be entirely browsers or mixed feeders on browse and grass, but they can never survive entirely as grazers. Thus, browse species are extremely important in the nutrition of elephants (Koch et al. 1995; Sukumar and Ramesh 1995).

Body size is obviously a key determinant of an elephant's foraging strategy. Elephants consume, in dry matter, 1.5–2.0% of their body weight daily. Dentition (Maglio 1973) and gut anatomy and physiology (Clemens and Maloiy 1982) are of equal importance in fostering elephants' strategy as nonruminant "megaherbivores" that subsist on coarse plants of low nutritional quality. They depend on high-crowned molars with a complex pattern of folded plates and occlusal surfaces to grind up the food and rely on hindgut fermentation to digest it.

The elephant's mineral and vitamin requirements are poorly understood. The nutrient requirements for other nonruminants, such as the horse, are usually viewed as adequate for elephants (Dierenfeld 1994). The elephant's appetite for bark may be partly a search for minerals such as calcium (Bax and Sheldrick 1963; Sukumar 1989) or for essential fatty acids (McCullagh 1973). Sodium may also be obtained from bark or from water and soil rich in this mineral (Weir 1972; Ruggiero and Fay 1994). Plant chemical defenses, such as tannins, alkaloids, and cyanogenic compounds also influence the choice of plant species and parts (Rosenthal and Janzen 1979). *Geophagy*—the consumption of soils and clays—may help bind or neutralize certain plant toxins such as alkaloids and thus may aid in digestion (Duquette 1991).

Elephants have been termed a *landscape species*. Because of their body size and food requirements, they have a tremendous impact on large areas, shaping the ecology of entire landscapes. By debarking, damaging, and pushing over trees, elephants can convert woodlands into grasslands (Laws 1970). This transformation favors the increase of grazing ungulate populations. Even in tropical moist forests elephants have created gaps in the forest canopy, and new plant regeneration in these openings may favor smaller ungulates. Ecologists have documented the role of elephants in seed dispersal (Lieberman, Lieberman, and Martin 1987; White et al. 1993), but few plants may have an obligate relationship with the elephant for their dispersal (Hawthorne and Parren 2000).

The dramatic transformation of woodland into grassland in parts of Africa triggered an intense debate over elephant population management through culling. One school of thought considered this habitat transformation as unnatural, caused mainly through human factors such as conversion of habitat for agriculture and compression of elephant populations (e.g., Buss 1977), while the other considered this as part of the natural ecology of the savanna woodland ecosystem (e.g., Caughley 1976; Norton-Griffiths 1979; Sinclair 1981). Ecologists proposed various hypotheses to explain the dynamics of elephants and vegetation. The "stable limit cycle" model of elephants and trees, a predator-prey feedback model for elephant-tree dynamics with oscillation periods of about 270 years, was the one that stimulated the most debate (Caughley 1976). Recent efforts at modeling the limit cycle for elephants and trees have shown that this model is unlikely to operate under realistic conditions of elephant and tree densities (Duffy et al. 1999). Nevertheless, Caughley's model served a very useful purpose; it spurred ecologists to explore the nonequilibrium nature of elephant-vegetation dynamics in semiarid ecosystems such as the savanna woodlands (Sukumar 2003).

Now the dynamic and stochastic nature of elephant-vegetation systems is much better appreciated. Several factors, including climatic fluctuations, fire, and changing populations of browsing and grazing herbivores drive the dynamics of grassland and woodland in semiarid Africa (Western and van Praet 1973; Norton-Griffiths 1979). Studies in the Serengeti-Mara region exemplify the complex nature of these dynamics. The potential of trees such as *Acacia tortilis* to regenerate, replacing those of their species killed by elephants, may be inadequately realized because of fire damage (Croze 1974), especially in combination with browsing by giraffes (Pellew 1983). Once these factors, in concert, convert enough woodland to grassland, the proliferation of grazers such as wildebeest potentially could lower the standing crop of grass and hence the frequency or intensity of fires. This would give these trees another opportunity to regenerate, or elephants could persistently feed on the tree seedlings, thus locking the system into the grassland stage (Dublin, Sinclair, and McGlade 1990). In fact, the semiarid regions of Africa could oscillate between two stable states, one where woodlands have relatively few elephants and the other where grasslands have a high abundance of elephants (Dublin et al. 1990). Similarly, I have elsewhere argued that elephant-vegetation dynamics can be expected to vary between a highly fluctuating condition in the markedly seasonal savanna woodlands to a relatively stable condition in aseasonal rain forests (Sukumar 1989, 2003).

The intrinsic dynamics of elephant populations are also of both theoretical and practical interest (Sukumar 2003). As can be expected of a large mammal, the elephant has a late age of sexual maturity (typically 10–15 years old), low reproductive rate (interbirth interval of four to six years), and low juvenile (5–15% per year) and adult mortality (2–3% per year in females). One consequence of these life history traits is the relative stability of elephant populations. When compared with other mammals, elephant populations generally do not intrinsically increase or decrease at high rates. On average, Asian elephant populations are unlikely to grow at greater than about 2% per year (Sukumar 1989); nor African elephant populations (in savannas), at greater than about 4% per year (Hanks and McIntosh 1973). Exceptions to this rule are the Addo and Kruger elephant populations in South Africa that have increased at more than 6% per year for over five decades (Whitehouse and Hall-Martin 2000; see also Whyte and Fayrer-Hosken, Chapter 20 in this volume). Similarly, decreases in population size are unlikely to be drastic, with the exception of the impact of a catastrophic drought, as in the case of Tsavo (Kenya) during 1970–1971 (Corfield 1973).

Elephants are adaptable to a range of conditions. I have argued that demographic traits in elephants, resulting from natural selection, can be expected to vary across environmental gradients. Thus, elephants in the relatively unstable semiarid habitats can be expected to show life history traits indicative of r-selection, favoring a reproductive strategy in which many offspring are produced (on a relative scale), while those in the more stable environments of aseasonal rain forests have traits indicative of K-selection, favoring a reproductive strategy in which few offspring are produced

(Sukumar 1989, 2003). Observed demographic variables in elephant populations broadly support this expectation; thus, female elephants in semi-arid regions not only have lower mean age at first reproduction and shorter interbirth intervals but also greater variance in these traits as compared with female elephants living in moist forest.

Human-Elephant Interactions

In recent centuries, human interactions with elephants have affected elephant population dynamics. In Asia, elephants have been captured in large numbers over the past 4,000 years, primarily by kings and chieftains for use in armies but also for other purposes, including logging and cultural festivities (Sukumar 1989). The African elephant has been hunted for ivory both historically (Spinage 1973) and in recent times (Douglas-Hamilton 1987). Various models of hunting of African elephants for ivory have clearly shown this to be the major factor in decline of populations over the past two centuries (Pilgram and Western 1986; Caughley, Dublin, and Parker 1990), though one model has also implicated habitat loss until about 1970 in this decline (Milner-Gulland and Beddington 1993). Poaching for ivory and other elephant products, such as hide and meat, has also decimated Asian elephant populations, especially in the region between Myanmar and Vietnam (Menon 2002) or resulted in highly female-biased sex ratios, as in southern India (Sukumar 2003).

One of the most significant effects of elephants on human interests is the destruction of agricultural crops, often accompanied by manslaughter in Asia and Africa (Blair, Boon, and Noor 1979; Sukumar 1989; Barnes 1996; Hoare 1999). These adverse interactions are now commonly referred to as *human-elephant conflict*. Not surprisingly, crops raided by elephants are usually the cultivated analogues of the wild-plant types that elephants consume. Cultivated grasses, palms, and legumes are commonly targeted (Sukumar 1989). There is also considerable selection for plant parts consumed in cultivated fields. Male elephants have a much greater propensity to raid crops than do the female-led family groups (Sukumar and Gadgil 1988; Hoare 1999) and may obtain significant proportions of their annual dietary requirements, both in quantitative and qualitative terms, from agricultural land (Sukumar 1989).

We need to gain a better understanding of the proximate and ultimate causes of crop-raiding behavior to mitigate human-elephant conflicts and the adverse consequences of these conflicts for elephants, specifically the capture or killing of the elephant perpetrators. Studies of human-elephant conflict should be considered within the framework of elephant ecology and should draw on landscape ecology, foraging and movement theory, and studies of social behavior (Sukumar 2003). Habitat loss is probably one of the main proximate causes of crop raiding. Given a certain fidelity to their home range (Baskaran et al. 1995), elephant family groups or solitary bulls may continue to move into land converted for cultivation. There could be opportunistic raiding related to the seasonal movement pattern of elephants observed in some elephant populations (Sukumar 1989). The same could be true of elephants having to traverse agricultural land to reach water sources (Allaway 1979; Sukumar 1989). One proximate factor that has a fairly clear relationship to raiding is the fragmentation of habitat (Sukumar 1989, 2003; Nath and Sukumar 1998). Habitat fragmentation increases the chances of elephants making contact with cultivated land, while the habitat fragments provide convenient daytime refuges for elephants that raid crops during the night.

In ultimate terms, however, crop raiding can be thought of as a strategy by elephants to maximize their forage intake both quantitatively and qualitatively (Sukumar 1989). Cultivated crops are usually more palatable and nutritious than wild-forage plants (Sukumar 1989; Osborn 1998). They provide more protein and minerals, such as calcium and sodium, as compared with analogous wild species. Raiding involves taking risks because the elephants may have to face hostile farmers. Some elephants may thus opt out of raiding even if they have access to crop fields. The higher propensity of male elephants to raid crops as compared with female elephants may be related to the higher variance in male reproductive success (Sukumar and Gadgil 1988). Selection pressures would favor the evolution of risky strategies by males for enhancing reproductive success. In this instance, the extra nutrition from crop plants would potentially benefit a male elephant through better body condition and more intense expression of musth, thereby increasing its chances of mating. Crop raiding by bulls could thus be a manifestation of a "high risk-high gain" strategy molded by natural selection to enhance reproductive success. Obviously, the realized reproductive success could depend on the interaction with humans; a bull could be injured or killed in the process of raiding.

If learning plays an important role in the development of behavior, including the choice of food plants, it follows that crop raiding could be learned behavior that is transmitted culturally from one generation to the next (Sukumar 1985, 1995a). Extreme climatic events such as severe drought may also trigger the long-distance movement or dispersal of elephants, thus escalating conflict (Sukumar 1995a).

Elephants in Human Culture

It is not surprising that these land giants have played a prominent role in human cultures both symbolically and more directly as beasts of burden (Lahiri Choudhury 1991, Chapter 7 in this volume; Ross 1992). Both Asian and African elephants have been tamed in historical times. With the former species, this culture has continued to flourish to present times, while with the latter the practice disappeared by about 2,000 years ago and was revived only marginally during the nineteenth century. The elephant has also been deified in Asia, while its role in religion has remained at the totemic level in Africa.

The earliest evidence for elephant taming comes from the seals of the Indus Valley civilization about 4,000 years ago (Carrington 1958), though non-Aryan people in the subcontinent may have tamed it even earlier (Lahiri Choudhury 1991, Chapter 7). The culture of capturing and training elephants in large numbers flourished with the rise of the Aryan republics and kingdoms in northern India and seems to have reached an early peak during Mauryan times (Lahiri Choudhury 1991; Sukumar 2003). Sanctuaries for the protection of wild elephants, presumably to supply the king's army, were set up in ancient India (see the translation of the *Arthashastra,* an ancient manual on statecraft, by Rangarajan 1992); the army of Chandragupta Maurya (late fourth-century BCE) included a 9,000–strong elephant force. During the early centuries of the modern era, we observe the imposition of a taboo on consuming elephant meat and also the rise of the classic elephant-headed deity Ganesha in the Indian subcontinent, reflecting the importance of this animal at least to certain powerful social classes (Sukumar 2003).

The African elephant, possibly a now extinct form from the Atlas Mountains in the north, was tamed by the Carthaginians and used in the Mediterranean wars from around the third century BCE. The elephants that Hannibal marched across the Alps in his much-celebrated campaign against Rome were mostly African (Scullard 1974). This ancient culture of capturing elephants, as well as their use in war, died out in Africa by about 2,000 years ago. On both continents the rise of the elephant culture—specifically the large-scale capture and use of the animal—probably was associated with the rise of major republics and kingdoms. The Carthaginians possibly overexploited a small elephant population already isolated in northern Africa. The Egyptians did not have access to Africa's large, essentially sub-Saharan elephant populations; the few elephants used by the Ptolemies came from Asia or were captured from the Carthaginians. The rulers of the Indian subcontinent, however, could exploit a ready source of abundant elephant populations, virtually at their doorsteps. African elephants could be tamed, as was shown in the erstwhile Belgian Congo during the late nineteenth century, but in modern times, the entire context has changed. It is unlikely that Africa would ever witness an Asian-type elephant culture; even in Asia this culture is on the decline.

Conservation of Elephants

The survival of elephants is linked not merely to biological or ecological factors but also to social, economic, and political forces. Conservation efforts to save Asian and African elephants face many similar problems, but there are also striking differences, especially with regard to the specific threats to these species (Sukumar 2003).

The population of wild African elephants (ca. 400,000) may be tenfold larger than that of the Asian elephant (ca. 45,000) with a corresponding difference in presently available habitat area; >5 million km^2 in Africa versus about 0.5 million km^2 in Asia. Extant Asian elephants are scattered in a series of isolated populations in fragmented landscapes. The majority of these populations may be small and not viable in the long term (Sukumar 1995b). Thus, the availability of sufficient habitat and the maintenance of large, integral landscapes are key to the survival of Asian elephant populations. This factor is also important in some regions of Africa, for instance, in western Africa, where the fragmented habitats and isolated elephant populations resemble the Asian situation. The fragmentation of habitat is not just due to human settlement and agricultural expansion. Roads, railway lines, dams, canals, and other developmental projects are also barriers to elephant movement. The maintenance or augmentation of existing "corridors" or creation of new corridors may thus be key to maintaining landscape integrity.

Human-elephant conflict is an issue of such magnitude that it must be resolved if local people are ever to consider and accept elephant conservation measures. In many regions, people who have traditionally coexisted with elephants have developed simple methods to chase them away from their settlements and crop fields. In the modern world, people's perception of the elephant and their tolerance to its crop raiding is changing. Barriers such as high-voltage electric fences are widely used in Asia and in Africa (Sukumar 1989; Thouless and Sakwa 1995; Nath and Sukumar 1998; O'Connell-Rodwell et al. 2000). These tactics can be reasonably successful if local communities actively participate in their maintenance (see Kiiru, Chapter 19 in this volume; Seneviratne and Rossel, Chapter 17 in this volume). Some form of elephant population management is also needed to contain crop raiding and loss of human lives (Sukumar 1991).

Poaching for ivory and other elephant products is linked to a hierarchy of markets, from local to regional and international (Barbier et al. 1990; Dublin et al. 1995; Menon, Sukumar, and Kumar 1997; Nash 1997; Martin and Stiles 2002; Duffy, Chapter 22 in this volume). Strategies to control poaching should thus include law enforcement by the range states, public education about the value of elephants, and the regulation of illegal international trade through effective enforcement of CITES, the Convention on International Trade in Endangered Species of Wild Flora and Fauna.

One additional factor has only marginal relevance for African elephants but is very important for Asian elephants, namely, the welfare and management of animals in captivity. About 16,000 Asian elephants, or about 30% of the global population, are kept in timber camps, zoos, or temples or are privately owned (Krishnamurthy and Wemmer 1995; Lair 1997). With a few notable exceptions (Kurt 1995; Sukumar et al. 1997; Taylor and Poole 1998), most of these captive stocks are declining because of the paucity or absence of captive births combined with excessive deaths and inadequate welfare.

Elephants can be powerful flagships for the conservation of the rich biodiversity of Asian and African tropical landscapes. It will be an insufficient conservation outcome if elephants merely survive as species. We must conserve elephant populations that actually represent their present diverse genetic makeup (Fernando et al. 2000; Fleischer et al. 2001; Eggert, Rasner, and Woodruff 2002) as well as the spectrum of landscapes they presently inhabit. The complex challenges to elephants' survival can be addressed only through a balanced, pragmatic, and ethical approach to conservation (Sukumar 2003).

References

Abe, E. L. 1994. The behavioural ecology of elephant survivors in Queen Elizabeth National Park, Uganda. DPhil diss., University of Cambridge.

Allaway, J. D. 1979. Elephants and their interactions with people in the Tana river region of Kenya. PhD diss., Cornell University.

Barbier, E. B., Burgess, J. C., Swanson, T. M., and Pearce, D. W. 1990. *Elephants, economics and ivory.* London: Earthscan Publications.

Barnes, R. F. W. 1996. The conflict between humans and elephants in the Central African forests. *Mammal Review* 26: 67–80.

Baskaran, N., Balasubramanian, M., Swaminathan, S., and Desai, A. A. 1995. Home range of elephants in the Nilgiri Biosphere Reserve, South India. In J. C. Daniel and H. S. Datye (eds.), *A week with elephants: Proceedings of the international seminar on Asian elephants, June 1993* (pp. 296–313). Bombay: Bombay Natural History Society and New Delhi: Oxford University Press.

Bax, P. N., and Sheldrick, D. L. W. 1963. Some preliminary observations on the food of elephants in the Tsavo Royal National Park (East) of Kenya. *East African Wildlife Journal* 1: 40–53.

Blair, J. A. S., Boon, G. G., and Noor, N. M. 1979. Conservation or cultivation: The confrontation between the Asian elephant and land development in peninsular Malaysia. *Land Development Digest* 2: 27–59.

Buss, I. O. 1977. Management of big game with particular reference to elephants. *Malayan Nature Journal* 31: 59–71.

Carrington, R. 1958. *Elephants: A short account of their natural history, evolution and influence on mankind.* London: Chatto and Windus.

Caughley, G. 1976. The elephant problem: An alternative hypothesis. *East African Wildlife Journal* 14: 265–283.

Caughley, G., Dublin, H., and Parker, I. 1990. Projected decline of the African elephant. *Biological Conservation* 54: 157–164.

Cerling, T. E., Harris, J. M., MacFadden, B. J., Leakey, M. G., Quade, J., Eisenmann, V., and Ehleringer, J. R. 1997. Global change through the Miocene/Pliocene boundary. *Nature* 389: 153–158.

Clemens, E. T., and Maloiy, G. M. O. 1982. The digestive physiology of three East African herbivores: The elephant, rhinoceros and hippopotamus. *Journal of Zoology* (London) 198: 141–156.

Corfield, T. F. 1973. Elephant mortality in the Tsavo National Park, Kenya. *East African Wildlife Journal* 11: 339–368.

Croze, H. 1974. The Seronera bull problem: Part II. The trees. *East African Wildlife Journal* 12: 29–47.

Dierenfeld, E. S. 1994. Nutrition and feeding. In S. K. Mikota, E. L. Sargant, and G. S. Ranglak (eds.), *Medical management of the elephant* (pp. 69–79). West Bloomfield, MI: Indira Publishing House.

Douglas-Hamilton, I. 1972. On the ecology and behaviour of the African elephant. DPhil diss., Oxford University.

Douglas-Hamilton, I. 1987. African elephants: Population trends and their causes. *Oryx* 21: 11–14.

Douglas-Hamilton, I., and Douglas-Hamilton, O. 1975. *Among the elephants.* New York: Viking Press.

Dublin, H. T., Sinclair, A. R. E., and McGlade, J. 1990. Elephants and fire as causes of multiple stable states in the Serengeti-Mara woodlands. *Journal of Animal Ecology* 59: 1147–1164.

Dublin, H. T., Milliken, T., and Barnes, R. F. W. 1995. *Four years after the CITES ban: Illegal killing of elephants, ivory trade and stockpiles.* Gland, Switzerland: International Union for Conservation of Nature and Natural Resources / Species Survival Commission African Elephant Specialist Group.

Duffy, K. J., Page, B. R., Swart, J. H., and Bajic, V. B. 1999. Realistic parameter assessment for a well known elephant-tree ecosystem model reveals that limit cycles are unlikely. *Ecological Modelling* 121: 115–125.

Duquette, J. T. 1991. Detoxification and mineral supplementation as functions of geophagy. *American Journal of Clinical Nutrition* 53: 448–56.

Edgerton, F. (trans.). 1931. *The elephant-lore of the Hindus: The elephant-sport (Matangalila) of Nilakantha.*: New Haven, CT: Yale University Press. Repr., New Delhi: Motilal Banarsidass, 1985.

Eggert, L. S., Rasner, C. A., and Woodruff, D. S. 2002. The evolution and phylogeography of the African elephant inferred from mitochondrial DNA sequence and nuclear microsatellite markers. *Proceedings of the Royal Society, London, B* 269: 1993–2006.

Eisenberg, J. F., McKay, G. M., and Jainudeen, M. R. 1971. Reproductive behaviour of the Asiatic elephant (*Elephas maximus maximus* L.). *Behaviour* 38: 193–225.

Fernando, P., and Lande, R. 2000. Molecular genetic and behavioral analysis of social organization in the Asian elephant (*Elephas maximus*). *Behavioural Ecology and Sociobiology* 48: 84–91.

Fernando, P., Pfrender, M. E., Enclada, S. E., and Lande, R. 2000. Mitochondrial DNA variation, phylogeography, and population structure of the Asian elephant. *Heredity* 84: 362–372.

Field, C. R., and Ross, I. C. 1976. The savanna ecology of Kidepo Valley National Park. II. Feeding ecology of elephant and giraffe. *East African Wildlife Journal* 14: 1–15.

Fleischer, C., Perry, E. A., Muralidharan, K., Stevens, E. E., and Wemmer, C. M. 2001. Phylogeography of the Asian elephant (*Elephas maximus*) based on mitochondrial DNA. *Evolution* 55: 1882–1892.

Folstad, I., and Karter, A. J. 1992. Parasites, bright males and the immuno-competence handicap. *American Naturalist* 139: 603–622.

Gadgil, M., and Nair, P. V. 1984. Observations on the social behaviour of free ranging groups of tame Asiatic elephant (*Elephas maximus* Linn). *Proceedings of the Indian Academy of Sciences (Animal Sciences)* 93: 225–233.

Garstang, M., Larom, D., Raspet, R., and Lindeque, M. 1995. Atmospheric controls on elephant communication. *Journal of Experimental Biology* 198: 939–951.

Gheerbrant, E., Sundre, J., and Cappetta, H. 1996. A Palaeocene proboscidean from Morocco. *Nature* 383: 68–70.

Guthrie, R. D. 1984. Mosaics, allelochemics, and nutrients: An ecological theory of late Pleistocene megafaunal extinctions. In P. S. Martin and R. G. Klein (eds.), *Quaternary extinctions: A prehistoric revolution* (pp. 259–298). Tucson: University of Arizona Press.

Hanks, J., and McIntosh, J. E. A. 1973. Population dynamics of the African elephant (*Loxodonta africana*). *Journal of Zoology* 169: 29–38.

Haynes, G. 1991. *Mammoths, mastodonts and elephants: Biology, behavior, and the fossil record.* Cambridge: Cambridge University Press.

Hawthorne, W. D., and Parren, M. P. E. 2000. How important are forest elephants to the survival of woody plant species in Upper Guinean forests? *Journal of Tropical Ecology* 16: 133–150.

Hoare, R. E. 1999. Determinants of human-elephant conflict in a land-use mosaic. *Journal of Applied Ecology* 36: 689–700.

Jainudeen, M. R., Katongole, C. B., and Short, R. V. 1972. Plasma testosterone levels in relation to musth and sexual activity in the male Asiatic elephant, *Elephas maximus*. *Journal of Reproduction and Fertility* 29: 99–103.

Kiltie, R. A. 1984. Seasonality, gestation time, and large mammal extinctions. In P. S. Martin and R. G. Klein (eds.), *Quaternary extinctions: A prehistoric revolution* (pp. 299–314). Tucson: University of Arizona Press.

Koch, P. L., Heisinger, J., Moss, C., Carlson, R. W., Fogel, M. L., and Behrensmeyer, A. K. 1995. Isotopic tracking of change in diet and habitat use in African elephants. *Science* 267: 1340–1343.

Krishnamurthy, V., and Wemmer, C. 1995. Timber elephant management in the Madras Presidency of India (1844–1947). In J. C. Daniel and H. S. Datye (eds.), *A week with elephants: Proceedings of the international seminar on Asian elephants, June 1993* (pp. 456–472). Bombay: Bombay Natural History Society; New Delhi: Oxford University Press.

Kurt, F. 1974. Remarks on the social structure and ecology of the Ceylon elephant in the Yala National Park. In V. Geist and F. Walther (eds.), *The behaviour of ungulates and its relation to management* (Vol. 2, pp. 618–634). Morges, Switzerland: International Union for Conservation of Nature and Natural Resources.

Kurt, F. 1995. The preservation of Asian elephants in human care: A comparison between the different keeping systems in South Asia and Europe. *Animal Research and Development* 41: 38–60.

Lahiri Choudhury, D. K. 1991. Indian myths and history. In S. K. Eltringham (ed.), *The illustrated encyclopedia of elephants* (pp. 130–147). London: Salamander Books.

Lahiri Choudhury, D. K. 1992. Musth in Indian elephant lore. In J. Shoshani (ed.), *Elephants: Majestic creatures of the wild* (pp. 82–84). Emmaus, PA: Rodale Press.

Lair, R. C. 1997. *Gone astray: The care and management of the Asian elephant in domesticity.* Bangkok: United Nations Food and Agriculture Organization Regional Office for Asia and the Pacific.

Langbauer, W. R., Jr., Payne, K. B., Charif, R. A., Rapaport, L., and Osborn, F. 1991. African elephants respond to distant playbacks of low-frequency conspecific calls. *Journal of Experimental Biology* 157: 35–46.

Laws, R. M. 1970. Elephants as agents of habitat and landscape change in East Africa. *Oikos* 21: 1–15.

Laws, R. M., Parker, I. S. C., and Johnstone, R. C. B., 1975. *Elephants and their habitats: The ecology of elephants in North Bunyoro, Uganda.* Oxford: Clarendon Press.

Leuthold, W. 1977. Spatial organization and strategy of habitat utilization of elephants in Tsavo National Park, Kenya. *Zeitschrift für Säugetierkunde* 42: 358–379.

Lieberman, D., Lieberman, M., and Martin, C. 1987. Notes on seeds in elephant dung from Bia National Park, Ghana. *Biotropica* 19: 365–369.

Lindeque, M., and Lindeque, P. M. 1991. Satellite tracking of elephants in northwestern Namibia. *African Journal of Ecology* 29: 196–206.

Lister, A. M. 1996. Dwarfing in island elephants and deer: Processes in relation to time of isolation. *Symposium of the Zoological Society of London* 69: 277–292.

MacPhee, R. D. E., and Marx, P. A. 1997. The 40,000–year plague: Humans, hyperdisease, and first-contact extinctions. In S. M. Goodman and B. D. Patterson (eds.), *Natural change and human impact in Madagascar* (pp. 169–217). Washington, DC: Smithsonian Institution Press.

Maglio, V. J. 1973. Origin and evolution of the Elephantidae. *Transactions of the American Philosophical Society of Philadelphia*, n.s., 63: 1–149.

Mahboubi, M., Ameur, R., Crochet, J. Y., and Jaeger, J. J. 1984. Earliest known proboscidean from early Eocene of north-west Africa. *Nature* 308: 543–544.

Martin, E., and Stiles, D. 2002. *The South and Southeast Asian ivory markets.* Nairobi: Save the Elephants.

Martin, P. S. 1984. Prehistoric overkill: The global model. In P. S. Martin and R. G. Klein (eds.), *Quaternary extinctions: A prehistoric revolution* (pp. 354–403). Tucson: University of Arizona Press.

Martin, P. S., and Klein, R. G. (eds.). 1984. *Quaternary extinctions: A prehistoric revolution.* Tucson: University of Arizona Press.

McComb, K., Moss, C., Durant, S. M., Baker, L., and Sayialel, S. 2001. Matriarchs as repositories of social knowledge in African elephants. *Science* 292: 491–494.

McCullagh, K.G. 1973. Are African elephants deficient in essential fatty acids? *Nature* 242: 267–268.

McKay, G. M. 1973. Behavior and ecology of the Asiatic elephant in southeastern Ceylon. *Smithsonian Contributions to Zoology* 125: 1–113.

Menon, V. 2002. *Tusker: The story of the Asian elephant.* New Delhi: Penguin Books India.

Menon, V., Sukumar, R., and Kumar, A. 1997. *A God in distress: Threats of poaching and the ivory trade to the Asian elephant in India.* Bangalore: Asian Elephant Conservation Centre and New Delhi: Wildlife Protection Society of India.

Milner-Gulland, E. J., and Beddington, J. R. 1993. The exploitation of elephants for the ivory trade: An historical perspective. *Philosophical Transactions of the Royal Society of London*, ser. B 252: 29–37.

Mosimann, J. E., and Martin, P. S. 1975. Simulating overkill by Paleoindians. *American Scientist* 63: 304–313.

Moss, C. J. 1983. Oestrous behaviour and female choice in the African elephant. *Behaviour* 86: 167–196.

Moss, C. J. 1988. *Elephant memories: Thirteen years in the life of an elephant family.* New York: William Morrow.

Nash, S. (ed.). 1997. *Still in business: The ivory trade in Asia, seven years after the ivory ban.* Cambridge: Traffic International.

Nath, C. D., and Sukumar, R. 1998. *Elephant-human conflict in Kodagu, southern India: Distribution patterns, people's perceptions and mitigation methods.* Bangalore: Asian Elephant Research & Conservation Centre, Indian Institute of Science.

Niemuller, C., and Liptrap, R. M. 1991. Altered androstenedione to testosterone ratios and LH concentrations during musth in the captive male Asian elephant (*Elephas maximus*). *Journal of Reproduction and Fertility* 91: 139–146.

Norton-Griffiths, M. 1979. The influence of grazing, browsing, and fire on the vegetation dynamics of the Serengeti. In A. R. E. Sinclair and M. Norton-Griffiths (eds.), *Serengeti: Dynamics of an ecosystem* (pp. 310–352). Chicago: University of Chicago Press.

O'Connell-Rodwell, C. E., Rodwell, T., Rice, M., and Hart, L. A. 2000. Living with the modern conservation paradigm: Can agricultural communities co-exist with elephants? A five-year case study in East Caprivi, Namibia. *Biological Conservation* 93: 381–391.

Olivier, R. C. D. 1978. On the ecology of the Asian elephant. DPhil diss., University of Cambridge.

Osborn, F. V. 1998. The ecology of crop-raiding elephants in Zimbabwe. DPhil diss., University of Cambridge.

Payne, K. B., Langbauer, W. R., Jr., and Thomas, E. M. 1986. Infrasonic calls of the Asian elephant (*Elephas maximus*). *Behavioural Ecology and Sociobiology* 18: 297–301.

Pellew, R. A. P. 1983. The impacts of elephant, giraffe, and fire upon the *Acacia tortilis* woodlands of the Serengeti. *African Journal of Ecology* 21: 41–74.

Pilgram, T., and Western, D. 1986. Inferring hunting patterns on African elephants from tusks in the international ivory trade. *Journal of Applied Ecology* 23: 503–514.

Poole, J. H. 1987. Rutting behavior in African elephants: The phenomenon of musth. *Behaviour* 102: 283–316.

Poole, J. H. 1989a. Announcing intent: The aggressive state of musth in African elephants. *Animal Behaviour* 37: 140–152.

Poole, J. H. 1989b. Mate guarding, reproductive success and female choice in African elephants. *Animal Behaviour* 37: 842–849.

Poole, J. H., and Moss, C. J. 1981. Musth in the African elephant, *Loxodonta africana*. *Nature* 292: 830–831.

Poole, J. H., and Moss, C. J. 1989. Elephant mate searching: Group dynamics and vocal and olfactory communication. *Symposium of the Zoological Society of London* 61: 111–125.

Poole, J. H., Payne, K., Langbauer, W. R., and Moss, C. J. 1988. The social context of some very low frequency calls of African elephants. *Behavioural Ecology and Sociobiology* 22: 385–392.

Rangarajan, L. N. (ed., trans.). 1992. Kautilya: The *Arthashastra*. New Delhi: Penguin Books.

Rasmussen, L. E., Buss, I. O., Hess, D. L., and Schmidt, M. J. 1984. Testosterone and dihydrotestosterone concentrations in elephant serum and temporal gland secretions. *Biology of Reproduction* 30: 352–362.

Rasmussen, L. E., Schmidt, M. J., Henneous, R., Groves, D., and Daves, G. D., Jr. 1982. Asian bull elephants: Flehmen-like responses to extractable components in female elephant estrous urine. *Science* 217: 159–162.

Rasmussen, L. E. L. 1998. Chemical communication: An integral part of functional Asian elephant (*Elephas maximus*) society. *Ecoscience* 5: 410–426.

Rasmussen, L. E. L., Haight, J., and Hess, D. L. 1990. Chemical analysis of temporal gland secretions collected from an Asian bull elephant during a four-month musth episode. *Journal of Chemical Ecology* 16: 2167–2181.

Rasmussen, L. E. L., Lee, T. D., Roelofs, W. L., Zhang, A., and Daves, G. D., Jr. 1996. Insect pheromone in elephants. *Nature* 379: 684.

Rasmussen, L. E. L., Riddle, H. S., and Krishnamurthy, V. 2002. Mellifluous matures to malodorous in musth. *Nature* 415: 975–976.

Rosenthal, G. A., and Janzen, D. H. (eds.). 1979. *Herbivores: Their interactions with secondary plant metabolites*. New York: Academic Press.

Ross, D. (ed.). 1992. *Elephant: The animal and its ivory in African culture*. Fowler Museum of Cultural History. Los Angeles: University of California.

Ruggiero, R. G., and Fay, J. M. 1994. Utilization of termitarium soils by elephants and its ecological implications. *African Journal of Ecology* 32: 222–232.

Scullard, H. H. 1974. *The elephant in the Greek and Roman world*. Ithaca, NY: Cornell University Press.

Short, J. C. 1981. Diet and feeding behaviour of the forest elephant. *Mammalia* 45: 177–185.

Shoshani, J. 1996. Evolution of the Proboscidea. In J. Shoshani and P. Tassy (eds.), *The Proboscidea: Evolution and palaeoecology of elephants and their relatives* (pp. 18–33). New York: Oxford University Press.

Shoshani, J., and Tassy, P. 1996. Summary, conclusions, and a glimpse into the future. In J. Shoshani and P. Tassy (eds.), *The Proboscidea: Evolution and palaeoecology of elephants and their relatives* (pp. 335–348). New York: Oxford University Press.

Sinclair, A. R. E. 1981. Environmental carrying capacity and the evidence for over abundance. In P. A. Jewell and S. Holt (eds.), *Problems in management of locally abundant wild mammals* (pp. 247–257). New York: Academic Press.

Spinage, C. A. 1973. A review of ivory exploitation and elephant population trends in Africa. *East African Wildlife Journal* 11: 281–289.

Sukumar, R. 1985. *Ecology of the Asian elephant (Elephas maximus) and its interaction with man in south India.* PhD diss., Indian Institute of Science, Bangalore.

Sukumar, R. 1989. *The Asian elephant: Ecology and management* (2nd rev. ed., 1992). Cambridge: Cambridge University Press.

Sukumar, R. 1991. The management of large mammals in relation to male strategies and conflict with people. *Biological Conservation* 55: 93–102.

Sukumar, R. 1994. *Elephant days and nights: Ten years with the Indian elephant.* New Delhi: Oxford University Press.

Sukumar, R. 1995a. Elephant raiders and rogues. *Natural History* 104: 52–60.

Sukumar, R. 1995b. Minimum viable populations for elephant conservation. In J. C. Daniel and H. S. Datye (eds.), *A week with elephants: Proceedings of the international seminar on Asian elephants, June 1993* (pp. 279–288). Bombay: Bombay Natural History Society and New Delhi: Oxford University Press.

Sukumar, R. 2003. *The living elephants: Evolutionary ecology, behavior and conservation.* New York: Oxford University Press.

Sukumar, R., and Gadgil, M. 1988. Male-female differences in foraging on crops by Asian elephants. *Animal Behaviour* 36: 1233–1235.

Sukumar, R., Krishnamurthy, V., Wemmer, C., and Rodden, M. 1997. Demography of captive Asian elephants (*Elephas maximus*) in southern India. *Zoo Biology* 16: 263–272.

Sukumar, R., and Ramesh, R. 1992. Stable carbon isotope ratios in Asian elephant collagen: Implications for dietary studies. *Oecologia* 91: 536–539.

Sukumar, R., and Ramesh, R. 1995. Elephant foraging: Is browse or grass more important? In J. C. Daniel and H. S. Datye (eds.), *A week with elephants: Proceedings of the international seminar on Asian elephants, June 1993* (pp. 368–374). Bombay: Bombay Natural History Society and New Delhi: Oxford University Press.

Taylor, V. J., and Poole, T. B. 1998. Captive breeding and infant mortality in Asian elephants: A comparison between 20 Western zoos and 3 eastern elephant centers. *Zoo Biology* 17: 311–332.

Thouless, C. R. 1996. Home ranges and social organization of female elephants in northern Kenya. *African Journal of Ecology* 34: 284–297.

Thouless, C. R., and Sakwa, J. 1995. Shocking elephants: Fences and crop raiders in Laikipia district, Kenya. *Biological Conservation* 72: 99–107.

Todd, N. E., and Roth, V. L. 1996. Origin and radiation of the Elephantidae. In J. Shoshani and P. Tassy (eds.), *The Proboscidea: Evolution and palaeoecology of elephants and their relatives* (pp. 193–202). New York: Oxford University Press.

Turkalo, A., and Fay, J. M. 1995. Studying forest elephants by direct observation: Preliminary results from the Dzanga clearing, Central African Republic. *Pachyderm* 20: 45–54.

van der Merwe, N. J., Lee-Thorp, J. A., and Bell, R. H. V. 1988. Carbon isotopes as indicators of elephant diets and African environments. *African Journal of Ecology* 26: 163–172.

Vartanyan, S. L., Garutt, V. E., and Sher, A. V. 1993. Holocene dwarf mammoths from Wrangel Island in the Siberian Arctic. *Nature* 362: 337–340.

Watve, M. G., and Sukumar, R. 1997. Asian elephants with longer tusks have lower parasite loads. *Current Science* 72: 885–889.

Webb, S. D. 1984. Ten million years of mammal extinctions in North America. In P. S. Martin and R. G. Klein (eds.), *Quaternary extinctions: A prehistoric revolution* (pp. 189–210). Tucson: University of Arizona Press.

Weir, J. S. 1972. Spatial distribution of elephants in an African national park in relation to environmental sodium. *Oikos* 23: 1–13.

West, R. M. 1980. Middle Eocene large mammal assemblage with Tethyan affinities, Ganda Kas Region, Pakistan. *Journal of Paleontology* 54: 508–533.

Western, D., and Lindsay, W. K. 1984. Seasonal herd dynamics of a savanna elephant population. *African Journal of Ecology* 22: 229–244.

Western, D., and van Praet, C. 1973. Cyclical changes in the habitat and climate of an East African ecosystem. *Nature* 241: 104–106.

White, L. J. T., Tutin, C. E. G., and Fernandez, M. 1993. Group composition and diet of forest elephants, *Loxodonta africana cyclotis* Matschie 1900, in the Lopé Reserve, Gabon. *African Journal of Ecology* 31: 181–199.

Whitehouse, A. M., and Hall-Martin, A. J. 2000. Elephants in Addo National Park, South Africa: Reconstruction of the population's history. *Oryx* 34: 46–55.

Zahavi, A. 1975. Mate selection: Selection for a handicap. *Journal of Theoretical Biology* 53: 205–214.

3 PERSONHOOD, MEMORY, AND ELEPHANT MANAGEMENT

GARY VARNER

In day-to-day speech, "person" is often used synonymously with "human being." In philosophical ethics, however, as in some political debates, the term is used to refer to individuals with certain cognitive or other characteristics, which, it is claimed, give them special moral status in comparison to, or even over, others that lack these characteristics.

In the abortion debate, for instance, if "person" just meant "human being," there would be no point in asking when a developing human being becomes a person. As used in this hot-button political debate, "persons" are individuals with moral status equivalent to, or at least approximating that of normal adults, including, presumably, a right to life. The debate over abortion and infanticide focuses largely on when developing human beings acquire whatever characteristics qualify them for personhood in this sense, and, of course, on what those characteristics are. Some claim that it is having a soul, or just being a member of our species that confers this special status, while others claim that it is the acquisition of consciousness or certain cognitive characteristics. Thus, there is a range of views about personhood when it comes to humans: some believe that an embryo is a person from the moment of conception, others believe that personhood arises at some fetal stage, and still others believe that human beings become persons well after birth and, sometimes, stop being persons before they die.

The term "person" is normally used in such a specialized sense by ethicists when discussing their views of animal welfare and animal rights. Species membership is generally dismissed as a criterion for personhood and instead the focus is on cognitive capacities. Thus, it could turn out that some nonhuman animals are persons, depending on which capacities are used to define personhood.

Although there is hardly a consensus on the issue, a number of authors have offered accounts of personhood that converge on what I call the "autonoetic consciousness paradigm." "Autonoetic" refers to self-knowledge, and some psychologists use the term "autonoetic consciousness" as a label for conscious awareness of one's own past, present, and future. Examples of the autonoetic consciousness paradigm are found in the works of R. M. Hare (1981, 1993), Jeff McMahan (2002), Peter Singer (1987, 1993), Michael Tooley (1983), and myself (1998), among others. These authors all agree that individuals with a robust, conscious sense of their own past and future deserve a special form of respectful treatment. This special respect is sometimes expressed in terms of "having rights," but it need not be so. Most generally, the idea behind the autonoetic consciousness paradigm is this: If an animal has a robust sense of its own past and future, then it deserves a special kind of respect over animals that lack this kind of consciousness, and it is not permissible to treat it in certain ways that it would be permissible to treat animals that, by contrast, "live entirely in the present."

This is how personhood, memory, and elephant management may be connected. The memory of elephants is legendary in certain traditional cultures and is celebrated in contemporary western pop culture. We can infer that elephants might be persons—or at least "near-persons"—under the autonoetic consciousness paradigm because the legendary memory of elephants is indicative of autonoetic consciousness. If this is true, then respectful treatment of elephants may entail forgoing management strategies that would be acceptable with other animals. For instance, assuming that deer are "merely sentient," it may be appropriate to cull them humanely for the sake of population control. However, if elephants might turn out to be persons—or even near-persons—could we still justify culling them as a means of population control?

In this chapter, I will not offer a complete defense of the autonoetic consciousness paradigm, but I will next try to articulate, in a clear and plausible-sounding way, two kinds of arguments in defense of it. Then, in discussing its application to elephants, I will first consider what science tells us or could someday tell us about whether elephants have autonoetic consciousness. Then I will discuss autonoetic consciousness from a practical perspective; if elephants turned out to be persons—or at least near-persons—what would this imply for elephant management?

Justifying the Autonoetic Consciousness Paradigm (ACP)

In a 1986 book, philosopher James Rachels used the term "biographical life" to shed light on life-and-death decisions in the context of human medicine. He argued that "the sanctity of life ought to be interpreted as protecting lives in the biographical sense, and not merely life in the biological sense." He characterized "biographical life" as "the sum of one's aspirations, decisions, activities, projects, and human relationships. The point of the rule against killing is the protection of *lives* and the interests that some beings, including ourselves, have in virtue of the fact that we are subjects of lives" (p. 5). Once an individual loses life in this biographical sense, Rachels argued, what makes human life particularly valuable has been lost, and there is little point in prolonging biological life. An irreversibly comatose person has already lost life in this sense, and a severely brain-damaged baby will never develop it, but so too people with advanced brain-wasting diseases may already have lost their biographical lives.

In what I call the autonoetic consciousness paradigm, a similar special status is afforded to any animals that have a robust, conscious sense of their own past and future. This is not to say that any animal has biographical life in the same sense that humans normally do. For a biography is a story, and storytelling is more likely to be a uniquely human characteristic than the more familiar candidates of tool making and language use. As individual humans, our identities are tied to a story we can tell about ourselves, which begins with where we came from, explains where we are now, and includes both aspirations for the future and an understanding of our own mortality. Communities, too, have biographies, and part of our identities as humans is a function of being able to tell and understand these stories. For this reason, we are unlikely ever to find full-blown *biographical* life in any nonhuman animal. Telling stories about ourselves and other beings requires far more than the simple syntax that has been taught to some great apes and dolphins. Without something very much like a human language, it may be impossible to represent one's life as a story, even to oneself.

Proponents of the autonoetic consciousness paradigm (or ACP) claim that a much more limited form of forward- and backward-looking consciousness would ground some kind of special respect for any animals that had such consciousness. If the paradigm of personhood is the normal adult human who is the subject of a biographical life, then we could say that animals with autonoetic consciousness are at least "*near*-persons," and as such deserve some higher form of respect than merely sentient animals.

The ACP thus implies a kind of moral hierarchy. At the bottom are *merely sentient* organisms—organisms that can consciously experience pain,[1] but have no robust sense of their own future and past—they "live entirely in the present." At the top of the ACP hierarchy are full-blown *persons*, with a biographical sense of their lives as wholes. Probably only human beings are persons in this strong sense, but among sentient nonpersons a further distinction can be drawn between merely sentient animals and those that qualify as *near-persons* because they have a robust, conscious sense of their own past and future, but fall short of having the normal human sense of their lives as complete biographies. Persons are claimed to deserve a special kind of respect over both near-persons and the merely sentient, and near-persons deserve some kind of special respect over the merely sentient.

Why think that an animal's having a robust sense of its own past and future should qualify it for some special kind of respectful treatment by humans? Here, I can give only sketches of two kinds of reasons that philosophers working within the autonoetic consciousness paradigm have given for endorsing the kind of hierarchy described above.

One reason appeals to common sense and consistency. This kind of argument begins from the premise that Rachels' account of biographical life makes good sense of commonsense judgments about the value of persons' lives. Then, if we learn that some nonhuman animals have a form of autonoetic consciousness that is somewhat similar to our biographical sense of self, shouldn't we treat those animals with some measure of special respect? That is, if common sense attributes special value to biographical consciousness, and if certain animals turn out to have a similar form of consciousness, then consistency requires that we treat them with a form of respect similar to, but not identical with, what persons are due according to common sense (for a more detailed version of this kind of argument, see Singer 1987 and Singer 1993, 127–131).

This argument will not appeal to those who believe humans are special simply because they are members of our species, but in the literature on animal welfare and animal rights, there is general agreement that species membership is irrelevant to personhood. At the same time, there is general agreement that consciousness mandates moral consideration. That is, if a being is incapable of suffering or enjoyment, then it has no direct moral standing, but, conversely, if it can suffer, then its suffering ought to be taken into consideration. From such a *sentientist* perspective, we can give the following, second kind of argument for the moral hierarchy endorsed by the ACP.

The sentientist holds that positive conscious states (pleasures and satisfactions) have intrinsic value and that negative conscious states (pain and dissatisfactions) have intrinsic *dis*value. Both persons and merely sentient animals are capable of experiencing momentary satisfactions in the pleasures of eating, drinking, and so on, but the merely sentient animal is capable only of this. The person is capable of these serial satisfactions but also has long-term and complicated desires, projects, and plans. Therefore, whatever the value of the momentary serial satisfactions of which each is capable, a person's life also includes the additional value that comes from the satisfaction of such long-term projects. When a person's life goes badly, it is also tragic in a sense that no merely sentient being's life can be because, in addition to losing all future opportunities for momentary satisfactions, the satisfactions a person achieves through long-term projects also go unfulfilled. From a sentientist perspective, then, persons' ability to project themselves into the future and past makes their lives especially important, in contrast to the lives of animals that live entirely in the present.

An analogous argument can be made that special respect is due animals that have a less robust sense of their own future. Near-persons have at least an analog of the profoundly long-term and complicated aspirations of persons. Their more limited sense of their own past and future allows them to form more complicated, longer-term desires than those of which merely sentient animals are capable, and the satisfaction of these desires, like the satisfactions persons receive from their long-term projects, adds some value to their lives that is missing from the lives of the merely sentient. Thus, near-persons are due some form of respect similar to but not identical with what persons are due (for a more detailed version of this kind of argument, see Varner 1998, Chapter 4).

These are two of the ways that the ACP value hierarchy has been defended. In the final section of this essay, I consider the characteristics of that "form of respect similar to, but not identical with what persons are due" with special attention to elephants. In the next section, I first consider what kinds of evidence could validate the claim that elephants have autonoetic consciousness, that is, a robust, conscious sense of their own past and future.

Looking for Autonoetic Consciousness in Elephants

For the reasons already discussed, it seems likely that only humans are persons with full-blown biographical lives. If autonoetic consciousness exists at all among animals, it probably differs in degrees. Perhaps no sentient animal lives *entirely* in the present. Some conscious sense of at least the immediate future seems necessary to have conscious desires or yearnings at all because to desire something is to want something to change. Therefore, between normal humans and animals that are merely sentient, there may lay a continuum of species having varying degrees of conscious awareness of the future and past. In discussing elephants' possible near-personhood, then, the question about elephants is, why should we believe that they are relatively close to us on this continuum?

In this section, I discuss three kinds of research relevant to determining whether elephants have a relatively robust, conscious sense of their own past, present, and future, in other words, an autonoetic consciousness. The three kinds of research concern: (1) episodic memory, which is the backward-looking portion of autonoetic consciousness, (2) mirror self-recognition, which is a sign of autonoetic consciousness in the present, and (3) the use of a theory of mind, which is relevant to the forward-looking element of autonoetic consciousness.

I will spend as much or more time describing these three lines of research as reporting results about elephants in particular. The reason is that, as Moti Nissani (a geneticist who has done behavioral studies of elephants) puts it, "Although elephants have been in close association with humankind for thousands of years, and although anecdotes about their wisdom or witlessness are many, their mentality has only been subject to a mere handful of controlled studies" (2004, 228). The dearth of data applies to each of the three types of research discussed in this section. No research at all exists on episodic memory in elephants. Only three studies of mirror self-recognition have been published, and only one very incomplete study of theory of mind in elephants has been published. The conclusion of this section will not be that elephants are or are not near-persons under the autonoetic consciousness paradigm. Discussing these three types of research will, however, illustrate that the question is amenable to scientific investigation, and, along the way, I give reasons for thinking that elephants are good candidates for autonoetic consciousness. Thus, it takes no great stretch of the imagination to think that scientific research may one day confirm that elephants are near-persons under the autonoetic consciousness paradigm, and this motivates the discussion in the final section of this essay.

Episodic Memory

Memory is the backward-looking third of autonoetic consciousness and striking anecdotal evidence suggests that elephants have really good memories. Field researchers (e.g., Moss 2000, caption to Agatha photo between 128 and 129, and text on 270–271) have reported that elephants linger among and closely examine the bones of long-deceased kin, and zoo elephants have been reported to remember keepers that they have not seen in dozens of years (Lewis 1978, 96). Some experimental evidence confirms that elephants have good memories. A classic study (Rensch 1957) showed that zoo elephants remembered which of up to twenty pairs of cards were the correct ones to choose for a reward, as much as a year after initially learning them. When field researchers in Africa broadcast recordings of the calls of elephant family members that had been absent from their groups for up to twelve years (either because they had died or emigrated to other groups), the remaining family members displayed "a strong affiliative response" (McComb et al. 2000, 1108).

Psychologists studying humans distinguish at least three kinds of memory. However, only one of these three definitely involves conscious awareness of the past. Psychologists draw a basic distinction between *procedural memory*, or "remembering how," and *propositional or declarative memory*, or "remembering that." The former is possible without the latter, as when a guitarist can play a complicated passage without being able to describe or in any other way answer questions about how to play it (besides just playing it, of course). Despite the terms used in this somewhat misleading label, "propositional or declarative memory" does not presume elucidating something by using mastery of a language; it only refers to the ability somehow to indicate answers to a problem or question, including, for instance, pointing out which of two images has been presented earlier in a lab trial, something that many animals can do. However, it is possible to have propositional or declarative memory of events without consciously remembering them at all. Therefore, since the 1970s, a further distinction has been drawn within propositional or declarative memory between *semantic memory*, or "symbolically representable knowledge that organisms possess about the world," and *episodic memory*, or the "remembering of personally experienced events" (Tulving 1985, 2 emphasis removed). For a prosaic illustration of the difference, think of a moviegoer who accurately recalls that she has seen a particular movie without having any conscious memory of when or where, or of what the plot or images in the movie were like. This person has no episodic memory of the event, but she has some semantic memory of it. Similarly, sometimes people can answer "forced recall" questions at far higher than chance accuracy about things of which they have no conscious memory. That is, they will "guess" right at a higher than chance rate, indicating that they do have relevant memories but not conscious ones. This is another example of semantic memory without episodic memory. (Sometimes such memories are called "implicit" memories, in contrast to "explicit" episodic memories.)

In 1985, psychologist Endel Tulving proposed a specialized account of episodic memory that he and colleagues have since expanded into a general account of what they call "autonoetic awareness":

> It occurs whenever one consciously recollects or re-experiences
> a happening from a specific time in the past, attends directly
> to one's present or on-line experience, or contemplates one's
> existence and conduct at a time in the future. Autonoetic aware-
> ness of the subjective past constitutes episodic retrieval. It repre-
> sents the major defining difference between episodic and semantic
> memory. (Wheeler, Stuss, and Tulving 1997, 350)

Elsewhere, they variously describe autonoetic awareness as "the kind of consciousness that mediates from the personal past through the present to the personal future," as "providing the characteristic phenomenal flavor of the experience of remembering" (Tulving 1985, 1), and as making possible "mental time travel" (Tulving 1985, 5; Wheeler et al. 1997, 331).

For the celebrated cases of elephant memory previously described to count in favor of the claim that elephants have autonoetic consciousness, we would have to be able to show that the elephants were having *episodic* memories. We would have to show that they were consciously recalling events rather than just "telling us," through their behavior, that those events had occurred. So, for instance, even if elephants clearly become emotionally aroused in the presence of a relative's carcass or an abusive trainer from long ago, they need not be consciously recalling events involving these former colleagues.

Tulving and associates have developed systematic ways of studying episodic memory, but none of their current research paradigms can be used on animals because they all involve language. Primarily, they have used word-list recall problems. When shown a list of words and asked to study them for later recall, subjects will respond with much higher than chance success when "cued" to recall words that they say they don't remember seeing. When given first a "free recall" test on the list, subjects will mention words that they say they consciously remember seeing. Then, if they are asked further "cueing" questions, such as "Do you remember any words similar to 'snake'?" they mention that similar words, such as "cake" or "snail," were on the list, and they are correct in the ones they suggest at a rate well above chance. In such cases, when asked "whether they actually 'remembered' its occurrence in the list or whether they simply 'knew' on some other basis that the item was a member of the study list," subjects will typically say that they "just knew it but don't actually remember it" (Tulving 1985, 8).

Certainly, it is hard to imagine how to distinguish episodic from semantic memory without relying on introspection and therefore on language. However, recent innovative research shows how a simple experiment can be used to show that some animals have at least episodic-like memory. Beginning in 1998, ethologist Nicola Clayton and colleagues developed a novel way of testing the hypothesis that scrub jays remember not only what happened and where but also *when* (see Clayton and Dickinson 1998). Scrub jays cache food in the wild, and Clayton et al. used this natural behavior to test their hypothesis by allowing the birds (in captive situations) to cache two kinds of food and retrieve them later. In one of the more sophisticated versions of the experiment, one of the foods was always peanuts, a stable, palatable, but less preferred food. The other was always an unstable but preferred food, either crickets or mealworms. The crickets and mealworms were treated to ensure that they would become unpalatable after specified periods of time, and the jays were allowed to learn this over a series of trials. Then the jays were tested to see which food they would try to retrieve after delays of various lengths, including some trials when the food items had been removed to ensure that the jays were not relying on visual or olfactory cues. Under two pairings:

1. peanuts paired with mealworms treated to be fresh through 4 hours but rotten by 28 hours and
2. peanuts paired with crickets treated to be fresh through 28 hours but rotten by 100 hours,

the birds were allowed to hide both kinds of food in each pair and then retrieve food after three lengths of delay: 4, 28, and 100 hours. If the birds remembered not only what food was hidden where but also *when*, then under pairing 1, they should have spent most of their time checking where they had cached mealworms after a 4-hour delay, but most of their time looking for peanuts after both 28- and 100-hour delays. Under pairing 2, they should have looked for mealworms after both 4- and 28-hour delays, switching to peanuts after 100 hours. The birds did exactly as predicted, indicating that they indeed had memory not only of what and where but *when* (Clayton et al. 2001, 1485).

Clayton et al. have always used, in the titles of their papers, the expression "*episodic-like* memory." Tulving originally described episodic memory as memory that "receives and stores information about temporally dated episodes or events" (1972, 385), suggesting that a memory's being somehow "time-stamped" suffices to make it an episodic memory. In subsequent papers, however, Tulving began including references to conscious reexperiencing of past events in his definitions of episodic memory.[2] Clayton et al.'s scrub jays clearly had time-stamped memories, but that still didn't show that in using those memories they were consciously recalling caching the foods. Indeed, as one commentary on their work observed, the time-stamp could even be incorporated in a perception of the present: "one could ask whether the birds remember the event of caching worms in a particular location, a memory that is oriented to the past, or simply know the current state of the world—that caches are in particular locations, and have been there for particular periods of time" (Hampton and Schwartz 2004, 3).

However, in normal human subjects, positron emission tomography (PET scanning) confirms that episodic memory is associated with intense activity in the prefrontal cortex (PFC). This suggests a way that scientists could strengthen the claim that some animals—at least some mammals—have episodic memories, using an argument by analogy. When subjects are asked to solve episodic memory problems (such as the word list problems described above) during a PET scan, intense activity occurs in the left PFC during the encoding of episodic memories, while intense activity occurs in the right side of the PFC during their recall (Wheeler et al. 1997, 335–343). The frontal cortex is the area of the cerebral cortex forward of the central gyrus. The frontal cortex as a whole is action oriented, whereas the rest of the cortex is sensory oriented (Fuster 1997, 3). The *prefrontal* cortex lies at the front tip of the frontal lobe, but it is defined in two different ways: in terms of how its cytoarchitecture (cellular architecture) differs from the rest of the frontal lobe or in terms of its connections to other parts of the brain. All mammals and only mammals have a PFC, defined as "the part of the cerebral cortex that receives projections from the mediodorsal nucleus of the thalamus" (Fuster 1997, 2). However, development of the PFC varies enormously across the mammals, with, proportionately, a strikingly larger PFC in primates than in all other mammals. No estimate of the relative size of the elephant's PFC is available, but a number of estimates of elephants' encephalization quotient (EQ) have been published. EQ is a systematic way of estimating how far growth in brain size outstrips growth in body size.

Specifically, the EQ expresses the relationship of a species' (or individual's) brain to body weight ratio to the overall average ratio for mammals. So an EQ of 2.0 means that the species' brain is twice the size predicted by the mammalian average, an EQ of 0.75 means that it is only three fourths of what would be expected, and so on. Published EQ estimates for elephants range widely:

> 0.97 for a male African elephant (Eisenberg 1981, 501t)
> 1.3 for an African elephant, sex unspecified (Jerison 1973, 344)
> 1.32 for an African elephant, sex unspecified (Cutler 1979, 77)
> 1.62–1.72 for Asian elephants, sexes unspecified (Eisenberg 1981, 501t)
> 1.87 for unspecified sex and species of elephant (Russell 1979, 138)
> 2.05 for a female African elephant (Eisenberg 1981, 501t)
> 2.28 for an Asian elephant, sex unspecified (Cutler 1979, 77)

However, on all but the first estimate listed, elephants compare favorably with primates, including the great apes:

> 1.40–1.68 for gorillas (Eisenberg 1981, 500t)
> 2.18–2.45 for chimpanzees (Eisenberg 1981, 500t)
> 2.49 for chimpanzees (Russell 1979, 138)
> 7.33–7.69 for humans (Eisenberg 1981, 500t)
> 7.44 for humans (Russell 1979, 138)

If elephants could be presented with problems that (like those the scrub jays faced) could only be solved using knowledge of *when* events occurred, and if, while solving those problems successfully, similar information on brain activity could be obtained as that provided from PET scans of human subjects in the verbal memory recall tests, a strong argument by analogy could be made for the claim that elephants have episodic (and not just episodic-like) memories. Such a study is impossible at present. Since subjects must lie still inside current PET scanning devices while being scanned, it is unclear how to give an animal subject a PET scan while it solves an episodic memory problem. In principle, however, other techniques for imaging brain activity could be developed and employed, and if the elephants both (a) solved the episodic-like memory problem and (b) displayed the same kind of lateralization of PFC activity as humans while doing so, this would provide a relatively strong argument by analogy for the claim that elephants do have episodic memory.

Even in the case of our fellow humans, we can never directly observe others' conscious states. We must, therefore, rely on arguments by analogy: we attribute similar conscious states to others based on similarities in (1) behavior, (2) neurophysiology, and (3) evolutionary history. Obtaining the results described in the preceding paragraph would show that similar problem-solving behaviors are correlated with similar neurophysiological structures and activities in humans and elephants.[3] The Proboscidea (elephants and their ancestors) diverged from other mammals about 55 million years ago (Shoshani 1991, 13), so humans and elephants do not share a close evolutionary history. For reasons that will be discussed later (in the theory of mind segment of this chapter), elephants do share with us a relevantly analogous evolutionary context. Therefore, if the previously described results are eventually forthcoming, the argument by analogy for episodic memory in elephants will be complete.

In summary, episodic memory is the backward-looking portion of autonoetic consciousness. Although it is still unclear how to study episodic memory without relying on introspection and therefore language, innovative techniques for studying memory of dated events, and therefore at least episodic-like memory in nonhumans, are being developed. In addition, in principle, these techniques could be applied to elephants and coupled with PET scan imaging to develop an argument by analogy to the conclusion that they possess this backward-looking third of autonoetic consciousness.

Mirror Self-Recognition

The well-known Gallup test of mirror self-recognition is commonly taken to be a sign of self-consciousness in the present, and here published studies of elephants do exist. In 1970, psychologist Gordon Gallup observed that when chimpanzees were first exposed to mirrors they reacted aggressively, as if confronting unfamiliar conspecifics (members of their own species), but they began, after further exposure, to act as if they recognized themselves in the mirrors:

> Such self-directed responding took the form of grooming parts of the body which would otherwise be visually inaccessible without the mirror, picking bits of food from between the teeth while watching the mirror image, visually guided manipulation of the anal-genital areas by means of the mirror, picking extraneous material from the nose by inspecting the reflected image, making faces at the mirror, blowing bubbles, and manipulating food wads with the lips by watching the reflection. (86)

That the chimpanzees recognized themselves in the mirror seemed obviously true, but Gallup proposed a simple, controlled test of the self-recognition hypothesis. He removed the mirrors and, placing these chimpanzees under anesthesia, applied nonirritating, nonodorous colored marks above one eyebrow and at the top of one of each of their ears. After they recovered from anesthesia, he counted the number of times they touched the marked areas, confirming that they did not touch these areas more often than other areas on their faces and ears. He then reintroduced the mirrors and counted again, finding that they now touched the marked areas at an enormously higher frequency than other areas. This, he argued, was clear proof that the chimpanzees recognized themselves.

He repeated the test on two species of macaques and found both (1) no mark-directed behavior involving the mirror after marking under anesthesia and (2) no decline in behavior indicating that they thought they were seeing an unfamiliar conspecific in the mirror. Thus was born the now familiar hypothesis that while the great apes are capable of self-recognition, monkeys are not. Thirty years later, Gallup, Anderson, and Shillito claimed this was confirmed by experiments on dozens of primate species (2002, 326).

The first two published experiments with elephants (Povinelli 1989; Hyatt et al. 2003) failed to find any mark-directed behavior, even though one of them (Povinelli) proved that elephants can use mirrors to retrieve food that they cannot see or smell. The third found significant mark-directed behavior, but only by one out of three elephants tested, and on only one of the four occasions, it was marked. That study used three female Asian elephants at the Bronx Zoo. After habituating them to the presence of a large (2.5 meter square) mirror, the elephants' handlers applied a mark on the right side of the elephants' heads and a sham mark on the left side.

> One elephant, Happy, passed the mark test on the first day of marking. Caretakers did not notice her touching either the mark or sham-mark before being released into the elephant yard. After being released into the yard, she walked straight to the mirror where she spent 10 seconds, then walked away. Seven minutes later she returned to the mirror, and over the course of the next minute she moved in and out of view of the mirror a couple of times, until she moved away again. In the following 90 seconds, out of view of the mirror, she repeatedly touched the visible mark but not the sham-mark. She then returned to the mirror, and while standing directly in front of it, repeatedly touched and further investigated the visible mark with her trunk. (Plotnik, de Waal, and Reiss 2006, 17054)

Happy was retested the next two days, and again two months later, but she never again showed mark-directed behavior. Neither of the other elephants (Maxine and Patty), who were tested twice and then again two months later, ever showed any interest in their marks. However, the authors emphasize that animals that have passed the spot test have generally progressed through four stages before facing the spot test:

1.   In the first phase, animals react to their mirror image as if to an unfamiliar conspecific.

2. In the second phase, they inspect the mirror in various ways, including (usually) looking behind it.

3. In the third, they engage in "repetitive mirror-testing behavior," things like systematically moving back and forth while studying the image in the mirror.

4. In the fourth, they exhibit various kinds of "self-directed behavior," such as inspecting their teeth or other areas of their bodies that they could not see without using the mirror. (Plotnik et al. 2006, 17053)

The authors emphasize that:

> All three elephants displayed behavior consistent with mirror-testing and self-directed behavior . . . such as bringing food to and eating right in front of the mirror (a rare location for such activity), repetitive, nonstereotypic trunk and body movements (both vertically and horizontally) in front of the mirror, and rhythmic head movements in and out of mirror view; such behavior was not observed in the absence of the mirror. . . . On more than one occasion, the elephants stuck their trunks into their mouths in front of the mirror or slowly and methodically moved their trunks from the top of the mirror surface downward. In one instance, Maxine put her trunk tip-first into her mouth at the mirror, as if inspecting the interior of her oral cavity, and in another instance, she used her trunk to pull her ear slowly forward toward the mirror. Because these behaviors were never observed in . . . the initial, "no mirror" control conditions . . . they indicate the elephants' tendency to use the mirror as a tool to investigate their own bodies. (Plotnik et al. 2006, 17054)

The authors characterize the fourth phase of mirror acclimation as "the beginning of mirror understanding" (17053), but it would be more accurate to say that animals who actively inspect otherwise invisible parts of their bodies using mirrors fully understand them, whether or not they pass the spot test. Various animals are known to acclimate to mirrors in the sense of progressing from stage one to two, when, after inspecting the mirror and looking behind it, they stop treating the mirror image as a conspecific. This is true of various species of monkeys that have failed the spot test and familiar animals such as cats. Although that kind of acclimation probably shouldn't count as understanding how mirrors work, when an animal enters phase three then it does appear to be testing hypotheses about how mirrors work. That, I suggest, is "the beginning of mirror understanding," which is fully achieved as soon as the animal exhibits self-directed behavior, whether or not it shows any interest specifically in spots that are applied to it by experimenters.

Both of the earlier published studies of elephants mentioned similar examples of apparently self-directed behavior, while being somewhat more cautious than Plotnik et al. in their interpretations of them. If the elephants understand that the mirror images are their reflections, then why don't they more consistently pass the spot test? Some general facts about them suggest that the Gallup spot test might not reliably reveal self-recognition in elephants. First, although elephants are known to use branches to scratch parts of their bodies, elephants do not extensively groom themselves the way primates do, and this may make elephants less likely to examine unusual spots on themselves and others. Also, elephants' eyesight is notoriously bad, they lack color vision, and their skin color changes noticeably as they spray themselves with water or mud, so they may simply disregard the kinds of marks traditionally used in the spot test (Nissani and Nissani 2001).

Although elephants' performance on the Gallup spot test has been mixed in the three studies published to date, the extent of self-directed behavior displayed by elephants exposed to mirrors still strongly suggests that elephants are capable of recognizing themselves in mirrors and hence that they have the present-time element of autonoetic consciousness.

Theory of Mind (ToM)

What about the forward-looking element of autonoetic consciousness? Here, long-term planning is the core of what we seek. This gives added significance to the large size of elephants' prefrontal cortex because it is confirmed that in humans the prefrontal cortex is heavily involved in long-term planning.

With long-term planning for the future, however, there is a problem analogous to that with long-term memory of the past: What we need to demonstrate is not just anticipation of the nonimmediate future but *conscious* anticipation. Just as it is possible to have semantic memory without episodic memory, it is possible to arrange things to serve one's long-term interests without any conscious awareness of one's long-term future. Presumably, this happens when a bird builds its first nest or a squirrel stores acorns before its first winter. When humans make complex plans for the long-term future, we know from introspection that this is done consciously. One way of underlining the difference between humans and all other animals is to emphasize the length and complexity of human projects (Varner 1998, Chapter 4). Probably no animal makes plans as complex and for as far into the future as humans when they arrange to provide for their children's education, their own retirement, and the interests of their heirs. It may be that language is necessary for this level of complexity in planning.

However, the use of a theory of mind (ToM) appears to involve conscious planning by animals for their own future, for several reasons. Theory of mind research is an area of great current interest to ethologists (scientists who study animal behavior, especially in natural settings), and one initial study has been published showing that elephants display at least one of the simpler components of a ToM.

To use a ToM is to interpret others' behaviors in light of their beliefs and desires. This constitutes "using a theory" insofar as the beliefs and desires are not observable but are theoretical constructs. All normal children acquire the human ToM in a developmentally stereotyped sequence and without explicit training, suggesting that humans are hard-wired to use ToM. According to Stone, Baron-Cohen, and Knight (1998, 640–641), at around 18 months, children first become able to understand what others are looking at in general and, in particular, that others are looking at the same objects as they. Between 18 and 24 months, they understand pretend play. By age 2, they understand and use attributions of desires to others. Between 3 and 4 years, they understand that others can hold different and sometimes false beliefs about the world. By 6 or 7, they understand that others are attributing mental states to various people, and thus they acquire an understanding of second-order false beliefs, that is, false beliefs about others' beliefs or desires. Finally, between 9 and 11 years, children understand the making of a faux pas, which requires a complex representation of multiple persons' beliefs and desires. To understand that "someone said something they shouldn't have said," one must understand both that the speaker did not know that he should not have said something and that the audience felt offended by what he said. Interestingly, girls develop this ability by 9 years, while boys take 11 years.

To suggest that some animals use ToM is to suggest that evolution has similarly programmed them to acquire at least some (but not necessarily all) of the abilities previously listed. Lacking language, no animal displays an understanding of faux pas, but both anecdotal field reports and some experiments suggest that some primates use an understanding of others' desires and beliefs to predict and manipulate others' behaviors (Whiten 1997). If they do use this relatively sophisticated and later-emerging element of ToM, it seems plausible, for a couple of reasons, to say that they do so consciously.

First, it seems that when we use ToM to predict another's behavior, we usually do so consciously. Introspectively, we describe our use of ToM in terms of "imagining future scenarios," and outside of Freudian psychologists' offices, to describe someone as "imagining" something is to describe them as consciously thinking about it. This is confirmed insofar as the prefrontal cortex is involved in ToM use (Stone et al. 1998) because our prefrontal cortex is involved in conscious planning for the future. Another reason for thinking that ToM use is conscious is that we sometimes attribute

beliefs and desires to others based on an understanding of what we ourselves would believe and desire under varying circumstances. In doing this,
we seem to be consciously projecting ourselves into various circumstances,
past, present, and future.

We can infer from some anecdotal field reports that elephants use a ToM
to understand and manipulate one another's behaviors. Joyce Poole, for instance, who has studied the elephants of Amboseli Park in Kenya since
1976, states that: "The evidence available on cognitive empathy, understanding death, sense of humour, imagination, the ability to teach, to imitate, and to deceive, are all very suggestive that elephants have a theory of
mind" (1998, 107).

The only experimental investigation of elephant ToM published to date
focused on one of the simpler and earlier-emerging portions of the human
ToM, specifically the ability to understand what others are looking at. Moti
Nissani (2004) applied an experimental paradigm first developed by Daniel
Povinelli, working with chimpanzees. In Povinelli's studies, chimpanzees
had the opportunity to beg from one of two experimenters who were holding something just out of reach in front of their cages. Only one of the two
experimenters could see the chimpanzee, either because only one was facing the cage, one had a bucket over her head, or one was blindfolded. On
analogous tasks, children aged 3 and up consistently solicit the attention
of the human who can see them, but Povinelli's chimps initially performed
at chance levels (Povinelli and Eddy 1996). Nissani applied the same paradigm to both elephants and chimpanzees older than Povinelli's and, in
some cases, wild reared. His results with the chimps were inconsistent with
Povinelli's, finding that six out of seven chimpanzees tested immediately
performed "significantly above chance" (Nissani 2004, 248).

Nissani's study involved only two elephants, Wanda and Winky, at the
Detroit Zoo. He therefore cautions that "the small sample, and the limited
number of trials . . . do not allow us to carry out meaningful statistical tests,"
but the results at least suggested the elephants understood who could and
could not see them. Although the elephants begged correctly less than two-
thirds of the time when a screen obscured one experimenter's gaze, or when
one experimenter had a bucket over her head (similar to the photo accompanying this chapter), they begged correctly 81% of the time when one was
lying face down or when one was standing sideways and looking away from
the elephant. These were much better performances than Povinelli's
chimps, whose best performance, in the bucket over the head condition,
was still only 58%. (All statistics from Nissani 2004, 244, table 1.)

Monitoring others' visual perspective is but one small part of what humans do with ToM. The ability to know that others are attending to the
same object emerges relatively early in development: Children acquire this
capacity around 18 months of age, well before they can attribute false be-

liefs to others. The various elements of theory of mind could have evolved at different times, and while the simpler elements, such as gaze monitoring, support the more sophisticated capacities, such as attributing beliefs and desires, the simpler elements could have evolved much earlier (Byrne and Whiten 1992; Povinelli and Eddy 1996, 14). Proving that elephants can—sometimes—monitor others' visual attention hardly shows that elephants have the kind of full-blown ToM we humans use to predict others' behaviors.

Although the first controlled experiment on elephants suggests they have one of the simpler elements of ToM, at present no decisive scientific evidence indicates that elephants use ToM to predict others' future behaviors the way humans consciously use their ToM. Unlike studies of episodic memory, however, more sophisticated means of studying theory of mind in animals have already been developed, and variations on these techniques could someday be applied to elephants. For instance, one way of testing for understanding of others' false beliefs involves asking young children where someone will look for an object after being out of the room while it was moved. Children over 3 or 4 years of age correctly predict that the person will look where the object was before that person left the room, indicating that they understand that the person has a false belief. Primate researchers have adapted this procedure as follows. A chimpanzee is presented with two experimenters and two boxes. The experimenters put food in one of the boxes, and then one of them leaves the room. The chimpanzee sees the other experimenter move the food to the other box and, when the first experimenter returns, the chimpanzee is allowed to beg one of them for food. Whom it chooses to beg from is taken to indicate whether it understands that one of the experimenters has been saddled with a false belief. This protocol could easily be adapted to test for elephants' understanding of false beliefs.

A reason for thinking elephants will pass such tests for ToM is that they are the kind of social animals that would benefit from mind reading. When Tulving and associates figuratively describe autonoetic consciousness as making possible "mental time travel," they speculate that the evolutionary value of this is the ability to play out future scenarios in the imagination (see Tulving 1985, 5). A similar adaptive significance has been claimed for ToM. As detailed in Joyce H. Poole's and Cynthia J. Moss's contribution to this volume (Chapter 4), elephants are long-lived, highly social organisms. They grow up in complex, changing social settings that involve interactions with multiple generations of individuals from variously related groups, over a life span up to 65 years. Such animals would benefit both from understanding others' intentions and being able to predict behaviors in light of them, and from being able to actively manipulate others' behaviors by manipulating their beliefs. So elephants are at least good candidates for ToM.

Summary

Episodic memory, mirror self-recognition, and use of a theory of mind reflect, respectively, autonoetic consciousness of the past, present, and future. Although no one has yet attempted to study their episodic memory, elephants clearly have very good, long-term semantic memory. Research on episodic-like memory problems, coupled with PET scan data, could someday provide a strong argument by analogy for the claim that elephants possess the backward-looking portion of autonoetic consciousness. Regarding self-consciousness in the present, elephants generally fail the Gallup spot test of mirror self-recognition, but there are reasons for thinking that the spot test will not reliably reveal self-recognition in elephants, and the extent of self-directed behavior displayed by elephants exposed to mirrors still strongly suggests that they are capable of recognizing themselves in mirrors. The one controlled study of theory of mind in elephants published to date confirms that they have one of the simpler elements of ToM. Elephants are the kind of animals ethologists think are good candidates for using a theory of mind. Future research could be designed to provide more definitive evidence about whether they do. Therefore, while there is not solid scientific evidence that elephants have autonoetic consciousness, the question can be scientifically investigated, and because elephants are good candidates for autonoetic consciousness, it is worth considering what would be the practical implications of near-personhood of elephants.

Near-Personhood and Elephant Management

According to the autonoetic consciousness paradigm, if elephants are near-persons, then they deserve some kind of special respect compared with merely sentient animals, although not necessarily the same kind of special respect full-blown persons are due. Here, I will focus on the question of culling overpopulated elephants, but what I say about culling will set a general framework for thinking about other issues, including working elephants, circuses, and zoos.

Regarding culling, what near-personhood implies will depend, first and fundamentally, on whether near-persons deserve rights in a strong sense. In the abortion literature, "persons" are sometimes defined in terms of having a very strong right to *life*, specifically. This right is strong in the sense that it trumps almost everything else, with proponents of fetal personhood sometimes going so far as to say that it prohibits abortion even in cases where the woman's health and life are threatened. Philosopher Tom Regan (1983) defends analogously strong rights for animals, arguing that if animals have rights in this sense, then we cannot justify culling them for the sake of population control. He writes:

> the goal of wildlife managers should be to defend wild animals in the possession of their rights, providing them with the opportunity to live their own life, by their own lights, as best they can, spared that human predation that goes by the name of "sport." . . . If, in reply, we are told that [not culling] does not guarantee that we will minimize the total amount of suffering wild animals will suffer over time, our reply should be that this cannot be the overarching goal of wildlife management, once we take the rights of animals seriously. (357)

The "reply" Regan imagines invokes utilitarian reasoning. Utilitarianism is the view that right actions, policies, and institutions are right because, ultimately, they maximize the aggregate happiness of those affected. Hunting for population control is commonly defended because it minimizes the opposite—unhappiness; however, Regan asserts that taking rights seriously means eschewing the utilitarian's ultimate goal of maximizing aggregate happiness.

If animals have rights in this sense of "trump cards against utilitarian arguments," then we cannot justify killing some of them on the utilitarian grounds that this will prevent more suffering than it causes in the culled animals, even if a *lot* of suffering would be prevented. In dramatic cases of overpopulation that have arisen among ungulates (especially deer and elk) in North America, this seems counterintuitive to most people. It is better, most people say, to humanely kill a smaller number of deer to save a larger number from starvation and improve living conditions for future generations of deer by preventing long-term habitat degradation.

Regan would condemn hunting for population control in the cases of both deer and elephants because he argues that all mammals equally deserve rights in the antiutilitarian sense. However, those who advocate on behalf of elephants commonly claim that they are special, that they deserve some kind of special respect in relation to the majority of other animals (see, for instance, Moss 2000, 317; Poole 1998, 107–108). They claim, in effect, that elephants are near-persons and that this rules out certain management practices that would be acceptable with other, cognitively simpler animals. So what if it turned out that only elephants and a few other animals are near-persons and a prohibition on culling for population control were imposed only for that small number of animal species, including elephants but excluding, for example, deer and elk? Would a prohibition on culling wild-animal herds be more plausible if restricted to just a few species, including elephants?

Unfortunately, elephants are remarkably like deer and elk when it comes to population dynamics, and this makes the consequences of not culling both predictable and terrible. If protected from human predation in currently available conservation areas, African elephant populations can grow at rates of up to 7% per year (Whyte and Fayrer-Hosken, Chapter 20 in this volume; here, I confine my discussion to African elephants because less field research has been done on Asian elephant populations). This implies a population doubling time that can be as short as a decade. At high densities, elephant populations cause dramatic habitat changes, transforming woodlands to grasslands. With human populations now boxing elephants into designated conservation areas, elephant populations cannot, as they might have done in ancient times, migrate to other woodland areas, allowing grasslands to revert to woodlands. As a result, unimpeded growth of elephant populations can cause the extinction of certain species of trees and animals. Relatedly, elephants rely on browse (bushes and trees) when grasses are dry (Sukumar 1991, 79), so grasslands cannot support as many elephants as can woodlands or savanna.

Like the ungulates of North America, then, African elephants have a tendency to overshoot the carrying capacity of their existing ranges, causing habitat degradation and diminished future carrying capacity. As Whyte and Fayrer-Hosken put it in Chapter 20, "as elephant densities increase, some other species will inevitably be lost, habitats will be transformed, and ultimately starvation will limit further elephant population growth." There are, they argue, only four options. Either we *actively* control the population by: (1) culling, (2) contraception, or (3) translocation, or we *passively* control the population by allowing (4) starvation. They conclude that the only currently feasible option for *actively* controlling elephant populations is culling because the areas available for translocation are limited, and currently available methods of contraception are prohibitively expensive, requiring massive programs of radio collaring and helicopter locating and darting of elephants in the larger African preserves.

I italicized the words "actively" and "passively" because some claim that when it comes to respecting persons or near-persons, it makes a difference whether we actively cause their deaths or passively allow their foreseeable deaths. To some extent, current human medical practice reflects this distinction by ruling out active euthanasia of the terminally ill while allowing withdrawal of life support. Regan holds a somewhat analogous position, that we should respect animals' rights by letting nature take its course rather than deciding who will die for the sake of population control. As he puts it, "Risks are not morally transferable to those who do not voluntarily choose to take them" (1983, 377), but in actively culling a population, we would be transferring risks from the animals likely to die in the natural course of events to those we choose to kill.

The killing / letting die distinction has some legitimate applications when it comes to persons, but Regan recognizes no distinction between the rights of human persons and animals. Within the autonoetic consciousness paradigm, however, even if the killing / letting die distinction were justified when it comes to full-blown persons, it could consistently be abandoned with regard to nonpersons. The ACP calls for special respect for near-persons compared with the merely sentient but not necessarily the same kind of respect due persons. So even if it were deemed appropriate to rule out actively killing persons for population control, we would not need to hold this view with regard to near-persons.

When it comes to culling for population control, one way of distinguishing between near-persons and merely sentient animals would be to raise the standard of justification by placing special restrictions on the culling of near-persons. I can think of at least two ways that contemporary policies already do something like this with regard to elephants compared with deer. First, while sport hunting of deer in North America is not uncontroversial, it is routinely used as a means of population control. In contrast, sport hunting of African elephants is not routinely permitted and even when it is used as a means of population control, it is always highly controversial. Therefore, we already distinguish between elephants and deer with regard to using sport hunters to cull populations. This can be understood as elephants requiring a higher level of respect as near-persons, which in turn mandates that only professionals are permitted to do the job. The most humane hunt will be unsporting. With deer, this might involve bait stations and automatic weapons (Varner 1998, 116). Today, in some African parks, elephants are culled by sharpshooters in low-flying helicopters (Whyte 2002, 303), while in others, elephants are herded by helicopters toward teams of marksmen who kill entire family groups at a time (Moss 2000, 315–316) to avoid orphaning some family members. While this sounds dreadful, these measures are appropriate if the goal is to cull humanely rather than rely on sport hunters to do the job on the cheap, as most areas in North America do with deer and elk.

Two other ways culling policies can reflect increased respect for near-persons are illustrated in the history of one African park's policies. When it comes to managing elephant populations, this story can be read as reflecting two different ways of erring on the side of caution. (The following details are based on Whyte 2002.) Kruger National Park in South Africa first placed a cap of 7,000 on its elephant population after a 1967 aerial census revealed the population was expanding quickly. This number was chosen because the park is about 8,000 square miles, and at the time, biologists at Tsavo National Park in Kenya were recommending a maximum density of one elephant per square mile. Although the population cap of 7,000 was somewhat arbitrary, it demonstrated a commitment to caution. Given that Tsavo's population has since expanded to five times that density, the fact that Kruger's biologists stuck to the lower number suggests their commitment to safeguarding the welfare of their park's elephants. That is, they seem to have chosen to maximize individual welfare. In both elephants and deer, on crowded ranges, density-dependent mechanisms, including nutritional stress, limit population to some extent by reducing the birth rate (Varner 1998, 116; Laws, Parker, and Johnstone 1975), but this drop indicates reduced vigor and health in the population.

In 1994, a lawsuit by an animal rights organization stopped Kruger's policy of annual culling. The park's response illustrates a second way of erring on the side of caution. Five years later, the park's management instituted a complex new policy, dividing the park into high- and low-impact elephant population zones. In the latter, plant and animal communities are closely monitored for evidence of elephant impacts. When a predetermined level of impact is measured (e.g., loss of some percentage of mature trees of a given species), a review process is triggered. Officials must then solicit input from all interested parties before deciding to cull and must give priority to whatever nonlethal means are available at the time. High- and low-impact zones will be rotated over long intervals, thus transferring the heaviest impact of elephants from area to area in the park. The new policy represents a second way of "erring on the side of caution" because it places the burden of proof on wildlife managers to justify culls on a case-by-case basis rather than allowing annual or routine culls.

Which policy adequately embodies special respect for elephants? Neither is obviously superior. With the newer policy, there may be more risk of "playing politics," and also of allowing a serious overpopulation problem to develop. In fact, in the first three years after it was instituted, the elephant population in Kruger grew by 14.8% (ABC Online 2002). However, a stronger anticulling stance may help improve the lot of elephants in the long run, even if it "goes over the brink" and harms elephants in the short run. By requiring that input be sought from interest groups, including animal rights and animal welfare advocates, and that nonlethal alternatives be considered every time population reduction is indicated, the new policy both simultaneously increases the need for nonlethal means of population control and ensures that they will be adopted as soon as they become practicable.

Elephants figure prominently in the popular lore of many cultures, including what I sometimes call the "Discovery Channel mentality" of the contemporary United States, which holds them up, along with cetaceans and the great apes, as examples of animals deserving special concern and respect. In the first section, I described the autonoetic consciousness paradigm, which holds, similarly, that near-persons deserve some kind of special respect in comparison to other animals that are merely sentient. Although popular culture is an unerring guide neither to values nor to facts, the discussion of autonoetic consciousness in section two suggested that common sense may have it right in this particular case: Science may someday confirm that elephants are at least near-persons. Yet, as the discussion in the section "Near-Personhood and Elephant Management" illustrates, this does not necessarily rule out culling as a population-control option. Special respect for elephants as near-persons requires that we work harder to tend to their physical and psychological needs. It may be possible for people to express respect in significantly different ways with regard to wild elephants, and these include culling under appropriate circumstances.

Similarly, it may be possible to express the special respect that near-personhood calls for in different ways when it comes to the question of holding elephants in captivity. We might say, as has been suggested for the great apes, that near-persons ought never to be held in captivity except "where it can be shown to be for their own good, or necessary to protect the public from [those]... who would clearly be a danger to others if at liberty" (Cavalieri and Singer 1993, 4). This standard would allow capture of wild elephants that pose a threat to villagers, although such elephants are often shot instead.

However, elephants, like dogs and horses, seem to be able to thrive under human control. For about 4,000 years, Asian elephants have worked with humans in agriculture, logging, construction, transportation, ceremonies, and warfare (de Alwis 1991). During that time, tamed Asian elephants have been allowed to interact and breed with wild ones, and some consider the Asian elephant semidomesticated. Although they are sometimes thought to be untameable, "elephants from North Africa were domesticated in antiquity and, in [the twentieth] century, elephant training stations were set up at a number of places in the then Belgian Congo" (Eltringham 1982, 189).

Although the methods traditionally used to capture, tame, and train Asian work elephants were harsh and brutal (de Alwis 1991, 117–120; Hart 1994, 299), traditional mahouts are devoted to their charges, and Lynette Hart notes that they spend more time each day caring for their animals than would any American or European person with any family member aside from a newborn (1994, 310). Modern training techniques can succeed without abuse, and it seems to me that a well-treated working elephant can lead a rich life in partnership with humans, just like working horses and dogs (Varner 2002, 456–459).

Most or all zoos provide more limited space and activities than working elephants enjoy. Still, I doubt that treating elephants as near-persons entirely rules out keeping them in zoos, and work- or performance-related training can contribute to an animal's physical and psychological well-being, suggesting that an idealized circus could be better than a zoo. The sheer size and power of elephants makes it amazing that they work smoothly with humans (Hart 1994, 298). Stephen Budiansky observes with regard to thoroughbreds and dressage horses that "training and learning may explain why a horse can be made to perform these tasks, but seem inadequate to explain the undeniable enthusiasm that many horses show for these pursuits" (1997, 99–101). Performing elephants sometimes appear similarly enthusiastic about their work (Kari Johnson, October 2002, interview by author). While studying confinement techniques for circus elephants, Ted Friend described elephants that were left on a picket line while the others went off to perform: "Those elephants were highly agitated, repeatedly oriented toward the tent and vocalized, and attempted to perform their act to music cues from the main tent while still chained" (Friend 1999, 85). Thus training elephants to work or perform can contribute to their physical and mental health, and a well-run circus could, perhaps, provide a better life for an elephant than most contemporary zoos.

I don't mean to suggest that keeping elephants in captivity is for their own good. That may literally be true for domesticated dogs (see Rollin 1992, 220, and 226–227) and maybe it's not a stretch to say that about horses. However, elephants are not fully domesticated animals, and the adult males do not live in hierarchically structured packs or herds in the same way that dogs and horses of both sexes do. For those two reasons, it seems less likely that human handlers of captive elephants can as effectively substitute for pack or herd leaders.

Nevertheless, the foregoing considerations lead me to think that holding elephants in captivity might be justifiable, at least in some narrow range of cases, for instance, if they are both (1) born in captivity and (2) their human keepers treat them like "domesticated partners." As I use the terms, "a *companion animal* is a pet who receives the affection and care owners typically give to pets, but who also has significant social interaction with its owner and would voluntarily choose to stay with the owner, in part for the sake of the companionship," and "a *domesticated partner* is a companion animal who works with humans in ways that emphasize and exercise the pet's mental and/or physical faculties in a healthy way" (Varner 2002, 463).

It may be a bit unnatural to refer to working elephants as "pets," but for captive-born elephants, life is better if they work with loving humans in ways that exercise the elephants' mental and physical faculties in healthy ways. So even though the adult males will sometimes pose special difficulties, keeping captive-born elephants in an enlightened zoo or circus environment may be consistent with respecting them as near-persons.

Acknowledgments
Several individuals and audiences provided me with helpful feedback on this chapter. As individuals I want to particularly thank Richard Byrne, whose incisive comments prevented some embarrassing blunders (all of the remaining blunders are, of course, entirely my own fault!); and Kate Chris-

ten and Chris Wemmer, whose detailed editorial comments made it all read much better. Helpful feedback also came from Joyce Poole, Adam Shriver, Colin Allen, and from audiences at the University of North Texas Philosophy Department and, of course, at the Front Royal conference out of which this anthology grew. Finally, research for this chapter was supported in part by National Science Foundation (NSF) grant no. 0620808, but the opinions expressed herein are my own and do not necessarily reflect the views of NSF.

Notes

1. Etymologically, "sentient" just means conscious of something or other. In the animal welfare and animal rights literature, however, "sentient" usually is assumed to mean capable of feeling pain specifically.

2. A memory's failing to be "stamped" with a specific date does not disqualify it as an episodic memory; it may just be dated as "before now." For instance, I may have no idea when I first heard a certain song, but when I imagine it, I'm having an episodic memory. I may even have heard the very same recording of it many times over many years, so that there just is no specific answer to the question, "How long ago did you hear that version of the song?"

3. Some philosophers (e.g., Allen 2004) and scientists (e.g., Povinelli, Bering, and Giambrone 2000) question the usefulness of arguments by analogy in discussions of animal minds, but this way of justifying attributions of conscious states to animals is widely accepted in the literature on animal welfare and animal rights. At a minimum, showing that the same parts of the brain are active in the same ways during a kind of problem solving that humans perform consciously puts the burden of proof on those skeptical about animal consciousness. To still maintain that humans solve the problem consciously while animals do not, the skeptic needs to explain how similar brain activity that produces similar behavior could be conscious in the one case but not in the other. (I owe this point to Adam Shriver, who earned his master's in philosophy at Texas A&M University.)

References

ABC Online. 2002. South African conservationists test elephant contraception. Accessed December 3, 2002, at http://abc.net.au/news/newsitems/s715364.htm.

Allen, C. 2004. Animal pain. *Noûs* 38: 617–643.

Budiansky, S. 1997. *The nature of horses: Exploring equine evolution, intelligence, and behavior.* New York: Free Press.

Byrne, R. W., and Whiten, A. 1992. Cognitive evolution in primates: Evidence from tactical deception. *Man* 27: 609–627.

Clayton, N. S., and Dickinson, A. 1998. Episodic-like memory during cache recovery by scrub jays. *Nature* 395: 272–274.

Clayton, N. S., Griffiths, D. P., Emery, N. J., and Dickinson, A. 2001. Elements of episodic-like memory in animals. *Philosophical Transactions of the Royal Society of London B* 356: 1483–1491.

Cutler, R. G. 1979. Evolution of longevity in ungulates and carnivores. *Gerontology* 25: 69–86.

de Alwis, Lyn. 1991. Working elephants. In S. K. Eltringham (ed.), *The illustrated history of elephants* (pp. 116–129). New York: Salamander Books.

Eisenberg, J. F. 1981. *The mammalian radiations: An analysis of trends in evolution, adaptation, and behavior.* Chicago: University of Chicago Press.

Eltringham, S. K. 1982. *Elephants.* New York: Sterling Publishing.

Friend, T. H. 1999. Behavior of picketed circus elephants. *Applied Animal Behaviour Science* 62: 73–88.

Fuster, J. M. 1997. *The prefrontal cortex: Anatomy, physiology, and neuropsychology of the frontal lobe.* 3rd ed. Philadelphia: Lippincott-Raven.

Gallup, G. G., Jr. 1970. Chimpanzees: Self-recognition. *Science* 167: 86–87.

Gallup, G. G., Jr., Anderson, J. R., and Shillito, D. 2002. The mirror test. In M. Bekoff, C.

Allen, and G. M. Burghardt (eds.), *The cognitive animal: Empirical and theoretical perspectives on animal cognition* (pp. 325–334). Cambridge: MIT Press.

Hampton, R. R., and Schwartz, B. L. 2004. Episodic memory in nonhumans: What, and where, is when? *Current Opinion in Neurobiology* 14: 1–6.

Hare, R. M. 1981. *Moral thinking: Its levels, method, and point.* New York: Oxford University Press.

Hare, R. M. 1993. Why I am only a demi-vegetarian. In R. M. Hare, *Essays on bioethics* (pp. 219–235). New York: Oxford University Press.

Hart, L. A. 1994. The Asian elephants-drivers partnership: The drivers' perspective. *Applied Animal Behaviour Science* 40: 297–312.

Hyatt, C. W., Metzler, T., French, B., and Fahrenbruck, D. 2003. Mirrors as enrichment for Asian elephants (*Elephas maximus*). *Journal of Elephant Management* 14: 12–16.

Jerison, H. J. 1973. *Evolution of the brain and intelligence.* New York: Academic Press.

Laws, R. M., Parker, I. S. C., and Johnstone, R. C. B. 1975. *Elephants and their habitats.* Oxford: Clarendon Press.

Lewis, G., and Fish, B. 1978. *I loved rogues: The life of an elephant tramp.* Seattle: Superior Publishing Company.

McComb, K., Moss, C., Sayialel, S., and Baker, L. 2000. Unusually extensive networks of vocal recognition in African elephants. *Animal Behaviour* 59: 1103–1109.

McMahan, J. 2002. *The ethics of killing: Problems at the margins of life.* New York: Oxford University Press.

Moss, C. J. 2000. *Elephant memories: Thirteen years in the life of an elephant family: with a new afterward.* Chicago: University of Chicago Press. Originally published 1988.

Nissani, M. 2004. Theory of mind and insight in chimpanzees, elephants, and other animals? In L. J. Rogers and G. Kaplan (eds.), *Comparative vertebrate cognition: Are primates superior to non-primates?* (pp. 227–261). New York: Kluwer Academic / Plenum Publishers.

Nissani, M., and Nissani, D. 2001. Absence of mirror self-recognition in two captive Asian elephants. Unpublished conference paper available online at: www.is.wayne.edu/mnissani/ElephantCorner/mirror.htm.

Plotnik, J. M., de Waal, F. B. M., and Reiss, D., 2006. Self-recognition in an Asian elephant. *Proceedings of the National Academy of Sciences* 103: 17053–17057.

Poole, J. 1998. An exploration of a commonality between ourselves and elephants. *Etica & Animali* 9: 85–110.

Povinelli, D. J. 1989. Failure to find self-recognition in Asian elephants (*Elephas maximus*) in contrast to their use of mirror cues to discover hidden food. *Journal of Comparative Psychology* 103: 122–131.

Povinelli, D. J., Bering, J. M., and Giambrone, S. 2000. Toward a science of other minds: Escaping the argument by analogy. *Cognitive Science* 24: 509–541.

Povinelli, D. J., and Eddy, T. J. 1996. What young chimpanzees know about seeing. *Monographs of the Society for Research in Child Development* 61(3): 1–152.

Rachels, J. 1986. *The end of life.* New York: Oxford University Press.

Regan, T. 1983. *The case for animal rights.* Berkeley: University of California Press.

Rensch, B. 1957. The intelligence of elephants. *Scientific American* 196: 44–49.

Rollin, B. E. 1981. *Animal rights and human morality.* Buffalo, New York: Prometheus Books.

Russell, I. S. 1979. Brain size and intelligence: A comparative perspective. In E. A. Oakley and H.C. Plotkin (eds.), *Brain, behaviour, and evolution* (pp. 126–153). London: Methuen.

Shoshani, J. 1991. Origins and evolution. In S. K. Eltringham (ed.), *The illustrated history of elephants* (pp. 12–29). New York: Salamander Books.

Singer, P. 1987. Life's uncertain voyage. In P. Pettit, R. Sylvan, and J. Norman (eds.), *Metaphysics and morality: Essays in honor of J. J. C. Smart* (pp. 154–172). New York: Oxford University Press.

Singer, P. 1993. *Practical ethics* (2nd ed.). Cambridge: Cambridge University Press.

Stone, V. E., Baron-Cohen, S., and Knight, R. T. 1998. Frontal lobe contributions to theory of mind. *Journal of Cognitive Neuroscience* 10: 640–656.

Sukumar, R. 1991. Ecology. In S. K. Eltringham (ed.), *The illustrated history of elephants* (pp. 78–101). New York: Salamander Books.

Tooley, M. 1983. *Abortion and infanticide.* New York: Oxford University Press.

Tulving, E. 1972. Episodic and semantic memory. In E. Tulving and W. Donaldson (eds.), *Organization of memory* (pp. 381–403). New York: Academic Press.

Tulving, E. 1985. Memory and consciousness. *Canadian Psychology* 26: 1–12.

Varner, G. 1998. *In nature's interests? Interests, animal rights, and environmental ethics.* New York: Oxford University Press.

Varner, G. 2002. Pets, companion animals, and domesticated partners. In D. Benatar (ed.), *Ethics for everyday* (pp. 450–475). New York: McGraw-Hill.

Wheeler, M. A., Stuss, D. T., and Tulving, E. 1997. Toward a theory of episodic memory: The frontal lobes and autonoetic consciousness. *Psychological Bulletin* 121: 331–354.

Whiten, A. 1997. The Machiavellian mind reader. In A. Whiten and R. W. Byrne (eds.), *Machiavellian intelligence:* II. *Extensions and evaluations* (pp. 144–173). New York: Cambridge University Press.

Whyte, I. J. 2002. Headaches and heartaches: The elephant management dilemma. In D. Schmidtz and E. Willott (eds.), *Environmental ethics: What really matters, what really works* (pp. 293–305). New York: Oxford University Press.

4 ELEPHANT SOCIALITY AND COMPLEXITY

THE SCIENTIFIC EVIDENCE

JOYCE H. POOLE AND CYNTHIA J. MOSS

Conventional wisdom places elephants among the more intelligent, socially intricate, and emotionally complex nonhuman species. Hancocks (Chapter 13 in this volume), citing Morris and Morris (1966) and Kellert (1989), notes that people rank elephants among the "most-liked" animals and consider the intelligence of elephants to be one of the most preferential factors. This popular conception reflects a long-term legacy of both legend and research about elephants. Scientists and philosophers, going back to Aristotle, have claimed elephants to be highly intelligent, and some have even viewed them as quasi-moral agents (Meredith 2001). This chapter examines what is currently known about the intelligence and complexity of elephants and attempts to answer whether they are deserving of special moral consideration.

Although the social complexity of elephants has been studied in depth, little systematic research on their cognitive abilities has been carried out until very recently (see Rensch 1956, 1957; Hart et al. 2001; McComb et al. 2001; Nissani 2004; Douglas-Hamilton et al. 2006; McComb, Bates, and Moss 2006; Plotnik, de Waal, and Reiss 2006). Though this body of research is growing rapidly, much of what is currently claimed regarding the superior intelligence of elephants is based on anecdotal evidence. Anecdotal evidence falls into several categories, and as long as the nature of this evidence is understood, some of it may be used to substantiate certain claims. For example, much critical data related to elephant cognition are based on rare behavior, which scientists have observed on an *ad libitum* (discretionary) basis. The observations of these scientists, who are trained in the interpretation of behavior, are an important data set (Byrne 1997a).

In this chapter, we provide empirical verification of elephant intelligence and social complexity. We examine such factors as social structure and flexibility, social network size, social learning, behavioral innovation, relative brain size and complexity, memory, and communication. We discuss Machiavellian intelligence, insight, and theory of mind, mirror self-recognition, and reaction to death. We argue that current and emerging evidence indicates we should err on the side of caution and treat elephants with special consideration. Evidence for these arguments comes in the form of scientific publications, previously unpublished data, and scientific records of rare behaviors. To reduce speculation as to the validity of the accounts of rare behavior, we rely on our own observations or those of our colleagues. The previously unpublished data presented in this essay come primarily from the Amboseli Elephant Research Project (AERP), a thirty-four-year study of known individuals in Amboseli National Park, Kenya.

Three recognized species of elephants exist: Asian elephants, *Elephas maximus*, African savanna elephants, *Loxodonta africana*, and African forest elephants, *L. cyclotis*. This chapter applies generally to all elephants. Although conclusions regarding the behavior of wild elephants come primarily from observations of African savanna elephants in Amboseli, and conclusions from experimental studies of captive elephants are primarily based on Asian elephants, it is generally acknowledged that there is broad similarity among the three species (Payne 2003; Sukumar 2003). Except where particularly relevant we have not attempted to compare elephants with other species.

The point of this chapter is not to argue that elephants deserve unprecedented rights that other species do not. Perhaps other animals also deserve such consideration, but we are not qualified to judge that. Many mammals may equal or even surpass elephants in their individual abilities, but we argue that it is the totality of elephant characteristics that makes them deserving of special consideration.

This chapter relies on a common understanding of certain basic concepts of elephant social structure. We define these here.

Family unit or *family:* One or more adult females and calves with a high frequency of association over time, who act in a coordinated manner and exhibit affiliative behavior toward one another (Moss and Poole 1983). This term does not exclude two or more adult females without offspring, or a single adult female and one or more juveniles who are not her immediate offspring, making up a family. The term family and family unit may be used interchangeably. In Amboseli, each family is referred to by a two-letter code, such as AA, AB, and so on.

Bond group: Two or more family units who associate with one another at high frequency relative to their associations with other family units in the population and whose members display affiliative behavior toward one another (Moss and Poole 1983).

Group: Any number of elephants of any age or sex moving together in a coordinated manner with no single member or subgroup at a distance from its nearest neighbor greater than the diameter of the main body of the group at its widest point.

Aggregation: A group specifically made up of more than one family unit with or without associating independent adult males.

Fission-fusion society: In relation to elephants the term fission-fusion society refers both to the slow changes in the structure of families or bond groups that occur over the course of years or decades and the very rapid changes that occur in social group composition over the course of hours.

In Amboseli, we have collected records on family group associations in two basic ways: as *sightings* records and as *census* records. *Sightings* record the presence of families, or portions of families, in any group that is encountered in the course of a day. Presence requires at least one member of a family to be sighted. We maintain the records in a Microsoft Access database (henceforth cited in this chapter as "AERP database") that contained more than 34,000 records as of the end of 2005. *Censuses* record the presence or absence of each family member at a sighting.

Social Complexity and Flexibility

Elephants live in a fluid, multitiered, fission-fusion society where group membership changes frequently, forming and dividing along lines that may be predicated on close social bonds, home range, and season (Douglas-Hamilton 1972; Moss and Poole 1983; Sukumar 2003; Wittemyer, Douglas-Hamilton, and Getz 2005; Archie, Moss, and Alberts 2005; Moss and Lee, forthcoming). Families are composed of a discrete, predictable composition of individuals, but over the course of hours or days, these groups may temporarily separate and reunite or they may mingle with other social groups to form larger social units. The close and lasting social relationships formed by elephants are remarkable in the context of their fluid social system (Archie et al. 2005; Moss and Lee, forthcoming). In fact, this combination of social qualities—close and enduring cooperative social relationships, and fission-fusion sociality—exists in only a small number of cooperatively hunting carnivores (e.g., hyenas, lions, and sperm whales) and also a few primates (e.g., chimpanzees and humans; Archie et al. 2005; Moss and Lee, forthcoming).

Relationships radiate out from the mother-offspring bond through members of the family, bond group, clan, subpopulation, to independent adult males and even beyond the population to strangers. Although special relationships between individual elephants may last a lifetime, the quality of relationships and the structure and degree of cohesion in an individual's social network may also change through time. With a maximum lifespan in the wild of sixty-five years (Moss 2001; AERP database), elephants are unusually long-lived mammals (Eisenberg 1981) and thus their relationships are extremely long term.

Like many sexually dimorphic mammals, adult male and female elephants live in very different social worlds (Moss 1988; Poole 1994). A complex network of bonds between individuals and families characterizes the lives of females and their offspring, while fluctuating sexual cycles distinguish the dynamic activities, associations, and relationships of males (Moss and Poole 1983). Elephants show strong individual personalities that affect

Families may temporarily separate and reunite or they may mingle with other social groups to form larger social units. The close and lasting social relationships formed by elephants are remarkable in the context of their fluid social system.
Photograph courtesy of Petter Granli / ElephantVoices.org

how they interact with other elephants and how well they are able to influence members of their group (Moss and Lee, forthcoming).

Elephant Social Structure

Members of an elephant family exhibit a high frequency of association over time, display strongly affiliative behavior, including a pattern of greeting ceremonies (Moss 1981) and are highly cooperative in group defense, resource acquisition, offspring care, and decision making (Douglas-Hamilton 1972; Dublin 1983; Moss and Poole 1983; Lee 1987; Moss 1988; Poole 1998; Payne 2003). A matriarch, usually the oldest female, leads each family, and most, though not all family members, are genetically related. Archie et al. have found complete uniformity of mitochondrial DNA (mtDNA) haplotypes in approximately 90% of Amboseli families. Average pair-wise relatedness between adults in families is 0.14, similar to the value expected in first cousins ($r = 0.125$) (Archie et al. 2005). Over thirty-four years of study in Amboseli, through 2004, there have been sixty-one families, seven of which no longer exist because of the deaths of their members. In the same period of time, new family units have been formed by five fissions and two fusions. Family membership has ranged from two to fifty-two individuals and, in 2002, averaged 18.7 individuals (Moss and Lee, forthcoming).

An African (*top*) and Asian (*bottom*) elephant family group walks in close proximity, providing care and reassurance to infants and calves.
Photograph of African elephants courtesy of Petter Granli / ElephantVoices.org; photograph of Asian elephants courtesy of Joyce Poole / ElephantVoices.org

An Asian male elephant in musth tests females for receptivity (*top*); two African elephant males in musth duel for supremacy and access to an estrous female (*bottom*).

Photograph of Asian elephants courtesy of Petter Granli / ElephantVoices.org; photograph of African elephants courtesy of Joyce Poole / ElephantVoices.org

Above the level of the family unit a second tier of relationships exists within bond groups (Douglas-Hamilton 1972; Moss and Poole 1983; Wittemyer et al. 2005). A bond group may include as many as five families (Moss 1988). Although the ties between individuals at this level are weaker than those within a family, bond group members also exhibit greeting ceremonies, form alliances against aggressors, assist in the care of one another's offspring, and defend one another in times of danger. Since the majority of bond groups are probably fission products of former families (Moss 1988) most carry the same mtDNA haplotype (Archie et al. 2005). Yet, rarely, there do exist bond groups (closely cooperating individuals) composed of unrelated families. Archie et al. (2005) have pointed out that even within families there are individuals who, through chance demographic events, have no close relatives. Yet, these individuals still benefit from the same cooperative behavior. Cohesion within families and bond groups varies significantly and depends upon a combination of factors, including individual personalities, the formation and dissolution of individual social bonds, the strength of the matriarch's leadership, historical events such as deaths of influential individuals, the type of habitat, and the season (Moss and Lee, forthcoming).

Clans are the next social level and have been defined as families who share the same dry-season home range (Laws and Parker 1968; Douglas-Hamilton 1972; Moss and Poole 1983). A clan is usually composed of several bond groups and numerous families, such that several hundred elephants may make up a clan. Playback experiments using individual vocalizations (McComb, Moss, Sayialel, and Baker 2000) indicate that elephants, too, distinguish between more familiar associates (typically clan members) and less familiar ones (typically nonclan).

Many of Amboseli's elephants are still in the same family, occupying the same home range that they were thirty-four years ago, while some families have shifted clan, or exchanged bond group. A few individuals whose families have gone extinct have been adopted by unrelated families, and in one case, individuals abandoned their natal family to create a new family with members of a previously unrelated family (Moss 1988; AERP database). Long-term records demonstrate both the extraordinary fidelity and unusual flexibility of elephant social relationships (Moss and Lee, forthcoming).

During dry conditions, families tend to move in relatively small groups in their clan area, but when resources are abundant, families may gather in large aggregations in an expanded range (Poole and Moss 1989). In these aggregations, several clans may intermingle. Far from being random, the fission and fusion of elephant groups follow predictable patterns, tending toward association according to hierarchical clustering into families, bond groups and clans, and breaking down in the reverse order. Elephants aggregate for different purposes (Western and Lindsay 1984), including an-tipredation (Laws, Parker, and Johnstone 1975), social benefits (Moss 1981), and improved mating opportunities (Poole and Moss 1989).

Although many mammals may follow some of the patterns exhibited by elephants, the combination of enormous social fluidity and durability of elephant associations and relationships is remarkable and rivals that of chimpanzees and humans in its complexity (McComb et al. 2000; Archie et al. 2005; Moss and Lee, forthcoming). Also notable is that kinship is not the only determinant of social bonds. Female elephants with few kin are not excluded from the benefits of sociality.

Young male elephants grow up in the tightly bonded society of females, maintaining close relationships with their relatives and participating in the many social events that affect their family, albeit at a lower intensity than their female agemates (Lee and Moss 1999; Poole, forthcoming). Males de-part from their natal families between 9 and 18 years of age (Lee and Moss 1999). A newly independent male must acquire a fresh set of behaviors to adapt to the society of males, where body size and fluctuating sexual state determine interactions and relationships (Poole 1989a). His transition from one society to the other occurs gradually, but dramatically, over a period of several years (Lee et al., forthcoming). During this time, a young male spends much of his time getting to know his agemates, and sparring and playing with novel partners from outside his natal family (Lee 1986). Thus he gathers information crucial to his longevity and reproductive suc-cess (Poole 1989a, 1989b; Lee et al., forthcoming; Poole, Lee, and Moss, forthcoming). Once fully independent, he forms relatively few close and long-lasting bonds with other elephants. Though males are often seen in small, all-male groups and may form lasting associations with certain in-dividuals, these are rather loose arrangements (Croze 1972; Lee et al., forth-coming). During sexually active periods males rove from one family group to the next, using olfactory and acoustic cues to search for receptive females (Poole and Moss 1989).

Elephants Have an Unusually Large Social Network

The social complexity hypothesis proposes that high intelligence has evolved to help individuals anticipate the plastic and often unpredictable behavior of group members (Jolly 1966). Social complexity has been cited as one indicator of intelligence (Byrne and Whiten 1988), and group size has been used as a measure of social complexity (Dunbar 1992). *Group* in this sense refers to long-term associates whose characteristics and histories are likely known to one another, rather than a temporary aggregation of strangers. This concept works well for species that live in stable, imperme-able groups, but less well for species like humans, chimpanzees, whales, and elephants, that do not. For fission-fusion societies the term "social network" encompasses such long-term associates. Just how large is an elephant's social network?

The Amboseli population is relatively small (1,417 at December 2005) compared with other elephant populations; yet, it is, nonetheless, a large society (McComb et al. 2000). In Amboseli, a female elephant may encounter literally hundreds of other individuals in the course of her daily range, and 34 years of records show that each family has been found with every other family in the population at least once (Moss and Lee, forthcoming).

Taking Amboseli's GB family as an example, we may estimate the size of one female's social network. Between 1974 and 2002, the GB family has grown from fourteen to thirty-seven members. Over this time, any given member of Amboseli's GB family found herself in a median group size of thirty-nine individuals, although at any particular moment she might (rarely) have been alone, with a small portion of her family, or in a group of 550 individuals (interquartile range 25–95; $N = 1,294$; AERP database). Members of the GB family have been sighted without other families on 428 occasions and with other families (in aggregations) 866 times. Of these 866 sightings, 59% of the aggregations included the IB family. The GB family was seen in aggregations with fifty other families, too, of which fourteen particular families were present in 15–30% of the sightings. Taking the population's current average family size, these close associates account for an estimated 261 elephants of which 125 (48%) are adult females at least ten years of age.

The question remains, however, do these animals know one another as individuals? To show that these are not just temporary aggregations of strangers, we must prove that elephants are able to distinguish between individuals, and to retain this knowledge over time. Work by biochemist Rasmussen and veterinarian Krishnamurthy (2000) found that offspring flehmen (a "chemical analysis" accomplished by touching the tip of the trunk on a liquid substance and placing it against the vomeronasal organ on the roof of the mouth) twice as often when presented with their mother's urine than with any other individual's urine, suggesting that they are able to distinguish their mother's urine. Even after a twenty-seven-year separation an offspring showed a flehmen response to its mother's urine (Rasmussen and Krishnamurthy 2000). Elephants answer the contact calls of close relatives, and, working in Amboseli, ethologist McComb et al. (2000) found that playing back the contact call of an elephant who had died elicited contact calling by her family unit almost a year after her death. McComb et al. (2000) also found that elephants recognized and usually answered the individual voices of their own family and bond group (ca. 40–60 individuals), and they also distinguished between familiar and less familiar elephants on the basis of how often they were encountered. McComb et al. (2000) predicted that the females would have to be familiar with the voices of at least 100 *adult* females to make the observed discriminations, indicating that elephants have one of the largest known social networks of any nonhuman species.

Males, too, make use of social knowledge in a large network. Complementary male postures in long-distance dyadic interactions (i.e., those between a paired group of individuals) indicate that males recognize one another visually, and that they are aware of one another's relative size, strength, and condition (Poole 1982, 1989a). An instance of such interaction would be a *higher-ranking male posturing with* Head-High; Ear-Spreading; Ear-Folding; Advance-Toward, while the corresponding *lower-ranking male is posturing with* Head-Low, Ear-Flattening; Turn-Away; Retreat-From; (Poole 1987, 1999a; Kahl and Armstrong 2000; Poole and Granli 2003). Males also use sound (Poole et al. 1988; Poole and Moss 1989; Langbauer et al. 1991) and scent (Rasmussen et al. 1982; Rasmussen and Schulte 1998), to remotely monitor the location of receptive females. Approximately, three weeks before true estrus and ovulation, female elephants exhibit a period of estrouslike behavior, referred to as "pre-estrus" (Poole and Moss 1989), that is associated with an anovulatory LH (lutenizing hormone) surge (Kapustin et al. 1996) and increasing pheromone secretions (Rasmussen, Krishnamurthy, and Sukumar 2005). Males accompanying a female during the pre-estrus are likely to be the same individuals who associate with her later during her true estrus (AERP database). During the intervening days, these same males may be many kilometers apart from that female, searching for, or consorting with, other females (AERP database). The median number of males in association with an estrous female in Amboseli is 5 (interquartile range 3–8; range 1–40; $N = 728$; AERP database). Because upward of 100 adult males may associate with female groups in the population at any given time, there is strong suggestion that individual males may relocate a female during true estrus using acoustic and olfactory cues as well as memory of her individual identity and clan area, and therefore of her likely location.

Although other mammals are also able to recognize the voices and scents of conspecifics (members of their own species), the sheer number and long-term retention of social information makes elephants unusual. McComb et al. (2001) showed that such vocal discriminations are learned through years of experience and older individuals are more discerning than younger ones. McComb et al. (2000) argue that an unusually large and fluid social network may be a phenomenon unique to long-lived mammals like elephants, equipped with long-range signaling capability and the mental capacity for extensive social recognition.

Rate of Change in Social Setting

Given a large and fluid social network in which group composition changes rapidly, elephants may encounter literally hundreds of individuals each day.

For example, of the 1,294 sightings of the GB family, 311 (24.0%) were of groups of 100 individuals or more; 190 (14.7%) were of groups of 150 elephants or more, and 122 (4%) were of groups of 200 elephants or greater. The number of elephants encountered by each female each day, and the rate of change, prove unusually large, especially when one considers that these data represent instantaneous sightings and that groups change in size and composition many times per day (Mutinda 2003).

Sexually active males, searching from group to group for receptive females, may also meet and interact with hundreds of different individuals, male and female, in the course of a day (Poole and Moss 1989). The nature of a male's interactions with those he meets is strongly influenced by his age, body condition, and sexual state (Poole 1982, 1989a).

Elephants are able to recognize the voices, scent, and appearance of many individuals, some of whom they encounter only relatively rarely. Although the social setting of many other species may also change during the course of a day, the sheer number of individuals involved and the hierarchical nature of the formation and dissolution of aggregations make elephants unusual. Under these circumstances, the importance of being able to distinguish genuine strangers from a wide range of more regular associates is not insignificant (McComb et al. 2001).

Social Learning

There is empirical evidence that social learning and behavioral innovation are positively correlated with brain size in mammals (Reader and Laland 2002). Both behavioral innovation and social learning are essential elements in the development and maintenance of elephant social complexity.

Learning What to Eat

Infant elephants must learn and practice complex coordination of the trunk and apply this gradually acquired skill to the procurement of appropriate foods (Lee and Moss 1999). Calves begin to sample potential food items when they are 1 to 2 months old. Sampling consists of grasping objects with their trunk, rolling food items in their trunk, and placing food matter in their mouths. At times young calves may chew, but sampling starts well before actual ingestion. Calves gradually acquire foraging knowledge by sampling what adults are eating. They attain this information by placing their trunks in the mouths of adults and pulling out food items, stealing food from their mothers and allomothers, as well as by eating the fresh dung of other elephants (Eisenberg 1981). Throughout their first five years, calves spend 15.8% of their social contact time exploring the food intake of others (Lee and Moss 1999). Elephants have many food sources, and social learning allows calves to exploit a wide range of these seasonally and geographically varying species (Lee and Moss 1999).

Allomothering increases calf survival and provides young females with an array of caretaking experiences that persist until they give birth to their own first calf. A juvenile female rescues her younger sibling from an attempted kidnapping by a nonfamily juvenile female and calls for the assistance of her mother (*top*); juvenile females shepherd infant Emily Kate back to her mother (*bottom*). *Photographs courtesy of Petter Granli / ElephantVoices.org*

Learning to Care for Calves

Juvenile females play a vital role in the care of young calves (Lee 1987). Allomothering both increases calf survival (Lee and Moss 1986) and provides young females with an array of caretaking experiences that persist until they give birth to their own first calf. Despite this practical social experience, first-born infants still have higher mortality rates than infants born to experienced mothers (Moss 1994), indicating that even by the age of first parturition (14–15 years) female elephants still have much to learn about mothering. The calves of inexperienced mothers show higher levels of distress than do calves born to experienced mothers, who appear to be more responsive to calf demands for food and protection, with obvious consequences for calf growth and survival (Lee and Moss 1999). In addition, first-time, or primiparous, mothers are given more assistance from family members than are older mothers. Knowledge gained from experiences over a succession of births plays an important role in calf survival.

Social Learning as an Aspect of Female Reproductive Behavior

Some behaviors essential to mate choice, such as consort behavior, appear to require a social context for learning. Young estrous females do not demonstrate the estrous behavior typical of older females and are frequently chased and mounted by a succession of young and nonmusth males (Moss 1983). As a female gains individual experience through successive estrous periods her responses to males change; she learns to avoid young males and to select and go into consort only with large musth males (Moss 1983; Poole 1989b). Both the acquisition of estrous behaviors and the choice of mates appear to be facilitated by the presence and behavior of the mothers of these young females (AERP database). Mothers of young estrous females are often observed exhibiting estrous postures and behaviors when not in estrus themselves (AERP database; J. Poole, personal observation). Mothers (and occasionally nonmother matriarchs) may be observed to approach and avoid males, run with their young estrous daughters (or young estrous members of their family) during long chases and occasionally make postcopulatory calls after the young female is mated (BBC Natural History Unit 1995). The behavior of mothers and daughters during a daughter's first estrous period not only indicates the importance of a social context for learning, but also suggests that mother elephants may be engaged in a rudimentary form of true teaching.

The birth of a female's first calf is another documented event where the presence and behavior of experienced females aids inexperienced mothers. Experienced family members assist young females to cope with the physical demands of birth, including helping a newborn to its feet, and with the immediate protection and socialization of the newborn calf (Moss 1988; Lee and Moss 1999).

Social Learning as an Element in Male Independence

In exploratory and play behavior, young males seek same-sex play partners outside the family, thereby socializing and bonding with their future peer group while simultaneously assessing the strength of their future rivals (Lee

1986). This male behavior can be linked to male emigration from the natal family, an event that occurs at an average age of 14 years (Lee and Moss 1999). The process of a male's departure from his family averages 16 months, but ranges anywhere from 5 to 35 months (Lee et al., forthcoming). The wide range in duration indicates that the process of male independence is probably a highly individualized event. Male mortality in the 10–20-year age bracket is also significantly higher than in females (Moss 2001). To survive on their own newly independent males require sufficient social experience with other males along with knowledge of male areas of residence and surrounding habitat (Lee and Moss 1999).

Social Learning and Role Models in Male Reproductive Behavior

Young males appear to learn to distinguish between reproductive odors. For example, young males, though not older males, often mistake the smell of a pre-estrous female or one who has just given birth, for that of a receptive female (J. Poole, personal observation). Successful mounting and intromission, too, appears to require considerable skill and experience. This learning process may, in part, be gained by watching the behavior of older, more experienced males. Young male elephants are often observed to follow older musth males, testing the same urine spots and females that they do. Musth males are extremely tolerant of these youngsters, allowing them to stand less than a meter from an estrous female while older males are kept at long distances (Poole 1982). Experience from southern Africa also highlights the importance of social learning in the acquisition of appropriate male reproductive behavior (Slotow et al. 2000). Juvenile male elephants that experienced their families killed in a culling operation and were then introduced to areas without adult role models exhibited abnormal reproductive behavior as young adults, including the mounting, tusking and killing of black rhinos. While trauma is likely to have been a causal factor in the development of this abnormal behavior (Bradshaw et al. 2005), it is also likely that the absence of adult males contributed to the inappropriate sexual reaction of these young males.

The Role of Social Learning in Tool Manufacture and Use

Elephants are unusual among nonprimates in that they use and even manufacture simple tools. The sensorimotor specializations of the trunk are extensive (Rasmussen and Munger 1996), allowing delicate manipulations of both large and small objects. Elephants pick up objects (logs, rocks) and throw them at opponents, use sticks to remove parasites, branches as fly switches, and logs to neutralize electric fences (Poole 1998). A study of Asian elephants by ethologist and sociologist Lynette Hart and colleagues (2001) indicates that elephants modify the length of long branches to make a suitable fly switch. That study also describes the role of social learning in the acquisition of tool use. Adult Asian elephants modify branches by pulling off leaves and shortening sticks to create fly switches of ideal length. Two young elephants, ages 18 and 9 months, imitated the fly switching behavior of adults. The 18-month-old calf was able to remove a side branch successfully and was coordinated in switching with the modified shorter branch. The 9-month-old showed only uncoordinated attempts to switch, but her movements appeared to imitate the fly switching behavior of older elephants.

Vocal Learning

Very few mammals are capable of producing or modifying sounds in response to auditory experience (McCowen and Reiss 1997; Tyack and Sayigh 1997; Janik and Slater 2000). Those that can include humans, some marine mammals (e.g., bottlenose dolphins, *Tursiops truncates;* harbor seals, *Phoca vitulina*; humpback whales, *Megaptera novaeangliae*) and bats. Despite the obvious significance of vocal imitation for human-language acquisition, there is scant evidence for vocal production learning in other primates. Recently, Poole et al. (2005) provided two examples of vocal imitation in African elephants: A young female elephant who imitated the sounds of trucks, and an African male zoo elephant who imitated the chirping sounds of the Asian elephants with whom he was raised. Earlier reports have described Asian elephants who have learned to produce whistling sounds by blowing air through their trunks, and still other elephants who have imitated this innovative whistling technique (Wemmer and Mishra 1982; Wemmer, Mishra, and Dinerstein 1985; I. Douglas-Hamilton, personal communication). Vocal production learning enables a flexible and open communication system, involving both behavioral innovation and social learning, in which animals may learn to imitate sounds not typical of their species. Poole et al. (2005) suggest that vocal learning should likely occur in other species in which long-lived social bonds are based on individual-specific relationships and involve fluid group membership, and where vocal communication functions in maintaining contact, in individual or group recognition and in mediating social interactions.

One of the hypotheses of the origin of human language is that it arose to encode increasingly complex information about social relationships (Worden 1998; Dunbar 2003; Pinker 2003). The association between species that show vocal production learning abilities and those with fission-fusion societies supports this hypothesis (e.g., cetaceans, elephants, chimpanzees, humans). In describing the social origin of language hypothesis, Pinker (2003) has suggested that sociality, knowledge gathering, and language co-evolved in humans, in that language reinforces social relationships and provides a mechanism for distributing knowledge to associates. These three characteristics of human societies appear to be evident, to a lesser extent, in elephants.

Members of an elephant family exhibit a high frequency of association over time and display strongly affiliative behavior, including a pattern of greeting ceremonies. Here mother and daughter greet one another.
Photograph courtesy of Petter Granli / ElephantVoices.org

Communication as a Measure of Social and Cognitive Flexibility

Species that make use of both *contextual* learning (behavioral context and sequential positioning) and *production* learning (modification based on experience) are theoretically capable of developing a more complex acoustic communication system than species utilizing only the former. Each additional form of vocal learning can increase the complexity of a communication system; the more forms present the greater the system's potential for openness and plasticity (Janik and Slater 2000).

Elephants make use of both contextual and production learning and, as predicted, their acoustic communication system includes an extensive vocal repertoire with a high degree of variability both within and between individuals (Poole et al. 1988; Langbauer 2000; McComb et al. 2000; Leong et al. 2003; Soltis, Leong, and Savage 2005a, 2005b; Poole, forthcoming). Elephants also have a wide range of visual and tactile gestures and displays (Kahl and Armstrong 2000; Poole and Granli 2003) and intricate chemical communication (Rasmussen and Krishnamurthy 2000). Together this complex suite of vocal, chemical, visual, and tactile signals mediates the intricate teamwork displayed by members of an elephant family (Poole, forthcoming; Poole and Granli 2003). Day to day decision making (e.g., deciding when and where to go; antipredator responses) involves broad vocal participation and often includes the building of consensus through vocal exchanges that may take up to an hour (J. Poole, personal observation; C. Moss, personal observation).

Indicators of Elephant Intelligence
Brain Size and Complexity

Until recently, only scant information was available on the brains of elephants, with the majority of the literature based on only a few specimens (Cozzi, Spagnoli, and Bruno 2001). Recent work has added to our knowledge (Hakeem et al. 2005; Shoshani, Kupsky, and Marchant 2006). The brains of Asian and African elephants rank among the highest of all animals for absolute and relative mass and cortical expansion and complexity, features comparable only to those of some of the cetaceans, the great apes, and humans. Weighing 4.5–6.5 kilograms, the elephant's brain is the largest in absolute mass among land mammals (Cozzi et al. 2001), with the brain of *E. maximus* weighing up to 5.5 kilograms, and that of *L. africana* being slightly heavier and larger. The temporal lobes of the elephant's brain, which are thought to function in recognition, storage and retrieval of information related to sight, touch, smell, and hearing, are especially large and extremely complex (Shoshani 1998; Shoshani et al. 2006).

The encephalization quotient (EQ), the ratio between the observed and expected brain weight for a defined body weight (Jerison 1973), implicitly holds that body size is taken as a "given" and selection operates only on brain size. An EQ equal to 1.0 is an "average" mammalian brain. Primates all have relatively high EQs, with *Homo sapiens* at 7.0+ demonstrating a quotient far above that of all other mammals. Elephant values, at between 1.7 (Shoshani 1998; Eisenberg 1981) and 2.3 (Cutler 1979) are comparable to those of larger primates (e.g., chimpanzee: 2.2–2.4; gorilla: 1.4–1.7; orangutan: 1.6–1.9; Eisenberg 1981). Where sexual dimorphism is pronounced, as in elephants, males have lower EQ values than females (EQ *L. africana*: male: ca. 1.0; female: 2.0; Eisenberg 1981). Clearly age and sex are important criteria in determining the importance of EQ, and for a variety of reasons many have argued that EQ may not be a particularly useful guide to relative intelligence (e.g., Byrne 1996; Roth 1999). Neurobiologist Roth (1999) and cognitive ethologist Byrne (1996) argue that what should be important for intelligence is the absolute number of nerve cells in the brain, as more nerve cells mean the potential for more complicated networks. Elephants have a very thick cerebral cortex, and although its cell density is lower than that of humans, elephants are estimated to have as many neurons as humans, namely between 10^{11} and 10^{12}. Shoshani et al. (2006) postulate that convergent evolution as seen in complex learned skills and be-havior may be responsible for the many similarities observed between the human and elephant brain.

Long-Term Memory

As a general rule vertebrate species with absolutely large brains have de-veloped the neocortex (the complicated parts of the cerebral cortex) to a greater degree, have greater capacity for learning, and seem to be able to learn more complicated tasks than vertebrates with absolutely smaller brains (Rensch 1956, 1957). In addition, it seems that larger animals with large brains also have the ability to retain information for longer than smaller animals with smaller brains. In other words they have better mem-ory (Rensch 1956, 1957). Exceptionally large and long-lived, elephants ac-cumulate and retain social and ecological knowledge, remembering other individuals (McComb et al. 2000; Rasmussen and Krishnamurthy 2000; McComb et al. 2001) and places (Viljoen 1990; Shoshani and Eisenberg 1992) for years.

In experimental trials, elephants show an excellent ability to learn and remember a large set of visual symbols and acoustic tones or commands over long periods (Rensch 1957). Playback experiments in Amboseli also provide good evidence for an elephant's exceptional memory. Companions use powerful calls to stay in contact when visually separated (Poole et al. 1988). As mentioned previously, elephants remember the contact calls of family and bond group members and distinguish them from those of fe-males outside these categories; moreover they can also discriminate be-tween the calls of family units farther removed than bond group members on the basis of how frequently they have encountered them (McComb et al. 2000). In fact, McComb et al. (2000) predicted that females would have to be familiar with (i.e., remember) the contact calls from at least 100 adult females in order to perform these discriminations. McComb et al. (2001) also found that older matriarchs possessed enhanced socially relevant, dis-criminatory abilities, which are learned and remembered over many years of experience. These older females acted as a repository of social knowledge that contributed to increased fitness for the entire family.

In addition, playback to her family unit of the contact call of a 15-year-old female that had died elicited contact calling three months after her death, and contact calling and approach to the loudspeaker twenty-three months after her death (McComb et al. 2000). Also, playback of the con-tact call of a female who had departed from her family to join another one, twelve years previously, also elicited contact calling (McComb et al. 2000).

Accounts of rare events by qualified observers also indicate that ele-phants are able to remember the voice and perhaps scent of individual people for over twelve years (R. Moore, personal communication; J. Poole, personal observation) and individual elephants for over 23 years (C. Buckley, personal communication). In the latter case, the specific protective behavior of elephant Shirley (ca. 53 years) toward Jenny (ca. 30 years), when reunited, suggested that they not only remembered one another, but also remembered the adult-juvenile relationship they had once shared. In the mid 1980s, one of us (J. Poole) established a relationship with Vladimir, a wild male elephant, who sometimes came to her car window allowing her to touch him (Poole 1996). In 2003, after a twelve-year separation, Poole met Vladimir, then aged 34 years, and called to him. He walked up to her window and allowed her to touch him. To the best of our knowledge, he has not interacted with any other person in this manner (AERP colleagues, personal communication).

Machiavellian Intelligence In Elephants

The term Machiavellian intelligence was coined to refer to the hypothesis that increased intelligence was linked to the evolution, through competitive tactics, of socially complex behavior predicated upon a good memory for socially relevant information (Byrne and Whiten 1988; Byrne 1996). The principles of Machiavellian intelligence were originally used to explain the socially complex behavior of monkeys and apes, but more recently, nonprimate species have been shown to exhibit similar traits (see Byrne 1996). Typical social traits include the formation of long-term social relationships, use of third party or triadic intervention to decide the outcome of competitive encounters, dependence on a network of allies, grooming used to build up a network of support, active reconciliation used to repair strained relationships, use of tactical deception, and the classification of others by dominance rank, affiliation or group membership. Successful use of these social tactics demands good perception and discrimination, attention to social attributes and good memory, all of which are displayed by elephants. The abilities ascribed to Machiavellian intelligence require neocortical enlargement and exceptional memory to retain and process socially relevant information (Byrne 1997b), qualities that are, again, characteristic of elephants.

Elephants are contenders for rank among those species possessing Machiavellian intelligence, in every respect. The formation of long-term relationships is the very essence of elephant society and a network of allies, who are usually though not always genetically related, defines the elephant family and bond group. These supportive relationships are maintained not by social grooming as in primates (Dunbar 1988), but through an elaborate system of vocal, visual, chemical and tactile signals and ceremonies between relatives and friends (Moss 1988; Poole et al. 1988; see also Poole and Granli 2003, 2004). The most well known of these is the greeting ceremony (Moss 1988), but reunions are by no means the only situation under which such a display of "bonding" occurs. Similar exuberant demonstrations of the importance of a social bond occur in a multitude of different contexts, including, among others: after a family member's infant or calf has been lost, frightened, or denied access to the breast; if a member of the family has been chased or harassed by another elephant; when a baby is born; or following a disturbing or exciting event (Poole, forthcoming).

When key relationships are put at risk by conflict, some primates make efforts to reconcile, and dyads that reconcile are more likely to continue to support one another (Cords 1997; de Waal and van Roosmalen 1979). Similarly, conflict between two individuals in an elephant family typically results in vocal reconciliation that—even more sociably—usually involves the intervention of a third party (J. Poole, personal observation). Third parties play an important role in deciding the outcome of conflicts among elephants, including support given to genetically unrelated individuals (Moss 1988; Payne 2003 Archie et al. 2005). These "friendships" are maintained by reciprocal altruism (Payne 2003). Work by McComb et al. (2000) shows clearly that elephants are able to recognize and respond appropriately to the voices of others and that they are able to classify others by group membership at the very least. Finally, there are many accounts of tactical deception by elephants, though these are mostly anecdotal reports by elephant keepers rather than data based on experimental trials or observations in the wild (e.g., see Rensch 1957; Chadwick 1992; Poole 1998). Many of these extraordinary accounts are of elephants deceiving their keepers to obtain food or freedom. Detailed documentation of such tactical deception is now needed.

Anticipatory Planning, Insight, and Theory of Mind

The combination of longevity, long-term memory, social learning, and behavioral innovation might lead us to expect elephants to display other cognitive abilities, such as insight and theory of mind. Insight refers to behavior that shows the understanding of relations between stimuli and events, while theory of mind is defined as an ability to understand that others see, feel, and know (see also Varner, Chapter 3 in this volume).

In the wild, anecdotal evidence certainly suggests the use of insight by elephants in response to electric fences. Elephants frequently disable electric fences by dropping logs onto them, or by uprooting and pushing trees onto them, causing live wires to sag onto ground wires, thus shorting the fence (Laikipia Ranch, Kenya, fence maintenance crew, personal communication; Seneviratne and Rossel, Chapter 17 in this volume). Reports exist of elephants lifting their infants over fences (Chadwick 1992; Ngulia Sanctuary, Tsavo National Park, Kenya, fence maintenance crew, personal communication). In captivity, elephants have been known to use sticks to pull out-of-reach food closer; open faucets, and, once these were bolted shut, use rocks to break the nuts loose from the bolt; throw tires onto nearby branches to weight them down to a level within reach; make a pile of tires to stand on so as to reach branches otherwise out of grasp (Chadwick 1992). Shoshani and Eisenberg (1992) recount an extraordinary tale of an elephant that ingeniously placed vegetation under his feet to prevent himself from sinking into muddy ground where he was tied, as he could not reach dry ground. Evolutionary biologist Bernhard Rensch's (1957) observations of the ability of domesticated elephants to work with minimal instruction and their talent to function as a team with extraordinary balance and coordination, pushing and dragging heavy logs up inclined bars onto a truck, caused him to credit elephants with true insight or "ideation"—the ability to anticipate what will come of certain actions.

Behavior that qualifies as insight cannot be genetically wired or acquired by trial and error. The problem must be new, or the animal must come up with a new solution. The correct behavioral sequence must be arrived at suddenly and completely and carried out relatively smoothly, with all its constituent elements purposefully aimed at a single goal (Nissani 2004).

In Amboseli one of us (J. Poole) also witnessed behavior that meets these criteria. In 1999 Poole witnessed Ella giving birth, an event accompanied by thirty minutes of excited vocalizations by her family and other females (Poole 1999b). The commotion attracted other elephants, including young and inexperienced Ramon, who upon sniffing Ella, mounted her, his body and feet suspended above the newborn. Matriarch Echo, and her adult daughter, Erin, rushed immediately to Ella's side, backing, in parallel, purposefully toward her and placing themselves resolutely on either side of her. In what appeared to be a very deliberate attempt to prevent the male from crushing the infant when he dismounted, their behavior appeared to be a case of true insight. Both authors also witnessed Victoria and Virginia (the two oldest females in a large family) rush to stand on either side of Vega, who had been darted with an immobilizing drug, and push into her, thus holding her up and preventing her from falling down. Only after forceful driving at them with vehicles would they give up.

Theory of mind accounts exist, too, with an often-related tale being that of Chandrasekhan, the elephant that would not lower a pillar of wood into a hole containing a sleeping dog until the dog was chased away (Shoshani and Eisenberg 1992). Many accounts are available from the wild, too. For example, the respected naturalist Tony Archer (personal communication) witnessed a young male repeatedly tusk a crippled female who had fallen behind her family. A large, older female (possibly a relative) suddenly ran from 40 meters away to chase the male and, having seen him off, returned to touch the other female gently on her withered leg. The specific behavior of the older female also indicated that she might have understood the particular feelings of the crippled female, possibly demonstrating her possession of theory of mind.

While it may be possible to explain some of these accounts as examples of genetically wired behavior or trial and error learning, without invoking insight or an ability of elephants to understand that others see, feel, and know (i.e., theory of mind) recent work by geneticist Moti Nissani (2004) indicates we should be cautious in drawing such conclusions. In three experiments using captive individuals, Wanda and Winky, Nissani explored the question of whether elephants are capable of theory of mind and insight, by focusing on whether elephants know that people use the sense of sight (see Varner, Chapter 3, for more details). Although Nissani's research is not conclusive, it adds to growing evidence supporting the view that elephants may be capable of both insight and theory of mind.

In the human brain, there exist multiple mirror neuron systems. Mirror neurons are cells in the brain that specialize in carrying out and understanding not only the actions of others but their intentions as well as the social meaning of complex behavior and secondary emotions such as disgust, shame, and guilt. By simulating a "mirror image" of another's actions in our brain we are able to empathize, imitate, and acquire language. Mirror neurons are, in effect, the source of our theory of mind. In a sense, mirror neurons absorb culture directly, with each generation teaching the next through sharing, imitation, and observation (cited in Blakeslee 2006).

Research by neuroscientist Giacomo Rizzolatti, on mirror neurons in the brains of monkeys have found that these cells, too, are active in situations that would inspire empathetic feelings (described in Blakeslee 2006). Some elephants have been observed attempting to raise an immobilized or dying elephant to its feet (J. Poole and C. Moss, personal observation; Douglas-Hamilton et al. 2006) or returning to the corpse of a dead companion many times over the course of days and weeks (Douglas-Hamilton et al. 2006). Taken together, these mirror neuron studies and elephant observations suggest that monkeys and probably other highly social species like apes, elephants, dolphins, and even dogs experience empathetic feelings and that the basis for theory of mind exists in a network of cells with the same origin as our own more complex capability.

Some years ago one of us (J. Poole) watched elephant Eliot flinch violently as Eudora, 10 meters away, reached out her trunk to test if an electric fence was connected. We suggest that Eliot's mirror neurons were actively firing as she experienced a moment of fear in anticipation of Eudora's potential shock. Vocal production learning and other imitative abilities in elephants may be yet another piece of evidence in the jigsaw puzzle pointing to elephant theory of mind.

Self-Recognition in Elephants

Self-directed behaviors in front of a mirror have been interpreted as indicative of self-recognition (Gallup 1970). Furthermore, researchers maintain that if an animal recognizes itself, then it must have, however rudimentary, a sense of self. Mirror self-recognition (MSR) has, thus, been expected to correlate with higher forms of both altruistic behavior and empathy. Dozens of primate species have been exhaustively tested for self-recognition using the mirror-mark test, and only the great apes have conclusively passed the test (Gallup, Anderson, and Shillito 2002). In 2001, mirror self-recognition was demonstrated in bottlenose dolphins, a species also noted for complex social behavior and empathetic traits (Reiss and Marino 2001). Recently, Plotnik et al. (2006) demonstrated that elephants, too, are capable of mirror self-recognition (see also Varner, Chapter 3). Plotnik et al. postulate that MSR in apes, dolphins and elephants offers persuasive evidence for convergent evolution in the self-other distinction that underlies the social complexity, cooperation and altruistic tendencies noted among these large-brained mammals.

Reaction to Trauma and Death

Recent work by trauma researchers Bradshaw et al. (2005) suggests that early disruption of attachment can result in social trauma that may affect the physiology, behavior, and culture of elephants over generations. Such disruption occurs during culls where adults are killed and infants are spared for capture; during abusive training for captive use; and during high-level poaching. An intact social order may buffer trauma, but as human populations increase more elephants are living in environments influenced by relentless human disturbances. The consequences of these experiences can be seen in abnormal levels of aggression in the wild (Slotow et al. 2000) and in captivity (see also Hancocks, Chapter 13 in this volume).

A discussion of elephant cognitive capacities would be incomplete without the mention of elephants' reaction to death. Elephants exhibit a variety of responses to dying or dead elephants and their bones, including touching with the trunk and feet, attempted lifting and carrying of the body or bones, mounting, feeding, body guarding, covering, and burying (Douglas-Hamilton and Douglas-Hamilton 1975; Moss 1975, 1988, 1992; BBC Natural History Unit 1992, 2005; Poole 1996; Payne 2003; Douglas-Hamilton et al. 2006; McComb et al. 2006). Although elephants exhibit this behavior primarily toward other elephants, they may also stand over or cover the bodies of humans or other animals that they kill (Poole 1996) or of those that they find killed by predators (N. Njiraini, S. Sayialel, and N. Sayialel, personal communication). People have known about the reaction of elephants to death for thousands of years (Meredith 2001) and as far as we are aware no one has yet come up with a plausible alternative to the explanation that elephants have, however basic, a concept of death.

In summary, relative to other mammals, including humans, elephants are unusually long-lived and exhibit a high degree of social complexity. Their

Early disruption of attachment such as occurred during culls where adults were killed and infants were spared can result in social trauma that may affect the physiology, behavior, and culture of elephants over generations.
Photograph courtesy of Oria Douglas-Hamilton

development includes social learning and behavioral innovation, both of which are manifested in the use and modification of rudimentary tools and in vocal learning. Elephants have extensive neocortical development, very good memory and are evidently adept users of Machiavellian intelligence. Mirror self-recognition by elephants indicates self-awareness (Plotnik et al. 2006), and numerous observations suggest elephants have a rudimentary theory of mind and anticipatory planning capabilities that may include imagining future events, such as pain inflicted on themselves and others and possibly their own deaths. Although many other species may rival elephants in one capability or another, there are few that equal or surpass elephants in the totality of their social and behavioral complexity.

The evidence presented in this chapter, including the recently published Bradshaw et al. 2005 report on the long-term effects of trauma, indicates that we must err on the side of caution when welfare issues are being weighed, and that individual elephant well-being, not commercial gain, must be the priority. This conclusion has broad implications for the management of elephants in the wild and in captivity. It is not within the scope of this essay to set guidelines for the treatment and care of elephants, but certain guiding principles, most of which are discussed in Clubb and Mason's report about European zoo elephants (2002) and in Kane, Forthman, and Hancocks (2005a, 2005b), should be put into practice.

As large, highly social, and intelligent animals, elephants require ample, environmentally complex space and a sufficient number of conspecifics for social contact and learning. We should be moving toward a situation in which institutions are permitted to keep elephants only if the captive situations provide sufficient space for adequate exercise and stimulation and allow the elephants to choose among social partners.

In seasonally cold climates, elephants are usually restricted to indoor barns when the outside temperature goes below about 4.4 degrees centigrade. In zoos that experience long, severe winters elephants may be indoors, with inadequate space, exercise, and social stimulation for months at a stretch. Because of the many negative consequences of restricted space on elephant physical and emotional well-being, zoos located in areas with prolonged cold winters should be precluded from keeping elephants.

Elephants (including males) should be allowed access to a social group, not kept in isolation. Males should be allowed to remain with their family until the age of natural dispersal; females should remain together for life. During parturition females should remain unchained and in the company of family members, particularly if experienced females are present. Infants and calves should not be removed from the care of their mothers and family members. Most elephant mothers in zoos were not themselves exposed to allomothering experiences, and their very poor success as mothers is a clear consequence. The tradition of removing an elephant from its social group for the purpose of exchange with other zoos or circuses should cease. All forms of physical discipline and punishment must be discontinued and chaining should stop unless absolutely necessary for veterinary care. Thus, on the basis of all these criteria, circuses are not an appropriate environment for elephants and most zoos will have to make substantial changes to meet the most basic needs of elephants.

In the wild the practice of abducting young elephants from their families should end. The culling of elephants should be avoided except where all other options have been exhausted. When culling is deemed essential it should include whole families; infants and calves should not be spared for export to zoos, circuses, safari parks, or private reserves. The parallel practice of introducing traumatized youngsters to new areas without adult role models should stop. Alternative practices to culling such as translocation and birth control also have welfare implications, and these must be carefully evaluated. Human-elephant conflict is the cause for increasing ethical dilemmas, and in cases where it is deemed that an elephant must be euthanized, it should be done efficiently and humanely.

What is the right way to treat beings such as elephants? As writer Douglas Chadwick (1992, 475) states, "If a continuum exists between us and such beings in terms of anatomy, physiology, social behavior, and intelligence, it follows that there should be some continuum of moral standards." Based on available evidence, the time has come for us to move beyond old patterns in the treatment of elephants. We must acknowledge and accept new standards for the future.

References

AERP Database. Data extracted from the long-term sightings, censuses, or field notes of the Amboseli Elephant Research Project.

Archie, E. A., Moss, C. J., and Alberts, S. C. 2005. The ties that bind: Genetic relatedness predicts the fission and fusion of social groups in wild African elephants. *Proceedings of the Royal Society B* 273: 513–522.

BBC Natural History Unit. 1992. *Echo of the Elephants.* Cinematography by Martyn Colbeck.

BBC Natural History Unit. 1995. *Echo of the Elephants: the Next Generation.* Cinematography by Martyn Colbeck.

BBC Natural History Unit. 2005 *Echo of the Elephants: the Last Chapter?* Cinematography by Martyn Colbeck.

Blakeslee, S. 2006. Cells that read minds. *New York Times,* January 10. http://query.nytimes .com/gst/fullpage.html?res=9900E3D81F30F9

Bradshaw, G. A., Schore, A. N., Brown, J. L., Poole, J. H., and Moss, C. J. 2005. Elephant breakdown. Social trauma: Early trauma and social disruption can affect the physiology, behaviour, and culture of animals and humans over generations. *Nature* 433: 807.

Byrne, R. W. 1996. Relating brain size to intelligence in primates. In P. A. Mellars and K. R. Gibson (eds.), *Modelling the early human mind* (pp. 49–56). Cambridge: Macdonald Institute for Archaeological Research.

Byrne, R. W. 1997a. What's the use of anecdotes? Distinguishing psychological mechanisms in primate tactical deception. In R. W. Mitchell, N. S. Thompson, and H. L. Miles (eds.), *Anthropomorphism, Anecdotes, and Animals* (pp. 134–150). Albany: State University of New York Press.

Byrne, R. W. 1997b. Machiavellian intelligence. *Evolutionary Anthropology* 5: 172–180.

Byrne, R., and Whiten, A. (eds.). 1988. *Machiavellian intelligence: Social expertise and the evolution of intellect in monkeys, apes, and humans.* New York: Oxford University Press.

Chadwick, D. H. 1992. *The fate of the elephant.* San Francisco: Sierra Club Books.

Clubb, R., and Mason, G. 2002. *A review of the welfare of zoo elephants in Europe.* Horsham, UK: Royal Society for the Prevention of Cruelty to Animals (RSPCA).

Cords, M. 1997. Friendships, alliances, reciprocity, and repair. In A. Whiten and R. W. Byrne (eds.), *Machiavellian intelligence II: Extensions and evaluations* (pp. 24–49). Cambridge: Cambridge University Press.

Cozzi, B., Spagnoli, S., and Bruno, L. 2001. An overview of the central nervous system of the elephant through a critical appraisal of the literature published in the XIX and XX centuries. *Brain Research Bulletin* 54: 219–227.

Croze, H. 1972. The Seronera bull problem: Part I. The elephants. *East African Wildlife Journal* 12: 1–27.

Cutler, R. G. 1979. Evolution of longevity in ungulates and carnivores. *Gerontology* 25: 69–86.

de Waal, F., and van Roosmalen, A. 1979. Reconciliation and consolation among chimpanzees. *Behavioral Ecology and Sociobiology* 5: 55–56.

Douglas-Hamilton, I. 1972. On the ecology and behaviour of the African elephant. DPhil diss., Oxford University.

Douglas-Hamilton, I., and Douglas-Hamilton, O. 1975. *Among the elephants.* New York: Viking Press.

Douglas-Hamilton, I., Bhalla, S., Wittemyer, G., and Vollrath, F. 2006. Behavioural reactions of elephants towards a dying and deceased matriarch. *Applied Animal Behaviour Science.* 100: 87–102.

Dublin, H. T. 1983. Cooperation and reproductive competition among female African elephants. In S. Wasser (ed.), *Social behavior of female vertebrates* (pp. 291–313). New York: Academic Press.

Dunbar, R. I. M. 1988. *Primate Social Systems.* London: Croom Helm.

Dunbar, R. I. M. 1992. Neocortex size as a constraint on group size in primates. *Journal of Human Evolution* 20: 469–493.

Dunbar, R. I. M. 2003. The social brain: Mind, language, and society in evolutionary perspective. *Annual Review of Anthropology* 32: 163–181.

Eisenberg, J. F. 1981. *The mammalian radiations: An analysis of trends in evolution, adaptation and behavior.* Chicago: University of Chicago Press.

Gallup, G. G., Jr. 1970. Chimpanzees: Self-recognition. *Science* 167: 86–87.

Gallup, G. G., Jr., Anderson, J. R., and Shillito, D. J. 2002. The mirror test. In M. Bekoff, C. Allen, and G. M. Burghardt (eds.), *The cognitive animal: Empirical and theoretical perspectives on animal cognition* (pp. 325–334). Cambridge, MA: MIT Press.

Hakeem, A. Y., Hof, P. R., Sherwood, C. C., Switzer, R. C., III, Rasmussen, L. E. L., and Allman, J. M. 2005. Brain of the African elephant (*Loxodonta africana*): Neuroanatomy from magnetic resonance images. *Anatomical Record Part A* 287A:1117–1127.

Hart, B. L., Hart, L. A., McCoy, M., and Sarath, C. R. 2001. Cognitive behaviour in Asian elephants: Use and modification of branches for fly switching. *Animal Behaviour* 62: 839–847.

Janik, V. M., and Slater, P. J. B. 2000. The different roles of social learning in vocal communication. *Animal Behaviour* 60: 1–11.

Jerison, H. J. 1973. *Evolution of the brain and intelligence.* New York: Academic Press.

Jolly, A. 1966. Lemur social behavior and primate intelligence. *Science* 153: 501–506.

Kahl, M. P., and Armstrong, B. D. 2000. Visual and tactile displays in African elephants, *Loxodonta africana:* A progress report (1991–1997). *Elephant* 2 (4): 19–21.

Kane, L., Forthman, D. L., and Hancocks, D. (eds.). 2005a. *Best Practices by the Coalition for Captive Elephant Well-Being.* Madison, WI: Coalition for Captive Elephant Well-Being. Accessed online at http://www.elephantcare.org/protoman.htm.

Kane, L., Forthman, D. L., and Hancocks, D. (eds.). 2005b. *Optimal Conditions for Captive Elephants: A Report by the Coalition for Captive Elephant Well-Being.* Madison, WI: Coalition for Captive Elephant Well-Being. Accessed online at http://www.elephantcare.org/protoman.htm.

Kapustin, N., Critser, J. K., Olsen, D., and Malven, P. V. 1996. Nonluteal estrous cycles of 3-week duration are initiated by anovulatory luteinizing hormone peaks in African elephants. *Biology of Reproduction* 55: 1147–1154.

Kellert, S. R. 1989. Perceptions of animals in America. In R. J. Hoage (ed.), *Perceptions of animals in American culture* (pp. 5–24). Washington, D.C.: Smithsonian Institution Press.

Langbauer, W. R., Jr. 2000. Elephant communication. *Zoo Biology* 19: 425–445.

Langbauer, W. R., Jr., Payne, K. B., Charif, R., Rapaport, L., and Osborn, F. 1991. African elephants respond to distant playback of low-frequency conspecific calls. *Journal of Experimental Biology* 157: 35–46.

Laws, R. M., and Parker, I. S. C. 1968. Recent studies on elephant populations in East Africa. *Symposia of the Zoological Society of London* 21: 319–359.

Laws, R. M., Parker, I. S. C., and Johnstone, R. C. B. 1975. *Elephants and their habitats.* Oxford: Clarendon Press.

Lee, P. C. 1986. Early social development among African elephant calves. *National Geographic Research* 2: 388–401.

Lee, P. C. 1987. Allomothering among African elephants. *Animal Behaviour* 35: 278–291.

Lee, P. C., and Moss, C. J. 1986. Early maternal investment in male and female African elephant calves. *Behavioural Ecology and Sociobiology* 18: 353–361.

Lee, P. C., and Moss, C. J. 1999. The social context for learning and behavioural development among wild African elephants. In H. O. Box and K. R. Gibson (eds.). *Mammalian social learning: Comparative and ecological perspectives* (pp. 102–125). Cambridge: Cambridge University Press.

Lee, P. C., Poole, J. H., Njiraini, N. W., and Moss, C. J. Forthcoming. Male elephant social dynamics: Independence and beyond. In C. J. Moss, and H. J. Croze (eds.), *The Amboseli elephants: A long-term perspective on a long-lived mammal*. Chicago: University of Chicago Press.

Leong, K. M., Ortolani, A., Burks, K. D., Mellen, J. D., and Savage, A. 2003. Quantifying acoustic and temporal characteristics of vocalisations for a group of captive African elephants *Loxodonta africana*. *Bioacoutics* 13: 213–232.

McComb, K., Bates, L., and Moss, C. 2006. African elephants show high levels of interest in the skulls and ivory of their own species. *Biology Letters* 2: 26–28.

McComb, K., Moss, C., Durant, S., Sayialel, S., and Baker, L. 2001. Matriarchs as repositories of social knowledge. *Science* 292: 491–494.

McComb, K., Moss, C., Sayialel, S., and Baker, L. 2000. Unusually extensive networks of vocal recognition in African elephants. *Animal Behaviour* 59: 1103–1109.

McCowan, B., and Reiss, D. 1997. Vocal learning in captive bottlenose dolphins: A comparison with humans and nonhuman animals. In C. T. Snowdon and M. Hausberger (eds.), *Social influences on vocal development* (pp. 178–207). Cambridge: Cambridge University Press.

Meredith, M. 2001. *Africa's elephants: A biography*. London: Hodder and Staughton.

Morris, R., and Morris, D. 1966. *Men and pandas*. New York: McGraw-Hill.

Moss, C. 1975. *Portraits in the wild: Behavior studies of East African mammals*. Boston: Houghton Mifflin.

Moss, C. 1981. Social circles. *Wildlife News* 16: 2–7.

Moss C. 1983. Oestrous behaviour and female choice in the African elephant. *Behaviour* 86: 167–96.

Moss C. 1988. *Elephant memories: Thirteen years in the life of an elephant family*. New York: William Morrow and Company.

Moss C. 1992. *Echo of the elephants: The story of an elephant family*. London: BBC Books.

Moss, C. 1994. Some reproductive parameters in a population of African elephants, *Loxodonta africana*. In C. S. Bambra (ed.), *Proceedings of the 2nd international NCRR (National Centre for Research in Reproduction) conference on advances in reproductive research in man and animals, held in Nairobi, Kenya, 3–9 May 1992* (pp. 284–292). Nairobi: Institute of Primate Research.

Moss, C. 2001. The demography of an African elephant (*Loxodonta africana*) population in Amboseli, Kenya. *Journal of the Zoological Society of London*. 255: 145–156.

Moss, C., and Lee, P. C. Forthcoming. Female elephant social dynamics: Fidelity and flexibility. In C. J. Moss, and H. J. Croze (eds.), *Amboseli elephants: A long-term perspective on a long-lived mammal*. Chicago: University of Chicago Press.

Moss, C. J., and Poole, J. 1983. Relationships and social structure in African elephants. In R. A. Hinde (ed.), *Primate social relationships: An integrated approach* (pp. 315–325). Oxford: Blackwell Scientific.

Mutinda, H. S. 2003. Social determinants of movements and aggregation among free ranging elephants (*Loxodonta africana*, Blumenbach) in Amboseli, Kenya. PhD diss., University of Nairobi.

Nissani, M. 2004. Theory of mind and insight in chimpanzees, elephants and other animals? In L. J. Rogers and G. Kaplan (eds.), *Comparative vertebrate cognition: Are primates superior to non-primates?* (pp. 227–261). New York: Kluwer Academic / Plenum Publishers.

Payne, K. 2003. Sources of social complexity in the three elephant species. In F. B. M. de Waal and P. L. Tyack (eds.), *Animal social complexity: Intelligence, culture, and individualized societies* (pp. 57–85). Cambridge, MA: Harvard University Press.

Pinker, S. 2003. Language as an adaptation to the cognitive niche. In M. H. Christiansen and S. Kirby (eds.), *Language evolution* (pp. 16–37). Oxford: Oxford University Press.

Plotnik, J. M., de Waal, F. B. M., and Reiss, D. 2006. Self-recognition in an Asian elephant. *Proceedings of the National Academy of Sciences* 103: 17053–17057.

Poole, J. H. 1982. Musth and male-male competition in the African elephant. PhD diss., University of Cambridge.

Poole, J. H. 1987. Rutting behaviour in African elephants: The phenomenon of musth. *Behaviour* 102: 283–316.

Poole, J. H. 1989a. Announcing intent: The aggressive state of musth in African elephants. *Animal Behaviour* 37: 140–152.

Poole, J. H. 1989b. Mate guarding, reproductive success and female choice in African elephants. *Animal Behaviour* 37: 842–849.

Poole, J. H. 1994. Sex differences in the behavior of African elephants. In R. Short and E. Balaban (eds.), *The differences between the sexes* (pp. 331–346). Cambridge: Cambridge University Press.

Poole, J. H. 1996. *Coming of age with elephants: A memoir.* New York: Hyperion Press.

Poole, J. H. 1998. An exploration of a commonality between ourselves and elephants. *Etica & Animali* 9: 85–110.

Poole, J. H. 1999a. Signals and assessment in African elephants: Evidence from playback experiments. *Animal Behaviour* 58: 185–193.

Poole, J. H. 1999b. Ella's Easter baby. *Care for the Wild News* 15: 24–25.

Poole, J. H. Forthcoming. The social contexts of elephant vocal communication. In C. J. Moss, and H. J. Croze (eds.), *The Amboseli elephants: A long-term perspective on a long-lived mammal.* Chicago: University of Chicago Press.

Poole, J. H., Payne, K. B., Langbauer, W., Jr., and Moss, C. J. 1988. The social contexts of some very low frequency calls of African elephants. *Behavioral Ecology and Sociobiology* 22: 385–392.

Poole, J. H., and Moss, C. J. 1989. Elephant mate searching: Group dynamics and vocal and olfactory communication. In P. A. Jewell and G. M. O. Maloiy (eds.), *The biology of large African mammals in their environment: the proceedings of a symposium held at the Zoological Society of London on 19th and 20th May, 1988* (pp. 111–125). Oxford: Clarendon Press.

Poole, J., and Granli, P. 2003. Elephant visual and tactile signals database. www.Elephant Voices.org.

Poole, J. H., and Granli, P. K. 2004. The visual, tactile and acoustic signals of play in African savannah elephants. In J. Jayewardene (ed.), *Endangered elephants, past, present and future: Proceedings of the symposium on human elephant relationships and conflicts, Sri Lanka, September 2003* (pp. 44–50). Colombo: Biodiversity and Elephant Conservation Trust.

Poole, J. H., Lee, P. C., and Moss, C. J. Forthcoming. Long-term reproductive patterns and musth. In C. J. Moss, and H. J. Croze (eds.), *The Amboseli elephants: A long-term perspective on a long-lived mammal.* Chicago: University of Chicago Press.

Poole, J. H., Tyack, P. L., Stoeger-Horwath, A. S., and Watwood, S. 2005. Elephants are capable of vocal learning. *Nature* 434: 455–456.

Povinelli, D. J. 1989. Failure to find self-recognition in Asian elephants (*Elephas maximus*) in contrast to their use of mirror cues to discover hidden food. *Journal of Comparative Psychology* 103: 122–131.

Rasmussen, L. E. L., and Schulte, B. A. 1998. Chemical signals in the reproduction of Asian (*Elephas maximus*) and African (*Loxodonta africana*) elephants. *Animal Reproduction Science* 53: 19–34.

Rasmussen, L. E., Schmidt, M. J., Henneous, R., Groves, D., and Daves, G. D., Jr. 1982. Asian bull elephants: Flehmen-like responses to extractable components in female elephant estrous urine. *Science* 217: 159–162.

Rasmussen, L. E. L., and Munger, B. 1996. The sensorimotor specializations of the trunk tip of the Asian elephant, *Elephas maximus. Anatomical Record* 246: 127–134.

Rasmussen, L. E. L., and Krishnamurthy, V. 2000. How chemical signals integrate Asian elephant society: The known and the unknown. *Zoo Biology* 19: 405–423.

Rasmussen, L. E. L., Krishnamurthy, V., and Sukumar, R. 2005. Behavioral and chemical confirmation of the preovulatory pheromone, (Z)-7–dodecenyl acetate, in wild Asian elephants: Its relationship to musth. *Behaviour* 142: 351–396.

Reader, S. M., and Laland, K. N. 2002. Social intelligence, innovation, and enhanced brain size in primates. *Evolution* 99: 4436–4441.

Reiss, D., and Marino, L. 2001. Mirror self-recognition in the bottlenose dolphin: A case of cognitive convergence. *Proceedings of the National Academy of Sciences* 98: 5937–5942.

Rensch, B. 1956. Increase of learning capability with increase of brain-size. *American Naturalist* 90: 81–95.

Rensch, B. 1957. The intelligence of elephants. *Scientific American* 196: 44–49.

Roth, G. 1999. Kleine Gehirne—grosse Gehirn. Evolutionare Aspekte und funktionelle Konsequnzen. *Naturwissenschaftliche Rundschau* 52: 213–219.

Shoshani, J. 1998. Understanding proboscidean evolution: A formidable task. *Trends in Ecology and Evolution* 13: 480–487.

Shoshani, J., and Eisenberg, J. 1992. Intelligence and survival. In H. Shoshani (ed.), *Elephants: Majestic creatures of the wild.* Singapore: Weldon Owen.

Shoshani, J., Kupsky, W. J., and Marchant, G. H., 2006. Elephant brain: Part I. Gross morphology, functions, comparative anatomy, and evolution. *Brain Research Bulletin* 70:124–157.

Slotow, R., van Dyke, G., Poole, J., Page, B., and Klocke, A. 2000. Older bull elephants control young males: Orphaned male adolescents go on killing sprees if mature males aren't around. *Nature* 408: 425–426.

Smolker, R. A., and Pepper, J. W. 1999. Whistle convergence among allied male bottlenose dolphins (Delphinidae, *Tursiops* sp.). *Ethology* 105: 595–617.

Soltis, J., Leong, K., and Savage, A. 2005a. African elephant vocal communication: I. Antiphonal calling behavior among affiliated females. *Animal Behaviour* 70: 579–587.

Soltis, J., Leong, K., and Savage, A. 2005b. African elephant vocal communication: II. Rumble variation reflects the individual identity and emotional state of callers. *Animal Behaviour* 70: 589–599.

Sukumar, R. 2003. *The living elephants: Evolutionary ecology, behavior, and conservation.* New York: Oxford University Press.

Tyack, P., and Sayigh, L. S. 1997. Vocal learning in cetaceans. In C. T. Snowdon and M. Hausberger (eds.), *Social influences on vocal development* (pp. 208–233). Cambridge: Cambridge University Press.

Viljoen, P. J. 1990. Daily movements of desert dwelling elephants in the northern Namib Desert. *South African Wildlife Research* 20 (2): 69–72.

Wemmer, C., and Mishra, H. R. 1982. Observational learning by an Asian elephant of an unusual sound production method. *Mammalia* 46: 557.

Wemmer, C., Mishra, H., and Dinerstein, E. 1985. Unusual use of the trunk for sound production in a captive Asian elephant: A second case. *Journal of the Bombay Natural History Society* 82: 187.

Western, D., and Lindsay, W. K. 1984. Seasonal herd dynamics of a savanna elephant population. *African Journal of Ecology* 22: 229–244.

Wittemyer, G., Douglas-Hamilton, I., and Getz, W. M. 2005. The socioecology of elephants: Analysis of the processes creating multitiered social structures. *Animal Behaviour.* 69: 1357–1371.

Worden, R. 1998. The evolution of language from social intelligence. In J. R. Hurford, M. Studdert-Kennedy, and C. Knight (eds.), *Approaches to the evolution of language: Social and cognitive bases* (pp. 148–168). Cambridge: Cambridge University Press.

5 ELEPHANTS, ETHICS, AND HISTORY
NIGEL ROTHFELS

In an old postcard of the elephant Betsie at the Hague Diergaarde from around 1905, Betsie stands in the doorway at the zoo's elephant house. Behind Betsie can be seen part of the elaborately decorated late nineteenth-century building with its keyhole-shaped doorways for the elephants, a threshold intended to suggest a building in India. The smallish Betsie stands next to her keeper, who is shown erect and stiff, with his arms to his side, in a classic zookeeper's buttoned-up tunic and small hat. The keeper's eyes are a little higher than Betsie's as he looks directly and rigidly at the camera. In front and to the side of Betsie and her keeper stands a small table holding three items—a hand-cranked music box, a corked bottle, and a bell. The objects are all classic props from elephant performances of the period. With these, Betsie could demonstrate the dexterity of her most notable feature, her trunk. Directly in front of Betsie, however, rests more bizarrely an infant's crib, and at the center of the picture, held up by the elephant in her curled trunk, is a swaddled baby or perhaps doll. Haloed by a white, fringed bonnet, the baby's or doll's face is directed to the audience. From under the table, we can see that Betsie has her left front leg casually draped over her right.

The picture captures a particular elephant, time, and place. Looked at from the perspective of another century, most people now involved in zoological gardens would either frown at the idea of an infant in a "free contact" setting with an elephant or smile at the "quaint innocence" or humor of the photo. Most everyone would agree, however, that such a scene would somehow be inappropriate in a progressive zoological garden of today. As a historian, I am interested in how and why scenes like this have become a part of the past of our zoological gardens and what that "pastness" can tell us about how we think about elephants and their ethical treatment *today*.

Current elephant debates revolve around a wide variety of questions. We wonder whether zoos and circuses should be allowed to keep elephants in captivity, and if so, under what circumstances. We argue about whether the ivory trade should be used as a mechanism to fund conservation and whether culling should be properly understood as just one among several effective management tools. We disagree about whether longstanding cultural traditions of working elephants in Asia should force us to pose a

Johann Elias Ridinger, "The Elephant and the Rhinoceros."
Reproduced from Clark (1986)

unique set of questions about "rights" and "wrongs" and elephants. We ask ourselves whether elephants (perhaps along with primates and cetaceans) deserve special legal or moral status. These discussions, and, frequently, heated arguments, do not spring from trivial or parochial intellectual interests. As the chapters in this volume make clear, these are often high-stakes debates in which international commerce and economic development, international trade in endangered species, international science and wildlife conservation, and international animal rights activism may play decisive roles in deciding the fate of a particular elephant or community of elephants in a particular zoo, circus, forest, veldt, or even elephant art academy in Thailand where the paintings of elephant artists are sold to tourists.

These debates are also, and importantly, historical debates. By this, I mean that they should be understood as rooted in specific historical moments and as part of a development of ideas about elephants and their place in human life. This is not, of course, an especially controversial point. Most of us would be willing to accept the position that animals, including elephants, have been thought about in different ways in the past. We do not imagine the lives of gorillas today, for example, in the same way people did a hundred years ago: images of the hideous, rapacious, and sexually violent creature seen repeatedly in nineteenth-century accounts have been largely supplanted by new images of a quiet, reticent, and peaceful animal (Rothfels 2002, 1–4). Nevertheless, is research into how animals have been thought about in the past relevant to discussions of current and future policies regarding the animals? Should historical accounts, for example, of the changing ways elephants have been described in the past, be considered as anything other than background noise in discussions directed to generating critically needed ethical management standards for elephants in the future? The answers to these questions become clear when we examine two pivotal changes in western ideas about elephants that occurred in the eighteenth and nineteenth centuries.

On June 12, 1748, the artist Johann Elias Ridinger completed a series of sketches of a rhinoceros that had arrived in the German city of Augsburg the month before. Having landed in Holland seven years earlier, this rhinoceros, known as Clara, would eventually travel for eighteen years across Europe, visiting hundreds of towns and cities throughout what we now call the Netherlands, Germany, Austria, England, France, Italy, Poland, and Sweden. One of the results of Clara's travels was the widespread dissemination of her image across Europe. In the many posters commissioned by the rhino's owner, in the drawings, paintings, and prints produced by Ridinger, Italian artist Pietro Longhi, and, most importantly, French artist Jean-Baptiste Oudry, Clara finally supplanted the image of the rhinoceros as a kind of ornate ramming machine, an image that had been stamped into European culture by the renowned engraver Albrecht Dürer 233 years earlier. Clara was a pachyderm heroine at the midpoint of the eighteenth century, and Ridinger, in an ensuing print, depicted her as a heroine in a battle to the death with an elephant.

The illustration recalls a story that can be traced at least to the second century before the Christian era (see Rookmaaker 1983; Clarke 1986). The ancient accounts claimed that whenever an elephant and a rhinoceros came across each other in the jungle, the latter would sharpen its horn on a special stone and a titanic battle would begin. Adopting a strategy also used by Eleazar in the Apocryphal battle against King Antiochus, the rhinoceros would eventually seek to attack the elephant's more vulnerable underbelly. The rhino would often perish under the crushing weight of the collapsing elephant (*Maccabees* 1949, 27). At the beginning of the sixteenth century, the old story was read with renewed interest when an Indian rhinoceros arrived in Lisbon on May 20, 1515, as a gift to Portugal's King Manuel. Eager to test the old stories, the king quickly organized a confrontation between the rhinoceros and a young elephant in a courtyard on June 3. After much anticipation, however, the results were disappointing. Although the rhinoceros apparently did show signs of aggression, the terrified elephant fled the scene, bursting through a wrought-iron gate (Bedini 1997, 117–120). Despite this ambiguous test, stories of the battles of the two colossi of nature continued to have remarkable staying power.

Before anything else, it should be immediately clear from the Ridinger print that while the artist had, in fact, had the opportunity to study a living rhinoceros, he had not seen an elephant. Indeed, his elephant resembles more nearly a large bull with massive legs, a tacked-on head, batlike ears, and funny little toes. This is an elephant of the human mind. It is a stylized embodiment—perhaps even an archetype—of brute power and strength. A poem accompanying Ridinger's print makes this point in its first two lines, stating a political truth about power: "No being in the world is so large and powerful / That it does not have an enemy which is its equal." Perhaps the best way to think of Ridinger's elephant is as one of the last representatives of a "species" of elephant we might designate as *Elephas horribilis*—a creature that was spectacularly terrifying, dreadful, and horrible but also astonishing, amazing, and tremendous. Like all elephants, a significant part of the "home range" of this creature was, in fact, the human imagination.

A surprising number of depictions of *E. horribilis* have survived the centuries. Their number, however, does not make it any easier to provide an unambiguous image of what the creature looked like. For example, although most depictions seem convinced that the creature was large, they differ significantly in their estimation of how large. An illustration in a fourteenth-century English bestiary suggests the creature was about the size of a large horse. Ridinger portrayed it as about the same heft as, though taller than, a rhino, and other descriptions imagine the creature at about the height of two people. Essentially, all the accounts agree that the creature had a long nose, and most agree that it had two elongated teeth or tusks. There nevertheless remained some confusion about whether these tusks grew from the top jaw, from the bottom, or from both. Beyond these more striking features, there was considerable disagreement about the conformation of the body, legs, and toes. Moreover, the *Physiologus*—the late antique compendium that sought to interpret classical accounts of animals according to Christian doctrine—created great confusion by insisting that the elephant had no knee joints and, therefore, could not lie down (*Physiologus* 1979, 29–32).

If the general physical descriptions of the animal varied, however, there was more consensus on a series of other points. First, this elephant was particularly useful in military engagements. This had been clear in the accounts of Alexander's conquests, the stories of Hannibal, and the *Books of the Maccabees*. Indeed, an elephant carrying a soldier-filled castle tower on its back is the dominant theme of most representations of the animal up through the seventeenth century. The animal was, moreover, generally understood to live as long as 200, and perhaps even 300 years (Pliny 1940, 23). It was intelligent and could be tamed to serve man. In one important way, though, this elephant contrasted sharply with all other domesticated animals. From classical times, it was known that this creature refused to reproduce in captivity and thereby perpetuate a state of slavery. This indignation for captivity—which the animal could accept for itself but not for its offspring—was further exacerbated, authorities were certain, because this kind of elephant needed utmost privacy for mating, a condition simply unattainable in captivity. When the time came for mating, it was thought that the male and female elephant would retire to a hidden clearing in the forest, and only when convinced that they were unobserved, would they let down their guard and mate. *Physiologus* took this story a bit further and, in a characteristic biblical parallel, claimed that the creature had no desire to mate until, once retired in the deep woods, the female ate the fruit of the mandrake tree and then convinced her prospective mate to do likewise. As to the particulars of mating, there was some wildly ungrounded speculation. Indeed, even as late as the beginning of the nineteenth century, these elephants were depicted mating with the female lying on the ground on her back (Robbins 2002, 227).

As should be clear from this brief account, *E. horribilis* was often little more than a literary or artistic convention. Of course, quite real elephants did occasionally show up in medieval and early-modern Europe. On July 20, 802, for example, the first elephant to arrive in Europe since classical times, Abul Abaz, walked into Aachen, a diplomatic gift to Charlemagne from Caliph Harun al Raschid in Baghdad. For two years, Charles the Great deployed Abul Abaz as the most suitable representation of his power in parades and on campaigns. Indeed, it was on an expedition against the Friesians that the elephant drowned attempting to cross the Rhine. According to Stephan Oettermann's 1982 history of elephants in Europe, in the almost 1,000 years between the arrival of Abul Abaz and the middle of the eighteenth century, around twenty-five elephants, variously from Africa and Asia, lived for a time in western Europe. Oettermann's figures have been expanded in recent years, but we are still talking about an animal that only very few people would ever see. The very rarity of *actual* elephants, of course, often meant that they left a substantial body of records behind. Clearly, alongside the more speculative elephants of Ridinger and others, a number of other animals were actually walking around that evidently looked a good deal more like the elephants we see today.

Even if, however, there were elephants walking around Europe in premodern times that physically resembled the elephants we know, we must nevertheless accept that they were understood quite differently in their day. Like Abul Abaz, these creatures were royal beasts and were meant to be the most spectacular embodiments of princely power. When Hanno, given to Pope Leo X, arrived in Rome in 1514, for example, his entrance to the city as part of a mission of obedience from the Portuguese king was the most memorable event anyone could recall. Ridden by a spectacularly dressed mahout and led by a young Saracen, Hanno's back was covered "with a saddlecloth of crimson velvet" and his "head masked in a caparison of gold brocade"(Bedini 1997, 47). According to historian Silvio Bedini, on his back Hanno carried "a great silver coffer, artfully disposed under a handsome brocaded covering which fell almost to the ground and was ornamented in the finest workmanship with the arms of the Portuguese king. Above the coffer rose a silver tower with many turrets," and under a "canopy at its centre [was] a large rock containing a gold tabernacle. . . . A gold chalice was displayed in one of its turrets, and the others contained silver cases displaying holy vestments of gold and silver cloth adorned with embroideries of pearls and unpolished rubies" (1997, 47–48). According to accounts that have been repeated for centuries, when the elephant finally saw the pope at an open window, he dropped to his knees and bowed his head reverently. Arising, he saluted the pope, trumpeting loudly three times. Hanno was the ideal diplomatic gift to Leo, the lion pope: He was both the rarest of treasures as well as a living representation of the expanding power of the papacy and the Catholic Church; power, through Portuguese conquests, moving even into the Indian subcontinent.

Against both the more legendary and occasionally quite real history of *E. horribilis*, the eleventh volume of the great French naturalist of the eighteenth century, Georges Louis Leclerc, comte de Buffon's *Histoire naturelle, générale et particulière* appeared in 1764. There are only a handful of works on natural history that have had as profound an influence on how people in the western countries have thought about the natural world. To put the point very simply, from 1749, when the first three volumes of Buffon's natural histories of quadrupeds, birds, and minerals appeared, to long after 1789, when the last supplement was published, the work was *the* European authority on nature.[1] In her book on exotic animals in eighteenth-century Paris, *Elephant Slaves and Pampered Parrots,* Louise Robbins reports, for example, a 1772 review of a new volume of Buffon's work. The announcement noted that "for a long time, the works of Buffon have needed only to be indicated; the simplest mention sufficed. . . . Everyone has read, reads and rereads the *Histoire naturelle;* what praise is worth more than that fact?" (2002, 172). Available in numerous editions, sizes, and prices, in abridged formats (some without the anatomical descriptions that many felt were dry and of interest only to scientists), and, of course, translated, copied, paraphrased, and quoted in publications across Europe and America for decades, Buffon's work was read for instruction, for pleasure, for moral edification, and because his scope seemed so encompassing. After completing an entry in the *Histoire naturelle,* a reader could have the sense that he or she knew everything that needed to be known about that animal. As Robbins makes clear, though, Buffon's goal was not to banish completely the allegorical significance attributed to animals in earlier centuries—indeed, part of the pleasure in reading his accounts was to be found precisely in the anecdotes that brought the animal closer to the human experience. At the same time, Buffon sought to include only those stories he believed were verified by reliable observers.

Thus, in his lengthy discussion of the elephant, the creature that he considered the "first and grandest of terrestrial creatures," Buffon insists that classical writers had erred in ascribing to elephants fantastic "intellectual powers and moral virtues." Ancient and more recent authors, he writes, "have given to these animals rational manners, a natural and innate religion, a kind of daily adoration of the sun and moon, the use of ablution before worship, a spirit of divination, piety towards heaven and their fellow creatures, whom they assist at the approach of death, and after their decease bedew them with tears, cover them with earth, &c." (1812, 139–40). Buffon concludes that, "after removing the fabulous credulities of antiquity, and the puerile fictions of superstition, which still exist, the elephant, even to the philosophers, possesses enough to make him be regarded as a being of the first distinction" (142). By "first distinction," Buffon meant that elephants were as close to humans as soulless matter could be. Indeed, of the four species Buffon felt were deserving of special note, the dog (for its ability to form attachments of affection), the ape (for its physical structure), the beaver (for its ability to act in society and cooperation), and the elephant, the last stood apart from the others. He writes: "The genius of the dog (if I may be permitted to profane this term) is borrowed; the ape has only the appearance of it; and the talents of the beaver extend no farther than to what regards himself and his associates. But the elephant is superior to all the three; for in him all their most exalted qualities are united" (137). In the elephant, Buffon saw the combination of "the sagacity of the beaver, the address of the ape, the sentiment of the dog, together with the peculiar advantages of strength, largeness, and long duration of life" (138).

The elephant, Buffon insists is, excepting humans, the most impressive creature in the world, the animal with "more memory and intelligence than any other animal" (188). In his listing of the abilities of the beast, Buffon notes the animal's size and strength, that it can carry an armed tower on his back, that it is courageous, prudent, moderate, impetuous in love, and, importantly, a vegetarian (139). Indeed, Buffon's description seems to be of an Enlightenment supercreature born out the pages of the philosopher Voltaire or the novelist Jonathan Swift. The important point is that Buffon's description marks the beginning of the end of the overwhelming *E. horribilis* and the entrance of a new creature that we might better designate as *E. sentiens*—an animal that feels, experiences, and is somehow physically, intellectually, and emotionally deeply sensitive to its surroundings.[2]

Buffon does not dispose of everything from earlier writers that might be considered a bit dubious. He accepts, for example, that elephants can live for 200 years, that the young suckle with their trunks, that elephants mate face-to-face, that the smell of hogs frightens them, that they are naturally modest, and that they refuse to breed in captivity and thus sustain an unnatural state of slavery. However, these points are accepted precisely because they fit so neatly within Buffon's broad conception of the elephant as a "miracle of intelligence and a monster of matter" (188). In a time when the issues of human slavery and subjugation were becoming increasingly important to pre-Revolutionary France, for example, it apparently made immediate sense to Buffon that elephants would abhor slavery so much— more, he felt than any other domesticated animal or perhaps even the human slaves from Africa—that they would deny themselves their deepest desires in order not to perpetuate the slavery of their kind (152–153). But if they despised their slavery, these elephants were both wise and moderate enough to accept their captivity; they would become model citizens, paying close attention to all instructions and obeying with enthusiasm but also with that calm dignity that marked the animal so much for Buffon. As he noted, "the elephant's character seems to partake of the gravity of his mass" (162).

What makes Buffon's discussion of the elephant such a milestone in our understanding of this animal, however, is his sense that a fundamental quality of the elephant is its capacity for sentiment. In a completely new discussion of elephant eyes, for example, we can read about a creature that, especially when compared with the monstrosities of earlier centuries, now seems strikingly familiar. Buffon writes:

> In proportion to the magnitude of his body, the eyes of the elephant are very small; but they are lively and brilliant; what distinguishes them from the eyes of other animals, is a pathetic expression of sentiment, and an almost rational management of all their actions. He turns them slowly and with mildness towards his master. When he speaks, the animal regards him with an eye of friendship and attention, and his penetrating aspect is conspicuous when he wants to anticipate the inclination of his governor. He seems to reflect, to deliberate, to think, and never determines till he has several times examined, without passion or precipitation, the signs which he ought to obey. The dog, whose eyes are very expressive, is too prompt and vivacious to allow us to distinguish with ease the successive shades of his sensations. But the elephant is naturally grave and moderate, we read in his eyes, whose movements are slow, the order and succession of his internal affections. (183)

If the soldier-filled tower was the most popular way to represent *E. horribilis*, an all-seeing eye—the eye that is ubiquitous in elephant photographs even today—marks the entrance of *E. sentiens*.

Throughout the nineteenth century, many of Buffon's ideas about elephants, especially those whose disproving required simply more contact with the animals, gradually fell by the wayside in European minds. However edifying it may have been for readers in post-Revolutionary France to think of the elephant as only willing to breed in a state of freedom, for example, such ideas were eventually dropped as more elephants made their way into the western world. Nevertheless, as new ideas were added to the elephant, Buffon's interest in the expression of sentiment in the elephant persisted among later commentators. When Buffon and other eighteenth-century natural historians discarded at least some of the unverified stories about elephants passed down through the centuries, we began looking into the eyes of a deeply emotional creature. It was in the nineteenth century, however, when a particular aspect of the emotional life of elephants emerged as especially important in western cultures. Whereas Buffon's ideas of elephant sentiment turned largely on the animal's capacity to show affection, by the late nineteenth century, authors increasingly began to focus on the elephant's capacity to suffer.

The interest in elephant suffering should at least partly be understood within the larger movement in humane ideas in the nineteenth century. These ideas are heralded in one of the most often quoted passages from the utilitarian philosopher Jeremy Bentham's 1789 *Introduction to the Principles of Morals and Legislation*. Bentham writes:

> The day *may* come, when the rest of the animal creation may acquire those rights which never could have been withholden from them but by the hand of tyranny. The French have already discovered that the blackness of the skin is no reason why a human being should be abandoned without redress to the caprice of a tormentor. . . . It may come one day to be recognized, that the number of the legs, the villosity of the skin, or the termination of the *os sacrum*, are reasons equally insufficient for abandoning a sensitive being to the same fate. What else is it that should trace the insuperable line? Is it the faculty of reason, or, perhaps, the faculty of discourse? But a full-grown horse or dog is beyond comparison a more rational, as well as a more conversable animal, than an infant of a day, or a week, or even a month, old. But suppose the case were otherwise, what would it avail? the question is not, Can they *reason*? nor, Can they *talk*? but, Can they *suffer*? (1789/1948, 311)

In general, the humane movement in the nineteenth century was an anti-cruelty movement, which was, as the modern-day animal rights philosopher Peter Singer has put it, "built on the assumption that the interests of non-human animals deserve protection only when serious human interests are not at stake" (1985, 4; see also Turner 1980). The movement thus tended to focus on the treatment of animals by the underclasses—people judged by the movement's leaders to have less serious human interests than their own. Thus, while dogfights could be seen as cruel, foxhunting rarely came up against sustained criticism. The dominant interest of anticruelty advocates was in animals in cities where cruel behavior toward animals could possibly lead to more disturbing antisocial activities directed against other humans. As for more exotic species, although animals in zoos did occasionally draw the attention of humane societies (again, largely for reasons based in concern about what witnessing suffering could do to people), wild animals rarely attracted much attention. However, if, as the British philosopher Mary Midgley pointed out (1983), animals can be deeply significant in human cultures even when that significance is outspokenly denied—if, in other words, "animals matter"—it seems that over the course of the nineteenth century, elephants, and especially their deaths, came to matter a great deal.

Indeed, by the end of the nineteenth century, elephants and their deaths had become almost unmatched metaphors for both power and suffering for Europeans and Americans. Thomas Edison understood this in 1903 when he filmed the electrocution of the Luna Park elephant, Topsy. For having killed several people, it was decided that Topsy should be put down. Edison, trying to prove the dangers of alternating current (AC) and the safety of direct current, signed on to handle the matter with AC. Filming the event, and then traveling around showing the film to prove to audiences the threat posed by AC, made clear sense to Edison. Even today, the film is riveting and the electrocution of Topsy remains probably the most known and frequently seen of all of Edison's films. As we watch, Topsy, in a peculiar conductive costume, is led to a special platform. She stands for a moment and then her entire body jerks up straight and stiff, sparks and smoke enshroud her, and then she falls to her right as a statue of an elephant might fall over, her legs sticking out as if rigor had already done its work. The film is so disturbingly compelling, I believe, because of Topsy's size; because we witness her obvious betrayal as she pliantly and unwittingly follows directions onto the platform; because of the obscene preposterousness of an elephant being electrocuted against a dismal backdrop resembling a garbage dump; because, simply, she's an elephant, and it is difficult for us to see her in a scene like this without recalling the many deeply resonant ideas that have become associated with the creature. Edison electrocuted stray dogs and cats to show the danger of AC; but he made a far more profound case when he electrocuted an elephant.

Relating a long elephantine necrology is not necessary to make it clear that by the end of the nineteenth century, the death of an elephant had become an event imbued with many layers of meaning. When, in 1885, the famous elephant Jumbo (the animal who marks the beginning of the use of the word "jumbo" in English to describe things of colossal scale) was accidentally hit and killed by a locomotive, P. T. Barnum immediately recast the event as a heroic clash between nature and technology wherein a faithful and brave elephant succeeded in saving the life of an elephant calf from an oncoming train only to lose his own. The fabricated story was immediately picked up in newspapers around the world and Barnum recouped his loss of the living Jumbo by traveling for years with two new Jumbos: the animal's mounted skeleton and mounted skin (Oettermann 1982, 88). Barnum made money on the death of Jumbo because the death of an elephant could come to mean more than the death of perhaps any other animal. Indeed, this is precisely the logic behind George Orwell's classic indictment of colonialism, "Shooting an Elephant." Though published in 1936, the expository piece recalls an episode from Orwell's earlier years when he served as a minor official in Burma. In the brief story, a young officer is called to investigate an elephant that had rampaged through a village and killed a man. Upon arriving, however, he finds the animal eating quietly in a field, no longer a threat to anyone. Despite his judgment that the elephant was no longer a concern, the officer soon realizes he will have to kill it anyway because if he were simply to walk away, refusing to do what the villagers expected him to do, they might laugh at him. Orwell's narrator notes that at that moment he realized his "whole life, every white man's life in the East, was one long struggle not to be laughed at" (1954, 8). In gruesome detail, "Shooting an Elephant" relates just how repellent, difficult, and *important* it can become to kill an elephant. "Shooting an Elephant" explains "the real nature of imperialism—the real motives for which despotic governments act" (4) by telling a story about what might otherwise be seen as the straightforward process of killing an animal.

Of all the stories of elephant killing and death in the nineteenth century, however, few were as widely read as that of the British hunter and adventurer Roualeyn Gordon Cumming, who described his hunting experiences most notably in his 1850 *Five Years of a Hunter's Life in the Far Interior of South Africa.* The two-volume account contains many elephant hunts that readers found riveting; great hunter-naturalists like Theodore Roosevelt and Frederick Courtney Selous read Cumming with unbridled enthusiasm. One story in particular, however, caught the attention almost twenty years later of Alfred Edmund Brehm, the celebrated encyclopedist of animal life in the second half of the nineteenth century in Germany. In his entry on elephants for his *Brehm's Thierleben* (*Brehm's Animal Life*), a new sort of animal encyclopedia in which the author hoped to move beyond sterile taxonomies of animals and uncover the lives of animals in the wild, Brehm related a slightly abridged version of an adventure told by Cumming. Cumming's original story reads as follows:

> On the 31st. . . . I came full in view of the tallest and largest bull elephant I had ever seen. He stood broadside to me, at upward of one hundred yards, and his attention at the moment was occupied with the dogs, which, unaware of his proximity, were rushing past him, while the old fellow seemed to gaze at their unwonted appearance with surprise.
>
> Halting my horse, I fired at his shoulder, and secured him with a single shot. The ball caught him high upon the shoulder-blade, rendering him instantly dead lame; and before the echo of the bullet could reach my ear, I plainly saw that the elephant was mine. The dogs now came up and barked around him, but, finding himself incapacitated, the old fellow seemed determined to take it easy, and, limping slowly to a neighboring tree, he remained stationary, eyeing his pursuers with a resigned and philosophic air.
>
> I resolved to devote a short time to the contemplation of this noble elephant before I should lay him low; accordingly. . . . I quickly kindled a fire and put on the kettle, and in a very few minutes my coffee was prepared. There I sat in my forest home, coolly sipping my coffee, with one of the finest elephants in Africa awaiting my pleasure beside a neighboring tree. . . .
>
> Having admired the elephant for a considerable time, I resolved to make experiments for vulnerable points, and, approaching very near, I fired several bullets at different parts of his enormous skull. These did not seem to affect him in the slightest; he only acknowledged the shots by a "salaam-like" movement of his trunk, with the point of which he gently touched the wound with a striking and peculiar action. Surprised and shocked to find that I was only tormenting and prolonging the sufferings of the noble beast, which bore his trials with such dignified composure, I resolved to finish the proceeding with all possible dispatch; accordingly, I opened fire upon him from the left side, aiming behind his shoulder; but even there it was long before my bullets seemed to take effect. I first fired six shots with the two-grooved, which must have eventually proved mortal, but as yet he evinced no visible distress; after which I fired three shots at the same part with the Dutch six-pounder. Large tears now trickled from his eyes, which he slowly shut and opened; his colossal frame quivered convulsively, and, falling on his side, he expired. The tusks of this elephant were beautifully

arched, and were the heaviest I had yet met with, averaging ninety
pounds weight apiece. (Cumming 1850, 2: 14–16)

What appears to have impressed readers in Cumming's day (and what still
stands out) are the vividness and detail of his descriptions. Whether he is
depicting the forest or the death throes of an animal he has just shot, he
takes his time to produce the maximum effect. And the effect, for Cumming,
had largely to do with the spectacular qualities of hunting large game—and
particularly elephants—in Africa. In an earlier hunting episode, for example, when he brought down a large giraffe, Cumming writes that had
elephants never existed, "I could have exclaimed, like Duke Alexander of
Gordon when he killed the famous old stag with seventeen tine, 'Now I can
die happy'" (Cumming 1850, 1: 259). But, of course, elephants did exist,
and the true measure of Cumming *and* his memoirs only becomes apparent when he finally faced his first elephant near the end of his first volume.
For Cumming—and, indeed, for most European and American hunters in
the nineteenth century—the elephant was just something altogether different from any other animal. Cumming writes: "The appearance of the
wild elephant is inconceivably majestic and imposing. His gigantic height
and colossal bulk, so greatly surpassing all other quadrupeds, combined
with his sagacious disposition and peculiar habits, impart to him an interest in the eyes of the hunter which no other animal can call forth"
(Cumming 1850, 1: 265–266). It appears that for most readers of Cumming in the nineteenth century, there was something deeply satisfying in
this kind of prose. Filled with ever-increasing tension and danger, Cumming's stories peak with the death of the animal, followed by a denoue-
ment of pondering during which the hunter finally marvels at his remarkable accomplishments.

Impressed as others may have been with Cumming, however, Brehm
was not. Beginning his critique at the point in Cumming's story when the
hunter began to shoot at different parts of the elephant's skull, he then
moves to a broader condemnation of unsportsmanly hunters. He writes:

> [Cumming] then tries to explain himself claiming that he had
> only entered into the experiment in order to shorten the suffering
> of other elephants. We can accept this excuse as unlikely, however,
> because an elephant hunter must know in advance to what point
> he should direct his shot. Also, in his book Cumming gives innu-
> merable proofs of such a savage and pointless bloodthirstiness that
> we would certainly see any excuse as only a recognition of his bru-
> tality. How infinitely higher stands the elephant over man, how
> wretched, how vile, the contemptible, treacherous enemy shows
> himself to be compared to the magnificent creature. On the occa-

sion of another elephant hunt, the little man relates that he shot a
large male elephant thirty-five times before it died. The hunters in
India fare little better. [James Emerson] Tennent [a mid-nineteenth
century natural historian of Ceylon] makes this abundantly clear.
They are just as shameless as our royals once were when they
would have hundreds of noble animals driven into a confined
space and then would massacre them from an elevated stand. The
bragging elephant hunters of India have shot a good portion of
their bags in the corrals or capture stations, of which we will soon
learn more. They cold-bloodedly shoot down the animals which
have been penned into a narrow space and then leave them to foul
simply in order to enter a few more numbers into their disgraceful
hunting registers. They have shot to pieces old and young without
being able to use their bodies. Among the so-called civilized peoples,
only the English are capable of such abominations.

The hunt after this noble game is also pursued cruelly and mer-
cilessly by the natives of the African Interior. (Brehm 1876–1879,
483–484)

Brehm is appalled that Cumming would chose to experiment on a lamed
animal by firing at its head and watching the effect of the different shots.[3]
Indeed, he is horrified. What drove his disgust, however, was neither the
easy jingoism of his concluding lines nor the fact that Cumming killed an
elephant. Much more importantly, Brehm's revulsion seems to be rooted
in a developing understanding in the period in Germany of the idea of suffering. When Brehm compares "shameless" behavior like that of Cumming
to both the hunting methods formerly employed by the German aristoc-
racy and the "cruel and merciless" methods of the English and the natives
of the African interior, he is arguing that modern, civilized Germans have
a different understanding of acceptable hunting behavior. For Brehm, using
this story of Cumming calculatingly prolonging the death of an elephant
allows him to address the problems of hunting, suffering, and killing in the
most dramatic way for his readers.

Near the end of his entry on elephants, Brehm turns to a seemingly different issue: how young elephants are captured in the wild in Africa. Not-
ing that the animals are often seen in tears as they suffer from their injuries
and that they have been known to die simply from the pain of having lost
their mothers and their freedom, he quotes briefly from Georg Schweinfurth, who explored central Africa in the late 1860s. Several days after having been given a young elephant as a gift, Schweinfurth watched it die from
the stresses of its capture. The explorer writes, "For me there was something
unendingly moving in watching the already quite large yet so helpless creature die while breathing with such difficulty. Whoever has looked into the

eye of an elephant will find that, despite its diminutiveness and the short-sightedness with which these animals are born, that eye holds a more soulful look than that of any other quadruped" (Brehm 1876–1879, 498). With Schweinfurth, Brehm, and many other writers from the second half of the nineteenth century, we can see the emergence of a new kind of elephant, one still with us, which might best be called *E. dolens*—an animal that both physically and mentally feels sorrow and pain; the elephant that suffers.[4]

Like *E. horribilis* and *E. sentiens,* the name *E. dolens* does not describe what an elephant is. I've used these terms to describe a dominant way that elephants have been imagined in different western historical periods. The terms describe *how we think* about elephants. I am not arguing that we only imagine elephants today in terms of suffering, but if we consider the main elements of most of the elephant management controversies over the last century and examine the range of recent popular literature about elephants (everything from Peter Beard's *End of the Game* [1977] and Jeffrey Moussaieff Masson and Susan McCarthy's *When Elephants Weep* [1995] to Barbara Gowdy's novel, *The White Bone* [1999], if we analyze the recent arguments brought forth by animal rights activists to move elephants from circuses and zoos to new elephant sanctuaries and to stop the importation of elephants into western countries, a key concern has been the belief that elephants can experience great emotional, mental, and physical pain. This *E. dolens,* a creature seen as a victim of the avarice, brutality, and disregard of modern man, came into existence during the nineteenth century.

What do these historical points about how elephants were imagined in the eighteenth and nineteenth centuries have to do with ethics and elephant management today? Quite simply, ethics has an important *historical* dimension that is often overlooked. On the face of it, it seems that an ethical position on a certain issue should be the same today, tomorrow, or yesterday. Irrespective of one's historical circumstances, for example, we should all know that it is clearly not right to use a position of power to force others to do things they would not otherwise want to do. The problem is that because we have thought about animals in different ways in different historical periods, our ethical responsibilities toward them has changed as well. In her 1983 book, *Animals and Why They Matter,* the practical ethicist Mary Midgley makes this historical point about animals and ethics when she writes:

> People in general have perhaps thought of animal welfare as they have thought of drains—as a worthy but not particularly interesting subject. In the last few decades, however, their imagination has been struck, somewhat suddenly, by a flood of new and fascinating information about animals. Some dim conception of splendours

and miseries hitherto undreamt of, of the vast range of sentient
life, of the richness and complexity found in even the simplest
creatures, has started to penetrate even to the least imaginative.
(13–14)

Simply because we think about animals differently today from how our ancestors may have thought about them two hundred or two thousand years
ago, we face a different set of ethical questions in living with them. For
Midgley, writing in the early 1980s, it was absolutely clear that there had
been a "marked change in the last few decades in the moral view that or-
dinary people take" of questions about animals. Although I would argue
that one should stretch her origin of the new ideas back another eighty or
ninety years, her basic conclusion still stands—that because we think differently about animals today, we should treat them differently as well.

Significantly for this discussion, Midgley illustrates her argument about
changing attitudes toward animals with an extended discussion of what she
calls "elephanticide." She writes: "To bring out this change, I quote here a
story about elephant-hunting. It comes from a book published in 1850, but
it would, I think, have passed without comment as normal at a much later
date" (14). Remarkably, not only has Midgley chosen to illustrate her argument with the prose of Roualeyn Gordon Cumming, but she has chosen precisely the same drawn-out story from his two-volume memoir that
Brehm did to exemplify a manifestly objectionable view of animals.[5] Like
Brehm, Midgley sees a gaping chasm between her own view of elephants
and that of Cumming. She concludes:

> I do not know whether there are still old gentlemen around today
> who can cheerfully look at that episode exactly as Cummings
> [sic] did, as a piece of perfectly natural civilized behaviour. There
> are plenty of records to show how freely this was done, notably
> the many photographs—taken regularly after a successful day's
> shooting—which show King Edward VII and similar heroes
> standing on lawns besides mountains of dead deer and birds. For
> most of us, however, the light seems somehow to have changed—
> indeed, it probably did so during the First World War. We cannot
> see things that way any longer. (15)[6]

When Midgley writes that "the light seems somehow to have changed," she
is talking about the changing ways people living in different historical periods have perceived the world and its animals. While it is true, that is, that at
a basic level an elephant is an elephant, it is also true that the human cultural
significance of an elephant can vary sharply across cultures and across time.

From my perspective, there are two central implications of this reasoning which indicate why it is useful for people interested in the ethical man-

agement of elephants today to be interested as well in the history of ideas about animals. First, by thinking carefully about history and animals, we can quickly see that although species themselves may not have changed much over the centuries, how we think about them has changed a great deal. With that recognition must come an acceptance that what may have seemed an ethical way to treat elephants 150 years ago, or fifty years ago, or twenty-five years ago, or even ten years ago may not, in fact, be ethical today. In his 1818 *The World as Will and Idea,* the German philosopher Arthur Schopenhauer noted that among certain animals, and he was discussing specifically elephants, "there certainly sometimes appears, always to our astonishment, a faint trace of reflection, reason, the comprehension of words, of thought, purpose, and deliberation" (1818/1907, 2: 232). Over 150 years later when Schopenhauer's words were echoed by Midgley when she described "splendours and miseries hitherto undreamt of," they rang even more true; and, with the work of people like Joyce Poole, Cynthia Moss, and many others, the words are even truer today. Understanding that our ideas about elephants have changed—an understanding that comes through historical research—forces us rationally to rethink our ethical positions. For example, just because a certain management technique "has always been used" or "has traditionally been used," or "has successfully been used" in the past, should not mean that such a technique has some assured status for the future. A technique developed in the past may have been acceptable and suitable in a particular time and place, but just because it was once acceptable should not necessarily mean it will continue to be so in the future.

Second, in thinking carefully about history and animals, it is also clear that to an important extent *we* make our elephants—they are part of *human* culture. I'm not saying that there are no such things as real elephants walking around in the world—that the only elephants are the ones we dream up. But when we see an elephant at the zoo, when we see an elephant performing in a circus, when we see an elephant in a nature documentary, when we read essays on the ethical management of elephants, it is important for us to recognize that an often significant part of what we think we are seeing as the *real thing* might just be a human cultural creation. Studying the history of how we have thought about elephants should make us cautious about what we think we know about the animals, because what we say about elephants might have a great deal more to do with what we think about ourselves than with what we know about elephants. When people claim that elephants never forget, when they still believe that elephants might be afraid of mice, when they spend thousands of dollars to purchase paintings made by elephants working in Thailand,[7] when they argue that an elephant having lived at a zoo for most of its life deserves "retirement" in a sunny sanctuary, they are thinking about both the lives of elephants and their own lives.

In the context of ongoing discussions about elephants and ethics, learning the lessons of history is not about reciting a routine formula about not wanting to repeat the mistakes of the past. Nor is it about being able to pat ourselves on the back at our own enlightenment in comparison to previous generations. In this case, the history of *E. horribilis, E. sentiens,* and *E. dolens* should be seen as a critical tool to be used to scrutinize ideas about elephants, ethics, and human culture.

Acknowledgments

I am indebted to the thoughtful comments on earlier drafts of this essay by Christen M. Wemmer, Catherine Christen, Mark V. Barrow, Jr., and Elizabeth Frank. I am pleased to thank my sister, Janet Rothfels, for her collaboration in naming the "species" of elephant described in this paper. Finally, I am grateful for the support of a fellowship from the Humanities Research Centre at the Australian National University in Canberra, Australia, which made the research for this project possible.

Notes

1. The standard edition of the *Histoire naturelle* runs to forty-four volumes. There were many other versions, however, including a fourteen-volume quarto and a twenty-seven-volume duodecimo. For a particularly compelling discussion of the importance of Buffon, see Robbins (2002).

2. Although the use of *sentiens* here might be somewhat confusing because of the animal rights literature on sentience, which refers most often to an animal's capacity to feel pain, *sentiens* has a much broader contextual use in Latin and refers to all kinds of abilities to feel, including emotional and intellectual capacities.

3. In Cumming's defense, it would be fair to point out that there was substantial disagreement well into the twentieth century about the most effective place to shoot an elephant. Indeed, there was essentially a constant dispute among elephant hunters in the nineteenth century about where to shoot and with what sort of weapon, and memoirs are replete with careful attention to ideal guns and shots. In general, two basic approaches had been recommended. Some authors/hunters felt that the best technique was to fire both barrels of a large caliber gun at the side of the animal near the shoulder with the hope of hitting the heart. The argument for that shot was that even if one missed the heart, the bullets might either hit the shoulder and disable the animal, which would then allow for further shots to the heart or brain, or hit the lungs, which would likely lead to a strong blood spoor and slow the animal for another shot. The other school held that the potentially fastest-working, but most difficult, shot would be one to the brain, either just behind the ear or at the beginning of the trunk. Theodore Roosevelt and many others were very keen on brain shots. The problem especially for a hunter like Roosevelt—was that one was often stuck having to "let fly" repeatedly with a large elephant gun at the brain before the animal could be brought to a stunned stand, at which point one could make a more careful shot, often with a smaller caliber rifle. At the same time, if the hunter did not have sufficient backup support and the first shot did not have the desired effect, the animal might well get away or, it was widely believed, kill the hunter. Cumming tended to shoot to lame elephants because, with his muzzle-loading weapons, it was very unlikely he could kill an elephant with a single shot.

4. In correspondence, Mark Barrow has argued that another important way of conceptualizing the elephant in the nineteenth and twentieth centuries is as a commodity, as, he argues, *E. mercabilis*—suggesting the idea of something that can be bought or purchased. While I would suggest that a better name might be *E. mercatorius*—suggesting more the idea of something seen as an object of commercial exchange—Barrow's point remains particularly important because I believe it points to a crucial element in the genealogy of *E. dolens.* Although it is beyond the scope of this essay, I think it can be argued that *E. dolens* arises out of *E. sentiens* through the apparent pathos of a creature of feeling being viewed primarily as a commodity (whether simply as a carrier of ivory, a trophy, or a lure to tourists and circus-goers). That is, *E. dolens* comes into existence at that moment when the commodification of elephants comes to be seen as brutal exploitation.

5. It appears that Midgley found her reference to Cumming in reading Richard Carrington's popular 1958 *Elephants.* Carrington uses the story as a particularly "revolting" and "repellent" foil to other hunters.

6. Midgley's choice of WWI as a watershed—and especially a *popular* watershed—makes sense. At the same time, most of her new orientation to questions about animals can be traced to the middle of the nineteenth century.

7. At a Christie's auction in 2000, buyers spent over $75,000 for elephant art (see Asian Elephant Art & Conservation Project).

References

Asian Elephant Art & Conservation Project. http://www.elephantart.com/catalog/history.php. Accessed July 4, 2005.

Beard, P. 1977. *The end of the game.* San Francisco: Chronicle Books.

Bedini, S. A. 1997. *The Pope's elephant.* Manchester: Carcanet.

Bentham, J. 1789/1948. *Introduction to the principles of morals and legislation.* Repr., New York: Hafner.

Books of the Maccabees. 1949. Repr., London: East and West Library.

Brehm, A. E. 1876–79. *Brehms Thierleben: Allgemeine Kunde des Thierreichs* (2nd ed.). Leipzig: Verlag des Bibliographischen Instituts.

Buffon, G. L. Leclerc, comte de. 1812. *Natural history: General and particular: Vol. 7. History of quadrupeds.* W. Smellie (Trans.). London: Cadell and Davies.

Carrington, R. 1958. *Elephants: A short account of their natural history, evolution, and influence on mankind.* London: Scientific Book Club.

Clarke, T. H. 1986. *The rhinoceros from Dürer to Stubbs: 1515–1799.* London: Southeby's.

Cumming, R. G. 1850. *Five years of a hunter's life in the far interior of South Africa. With notices on the native tribes, and anecdotes of the chase, of the lion, elephant, hippopotamus, giraffe, rhinoceros, &c.* (2 vols.), New York: Harper Brothers.

Gowdy, B. 1999. *The white bone.* New York: Metropolitan.

Masson, J. M., and McCarthy, S. 1995. *When elephants weep: The emotional lives of animals.* New York: Delta.

Midgley, M. 1983. *Animals and why they matter.* Athens: University of Georgia Press.

Oettermann, S. 1982. *Die Schaulust am Elefanten: Eine Elephantographia curiosa.* Frankfurt: Syndikat.

Orwell, G. 1954. *Shooting an elephant and other essays.* New York: Harcourt, Brace.

Physiologus. 1979. Trans. M. J. Curley. Austin: University of Texas Press.

Pliny. 1940. *Natural History.* H. Rackham (Trans.). Cambridge, MA: Harvard University Press.

Robbins, L. E. 2002. *Elephant slaves and pampered parrots: Exotic animals in eighteenth-century Paris.* Baltimore: Johns Hopkins University Press.

Rookmaaker, L. C. 1983. *Bibliography of the rhinoceros: An analysis of the literature on the recent rhinoceroses in culture, history and biology.* Rotterdam: Balkema.

Rothfels, N. 2002. *Savages and beasts: The birth of the modern zoo.* Baltimore: Johns Hopkins University Press.

Schopenhauer, A. 1818/1907. *The world as will and idea.* R. B. Haldane and J. Kemp (Trans.). London: Kegan.

Singer, P. (ed.). 1985. *In defense of animals.* New York: Basil Blackwell.

Turner, J. 1980. *Reckoning with the beast: Animals, pain, and humanity in the Victorian mind.* Baltimore: Johns Hopkins University Press.

6 PAIN, STRESS, AND SUFFERING IN ELEPHANTS

WHAT IS THE EVIDENCE AND HOW CAN WE MEASURE IT?

JANINE L. BROWN, NADJA WIELEBNOWSKI, AND JACOB V. CHEERAN

The past decade has seen a substantial increase in research on how to identify and measure pain, stress, suffering, and other psychological states, such as pleasure, fear, and grief, particularly in farm and laboratory animals (see, for example, Broom and Johnson 1993; Webster and Main 2003; Kirkwood, Roberts, and Vickery 2004; Touma and Palme 2005). In part because of their charismatic nature, the welfare of captive elephants is now being intensely debated among the general public, scientists, animal welfare/rights groups, animal managers, and the media. Increasingly, questions are raised about the ethics of using elephants for research, economic, education, or entertainment purposes. Animal rights groups often maintain it is cruel and inhumane to keep elephants in most captive situations (see Clubb and Mason 2002). Yet, the criteria used as indicators of poor animal welfare often are subjective, based on anecdotes, and lack a scientific basis. Thus, as scientists we need to communicate clearly the challenges to evaluating psychological states in animals objectively, and we need to develop reliable tools to measure stress, pain, and suffering for elephants.

Stress can manifest itself in a variety of ways. These differ depending on an animal's perceptions and coping strategies—its adaptability—and whether the stressor is acute, that is, short lived, or chronic, that is, ongoing. Reactions also differ depending on the individual's life history and previous experiences. Not all stress is bad. Stress is a natural response that allows the body to adapt quickly to challenging situations. Indeed, a lack of stimuli can lead to boredom, with negative effects on health, behavior, and reproduction (van Rooijen 1991). This could be termed the *golden cage phenomenon*, where individuals are provided with all basic needs and are sheltered from predators and diseases and other challenging life events but are not given the opportunity to make choices or to establish a certain element of control over their lives.

Undeniably, elephants face enormous challenges both *ex situ* and *in situ*. Free-ranging elephants are exposed to numerous stressors. Natural stressors can include efforts to locate food and water resources and dealing with scarcities of food and water, environmental change, encounters with predators, and antagonistic social interactions. Poaching, habitat destruction, translocation, and tourism are stressors of human or anthropogenic origin. Many wild populations are now intensively managed because of limited space and conflict with humans.

Captivity liberates elephants from many of these "natural" stressors, but at what cost? Captive herds differ significantly from those in the wild. Groups often are biased toward one sex, with some private temples holding primarily male elephants and zoos and circuses preferring females. Most "herds" are small; for example, three-quarters of North American facilities maintain three or fewer individuals (Taylor and Poole 1998; Schulte 2000). Such herds are not multigenerational, and are not based on family groups. Calves are rare, and infant mortality is often high (Taylor and Poole 1998). Under these conditions, social bonding may not occur, and at worst, antagonism can create management problems and lead to chronic social stress.

The challenge to scientists and animal managers is to determine which environmental factors are harmful, are benign, or actually improve the welfare of individual animals. Elephants in zoos are exposed to a variety of management styles, training methods, climates, and social situations. Even in range countries, lack of suitable habitat forces free-ranging elephants into small areas where conflict with neighboring human populations is inevitable. In parts of Asia, "domesticated" elephants often still work under what can only be considered unacceptably harsh conditions in timber camps, during festivals, and in conjunction with nature tourism (Ramanathan and Mallapur, in press). When captive elephants are not engaged in activities coordinated by people, they may be chained for extensive periods of time because of lack of space and to maintain control. Last, there is consensus among elephant conservationists that replenishing captive elephant populations from wild specimens is undesirable and perhaps even unethical. This has been identified as a reason for breeding elephants in captivity. However, some contend that it is also unethical to breed and keep elephants in captivity until management problems have been resolved (Clubb and Mason 2002). If we are to continue maintaining elephants *ex situ*, we no longer can ignore our obligation to assess the adequacy of environmental and husbandry conditions and to provide conditions that foster normal behavior, health, reproduction, and overall well-being. Only by carefully gathering the necessary scientific evidence can we identify factors associated with physical pain and psychological suffering and take concrete steps to improve the welfare of elephants in human care.

Defining the Terms
Stress

Stress has become an increasingly popular term in research and management circles in recent decades and is often associated with negative events.

Definitions of stress vary, but in general, it is considered a syndrome occurring in response to any stimulus that threatens homeostasis (the body's physiological or psychological equilibrium state; Selye 1936, 1984). These stimuli are called *stressors*, and the response is believed to have evolved as an adaptive mechanism that allows animals to respond quickly to changes in their environment. Therefore, stress and the stress response are natural elements of life. Consequently, some argue that the term *stress* should be used only for describing normal, adaptive responses, whereas the term *distress* may be more appropriate for defining responses that result in negative effects on animal health and well-being. To avoid confusion, we have chosen to use the generic term *stress* to refer to the various responses described in this chapter.

Pain and Suffering

Pain as an unpleasant sensory or emotional experience; it may or may not be associated with tissue damage (Mellor, Cook, and Stafford 2000). Physical pain, while not necessarily a result of stress per se, can lead to suffering and thus may elicit a physiological response. *Suffering* has been defined as an unpleasant subjective feeling that is prolonged or severe (Broom 1998). Continued exposure to pain and stress can therefore cause suffering. There is little disagreement that many vertebrates have the biological capacity to experience subjective states (in other words, feelings) of fear and anxiety indicative of suffering. Dawkins (1990) clearly outlined the evolutionary advantage of feelings and emotions. They may allow for more flexible and varied responses to the everyday challenges facing species living in highly interactive social and physical environments. Scientists infer advanced cognitive abilities in elephants from their relatively long maturation period (Sukumar 1989), tool use similar to that of higher apes (Chevalier-Skolnikoff and Liska 1993; Hart, Hart, and McCoy 2001), advanced problem-solving skills (Chevalier-Skolnikoff and Liska 1993), long-term memory (see, for example, Rasmussen 1995; Bekoff 2001; McComb et al. 2001), and similarity in certain brain structures to those of humans (Eisenberg 1981; Eltringham 1982; Poole and Moss, Chapter 4 in this volume).

Welfare and Well-Being

Physiologists agree that fear, suffering, and prolonged exposure to stress and pain clearly threaten an animal's welfare and decrease individual well-being. *Welfare* refers to both subjective and objective aspects of an animal's condition of life. It is not solely a scientific concept but also reflects ethical, societal, and political views (Broom 1991; Fraser 1995; Appleby 1997; Fraser et al. 1997). Animal well-being is usually regarded as one of the measurable aspects of animal welfare and is not usually associated with the ethical and political issues that the term *animal welfare* conjures up among various audiences. Assessing animal well-being is key for animal welfare science. Both the physical and psychological aspects must be considered. Poor well-being can be caused by social or physical environments that induce stress, fear, or frustration and by physiological changes that cause pain or discomfort. Both of these can result in animal suffering.

Animal managers should have three major goals. They should ensure that all animals in their care are given the ability to cope effectively with day-to-day changes in their social and physical environment, are allowed to engage in a broad range of species-typical activities, and are generally free from thirst, hunger, physical discomfort, injury, disease, and fear (Spedding 1993; Wolfle 2000). Animal managers need valid scientific criteria to determine a species' welfare requirements and to ascertain effectively when they are not being met. They should carefully quantify conditions that cause a decline in animal well-being and health and lead to suffering. Then they can take corrective actions to modify the environment to provide species-appropriate stimulation or whatever other conditions are needed for improved welfare.

The Physiological Stress Response and How It Relates to Well-Being

Physiological changes occur when an animal is exposed to a stressor. Organic compounds called *catecholamines* are released from the sympatho-adreno-medullary (SAM—"emergency response") system. This increases heart rate and energy allocations to muscles and the brain, while decreasing blood flow to internal organs and the gastrointestinal system. This also stimulates the hypothalamo-pituitary-adrenal (HPA) axis within the body, resulting in adrenal activation. During the HPA response, corticotropin-releasing hormone (CRH) from the hypothalamus stimulates the pituitary gland to release adrenocorticotropin (ACTH), inducing the adrenal gland to secrete glucocorticoids such as cortisol or corticosterone—often referred to as *stress* hormones (Axelrod and Reisine 1984). Glucocorticoids facilitate an ongoing stress response, also preparing the animal for a subsequent stressor. Events such as mating, fighting, hunting, or being hunted all can evoke a stress response (Colborn et al. 1991; Moberg 2000) and activate the adrenal cortex. Glucocorticoids modulate increases in carbohydrate and lipid metabolism, while protecting the animal from adverse inflammatory or immune responses. Together, activation of the SAM and HPA systems result in the "fight or flight" response, allowing an animal to react appropriately to challenges in its environment.

When stressors are short term, the associated physiological processes usually are beneficial. Rapid changes in heart rate and blood pressure allow the organism to react quickly to a new or dangerous situation. However, when acute stress occurs repeatedly or when the stress response is chronically activated, significant biological costs can arise (Ladewig 2000; Moberg 2000). Many stress-related problems have been linked to chronically elevated glucocorticoids; these include immune deficiency, reproductive suppression, growth reduction, muscle and bone wasting, gastrointestinal dysfunction, and impaired brain function (Munck, Naray-Fejes Toth, and Guyre 1987; Sapolsky 1992; Capitanio et al. 1998; Balm 1999; Ferin 2000; Murphy 2000). In African elephants, two diseases of the heart and arteries were associated with habitat degradation as a result of overcrowding in Kenya's Tsavo National Park (Sikes 1968).

Initially, it was thought that the stress response was nonspecific, resulting in the same physiological reactions regardless of type or intensity of the stressor (Selye 1936). We now know that these processes are quite complex and may occur in a graded fashion rather than in an all-or-none response (Matteri, Carroll, and Dyer 2000). Furthermore, feelings such as fear, guilt, and happiness arise through fairly complex neural processing, whereas recognition of pain and discomfort probably rely on more instinctual processes (Broom 1998). The pain detection apparatus appears to function equally across species; however, pain tolerance and perception may vary because of differences in genotype and past experiences (Broom 1998; Mellor et al. 2000). For these reasons, it is essential to identify early symptoms of excessive stress and pain and pinpoint chronic as well as repetitive acute stressors to evaluate animal well-being.

Identifying and Measuring Stress

Studying animal welfare is difficult because no single biochemical or behavioral parameter provides a conclusive measure of stress (Mason and Mendl 1993; Fraser 1995; Möstl and Palme 2002). Behavioral observations may indicate an animal is stressed, but they can be misleading (Henry 1986; Von Holst 1998; Rushen 2000). Similarly, measures of reproductive success, growth rate, and general health, although important for examining certain aspects of animal well-being, may not provide reliable early indicators of stress when examined independently. Rather, understanding the influence of stress on animal well-being will require examining a number of factors, including measures of behavioral and physiological responses.

Glucocorticoid concentrations in blood samples are one indicator of adrenal activity in response to stressors in mammals (Broom and Johnson 1993; Möstl and Palme 2002) and have been measured in wild and captive elephants (Brown, Wemmer, and Lehnhardt 1995). However, there are limitations to using blood corticoids as an index of stress. The act of collecting the sample can elicit a cortisol response (Reinhardt et al. 1990; Cook et al. 2000). Cortisol is secreted in regular pulses following diurnal rhythms. It can be difficult to determine an accurate measure of baseline levels by obtaining occasional samples (see Monfort, Brown, and Wildt 1993). Development of techniques to measure salivary, urinary, and fecal corticosteroids or their metabolites has provided us with new tools for wildlife studies, including in elephants (Dathe, Kuckelkorn, and Minnemann 1992; Brown et al. 1995; Wasser et al. 2000; Foley, Papageorge, and Wasser 2001; Ganswindt et al. 2001; Kalthoff, Schmidt, and Sachser 2001; Schmid et al. 2001; Touma and Palme 2005; Palme et al. 2005). Analysis of excreted urinary and fecal glucocorticoid metabolites also provides a measure of endocrine activity pooled over time. That is, one sample contains the sum of metabolites produced over several hours, depending on metabolic rates and the excretion process. Minor fluctuations in hormone concentrations due to regular pulses are dampened, thus improving the ability to distinguish between normal secretory dynamics and physiological responses to external stimuli (Brown et al. 2001). The noninvasive nature of data collection also allows for continued sampling over long time periods, which in turn enables better detection of glucocorticoid elevations as possible physiological evidence of stress. When used in combination with quantitative behavioral observations and other measures of overall health and well-being, these hormone evaluations can provide a scientific baseline to understand an individual's reactions to potential stressors. They also provide a starting point for ascertaining species-specific needs.

Ensuring the Validity of Stress Hormone Measures

New tools for assessing physiological aspects of stress responses must be carefully tested to ensure both methodological and biological validity, so that the measured results actually reflect true physiological status. Standard laboratory experiments indicate whether an assay system is measuring the appropriate hormone in a proportional manner. Biological validations allow researchers to assess hormone changes in animals exposed to known stressors or to stimuli expected to induce a stress response, for instance, translocation, veterinary procedures, and illness or injury. Increases in measured stress responses should occur after these manipulations.

Hormone responses also can be induced using hormone challenges, relatively simple procedures that rely on intramuscular or intravenous injections of substances that alter hormone secretion. The biological and physiological validity of the assay are assumed to have been demonstrated if hormone concentrations increase after injection of a stimulating agent or decrease following administration of an inhibitory factor. The process of validating corticoid assays involves injecting ACTH, a stimulatory factor that induces the release of cortisol from the adrenal gland, or giving dexamethazone, a synthetic corticosteroid drug that blocks adrenal corticoid production (see Touma and Palme [2005] for review). For noninvasive methods, these changes should be measurable in the urine or feces within the expected excretory lag time, usually a few hours for urine and up to a couple of days for feces. Invasive (blood) and noninvasive (urine, feces) corticoid methods to assess acute "stress" responses through adrenal activation have been validated by ACTH challenge tests in elephants (Brown et al. 1995; Wasser et al. 2000). Now studies are needed to test how effective these methods are at measuring chronic stress and changes in well-being of elephants.

Examples of Measuring Stress in Response to Captive Management in Wildlife Species

Noninvasive glucocorticoid metabolite monitoring is gaining in popularity as a means to investigate animal welfare issues, and when used in combination with other techniques, including behavioral observations and health assessments, such monitoring can produce results relevant to animal management. A study on leopard cats showed that urinary cortisol concentrations were reduced and pacing behavior was decreased when hiding spaces were added to the enclosure and foliage enrichment was provided (Carlstead, Brown, and Seidensticker 1993a). A study on capuchin monkeys using fecal cortisol and behavioral measures also indicated a positive effect of enrichment on stress responses, with a decline in cortisol (Boinski et al. 1999). In cheetahs, the occurrence and severity of gastroenteritis was related to the captive environment and associated with increased fecal glucocorticoid concentrations (Terio and Munson 2000). And in clouded leopards, several husbandry variables correlated significantly with fecal glucocorticoid concentrations (Wielebnowski et al. 2002b). Most notably, lower glucocorticoid concentrations were associated with tall enclosures where cats often rest above and out of the view of the public, whereas those on display or visibly exposed to other large cats had higher glucocorticoid concentrations. These are just a few examples of an ever-growing number of vertebrate taxa in the wild and in captivity among which noninvasive glucocorticoid metabolite monitoring has been used to study stress responses under a variety of circumstances (see also, Wasser et al. 2000; Goymann, Möstl, Gwinner 2002; Monfort 2003).

Challenges of Interpreting Stress Hormone Data

Sometimes welfare indicators are contradictory or the data they provide are difficult to interpret. For example, stereotypies are used as indicators of poor welfare because animals that exhibit these behaviors often are housed under poor or unstimulating conditions (Hediger 1955; Morris 1964; Carlstead et al. 1993a). Although stereotypic behavior may reflect a response to inadequate conditions, the performance of these behaviors may allow some individuals to cope better, presumably offering an outlet, albeit suboptimal, for expressing highly motivated natural behaviors

(Clubb and Mason 2002). Physiologically, the relationship between stereotypies and measures of stress is unclear. In some cases, stereotypies have been associated with factors suggestive of decreased reactivity, like lowered heart rates, increased internal secretion of opioids, which are brain endorphins that modulate pleasure or pain control responses, and reduced cortisol levels (Dantzer 1986; Mason 1991; Rushen 1993; Bettinger et al. 1997; Shepherdson, Carlstead, and Wielebnowski 2004). In other studies, elevated corticoid concentrations were observed in animals exhibiting more stereotypic or self-injurious behaviors (Schmid et al. 2001; Wielebnowski et al. 2002a). A recent study on three African elephants revealed no clear relationship between stereotypies and serum cortisol concentrations but rather highlighted that each individual responded differently to environmental change (Wilson, Bloomsmith, and Maple 2004). Regardless of the physiological impact, the occurrence of unnatural behavior patterns points to the widely held view that stereotypies emerge as a result of environmental inadequacies (Toates 2000). Also, if carried out excessively, many stereotypic behaviors, such as obsessive licking, swimming, or pacing, can become self-injurious or debilitating. That is why proximate, or quick-fix, approaches such as medication, which may eliminate these behaviors, nonetheless do not represent an adequate solution. Instead, the goal should be to eliminate the *conditions* that cause abnormal behavior.

Lack of a physiological stress response, as measured by glucocorticoid levels, does not always mean a lack of stress (Smith and Dobson 2002). For example, in a study of incompatible cheetah pairs, "forced" social housing was associated with suppressed ovarian activity, whereas separation of these pairs resulted in a resumption of normal ovarian cycling (Wielebnowski et al. 2002b). Previously, some researchers have hypothesized that ovarian inhibition might be stress related because elevated glucocorticoids can suppress reproductive activity (Moberg 1985; Sapolsky 1987). But the suppressed female cheetahs and the cycling female cheetahs exhibited no consistent differences in fecal glucocorticoid excretory patterns, either before or after pairing with the males (Wielebnowski et al. 2002b). These results do not mean that the paired cheetahs were not stressed or that adrenal activity was not altered. Chronically stressed individuals can produce paradoxically low corticoid concentrations that are indistinguishable from the concentrations in unstressed subjects. Changes in baseline corticoid production and physiological consequences of chronic stress might not become apparent until the animal is exposed to additional stressors, such as handling or translocation. Chronic stress-induced changes in HPA function may be tested using CRH, to examine pituitary activity, or ACTH, to test adrenal responsiveness. For example, in tethered cattle adrenal response to ACTH was less than that of cattle maintained in paddocks (Ladewig and Smidt 1989). In pigs, repeated social isolation resulted in decreased ACTH and cortisol responses, while adrenaline, noradrenalin, or heart rate responses did not change (Schrader and Ladewig 1999).

The Importance of Comparing Wild versus Captive Populations

Determining how the welfare of captive and free-living animals differs will require comparative analyses, keeping in mind that conditions in the wild vary regionally depending on weather and microclimates, population size, food availability, habitat quality, and extent of human activity. Free-ranging animals undisturbed by humans frequently experience an array of stressors, including predatory defense, social interactions, diseases, and inclement weather. Their responses to these stressors can vary greatly depending on the individual's gender, age, rank, genetics, nutritional status, overall health, and early rearing environment, and even prenatal experiences (Mason 2000; van Rooijen 1991).

Animals born or raised in captivity may react differently than their wild counterparts even when faced with the same stressors because their life experiences differ. If captivity removes too many stimuli, then the lack of stimulation may present a source of stress and result in development of adverse stereotypies (van Rooijen, 1991). Captive housing should take into account social groupings in the wild, for instance, the solitary nature of tigers as compared with the herd and dominance hierarchy characterizing elephants. The challenge is to identify appropriate measures for assessing the efficacy of captive management practices like environmental enrichment, keeper interactions, and social groupings, so as to ensure a state of well-being for particular captive individuals in light of their individual circumstances, experiences, and rearing histories. Urine and feces collected during field studies of wild populations could yield hormonal data for a normative database on wild elephants' variability in stress responses and associated behavioral and biological changes. These norms could then be compared with stress responses observed in captivity.

Studies Measuring Stress Responses in Captive Elephants

Surprisingly few studies have scientifically assessed the welfare of captive elephants in light of their complex social system, intelligence, and charisma. Mostly, researchers have drawn conclusions about the welfare status of elephants based on factors assumed to be related to stress in other species. Stereotypies have been observed in captive elephants, especially those confined in small spaces or chained (Hediger 1950; Meyer-Holzapfel 1968; Kurt and Hartl 1995; Kurt and Garaï 2001). Problems such as aggression that poses a danger to cohorts or handlers, poor maternal care, and infanticide also have been quantified in surveys and found to be higher in captivity than in wild populations (Clubb and Mason 2002). Morbidity, including the prevalence of illnesses that might be related to stress, and mortality also appear to be higher in captive than in wild elephants. Mortality is greater in intensive management systems, especially in Europe and North America, compared with extensive management systems in range countries (Clubb and Mason 2002). The greater prevalence of these problems has been cited as indicating poor welfare conditions for captive elephants (Clubb and Mason 2002). But again, the degree to which these problems actually are related to stress, to pain, or to suffering has not been critically studied.

Separation from herd mates may be stressful for female elephants and young offspring because of their highly social nature. However, studies in this area are limited, and the data are somewhat contradictory. In zoo elephants, Garaï (1992) reported that the temporary removal (for thirty minutes) of a female from a group elicited locomotion and vocalization in all remaining elephants. After separation, aggression directed toward herd mates or inanimate objects and stereotypic behavior was also observed. Kurt and Garaï (2001) noted that the frequency of stereotypies increased in elephants at an orphanage in Sri Lanka after separation. By contrast, transfer of three female Asian elephants into an existing group of four revealed no significant change in behavior or in the concentration of urinary cortisol measured in the two older females one week after transfer (Schmid et al. 2001). Similarly, introduction of an Asian elephant female into an existing herd of six resulted in elevated salivary cortisol for only two days following introduction, and concentrations quickly returned to baseline thereafter (Dathe et al. 1992).

Other studies have focused on the effect of chaining on stereotypies such as weaving, head bobbing, and trunk tossing. Circuses, for example, often chain the diagonal legs of elephants in picket lines when they are not performing. As reported by Friend (1999), the amount of stereotypic behavior exhibited by individual elephants varied but seemed to be consistent within each individual. By contrast, maintaining the same circus elephants unchained in an electrified paddock reduced stereotypic behavior (Friend and Parker 1999). Schmid (1995) also reported a decline in the frequency of stereotypic behavior when circus elephants were kept in paddocks instead of being chained. At an orphanage in Sri Lanka, Kurt and Garaï (2001) found that young Asian elephants rapidly developed stereotypies when chained. Interestingly, Wilson, Bloomsmith, and Maple (2004) found no consistent relationship between serum cortisol levels and stereotypic swaying in three captive African elephants, whereas Bettinger et al. (1997) reported lower serum cortisol concentrations in an Asian elephant that swayed during periods of inactivity. These examples serve to highlight the difficulty of only using one indicator, even a popular one such as stereotypic behavior, to assess stress.

A major problem with most studies is that they were based on relatively small animal numbers with confounding variables that cannot be controlled, that only limited physiological data were collected, and often that only one measure of health and well-being was obtained. Thus, we have to learn about what factors cause stress and possibly suffering in elephants and what constitutes optimal well-being for individuals of these species.

Future Directions for Welfare Research in Elephants

Integration of cross-disciplinary research, including veterinary medicine, nutrition, genetics, reproductive physiology, endocrinology, behavior, and psychology, is needed to accomplish the task of identifying and evaluating pain, stress, and suffering in elephants. To obtain sufficiently large sample sizes, studies need to be multi-institutional, taking care to standardize data collection and minimize extraneous variability whenever possible. Researchers must discriminate between different sources of stress, whether of psychological, physiological, or environmental origin. Psychological stressors would, for instance, be aversive stimuli that stimulate fear responses. Physiological stressors would include dietary deficiencies and diseases. Environmental stressors might include poor enclosure substrates or lack of exercise as causes of foot problems and arthritis. Some stressors related to aversive physical or environmental stimuli, such as enclosure design, visitor viewing access, and nutrition, may be relatively easy to rectify. The subjective, psychological factors are the ones most challenging to study and remediate, especially in the case of the highly intelligent and social elephants. In the following pages, we highlight several areas for which comprehensive studies are currently lacking but are urgently needed.

Breaking

Training an untamed elephant is referred to as *breaking*. It is most difficult in older animals recently captured from the wild because they have already established some level of dominance status in their own hierarchy. Hence, it is more difficult for a trainer or mahout to establish dominance over an older elephant than over a calf. For that reason trainers and mahouts prefer breaking wild-born calves, as well as those born in Asian work camps, at around 4 or 5 years of age.

The trainer seeks to establish dominance over the elephant as quickly as possible. In extreme cases, breaking involves repeated exposure to aversive stimuli (beatings with sticks or hooks, restriction of movement, and denial of food, water, and sleep; see Clubb and Mason 2002). Under these conditions, acute stress responses such as severe diarrhea and vocalizations undoubtedly indicate suffering. To reduce the trauma of breaking, calves in a Burmese working camp were treated repeatedly with xylazine, a sedative (Aik 1992). Animals became calmer and more tolerant of humans during the three-week training process, so punishments were gradually reduced.

However, xylazine and similar sedatives are powerful analgesics that may numb the animal, preventing it from responding readily to painful punishments. Trainers unaware of this effect of analgesics may cause unnecessary wounds and injuries.

In southern India, wild elephants are initially trained by keeping the animals inside a large cagelike enclosure called a *kraal*. Injuries sustained by the animal are fewer than with other methods and only occur if the animal attempts to break free from the kraal. Mild sedatives will be effective in reducing the frightened and aggressive behavior exhibited by the captive elephant. Still, even gentler breaking methods differ considerably from the way elephants establish and maintain dominance in the wild (Moss 1988).

Female social status within a herd is mainly determined by age, size, kinship, and individual disposition (Dublin 1983; Sukumar 1989; Thouless 1996). Generally, the largest, eldest female—the matriarch—is dominant, playing a critical role in elephant herd survival (Douglas-Hamilton 1972; Douglas-Hamilton and Douglas-Hamilton 1975; Dublin 1983; Poole and Moss 1989; Sukumar 1989; McComb et al. 2001). Adult males are more solitary but will compete with each other for access to breeding females. Losers of aggressive interactions retreat to indicate their subordinate position.

It is not natural for an individual of lesser stature, such as a human, to dominate an elephant—and when this does happen, the animal is placed in the unusual situation of being deprived of the option to retreat from an aggressor, namely, the trainer (Clubb and Mason 2002). Because elephants in the wild live in social groups with a dominance hierarchy, a trainer's dominance is more likely to be accepted when it is established gradually and tactfully. However, during the period of musth, a reversal of dominance occurs, with the bull becoming aggressive and often refusing to obey. Then a mahout may punish the animal by cruel and painful means. If the animal gets a chance, it may become more aggressive and attack the mahout, causing injury or death. At this point, we do not yet understand the psychological impact of dominance-oriented breaking methods, whether harsh or not, on the psychological welfare of elephants.

Training and Management

Elephants have thick skin; yet, they perceive pain through receptors in the skin just as other mammals do (Shoshani 2000). Handlers may use an ankus (elephant hook) or other devices on sensitive body areas to encourage desired responses or impart discipline. These are also used to punish recalcitrant elephants and may impart fleeting discomfort or long-lasting pain. Sometimes electric prods are used on especially difficult elephants. The elephant no longer has control over its environment, and after a time, it submits and no longer attempts to escape, retaliate, or resist being handled. It succumbs to what has been termed *learned helplessness* (Maier and Seligman 1976; Job 1987; Mazur 1998; Seligman and Isaacowitz 2000).

When does physical punishment have a detrimental effect on welfare? Is any form of physical discipline an acceptable management strategy? Assessing the long-term effects of harsh training methods or of less severe forms of discipline is difficult. Little is known about the consequences of maintaining subordinate behavior in elephants, as acute stress during breaking may be replaced by chronic stress associated with habituation. Psychosocial stress has been described as one of the most potent stressors (McCobb et al. 2003), with detrimental effects on behavior, physiology, and overall health, including immune responses (see Mendoza, Capitanio, and Mason 2000; McCobb et al. 2003). Rhesus macaques in unstable social conditions had a significantly shorter survival rate when exposed to the simian immunodeficiency virus than macaques in stable social conditions (Capitanio et al. 1998). Chronic subordination in tree shrews resulted in depressed gonadal function and other symptoms paralleling those of severe human depression, such as reductions in locomotor activity, autogrooming and scent marking, and adrenal hypertrophy (see McCobb et al. 2003). In a highly communal species like the elephant, prolonged social stress and subordination could affect behavior, reproduction, disease resistance, and overall health.

Whether continued harsh training elicits reactions, or feelings, akin to resentment is another concern. Such responses may explain why elephants attack or kill elephant keepers and mahouts (Hart 1994; Kurt 1995; Schmid 1998). We must examine the effectiveness of training based on positive reinforcement intended to build better relationships between elephant and handler (Fernando 1990). Indeed, studies in domestic species have convincingly linked handler attitude and behavior with animal stress levels, including cortisol levels and heart rate, as well as with their reproduction, milk production, and growth levels and their overall health (Hemsworth and Coleman 1998; Waiblinger, Menke, and Coleman 2002).

Objective measures are needed to determine what handling strategies—protected, free, or no contact—ensure the well-being of elephants in captivity. Some believe positive hands-on training interactions in free contact provide an enriching experience of mental and physical stimulation for the elephants (Hediger 1955; Schmidt and Markowitz 1977; Molter 1980; Dudley 1986; Koehl 2000). Others state that elephants in free contact are controlled using negative reinforcement and occasional harsh discipline so that handlers can maintain dominance and that it is a cruel management technique. Experts disagree about whether elephants in free-contact situations can be managed using positive reinforcement only. In free contact, when an elephant is subjected to long or painful veterinary interventions, physical restraint, negative reinforcement, or punishment may be required. If the mahout or keeper has an excellent relationship with the elephant in question, physical restraints plus oral commands—even oral commands alone—often suffice.

With protected contact the handler does not enter the elephant's enclosure so there is no need for dominance. Protected contact relies on operant conditioning and positive reinforcement, and the elephant's participation is voluntary (Desmond and Laule 1991). Operant conditioning is also used to train performing animals (Clubb and Mason 2003). This system is believed to offer a sense of security and safety to the elephant because it is able to retain some control over its environment. In other mammals, this response has been known to reduce stress-related pathologies (see Weiss 1971; Weiss et al. 1981; Broom and Johnson 1993; Holmer et al. 2003). Increasing evidence suggests that by using positive reinforcement and protected contact even older animals can be trained successfully to participate in a variety of required activities, such as veterinary exams, blood draws, and training (Del Castillo 2003). There is no valid reason for handlers to use an ankus in protected contact, although many still do (Clubb and Mason 2002).

Caretaking

In general, increased predictability and quality of caretaking has been shown to reduce stress responses in many species. For example, pigs managed using positive handling as compared with minimal handling and negative handling procedures demonstrated decreased serum cortisol concentrations and increased pregnancy rates (Hemsworth et al. 1986). Even individual pigs exposed to aversive housing practices such as tether-stalling showed reduced physiological stress responses when they received positive handling and caretaking experiences (Pedersen et al. 1997). In cows, positive handling produced lowered cortisol concentrations and increased milk production (Hemsworth et al. 1995; Breuer, Coleman, and Hemsworth 1998).

Improved caretaking has been associated with decreased rates of domestic cat stereotypic behavior (Carlstead, Brown, and Strawn 1993b), decreased morbidity and mortality of rhinoceros (Carlstead et al. 1999), increased reproductive rates of several small felids (Mellen 1991) and lowered corticoid concentrations in clouded leopards (Wielebnowski et al. 2002a). Friendly interactions with humans have shown direct or indirect physiological effects, including decrease in heart rate and arterial blood pressure in rodents and various pet species (McMillan 1999). By contrast, aversive handling can increase cortisol secretion and negatively affect other physiological indicators of health and well-being, including growth rate in young pigs and calves, milk production in cows and immune response in cows and pigs (Hemsworth et al. 1986; Hemsworth and Barnett 2000).

In complex social mammals such as elephants we must assume that the quality and consistency of the animal-handler relationship is of high importance. Strong bonds between elephants and keepers in free contact are common, and increased stereotypies, behavioral problems, and disease susceptibility have sometimes followed the departure of a favored handler. In protected contact, elephants and keepers do not share the same space. Yet, it is just as important to establish good relations to achieve similar benefits. A somewhat more difficult but not insurmountable challenge is to ensure the protected-contact environment provides adequate enrichment opportunities and encourages exercise to make up for any lack of keeper-directed activities (see, for example, Del Castillo 2003).

Regardless of which type of "contact" is employed, operant conditioning and positive reinforcement should constitute the only acceptable captive elephant management method. Many elephants currently in western zoos were trained using harsh methods or obtained from range countries during highly stressful captures. It is difficult to undo this damage. With the increase in captive births now occurring in western zoos, we have a unique opportunity to examine the effectiveness of continued positive reinforcement training. Likely there is not a one-size-fits-all solution to managing elephants. On an individual basis, some elephants may do better in free contact, while others clearly should be managed in protected contact. How elephants respond to different management options will differ among facilities, situations, and individuals. For instance, captive-born calves not raised in a normal herd structure may tend to become more mischievous than otherwise, owing to lack of disciplining by herd mates combined with pampering by owners or managers. We must be willing to critically and scientifically evaluate all options and choose the one that is best for each elephant. Implementing a protected contact system should never be used as a managerial excuse to provide less care and attention to elephants in an effort to lower the risk of staff injury.

Chaining and Confinement

How long is too long for an elephant to be chained? How do we know which enclosure sizes are adequate for different-sized groups of elephants, even when not chained? Elephants display more stereotypies while chained or confined in small spaces, but there is disagreement about what conditions are inhumane. Chaining for an hour to do foot care may not be stressful, but it may be if the chaining occurs for fourteen hours a day, even when the measure is used to prevent intragroup aggression. What about access to the outdoors? Depending on the climate and time of year, elephants in captivity can spend 60% or more of their time indoors, often chained, at least according to relatively old studies (Green 1989; Galloway 1991). During severe winter weather, elephants may be inside continually for days or even weeks. Indoor enclosures generally are smaller and provide less enrichment than outdoor enclosures. More studies should be conducted to examine how confinement affects well-being in the short and long term, while taking into account possible seasonal changes in stress responses (e.g., Touma and Palme 2005).

Enrichment and Exercise

Elephants in zoos should be provided with environmental enrichment and exercise opportunities designed to fulfill their physical and psychological needs for social companions, sufficient space, wallowing and bathing pools, rocks or trees for rubbing, dust or sand for dust-bathing, and various browse options. But these alone may not provide the level of stimulation required by highly social and intelligent animals like elephants.

The daily choices wild animals have to make constitute a key element in their lives. Providing options and allowing for choices whenever possible through enrichment can be of great importance for reducing stress. The amount of control an animal has over its environment can be a key factor in reducing the severity of stressful stimuli (e.g., Weiss et al. 1981). Choices allow the animal to find ways to cope with perceived or actual stressors and respond in an adaptive manner. Behavioral and physiological effects of different types of environmental enrichment should be evaluated. Zoo biologists should conduct experiments to determine the effects of modifying the size and complexity of the enclosure environment, varying presentation and choices of food (Gilbert and Hare 1994; Harnett 1995), introducing novel objects or scents (Green 1989; Haight 1993; Hare and Gilbert 1995; Leach, Young, and Warren 1998), ensuring an affiliative social environment and differing the amounts and kinds of physical activity.

Other Management Issues

Many other questions about potentially stressful husbandry and management practices could benefit from thorough physiological and behavioral investigations. Among these are the following: Should female offspring be separated from their dams, and if so, at what age? How should elephants be introduced to a new herd? How can we better manage bulls? What should the minimum herd size be and how do we prevent intragroup aggression? And, should elephants be used for rides, painting pictures, performing in shows, participating in festivals, or interacting directly with the public in any other manner?

Recommendations for Assessing Welfare in Elephants

To improve the way we care for elephants in captivity, the goal should not be to eliminate all stressors, just those that are psychologically or physically harmful. We must provide opportunities that allow elephants to express a full complement of behavioral responses similar to those in nature (Markowitz 1978). Detailed behavioral observations coupled with longitudinal physiological analyses are needed to identify factors that correlate with measures of elephant welfare. Whenever possible, data from captive elephants should be compared with those collected from free-living populations. Behavioral data can be collected by direct observation or by keeper and facility evaluation surveys, or preferably by both means (Carlstead et al. 1999; Wielebnowski et al. 2002a). Stress hormones can be assessed using serum, or by noninvasive analysis of metabolites in urine, saliva, or feces. Samples for corticoid analysis should be collected at approximately the same time each day to control for circadian rhythms (T. Wagener and J. Brown, unpublished data), as found in other species (Buckingham 2000; Murphy 2000; Touma and Palme 2005). For studying chronic stress, periodic CRH or ACTH challenge tests could provide a more meaningful evaluation of HPA responsiveness than other approaches. Biochemical analyses of other stress factors, like catecholamines, and tests of immune function could add another level to our understanding of how the captive environment affects animal health. Periodic studbook and survey analyses should also be conducted to establish long-term relationships between potential stress-inducing factors and reproductive health, as well as with morbidity and mortality data.

One of the biggest challenges, at least in the short term, will be obtaining enough physiological and behavioral baseline data on captive and free-ranging elephants to allow proper interpretation of individual observations. Study designs should provide for collecting data before, during, and after changes in environment or management to facilitate evaluation of individual responses. To date, very few environmental or management parameters have been extensively studied under the variety of conditions needed to make informed decisions. A large number of animals should be evaluated to control for variability among facilities, especially when examining the relative importance of different factors. This would apply, for example, to comparing stress levels of elephants in protected versus free contact, or in seeking correlations between stress and enclosure size. Ideally, multiple samples should be collected to ensure an adequate baseline estimation, with standardized parameters facilitating proper assessment and meaningful comparison. To rule out seasonal influences, data should be collected for a minimum of one year. For example, it would not be appropriate to make inferences about stress levels in elephants by analyzing a single blood sample for cortisol because too many intrinsic and extrinsic factors can affect circulating levels.

In their 2002 Royal Society for the Prevention of Cruelty to Animals report, Clubb and Mason identified potential causes of poor welfare for captive elephants. These included restricted space or opportunities for exercise; cold and wet climates; extended periods of confinement; hard or wet flooring substrates; inappropriate diets; the lack of opportunities to perform various natural behaviors; small social groups; isolation-housing; lack of relatedness or stability within social groups; and early weaning. Also indicted were training in free contact systems only, breaking when young, and the exposure to aversive stimuli during training. They admitted, however, that "In rare instances, scientific data allowed us to make a link between specific aspects of husbandry and welfare—but for the most part, these issues remain unresolved due to a lack of research" (p. 247). Clearly, a serious commitment by researchers and management organizations alike will be needed to improve the welfare of elephants. By applying a scientific approach to the implementation of conservation strategies, we become accountable for ethical decisions. We must be proactive in finding new and better ways to assess and improve elephant welfare, while recognizing there may be more than one solution. And we must also be prepared to admit that some welfare issues cannot be resolved by science alone.

To begin, organizational groups such as the Association of Zoos and Aquariums (AZA—formerly, the American Zoo and Aquarium Association), the European Association of Zoos and Aquaria, the World Association of Zoos and Aquariums, the Asian and African Elephant Specialist Groups, and nongovernmental organizations such as the Elephant Welfare Association should strongly encourage and endorse the initiation of large scale, multi-institutional, and multidisciplinary studies on captive elephants to investigate welfare issues. Funding entities like the AZA's Conservation Endowment Fund, Disney's Conservation Fund, and the International Elephant Foundation should make a high priority of funding well-designed studies on *ex situ* elephant welfare, and equally they must insist that management decisions are based on the scientific evidence, regardless of outcome. If findings suggest that elephants require more space and larger social groupings, then the standards should be changed to reflect that. If chaining is found to cause undue stress, it should be prohibited. In range countries, government and private wildlife organizations should insist that mahouts be better trained to ensure humane treatment of elephants and that their facilities meet the social and physical requirements of the animals.

We need not wait for scientific data to effect some of these changes. In some cases, an accumulation of documented anecdotal evidence and plain common sense may already help us make decisions toward improving elephant welfare. Any facility or circus that provides insufficient care for health and other daily needs should not be permitted to have elephants. Use of painful and psychologically cruel breaking techniques should be forbidden. The beating of elephants under any circumstance should not be tolerated.

We can no longer afford to ignore these problems. Given our currently available research tools we are obligated to take a close look at harmful practices and proactively address valid welfare concerns to ensure that elephants under human care are managed in the best way possible. Some welcome signs of reform are already under way, such as the Asian Elephant Specialist Group's work on creating a protocol for captive elephant management. Recently the southern Indian state of Kerala published a set of captive elephant management rules, enforceable under the provisions of India's statutory 1972 Wildlife Protection Act. Let us hope that other states and countries will soon devise and enforce similar rules that will help improve the lives of all elephants in human care.

References

Aik, S. S. 1992. Preliminary observations on the training of Burmese elephants using xylazine. *New Zealand Veterinary Journal* 40: 81–84.

Appleby, M. C. 1997. Life in a variable world: Behaviour, welfare, and environmental design. *Applied Animal Behaviour Science* 54: 1–19.

Axelrod, J., and Reisine, T. D. 1984. Stress hormones: Their interaction and regulation. *Science* 224: 452–459.

Balm, P. H. D., ed. 1999. *Stress physiology in animals.* Sheffield, UK: Sheffield Academic Press.

Bekoff, M. 2001. Thick skins, tender hearts: Incidents like those at the Denver Zoo in June remind us that elephants are complex creatures whose feelings and instincts are ignored at our own peril. *Rocky Mountain (Denver, CO) News,* July 7.

Bettinger, T., Larry, M., Goldstein, M., and Laudenslager, M. 1997. Plasma cortisol concentrations and behavioral traits of two female Asian elephants. In *Proceedings of the American Zoo and Aquarium Association Annual Conference, September 13–17, Albuquerque, New Mexico* (pp. 88–90). Bethesda, MD: American Zoo and Aquarium Association.

Boinski, S., Swing, S. P., Gross, T. S., and Davis, J. K. 1999. Environmental enrichment of brown capuchins (*Cebus capella*): Behavioral and plasma and fecal cortisol measures of effectiveness. *American Journal of Primatology* 48: 49–68.

Breuer, K., Coleman, G. J., and Hemsworth, P. H. 1998. The effect of handling on the stress physiology and behaviour of nonlactating heifers. In *Proceedings of the Australian Society for the Study of Animal Behaviour, 29th Annual Conference, Palmerston North, New Zealand* (pp. 8–9).

Broom, D. M. 1991. Animal welfare: concepts and measurement. *Journal of Animal Science* 69: 4167–4175.

Broom, D. M. 1998. Welfare, stress, and the evolution of feelings. In A. P. Møller, M. Milinski, and P. J. B. Slater (eds.), *Stress and behavior: Advances in the study of behavior* (pp. 371–404). San Diego, CA: Academic Press.

Broom, D. M., and Johnson, K. G. 1993. *Stress and animal welfare.* London: Chapman and Hall.

Brown, J. L., Graham, L. H., Wielebnowski, N., Swanson, W. F., Wildt, D. E., and Howard, J. G. 2001. Understanding the basic reproductive biology of wild felids by monitoring faecal steroids. In P. W. Concannon, G. C. W. England, W. Farstad, C. Linde-Forsberg, J. P. Verstegen, and C. Doberska (eds.), *Advances in reproduction in dogs, cats and exotic carnivores. Reproduction Supplement* 57: 71–82. Nottingham, UK: Nottingham University Press.

Brown, J. L., Wemmer, C. M., and Lehnhardt, J. 1995. Urinary cortisol analyses for monitoring adrenal activity in elephants. *Zoo Biology* 14: 533–542.

Buckingham, J. C. 2000. Effects of stress on glucocorticoids. In G. Fink (ed.), *Encyclopedia of stress* (Vol. 2, pp. 229–237, 261–269). New York: Academic Press.

Capitanio, J. P., Mendoza S. P., Lerche, N. W., and Mason, W. A. 1998. Social stress results in altered glucocorticoid regulation and shorter survival in simian acquired immune deficiency syndrome. *Proceedings of the National Academy of Sciences of the United States of America* 95: 4714–4719.

Carlstead, K., Brown, J. L., and Seidensticker, J. 1993a. Behavioral and adrenocortical responses to environmental changes in leopard cats (*Felis bengalensis*). *Zoo Biology* 12: 321–331.

Carlstead, K., Brown, J. L., and Strawn, W. 1993b. Behavioral and physiological correlates of stress in laboratory cats. *Applied Animal Behaviour Science* 38: 143–158.

Carlstead, K., Fraser, J., Bennett, C., and Kleiman, D. G. 1999. Black rhinoceros (*Diceros bicornis*) in U.S. zoos: II. Behavior, breeding success, and mortality in relation to housing facilities. *Zoo Biology* 18: 35–52.

Chevalier-Skolnikoff, S., and Liska, J. 1993. Tool use by wild and captive elephants. *Animal Behaviour* 46: 209–219.

Clubb, R., and Mason, G. 2002. *A review of the welfare of zoo elephants in Europe.* Horsham, UK: Royal Society for the Prevention of Cruelty to Animals (RSPCA).

Colborn, D. R., Thompson, D. L., Jr., Roth, T. L., Capehart, J. S., and White, K. L. 1991. Responses of cortisol and prolactin to sexual excitement and stress in stallions and geldings. *Journal of Animal Science* 69: 2556–2562.

Cook, C. J., Mellor, D. J., Harris, P. J., Ingram, J. R., and Matthews, L. R. 2000. Hands-on and hands-off measurement of stress. In G. P. Moberg and J. A. Mench (eds.), *The biology of animal stress: Basic principles and implications for animal welfare* (pp. 123–146). Wallingford, Oxon, UK: CABI.

Dantzer, R. 1986. Behavioural, physiological and functional aspects of stereotypic behaviour: A review and reinterpretation. *Journal of Animal Science* 62: 1776–1786.

Dathe, H. H., Kuckelkorn, B., and Minnemann, D. 1992. Salivary cortisol assessment for stress detection in the Asian elephant (*Elephas maximus*)—a pilot study. *Zoo Biology* 11: 285–289.

Dawkins, M. S. 1990. From an animal's point of view: Motivation, fitness, and animal welfare. *Behavioral and Brain Sciences* 13: 1–61.

Del Castillo, G. M. 2003. Psychological and physical benefits in a clinical session. In *Abstracts for the 6th International Conference on Environmental Enrichment.* November 2003, Johannesburg, South Africa, Johannesburg Zoo, p. 10.

Desmond, T., and Laule, G. 1991. Protected contact elephant training. *Proceedings of the American Association of Zoological Parks and Aquariums Annual Conference 1991* (pp. 606–613). Wheeling, WV: American Association of Zoological Parks and Aquariums.

Douglas-Hamilton, I. 1972. On the ecology and behaviour of the African elephant. DPhil diss., Oxford University.

Douglas-Hamilton I., and Douglas-Hamilton, O. 1975. *Among the elephants.* New York: Viking Press.

Dublin, H. T. 1983. Cooperation and reproduction competition among female African elephants. In S. K. Wagner (ed.), *Social behavior of female vertebrates* (pp. 291–313). New York: Academic Press.

Dudley, J. P. 1986. Notes on training of captive elephants. *Elephant* 2: 15–18.

Eisenberg, J. F. 1981. *The mammalian radiations: An analysis of trends on evolution, adaptation and behavior.* Chicago: University of Chicago Press.

Eltringham, S. K. 1982. *Elephants.* Dorset, England: Blandford Press.

Ferin, M. 2000. Effects of stress on gonadotropin secretion. In G. Fink (ed.), *Encyclopedia of stress*, (Vol. 2, pp. 283–288). New York: Academic Press.

Fernando, S. B. U. 1990. Training working elephants. In *Animal training: A review and commentary on current practice: Proceedings of a symposium organized by Universities Federation for Animal Welfare, held at the Peterhouse Theatre, University of Cam-*

bridge, 26–27 September 1989 (pp.101–113). Potters Bar, Hertfordshire, England: Universities Federation for Animal Welfare.

Foley, C. A. H., Papageorge, S., and Wasser, S. K. 2001. Noninvasive stress and reproductive measures of social and ecological pressures in free-ranging African elephants. *Conservation Biology* 15: 1134–1142.

Fraser, D. 1995. Science, values and animal welfare: Exploring the "inextricable connection." *Animal Welfare* 4: 103–117.

Fraser, D., Weary, D. M., Pajor, E. A., and Milligan, B. N. 1997. A scientific conception of animal welfare that reflects ethical concerns. *Animal Welfare* 6: 187–205.

Friend, T. H. 1999. Behavior of picketed circus elephants. *Applied Animal Behaviour Science* 62: 73–88.

Friend, T. H., and Parker, M. L. 1999. The effect of penning versus picketing on stereotypic behavior of circus elephants. *Applied Animal Behaviour Science* 64: 213–225.

Galloway, M. 1991. Update on 1990 chaining survey. In *Proceedings of the 12th International Elephant Workshop, held at Burnet Park Zoo, Syracuse, New York, October 16–19, 1991* (pp. 63–64). Syracuse, NY: Elephant Managers Association.

Ganswindt, A., Palme, R., Heistermann, M., Borragan, S., and Hodges, J. K. 2003. Noninvasive assessment of adrenocortical function in the male African elephant (*Loxodonta africana*) and its relation to musth. *General and Comparative Endocrinology* 134:156–166.

Garaï, M. E. 1992. Special relationships between female Asian elephants (*Elephas maximus*) in zoological gardens. *Ethology* 90: 187–205.

Gilbert, J., and Hare, V. J. 1994. Elephant feeder balls. *The Shape of Enrichment* 3: 3–5.

Goymann, W., Möstl, E., and Gwinner, E. 2002. Non-invasive methods to measure androgen metabolites in excrements of European stonechats, *Saxicola torquata rubicola. General and Comparative Endocrinology* 129: 80–87.

Green, C. 1989. Environmental enrichment: The elephant. *Ratel* 16: 73–75.

Haight, J. 1993. Reverse perspective: Basic elephant management or enrichment? *The Shape of Enrichment* 2: 5–6.

Hare, V. J., and Gilbert, J. 1995. The shape of enrichment: The first generation. *Proceedings of the American Zoo and Aquarium Association Annual Conference, Seattle, Washington, 1995* (pp. 88–90). Bethesda, MD: American Zoo and Aquarium Association.

Harnett, G. 1995. Enrich one, empower the other. *The Shape of Enrichment* 4: 5–6.

Hart, B. L., Hart, L. A., and McCoy, M. 2001. Cognition and tool use in elephants. In H. M. Schwammer, T. J. Foose, M. Fouraker, and D. Olson (eds.), *Recent research on elephants and rhinos. Scientific progress reports: Abstracts of the International Elephant and Rhino Research Symposium, Vienna, June 7–11, 2001* (p. 17). Münster, Germany: Schüling.

Hart, L. A. 1994. The Asian elephant-driver partnership: The drivers' perspective. *Applied Animal Behaviour Science* 40: 297–312.

Hediger, H. 1950. *Wild animals in captivity.* London: Butterworth Scientific.

Hediger, H. 1955. *Studies of the psychology and behaviour of animals in zoos and circuses.* London: Butterworth Scientific.

Hemsworth, P. H., and Barnett, J. L. 2000. Human-animal interactions and animal stress. In G. P. Moberg and J. A. Mench (eds.), *The biology of animal stress: Basic principles and implications for animal welfare* (pp. 309–336). Wallingford, Oxon, UK: CABI.

Hemsworth, P. H., Barnett, J. L., Breuer, K., Coleman, G. C., and Matthews, L. R. 1995. An investigation of the relationship between handling and human contact and the milking behaviour, productivity and welfare of commercial dairy cows. Research Report on Dairy Research and Development Council Project, Attwood, Australia.

Hemsworth, P. H., Barnett, J. L., Hansen, C., and Gonyou, H. W. 1986. The influence of early contact with humans on subsequent behavioral response of pigs to humans. *Applied Animal Behaviour Science* 15: 55–63.

Hemsworth, P. H. and Coleman, G. J. 1998. *Human-livestock interactions: The stockperson and the productivity and welfare of intensively farmed animals.* Wallingford, Oxon, UK: CABI.

Henry, J. P. 1986. Neuroendocrine patterns of emotional response. In R. Plutchik (ed.), *Emotion: Theory, research and experience.* (Vol. 3. pp. 37–60). New York: Academic Press.

Holmer, H. K., Rodman, J. E., Helmreich, D. L., and Parfitt, D. B. 2003. Differential effects of chronic escapable versus inescapable stress on male Syrian hamster (*Mesocricetus auratus*) reproductive behavior. *Hormones and Behavior* 43: 381–387.

Job, R. F. S. 1987. Learned helplessness in chickens. *Animal Learning and Behaviour* 15: 347–350.

Kalthoff, A., Schmidt, C., and Sachser, N. 2001. Influence of zoo visitors on behaviour and salivary corticosterone concentrations of zoo animals. *KTBL-Schrift* 403: 104–112.

Kirkwood, J. K., Roberts, E. A., and Vickery, S., eds. 2004. *Proceedings of the UFAW International Symposium, Science in the Service of Animal Welfare, Edinburgh 2003,* Published as a Supplementary Issue to *Animal Welfare.* February 13, 2004. Wheathampstead, UK: Universities Federation for Animal Welfare.

Koehl, D. 2000. Elephant Training. Available online at The Absolut Elephant website, www.elephant.se/elephant_training.php?open=Elephant%20training/.

Kurt, F. 1995. The preservation of Asian elephants in human care: A comparison between the different keeping systems in South Asia and Europe. *Animal Research and Development* 41: 38–60.

Kurt, F., and Garaï, M. E. 2002. Stereotypies in captive Asian elephants: A symptom of social isolation. In H. M. Schwammer, T. J. Foose, M. Fouraker, and D. Olson (eds.), *A research update on elephants and rhinos*: *Proceedings of the International Elephant and Rhino Research Symposium, Vienna, June 7–11, 2001* (pp. 57–63). Münster, Germany: Schüling.

Kurt, F., and Hartl, G. B. 1995. Asian elephants (*Elephas maximus*) in captivity—a challenge for zoo biological research. In U. Gansloßer, J. K. Hodges, and W. Kaumanns (eds.), *Research and Captive Propagation* (pp. 310–326). Furth, Germany: Filander.

Ladewig, J. 2000. Chronic intermittent stress: A model for the study of long-term stressors. In G. P. Moberg and J. A. Mench (eds.), *The biology of animal stress: Basic principles and implications for animal welfare* (pp. 159–169). Wallingford, Oxon, UK: CABI.

Ladewig, J. and Smidt, D. 1989. Behavior, episodic secretion of cortisol, and adrenocortical reactivity in bulls subjected to tethering. *Hormones and Behavior* 23: 344–360.

Leach, M., Young, R., and Warren, N. 1998. Olfactory enrichment for Asian elephants: Is it as effective as it smells? *International Zoo News* 45 (5): 286.

Maier S. F., and Seligman, M. E. P. 1976. Learned helplessness: Theory and evidence. *Journal of Experimental Psychology: General* 105: 3–46.

Markowitz, H. 1978. Engineering environments for behavioral opportunities in the zoo. *Behavior Analyst* 1: 34–47.

Mason, G., and Mendl, M. 1993. Why is there no simple way of measuring animal welfare? *Animal Welfare* 2: 301–319.

Mason, G. J. 1991. Stereotypies: A critical review. *Animal Behaviour* 41: 1015–1037.

Mason, W. A. 2000. Early developmental influences of experience on behaviour, temperament and stress. In G. P. Moberg and J. A. Mench (eds.), *The biology of animal stress: Basic principles and implications for animal welfare* (pp. 269–290). Wallingford, Oxon, UK: CABI.

Matteri, R. L., Carroll, J. A., and Dyer, C. J. 2000. Neuroendocrine responses to stress. In G. P. Moberg and J. A. Mench (eds.), *The biology of animal stress: Basic principles and implications for animal welfare* (pp. 43–76). Wallingford, Oxon, UK: CABI.

Mazur, J. E. 1998. *Learning and Behavior.* Upper Saddle River, NJ: Prentice Hall.

McCobb, D. P., Hara, Y., Lai, G., Mahmoud, S. F., and Flügge, G. 2003. Subordinations stress alters alternative splicing of the Slo gene in tree shrew adrenals. *Hormones and Behavior* 43: 180–186.

McComb, K., Moss, C., Durant, S. M., Baker, L., and Sayialel, S. 2001. Matriarchs as repositories of social knowledge in African elephants. *Science* 292: 491–494.

McMillan, F. D. 1999. Effects of human contact on animal health and well-being. *Journal of the American Veterinary Medical Association* 215: 1592–1598.

Mellen, J. D. 1991. Factors influencing reproductive success in small captive exotic felids (*Felis* spp.): A multiple regression analysis. *Zoo Biology* 10: 95–110.

Mellor, D. J., Cook, C. J., and Stafford, K. J. 2000. Quantifying some responses to pain as a stressor. In G. P. Moberg and J. A. Mench (eds.), *The biology of animal stress: Basic principles and implications for animal welfare* (pp. 171–198). Wallingford, Oxon, UK: CABI.

Mendoza, S. P., Capitanio, J. P., and Mason, W. A. 2000. Chronic social stress: Studies in non-human primates. In G. P. Moberg and J. A. Mench (eds.), *The biology of animal stress: Basic principles and implications for animal welfare* (pp. 227–247). Wallingford, Oxon, UK: CABI.

Meyer-Holzapfel, M. 1968. Abnormal behaviour in zoo animals. In M. W. Fox (ed.), *Abnormal behaviour in animals* (pp. 476–503). London: Saunders.

Moberg, G. P. 1985. Influence of stress on reproduction: a measure of well-being. In G. P. Moberg (ed.), *Animal stress* (pp. 245–267). Bethesda, MD: American Physiological Society.

Moberg, G. P. 2000. Biological response to stress: implications for animal welfare. In G. P. Moberg and J. A. Mench (eds.), *The biology of animal stress: Basic principles and implications for animal welfare* (pp. 1–21). Wallingford, Oxon, UK: CABI.

Molter, R. 1980. The taming of elephants. *Elephant* 1: 188–190.

Monfort, S. L. 2003. Non-invasive endocrine measures of reproduction and stress in wild populations. In W. V. Holt, A. R. Pickard, J. C. Rodger, and D. E. Wildt (eds.), *Reproductive Science and Integrated Conservation* (pp. 147–165). Cambridge: Cambridge University Press.

Monfort, S. L., Brown, J. L. and Wildt, D. E. 1993. Episodic and seasonal rhythms of cortisol secretion in male Eld's deer (*Cervus eldi thamin*). *Journal of Endocrinology* 138: 41–49.

Morris, D. 1964. The response of animals to a restricted environment. *Symposium of the Zoological Society, London* 13: 99–120.

Möstl, E., and Palme, R. 2002. Hormones as indicators of stress. *Domestic Animal Endocrinology* 23: 67–74.

Moss, C. J. 2000. *Elephant memories: Thirteen years in the life of an elephant family: With a new afterward.* Chicago: University of Chicago Press.

Munck, A., Naray-Fejes Toth, A., and Guyre, P. M. 1987. Mechanisms of glucocorticoid actions on the immune system. In I. Berczi and K. Kovacs (eds.), *Hormones and immunity* (pp. 20–37). Lancaster, UK: MTP Press.

Murphy, B. E. P. 2000. Overview of glucocorticoids. In G. Fink (ed.), *Encyclopedia of stress* (Vol., pp. 244–260). New York: Academic Press.

Palme, R., Rettenbacher, S., Touma, C., El-Bahr, S. M., and Möstl, E. 2005. Stress hormones in mammals and birds: Comparative aspects regarding metabolism, excretion, and noninvasive measurements in fecal samples. *Annals of the New York Academy of Sciences* 1040: 162–171.

Pedersen, V., Barnett, J. L., Hemsworth, P. H., Newman, E. A., and Schirmer, B. 1997. The effects of handling on behavioural and physiological responses to housing in tether-stalls in pregnant pigs. *Animal Welfare* 7: 137–150.

Poole, J. H., and Moss, C. J. 1989. Elephant mate searching: Group dynamics and vocal and olfactory communication. In P. A. Jewell and G. M. O. Maloiy (eds.), *The biology of large African mammals in their environment: The proceedings of a symposium held at the Zoological Society of London on 19th and 20th May, 1988* (pp. 111–125). Oxford: Clarendon Press.

Ramanathan, A., and Mallapur, A. In press. A visual health assessment of captive Asian elephants (*Elephas maximus*) housed in India. *Journal of Zoo and Wildlife Medicine*.

Rasmussen, L. E. L. 1995. Evidence for long-term chemical memory in elephants. *Chemical Senses* 20: 237.

Reinhardt, V., Cowley, D., Scheffler, J., Vertein, R., and Wegner, F. 1990. Cortisol response of female rhesus monkeys to venipuncture in homecage versus venipuncture in restraint apparatus. *Journal of Medical Primatology* 19: 601–606.

Rushen, J. 1993. The coping hypothesis of stereotypic behavior. *Animal Behaviour* 45: 613–615.

Rushen, J. 2002. Some issues in the interpretation of behavioural responses to stress. In G. P. Moberg and J. A. Mench (eds.), *The biology of animal stress: Basic principles and implications for animal welfare* (pp. 23–42). Wallingford, Oxon, UK: CABI.

Sapolsky, R. M. 1987. Stress, social status and reproductive physiology in free-living baboons. In D. Crews, (ed.), *Psychobiology of Reproductive Behavior: An Evolutionary Perspective* (pp. 292–322). Englewood Cliffs, NJ: Prentice Hall.

Sapolsky, R. M. 1992. *Stress, the aging brain, and the mechanisms of neuron death.* Cambridge, MA: MIT Press.

Schmid, J. 1995. Keeping circus elephants temporarily in paddocks—the effects on their behaviour. *Animal Welfare* 4: 87–101.

Schmid, J. 1998. Hands off, hands on: Some aspects of keeping elephants. *International Zoo News* 45: 478–486.

Schmid, J., Heistermann, M., Gansloßer U., and Hodges, J. K. 2001. Introduction of foreign females Asian elephants (*Elephas maximus*) into an existing group: Behavioural reactions and changes in cortisol levels. *Animal Welfare* 10: 357–372.

Schmidt, M. J., and Markowitz, H. 1977. Behavioral engineering as an aid in the maintenance of healthy zoo animals. *Journal of the American Veterinary Medical Association* 171: 966–969.

Schrader, L., and Ladewig, J. 1999. Temporal differences in the responses of the pituitary adrenocortical axis, the sympathoadrenomedullar axis, heart rate, and behavior to a chronic intermittent stressor in domestic pigs. *Physiology and Behavior* 66: 775–783.

Schulte, B. A. 2000. Social structure and helping behavior in captive elephants. *Zoo Biology* 19: 447–459.

Shepherdson, D. J., Carlstead, K., and Wielebnowski, N. 2004. Cross-institutional assessment of stress responses in zoo animals using longitudinal monitoring of faecal corticoids and behaviour. *Animal Welfare* 13 (Suppl.): 105–113.

Seligman, M. E. P., and Isaacowitz, D. M. 2000. Learned helplessness. In G. Fink (ed.), *Encyclopedia of stress* (Vol. 2, pp. 599–603). San Diego, CA: Academic Press.

Selye, H. 1936. A syndrome produced by diverse nocuous agents. *Nature* 138: 32–34.

Selye, H. 1984. *The stress of life.* New York: McGraw-Hill.

Shoshani, J. 2000. A compilation of frequently asked questions about elephants. Part I. *Elephant* 2: 78–84.

Sikes, S. K. 1968. Habitat stress and arterial disease in elephants. *Oryx* 9: 286–292.

Smith, R. F., and Dobson, H. 2002. Hormonal interactions within the hypothalamus and pituitary with respect to stress and reproduction in sheep. *Domestic Animal Endocrinology* 23: 75–85.

Spedding, C. R. 1993. Animal welfare policy in Europe. *Journal of Agricultural and Environmental Ethics* 6: 110–117.

Sukumar, R. 1989. The Asian elephant: Ecology and management (2nd rev. ed., 1992). Cambridge: Cambridge University Press.

Taylor, V. J., and Poole, T. B. 1998. Captive breeding and infant mortality in Asian elephants: A comparison between twenty western zoos and three eastern elephant centers. *Zoo Biology* 17: 311–332.

Terio, K. S., and Munson, L. 2000. Gastritis in cheetahs and relatedness to adrenal function. In B. Pukazhenthi, D. Wildt, and J. Mellen (eds.), *Felid Taxon Advisory Group*

Action Plan (p. 36). Columbia, SC: Riverbanks Zoological Park and Botanical Garden and American Zoo and Aquarium Association.

Thouless, C. R. 1996. Home ranges and social organization of female elephants in northern Kenya. *African Journal of Ecology* 34: 284–297.

Toates, F. 2000. Multiple factors controlling behaviour: implications for stress and welfare. In G. P. Moberg and J. A. Mench (eds.), *The biology of animal stress: Basic principles and implications for animal welfare* (pp. 199–226). Wallingford, Oxon, UK: CABI.

Touma, C., and Palme, R. 2005. Measuring fecal glucocorticoid metabolites in mammals and birds: The importance of validation. *Annals of the New York Academy of Sciences* 1046: 54–74.

van Rooijen, J. 1991. Predictability and boredom. *Applied Animal Behaviour Science* 31: 283–287.

Von Holst, D. 1998. The concept of stress and its relevance for animal behavior. In A. P. Møller, M. Milinski, and P. J. B. Slater (eds.), *Stress and behavior: Advances in the study of Behavior.* (Vol. 27, pp. 1–132). London: Academic Press.

Waiblinger, S., Menke, C., and Coleman, G. 2002. The relationship between attitudes, personal characteristics and behaviour of stockpeople and subsequent behaviour and production of dairy cows. *Applied Animal Behaviour Science* 79: 195–219.

Wasser, S. K., Hunt, K. E., Brown, J. L., Cooper, K., Crockett, C. M., Bechert, U., Millspaugh, J. J., Larson, S., and Monfort, S. L. 2000. A generalized fecal glucocorticoid assay for use in a diverse array of non-domestic mammalian and avian species. *General and Comparative Endocrinology* 120: 260–275.

Webster, A. J. F., and Main, D. C. J., eds. 2003. Proceedings of the 2nd international workshop on the assessment of animal welfare at farm and group level. *Animal welfare* 12(4), special issue. Wheathampstead, UK: Universities Federation for Animal Welfare.

Weiss, J. M. 1971. Effects of coping behaviour in different warning signal conditions on stress pathology in rats. *Journal of Comparative Physiology and Psychology* 77: 1–13.

Weiss, J., Goodman, P., Losito, B., Corrigan, S., Charry, J. and Bailey, W. 1981. Behavioral depression produced by an uncontrollable stressor: relationship to norepinephrine, dopamine, and serotonin levels in various regions of the rat brain. *Brain Research Review* 3: 167–205.

Wielebnowski, N. C., Fletchall, N., Carlstead, K., Busso, J. M., and Brown, J. L. 2002a. Noninvasive assessment of adrenal activity associated with husbandry and behavioral factors in the North American clouded leopard population. *Zoo Biology* 21: 77–98.

Wielebnowski, N. C., Ziegler, K., Wildt, D. E., Lukas, J., and Brown, J. L. 2002b. Impact of social management on reproductive, adrenal and behavioural activity in the cheetah (*Acinonyx jubatus*). *Animal Conservation* 5: 291–301.

Wilson, M. L., Bloomsmith, M. A., and Maple, T. L. 2004. Stereotypic swaying and serum cortisol concentrations in three captive African elephants (*Loxodonta africana*). *Animal Welfare* 13: 39–43.

Wolfle, T. L. 2000. Understanding the role of stress in animal welfare: Practical considerations. In G. P. Moberg and J. A. Mench (eds.), *The biology of animal stress: Basic principles and implications for animal welfare* (pp. 355–368). Wallingford, Oxon, UK: CABI.

PART II

ELEPHANTS IN THE SERVICE OF PEOPLE

CULTURAL DIFFERENCES AND ETHICAL RELATIVITY

7 ELEPHANTS AND PEOPLE IN INDIA
HISTORICAL PATTERNS OF CAPTURE AND MANAGEMENT
DHRITI K. LAHIRI CHOUDHURY

The relationship between people and captive Asian elephants is ancient. Cave paintings, ancient texts, and other artistic depictions testify to the antiquity of elephant culture—the customs, and technology related to the use of elephant—but knowledge of the subject is incomplete. The largest body of ancient information comes from the Indian subcontinent. We can infer a great deal about the ethical dimension of elephant culture from historical evidence and more detailed firsthand accounts of the recent past.

Tribal people, probably hunters and gatherers, were the first "elephant people" who captured, trained, and cared for wild elephants for their own transportation and domestic use, as well as for elephant warfare (Lahiri Choudhury 1991). It is likely that elephant culture emerged at several points of its geographical range independently, but as communication improved, knowledge dispersed along trade routes across the Bay of Bengal and South China Sea. Invading civilizations, such as that of the Moguls also encountered these people, appropriated their folk knowledge, and apparently employed them directly in military causes. Ancient stone reliefs clearly show people with the ethnic features seen in present-day tribal elephant handlers in India (Lahiri Choudhury 1991). Many of these ancient practices have survived to the present. The elephant also figures prominently in the Hindu and Buddhist scriptures, which have always represented it as a creature of sagacity. However, the history of the elephant in human culture predates modern religion, and the practices of pre-Hindu and pre-Buddhist cultures apparently persisted into more recent millennia.

The following pages will review these practices through time. Where possible, I have made reference to ethical issues, and the effect of these practices on the health and welfare of the animals. As will be seen, the primary concern of people in the past was capturing and using the elephant, and regard for the creature's welfare was based on a standard that differed significantly from today.

Early Records of Human-Elephant Relations

In ancient India, records of captive elephants go back to ca. 4500 BCE, and perhaps even as early as the Mesolithic age, which started around 8000 BCE, though the evidence is scant. Cave paintings of the period have not yet been scientifically dated, but benchmarks have been estimated. Much can be deciphered about the use and treatment of captive elephants from these ancient records, and there is ample evidence that the domestic use of elephants was a continuing process. We know, for example, that men were probably riding elephants by ca. 2500 to 1500 BCE. Elephants in captivity were clearly depicted on the Harappan stone seals of the Indus Valley, which date to around 2500 BCE. Indications of elephants in domesticity are recorded in the three classics of Indian literature, the *Rigveda* (ca. 1500 BCE), the *Atharbaveda* (ca. 600 BCE), and Kautilya's *Arthashastra* (ca. 321 BCE to 300 BCE). Palakapya's *Gajashastra*, a text on elephant management, was probably written in the early medieval period. The ambassador to the court of Emperor Chandragupta Maurya, Megasthenês, wrote his famous travelogue *Indica* around the outset of the third century BCE.

The final revised text of the *Ramayana*, dating to about 400 CE, mentions captive and domestic elephants as the property and proper mounts of noblemen. Perhaps its most revealing remark, however, is that elephants are fearless of fire, weapons, and spears but harbor fear of their own relations. The only logical explanation for such a comment is that elephants feared their relations because domestic elephants were used to capture wild ones. From this we infer that *koonkies* (domestic elephants trained in the capture and training of wild animals) were used even then to capture wild elephants. The *Mahabharata*, an epic that has existed in various forms for about 2,000 years, graphically records battlefield engagements of elephants and many thousands of casualties. In the *Mahabharata*, the character Bhishma makes reference to an elephant goad or ankus, confirming the antiquity of this tool.

Megasthenês, the ambassador of Seleukos of Babylonia to the Indian court of Chandragupta Maurya, was surely impressed with what he saw 2,300 years ago, for he described the method of capture and subjugation of wild elephants in detail (McKrindle 1926; Govinda Rao 1931). Clearly, elephant catchers of the period understood the disposition of wild elephants and used that knowledge to advantage. Megasthenês described how tame elephants were used as decoys to entice elephant herds or bulls into the stockade. From protected positions, the elephant-men noosed the legs of the wild elephants. The new captives were controlled with koonkies. The men used a variety of methods to break the animals' spirit and gain control. They created open wounds around the animals' necks with knives and sharp weapons to make the ropes tied around their necks bite painfully. They beat the elephants and deprived them of food. But they also used words, songs, and music to sooth the new captives. The latter inducements are a tradition that lives on in northeastern India even today. Not surprisingly, Megasthenês also observed that captive elephants were subject to sores and also recorded inflammation of the eye.

During the classical period of the Imperial Guptas (319 to 454 CE), we find reference to the use of chains. Apparently, some elephants were kept free of chains when at work. However, cane or rattan hobbles, as employed even today in remote parts of Myanmar and northeast India, have probably been used ever since elephants were kept in captivity.

At this time in human history, when war and conquest were primary human activities, cruelty to animals was a lesser concern than the slaughter of men on the battlefield. People expected elephants, as a medium of war, to be injured and lost as casualties. Even as early as the second century BCE, these elephants were trained to hold their heads high to shield the mahouts on their necks from the archers. They were valuable as war machines and prized as booty of war; ultimately, like soldiers, they were considered expendable and replaceable. Elephants clearly lived in abundance throughout the forests of India and Southeast Asia during this period; in the *Arthashastra*, elephant ranges were protected as a resource and superintended by officers and guards, and offenders were punished severely (Kautilya 1972).

Nonetheless, the welfare of elephants during this early period was not ignored altogether. In Kautilya's *Arthashastra* (1972), we find detailed instructions regarding the management of captive elephants, including housing, diet, protection, and even preservation of their habitat. Megasthenês and Palakapya also had much to say regarding the classification of elephants into castes based on body type (Lahiri Choudhury 1991). We find less information about care and treatment. Most literature on the topic goes into much more detail about the design and procedure of the capture method rather than on training and care of the charges following capture.

Records from a somewhat later period also discuss management, and indicate the longevity of the systems they were using. According to Abū'l-Fazl (1551–1602), court historian to Mughal ruler Akbar, "an elaborate system was devised for the management of the royal elephant stable," and Akbar developed a sevenfold system of classification of elephants, based no doubt on the ancient Hindu system. The duties, number, and rank of attendants were prescribed for each type and different diets were specified. "An elaborate system of fines was introduced to enforce strict discipline among the elephant keepers" and records were kept of every animal (Lahiri Choudhury 1991, 146).

The use of the stockade as a method of capture in India continued, but regional variations evolved in the details of construction and operation. Elephant capture was an activity that relied on the people's knowledge of the landscape and wildlife. It is clear that the practitioners of elephant capture were acutely aware of the elephant's natural history and perceptual capacities. The geographic variation attests to the elephant catchers' inventiveness. For example, the *pung garh* (literally salt-lick stockade) of northeast India capitalized on the elephants' habitual use of salt licks. This key resource determined the stockade's location. The *dandigarh* (or elephant-track stockade) was also used in the same area; here heavily used elephant trails determined the stockade's placement. Similar differences were manifested in the construction of the stockade. In Karnataka State, for example it was fashioned with a small accessory corral for roping of selected animals after initial capture. British Army captain and hunting chronicler Thomas Williamson (1808), one of the first modern-day authors to study wild elephants, showed a drawing of this kind of a stockade, probably from his experience somewhere in Orissa. Colonial forestry officer and wildlife authority P. D. Stracey (1963) pointed out that such roping stockades were used in Mysore (now Karnataka) and Orissa but not in northeast India.

The Diversity of Capture Methods

The diversity of documented capture methods indicates that the art of capturing wild elephants in Asia probably emerged separately at widely divergent points of the subcontinent's geographical range. Each method had its advantages and disadvantages in terms of physical risks to man and elephant. Even today, in the age of tranquilizer guns and drugs, traditional methods have applications when wild elephants have to be removed from encroached wild lands for their own safety. Let us briefly review the main themes of elephant capture practiced in more recent times.

Hand-Noosing

This was the specialty of traditional elephant noosers in Sri Lanka, an ethnic group of Muslims called Panikkans. Their capture method depended entirely on stealth and they used a strong flexible rope of buffalo or sambar deer hide for the purpose. Noosing was effected by hiding in a tree above the noose, which was concealed on the trail. Alternatively, "by stealing up close to it [the elephant] when at rest, and availing themselves of its well known propensity at such moments to swing the [back] feet backwards and forwards, they contrived to slip a noose over the hind leg" (Tennent 1861, 160). Stracey (1963) described the method in detail and mentioned meeting one such catcher in Gal Oya Valley about 300 kilometers east of Colombo, in what is today called North-East Province. In 1959, this elephant catcher was still vigorously pursuing his vocation at age 70 and had up to then captured more than 400 elephants.

Mela Shikar

Noosing wild elephants from the back of a domestic elephant (*mela shikar*) was a traditional practice in northeast India, Nepal, and in some parts of Southeast Asia, but not in south India (Govinda Rao 1931; Williams 1954; Baze 1959; Stracey 1963). The first recorded account of mela shikar seems to occur in Arrian's *Anabasis*, where Alexander is described as pursuing the sport. G. P. Sanderson, an expert elephant hunter and administrator for the British colonial government in India during the nineteenth century, is reported to have called it the most spirited and exciting, though by no means advantageous, manner of hunting the wild elephant (Govinda Rao 1931), and Stracey (1963) said it was a capture method that called for both great skill and great daring, on the part of both the rider and the koonki.

The practitioners of mela shikar preferred to work in groups for mutual safety. The usual team on a koonki consisted of three persons: the *mahout* (elephant handler) driving the koonki, the *phandi* or nooser, and a *lohatia,* who spurred the koonki to action with a spiked mallet called a *lohat.* During the chase, the lohatia stood on the crupper (the elephant's rump) and hit the buttocks of the koonki. Temple reliefs in south India of the Hoysala period (eleventh to thirteenth century AC) show such lohatias in action on war elephants.

The koonki party approached the quarry with stealth, and targeted a young animal from a family group, usually a juvenile or subadult measuring between one-and-a-half to two meters in height at the shoulder. When the animal fled, the koonki team gave chase, and isolated it from the herd. From the back of the furiously running koonki, the phandi then dangled the noose in front of the fleeing elephant's head. When the noose touched the quarry elephant's head and face, the animal instinctively curled its trunk up, and the noose slipped back round its neck. Next they had to tie the noose quickly with a slipknot so that the struggling captive was not strangled.

Etiquette allowed the phandi who put the first noose on an elephant to "claim" it without interference from others in the party. However, in difficulty he could summon help, calling for a second and even a third noose in the case of an unmanageably large animal. Stracey (1963) recorded a case, as reported by Indian Forest Service employee and elephant management pioneer A. J. W. Milroy, of as many as seven phandis being used to capture one elephant. Fees received for the capture were then divided among the captors. The ever-present danger that hung over a mela shikar party was the possible presence of a *goonda* (a large aggressive bull) nearby. To a mela shikar party in the forest, shouts of "goonda" were the most dreaded sounds. The late Rajkumar P. C. Barua of Gauripur, Assam, also known as Lalji, organized the capture and training of more than 1,200 elephants in his lifetime. Lalji once recalled that Tirath Singh, a big left-tusker koonki belonging to him, used to bolt if he even heard shouts of "goonda" (P. C. Barua, personal communication).

Elephant noosers sometimes tackled especially large animals that could only be overpowered by two or more koonkies, but this frequently resulted in the captive's strangulation. Attempting to capture large animals was therefore discouraged. Present regulations framed under India's Wildlife Protection Act of 1972 restrict capture by mela shikar to the juvenile class of elephants, one-and-a-half to two in height at the shoulder. Large koonkies are not preferred in such operations because they are difficult to use to chase a herd through dense cover. Accidents are common among the elephant catchers. For instance, Tirtha Phandi, in the employ of Lalji, had captured more than 100 elephants in mela shikar already in his career, when, on chase one day, a hanging creeper caught him around the midriff and fractured his spine. Even after six months in the hospital, he was left a lifelong invalid, fit only for depot (training camp) work (P. C. Barua, personal communication).

In the early 1930s, in his capacity as assistant conservator of Forests in Assam, A. J. W. Milroy modified the rules of mela shikar to prevent illegal capture, particularly on the north bank of the Brahmaputra River. He called this modified version "anchored" mela shikar. At that time, professional mela shikar parties holding leases on an area on the Brahmaputra's north bank would sneak into Manas Wildlife Sanctuary in the foothills of the Bhutan Himalayas, first designated a sanctuary in 1928, and hence legally out of bounds to them. There, these mela shikar parties would catch elephants illicitly. Carrying some dry rations, they would operate within the sanctuary for days on end, and then eventually return to base with their capture. Milroy made it obligatory for all such capture parties to report back to camp every evening, thereby restricting their area of operation to a radius of ten to fifteen square kilometers from the base camp. This anchored method has been used more recently in North Bengal, from 1980 to 1996, as an antidepredation measure to frighten wild herds away from a locality especially vulnerable to crop depredation by elephants.

Mela shikar has been the only method of large-scale capture practiced in northeast India since the mid-1950s. It cannot be used to catch large aggressive bulls or to capture whole herds or family groups, nor is it an effective management tool to remove large numbers of elephants in a planned manner. But in modern-day use it is an effective way to keep herds away from crop fields. It is also used sometimes today in Myanmar (Burma) to capture problem elephants. Its costs in terms of disturbance and stress to wild elephants are definitely less than the costs incurred by capturing and removing wild elephants from an area.

Jangi Shikar

This variation on the mela shikar theme of capture translates as "fighting capture." In *jangi shikar*, a full-grown wild male elephant was chased to exhaustion and then noosed. This is well described in Stebbing (1922) and Stracey (1963). Sportsman-naturalist E. P. Stebbing noted that the last operations carried out were in 1904–1905; but no details of the license issued are available. The Balarampur Estate did the catching with their own stud of elephants. The system in the United Provinces [of North India] and adjoining Nepal is to track and locate the wild elephants and then drive them out with tame ones, chase them to a standstill and then noose and tie them up. No stockade is built, and if the elephants cannot be driven away from water, they keep refreshing themselves and are very difficult to catch. It is somewhat dangerous work. (3: 662)

Stebbing's warning had sound basis in the account of an unfortunate accident resulting in the death of Mrs. Anson, the wife of the agent of the Balarampur Estate. Stracey (1963, 90) recounts that incident:

On this hunt two hundred men and ninety-six elephants were used and the line of elephants had not yet got into position when a large rogue tusker charged without warning as they were emerging from a narrow defile. It knocked down the elephant on which husband and wife were seated, gored it viciously and then fled. Mrs. Anson, who was half crushed by the fallen elephant, was extricated by the husband and taken back to the camp on an elephant, but she died there. The rogue elephant persistently followed them home and Major Anson had to take his wife off the [second transport] elephant and up a steep bank for safety.

The provincial government, acting on a 1912 note written by then-governor Sir John Hewett, stopped issuing elephant-capture licenses in that province. Hewett considered the Balarampur captures responsible for the sharp decline in the number of elephants in these forests. His writings did not mention, however, that these captures were made under the patronage of previous governors for the purpose of entertaining their guests from Delhi and abroad (Hewett 1938). The method is now obsolete, not only because of its inherent risks but also given the easy availability of tranquilizing and sedating drugs.

Pitha Shikar

"Hunting capture" is another variation on mela shikar. It consists of the chase and capture of a single large elephant by a relay of koonkies. The quarry is worn out by the chase and finally noosed. The large loner males captured by this method were prize catches because such bulls were not normally captured either in kheddah or in mela shikar. E. A. Smythies, a forester who practiced in both India and Nepal, provides an account and photographs of such a chase in Nepal (1942). Pitha shikar, mela shikar, and jangi shikar were all in wide use at the beginning of the nineteenth century; all three, therefore, can be taken to have been extensions of earlier practices.

Partala

This was the term used in East Bengal (now Bangladesh) to denote capture of single bull elephants with the help of female elephants as decoys. A female elephant used as a decoy was either not ridden or ridden by mahouts hiding under dung-and-urine smeared blankets. If the decoy was used without a rider, the mahouts watched from a distance and mounted their koonkies when the bull showed interest in the cow. Relays of koonkies were used round the clock to pursue and eventually tire out the bull, which was not given a chance to feed or drink. Finally, the mahouts would dismount from their koonkies and tie the legs of the exhausted bull. The koonkies then maneuvered the bull against a standing tree, and the mahouts lashed its legs to the tree (Williamson 1808; Sanderson 1912; Stracey 1963). The method is now obsolete because such koonkies and mahouts are no longer available, and in any case, sedatives and muscle relaxants have made their jobs redundant.

Pitfall

The use of a deep pit camouflaged with vegetation, placed on an elephant trail, was the traditional method of capture in both south India and on the north Indian border with Nepal (Abū'l-Fazl 1993). The English recorded its use in the early nineteenth century (Williamson 1808; Fayrer 1875; Sanderson 1912; Stebbing 1922; Govinda Rao 1931; Stracey 1963). As British civil servant G. P. Sanderson put it:

> A most barbarous method of catching wild elephants . . . an immense majority of elephants have their limbs dislocated or broken, or receive permanent internal injury, even if they are not killed on the spot, as some times happens. . . . An immense number died from the effects of this violent mode of capture, and those that lived were only small ones. (1912, 75)

To effect a single successful capture, between ten and twenty elephants were killed. Not uncommonly, other animals, including tigers, also fell into the pit and eventually perished (Govinda Rao 1931).

This method of capture was practiced throughout the nineteenth century in Mysore, largely because of the patronage and insistence of the then-Maharaja of Mysore and against the advice of the forest administration. The practice was abandoned there only after the Maharaja's death (Govinda Rao 1931). It was apparently also the only method of capture from the wild ever practiced in the states of Travancore and Cochin (present-day Kerala). It was practiced also in parts of Madras (now Tamil Nadu) and in fact went on in Madras at least up to the middle of the twentieth century. In that region, the effect of the fall was supposed to have been reduced, if not eliminated, by cushioning the floor of the pit with vegetation (Krishnamurthy and Wemmer 1995a).

Kheddah

A version of the *kheddah*, or stockade method of elephant capture, is described by Arrian in *Indica,* with Megasthenês referred to as the source (1983). This account differs from more modern practices in one detail; a ditch encloses the capture area, instead of palisades.

The driving of wild elephants into a stockade was apparently the main traditional practice of capture followed in northeast India and what is now Bangladesh. The word kheddah is sometimes used loosely to describe any capture in a stockade, but literally, it means to chase or drive. The pung garh and dandigarh were the commonest and cheapest forms of stockade used in Assam. For dandigarh, a stockade with two doors was built around a well-used elephant track. This allowed the elephants to move freely through and become accustomed to the stockade area. Elephants entered the stockade on their own by either gate, and then men, who were hiding from the elephants, would close both the gates. Dandigarh can be useful even now for the low cost incurred when capturing entire herds or family groups. Chemical tranquilization is useless in this situation because each case of darting puts the herd to flight and increases risk of injury due to stress and excitement.

The Kheddah in Practice

The kheddah method was relatively nonselective. Expert trackers and guides called *kukur sunganias*—Bengali for people who track like hounds—supervised the whole operation (Sanderson 1912; Shebbeare 1958). Its effectiveness depended on a division of labor. The capture area was from eight- to nine-and-a-half-square kilometers in circumference and situated in the heart of an active elephant range. Importantly, the area had to have water, fodder, and cover, enough for the elephants to linger naturally for a few days. On the perimeter, huts of leaves and reeds were spaced out every fifty meters or less. Two men were stationed at each post to prevent elephants from breaking out of the circle at night. When a herd of elephants entered the area, the capture operation commenced. Initially, the men used minor disturbances to coax the animals toward the stockade slowly and methodically. This was the passive phase of the operation, and the guiding principle was "easy does it." In a sense, the elephants were contained psychologically, and, apparently, fear was not used as a factor.

In due course and barring premature alarm, the elephants approached the wooden palisades that formed the funnel leading to the stockade. The herd crossed a strip of land that had been cleared of vegetation (called a *chhiuniabat* in northeast India), located just at the edge of the palisades. At that moment, the active part of the drive began. Full-throated chase started when the herd crossed the chhiuniabat toward the stockade gate. The guards in the outer ring of huts abandoned their posts and provided the main manpower. If the matriarch stampeded into the stockade, her followers could be trusted to follow. The stockade was originally furnished with a drop-gate, but later a swing gate came to be preferred (Sanderson 1912). The success of the operation often depended on the abilities of the porter, who had the onerous task of closing the gate at the appropriate moment. Ill-timed closure could exclude a part of the herd, allowing some entrapped elephants to escape, or it could injure an elephant. A trapped big bull accompanying a herd could cause havoc. It was permissible under the prevailing rules to shoot down such a bull in the interest of the capture.

The Spectator Phenomenon

The spectacle of the kheddah event obviously was compelling. This form of capture became a spectator sport in south India, where the Maharajah of Mysore invited spectators, especially other dignitaries, to witness the event, and the geography of the capture was planned to best advantage for those watching for entertainment (Govinda Rao 1931). In the Mughal court, kheddah also provided court entertainment, apart from its more practical uses, as Abū'l-Fazl recounted (1993). In nineteenth-century Thailand (then Siam), elaborate drives into stockade were also practiced, similarly providing great attraction as a gorgeous spectator-show (Whitney 1905; Bock 1985).

Training Wild Elephants

There are few extant historical records concerning the training of elephants. As noted by Stracey, the realm of the elephant trainer is shrouded with mystery, and to a certain extent, reverence for the art of training: "Laymen fear to tread where professional elephant men walk boldly, secure in the inherited and acquired knowledge of centuries" (1963, 116). We know there were regional variations in training, but in general, wild elephants were trained in two phases. Before the animal could be trained to respond to commands it had to be "broken" of its attempts to escape and habituated to the proximity of people and handling by them.

The traditional means started with starvation, no longer considered acceptable. In Assam for example, until early in the twentieth century, newly captured elephants were left in the stockade, without food, until buyers could be found. Iron-tipped spears, knives, and fire were used to intimidate the captives and render them submissive to their handlers. In particular cases, phandis would create an open wound on the neck of particularly troublesome elephants or put damp sand under the neck ropes, "so the captive could be brought under control more speedily" (Stracey 1963, 118). Cutting a small vein on the back of the ear was also "very effective in controlling a fractious captive—it was as if the elephant feared the smell of its own blood" (Stracey 1963, 118). Strangulation and heart failure, which sometime occurred in captivity, were attributed to harsh handling.

In addition to these cruelties, forced proximity and tactile stimulation were also part of the process. Once out of the stockade, the men hobbled the elephant's fore and hind legs and tethered it by the neck to a post, technically known, in northeast India, as *galakhamari.* In south India, the elephant was confined to a *kraal*—a confining cage of heavy timber. Soon after, a man mounted the animal's neck and men besieged the elephant. But as Stracey (1963, 125) noted, "The monotonous musical chanting eventually lulls the animal as much as the continuous patter of the human attendants, who take every opportunity to rub and stroke the elephant." The process of domination and acclimation continued after dark, when the men tethered the animal and subjected it to further tactile stimulation. In the light and smoke of burning torches, "the men gather round the captive and sing to it, rubbing its body down with straw and branches of leaves, while a rider perched on its back clings on for all his worth while the struggling captive throws itself about. When it reaches out with its trunk, it is beaten and intimidated by a torch thrust in its face" (Stracey 1963, 125).

When the animal's spirit was sufficiently broken the training began in earnest. Men and koonkies constantly attended the elephant. Then began the process of training the elephant to respond to a small number of commands—go forward, go backward, turn to one side, and halt. The captive, tied tightly between two koonkies on short lead, was marched backward, forward, and sideways with abrupt halts. The men's verbal commands were intoned loudly and reinforced with jabs from pointed wooden goads and by whacking with sticks. But more gentle stimuli were used as well. For example, a touch to the foot elicited lifting. The men reinforced their persuasion with words of appreciation like "well done!" In time, the animal responded to a basic set of commands.

In the northeastern India system, the last step was to make the animal sit on all fours, an unnatural position for an elephant, the training for which might take up to three months. The ultimate test of a captured elephant's training, apart from its behavior and temper in the field, would be the length of time an elephant would sit still on all fours like a dog. A good riding elephant was expected to be able to sit still for half an hour or more to allow the *howdah,* the riding box, to be tied securely on its back. To put all the accoutrements needed for a ceremonial appearance onto an elephant would have been (and still is) an even more time-consuming process. Elephants were sold when they had learned the four basic commands: go forward, halt, turn left, turn right, and when they had learned to carry their own fodder. The new owner had the task of training the animals to respond to additional commands, a process that might go on for several years.

In the kraal method of training practiced in Tamil Nadu, the *first* step was to make the elephant kneel by offering it food through the lowest bar of the kraal. The next step was to make it sit, forcibly done by putting the pressure of a heavy log of timber on its back. The kraal method of training had an advantage over the northeast India method in that it did not result in rope gall, which causes more struggling and can be a troublesome in hot, humid weather. However, in the southern Indian system, the period of training and stress would be much longer than in the northeastern India method: about three months. In the northeastern India method, the first part of the job, the "breaking" and initial training, were completed in two weeks and then further training could go on.

In summary, the training process began with heavy use of the "stick," succeeded by increasing use of the "carrot." There can be little debate that under any circumstances the breaking in of wild elephants is stressful, at least as much so as the breaking in of wild horses. In the early stage, the animals respond with aggression and attempt to escape. These behaviors subside and disappear in a couple of weeks. At the same time, the animals habituate to the proximity of humans and the domination of individual people, in particular the mahouts. Captive elephants do not exhibit a universal and consistent response to all people, and the familiarity of an individual person may well be of central importance to the animal's response.

Regulating Ethical Treatment of Elephants

Provincial forest administrators in the British colonial regime in India during the nineteenth and early twentieth centuries reacted strongly to the inhumane nature of a number of traditional methods and in their place introduced humane practices for elephant management. Several key figures played roles in reforming these practices. G. P. Sanderson, who introduced kheddah operations to south India from Bengal in the 1870s, wrote that "the proper management of elephants attached to the military and other [government] departments in India is a subject of much importance, both financially and from a humane point of view" (1912, 96).

The overall system of elephant capture in Assam was revised around 1918. Until then, the Dhaka Kheddah Department (under the Commissariat, or Army supply department) had used the traditional lease system, in which *mahaldars,* local leaseholders, relied on kheddah and mela shikar to capture elephants in *mahals,* or forest blocks. Chief Conservator of Forests A. V. Munro (Krishnamurthy and Wemmer, 1995a, 470) observed the operations and reported that "a good deal was learnt of the inner workings of the elephant catching business, including both the unbusinesslike methods of the mahaldars and their unreliability except under the closest personal supervision, while the needless cruelty inflicted on the animals was a revelation." The system had its good points; there were rewards for local communities, but it abused the resource from the standpoint of humane treatment. Stracey (1963, 119) noted that the greed of elephant catchers, driven by the profit motive, "can play a large part in cruelty in the elephant catching business." Stracey recorded his encounter with some of these professional European entrepreneurs, who made no bones that the profit motive was indeed driving them.

Many of the inhumane ways of postcapture management went on until a particularly enlightened employee of the Indian Forest Service, A. J. W. Milroy, intervened in 1929–1930 (Wemmer 1995; see also Milroy 2002). Milroy was trained as a forester in Germany, as were many British foresters of his time, but his work as assistant conservator of Assam Forests was primarily concerned with the capture of wild elephants. The Indian Forest Service considered elephants to be harvestable forest products, as dictated by precolonial tradition, and conducted annual capture operations to maintain domestic elephant stocks for use by the government, timber extraction, and transport. However, postcapture casualties were common because of starvation and lingering illness and debilitation afflicted captive elephants that were incarcerated within the stockade. As Wemmer noted in his historical profile of Milroy, the assistant conservator

> was a self-appointed defender of the right of domestic elephants to humane treatment, and published his convictions in *A Short Treatise on the Management of Elephants* (1922). He was keenly attuned to the fact that elephants can be pushed to great extremes, and was equally aware that men take ignorant or insensitive advantage of the creature's endurance. His humane philosophy was based on common sense and a penetrating insight into the nature of man and elephant. (1995, 490)

Milroy ended the practice of starvation by insisting that elephants should be fed and watered soon after capture, brought out of the stockade preferably within forty-eight hours, and tied in separate stalls. One of his standing orders was that captured elephants should be given basic training within fifteen days of the capture, so that they would learn four basic command words and learn the feeding and watering routine (P. C. Barua, personal communication). He advocated that the sale of captives take place as quickly as possible because, in his words, "an ownerless elephant is an elephant without a future" (Stracey 1963, 118).

Milroy was apparently quick to assess the risks of loss to newly caught animals and changed the prevailing practices that were causing high rates of injury and death. Injury to calves in the milling herd of the stockade was high because bulls wreaked havoc among the captives; the extant policy was to shoot the bulls immediately. Milroy instructed the elephant men instead to avoid capturing bulls, and if captured, to make every effort to let them out of the stockade quickly. He also ruled that pregnant cows and cows with calves at heel were to be released. Also to be released were elephants that showed "dangerous symptoms" during breaking. Because of these changes, causalities associated with capture and training were reduced from nearly 30% to less than 5% (Stracey 1963). Milroy also banned the use of metal-tipped spears and knives and replaced them with sharp sticks, which were much less injurious to the captive elephants.

Mahouts and "native" doctors practiced their own traditional medicine on elephants, but veterinary medicine began to play a role early in the history of government-managed elephant camps. The colonial British did not dismiss the traditional remedies, but they questioned the rationale and efficacy of many treatments. Economics were also a driving force to change practices that were evidently leading to injuries and death. One British captain noted in 1879 that from an economic standpoint, tolerating ignorance about so costly an animal as a captive elephant was itself very costly because that ignorance led to long periods of illness and injury for the elephants, as well as high mortality (Krishnamurthy and Wemmer 1995b).

The first professional veterinary work on elephants in India was carried out by British veterinarian W. Gilchrist (1841). In Madras Presidency, now Tamil Nadu State, two veterinary posts were mandated in 1906. The inspector of livestock was responsible for all veterinary matters in the Forest Department, including prescription of workload and diet for individual elephants, and the allocation of elephant power to the various timber extraction ranges. Elaborate commissariat rules for the use of elephants much predated the Forest Department's practice.

Forest administrators imposed a rigorous system of recordkeeping and reportage to ensure that work was done. Those in charge of elephants, the *jemadar* (platoon or troop commander) and later the forester, maintained a service register for each animal as long as it was employed in the camp, and recorded, among other things, temperament, diet, prescribed workload, and veterinary treatments. Semiannual and annual reports were submitted to the inspector of livestock, and visiting inspectors reviewed the registers and noted their own remarks.

Reflections on Ethics

As Indian culture has changed over time, so has elephant culture. Ethical concerns regarding capture and husbandry have also changed, though we will never be certain of the details of these transitions. In modern times, chemical sedation has revolutionized the capture and postcapture management of adult animals, but many vestiges of past practices remain. Elephant-human bonding has a millennia-old history in India and still is one of the pillars supporting the future of the species in the country. Domestic elephants have helped to create this bonding. However, the survival of the Asian elephant as a species is the ultimate ethical issue, and domestic elephants are linked to the species' survival in many ways. In India, improving management practices of domestic elephants can help to maintain the ancient connection between people and elephants.

At least since Palakapya's time, certain elements of the relationship between man and elephant have been inherently antagonistic. They compete for the same vital natural resources: water, good soil for the growth of vegetation, and shade. The object is to strike a balance between the needs of each species. As the British colonizers realized, cruelty is not only inhumane, it is also poor economics and poor management because it wastes a valuable natural resource. Cruelty cannot be countenanced; yet, the concept of "optimum" numbers must not be forgotten. Some dangerous animals should be eliminated, and in some places, there is a need to keep a watchful eye on elephants' growing numbers.

In some instances, lobbying by animal lovers has hindered scientific conservation, inducing inertia in forest departments of many Asian countries. For example, in India, a 1977 amendment of the country's 1972 Wildlife Protection Act added elephants to Schedule 1 (absolute protection). From 1980 to 2002, elephants in India increased in number from 15,600 to about 26,000; of these, only about 3,500 were captive elephants. Between 1991 and 2005, at least 769 elephants were killed illegally; during the same period, 3,140 humans were killed by elephants (Government of India 2005).

Ivory poachers are not the only people killing elephants in India. A new face of human-elephant conflict has emerged: retaliatory and often indiscriminate killing of dangerous elephants by private citizens in the face of government inaction. By 2005, only nine elephants had been legally eliminated under section 11 of the Wildlife Protection Act, and only two elephants were allowed to be captured in three states. A tusker was ordered destroyed only after it had killed thirty people. In contrast, between 1991 and 2002, in only one district in Orissa State, over thirty animals were killed illegally, among them tuskers, females, and calves (Government of India 2005).

These numbers indicate the unfortunate downside of an emotional approach to elephant conservation. The present situation has emerged largely because of pressure from vocal animal rights groups. Absolute protection laws lead to de facto vigilante action in place of scientific management. This does not help with conserving this species, which, in nature, unlike in zoos, does not live isolated from humans.

References

Abū'l-Fazl, Allāmī. 1993. *Āin-i Akbarī.* Calcutta: The Asiatic Society. Repr. of 1873 ed.

Arrian. 1983. *Anabasis of Alexander. Books 5–7. Indica.* Trans. P. A. Brunt and E. Iliff Robson. Cambridge, MA: Harvard University Press.

Baze, W. 1959. *Just elephants.* London: Corgi.

Bock, Carl. 1985. *Temples and elephants: The narrative of a journey of exploration through upper Siam and Lao.* Bangkok: White Orchid Press. (Originally published 1884.)

Fayrer, J. 1875. *The royal tiger of Bengal: His life and death.* London: J. & A. Churchill.

Gilchrist, W. 1841. *A practical memoir of the history and treatment of the diseases of the elephant.* Honsur, Madras Presidency: Government Cattle Farm.

Government of India. 2005. *Report of Project Elephant, 2004–05.* Delhi: Ministry of Environment and Forests.

Govinda Rao, P. S. 1931. *Elephant catching: Ancient and modern.* Mysore: Government Branch Press.

Hewett, J. 1938. *Jungle trails in northern India.* London: Methuen.

Kautilya. 1972. *Arthashastra.* Trans. and ed. R. P. Kangle. 3 vols. Delhi: Motilal Banaradass.

Krishnamurthy, V., and Wemmer, C. 1995a. Timber elephant management in the Madras Presidency of India (1844–1947). In J. C. Daniel and H. S. Datye (eds.), *A week with elephants: Proceedings of the international seminar on Asian elephants, June 1993* (pp. 456–472). Bombay: Bombay Natural History Society and New Delhi: Oxford University Press.

Krishnamurthy, V., and Wemmer, C. 1995b. Veterinary care of Asian timber elephants in India: Historical accounts and current observations. *Zoo Biology* 14: 123–133.

Lahiri Choudhury, D. K. 1991. Indian myths and history. In S. K. Eltringham (ed.), *The illustrated encyclopedia of elephants* (pp. 130–147). London: Salamander Books.

McKrindle, J. W. 1926. *Ancient India as described by Megasthenês and Arrian.* Repr., Delhi: Munshiram Manoharlal, 2000.

Milroy, A. J. W. 2002. *Management of elephants in captivity.* S. S. Bist (ed.). Dehradun: Nataraj. (Originally published 1922.)

Sanderson, G. P. 1912. *Thirteen years among the wild beasts of India: Their haunts and habits from personal observations; with an account of the modes and capturing and taming elephants.* Edinburgh: John Grant. (Originally published 1879.)

Shebbeare, E. O. 1958. *Soondar Mooni: The life of an Indian elephant.* London: Victor Golanz.

Smythies, E. A. 1942. *Big game shooting in Nepal.* Calcutta: Thacker Spink.

Stebbing, E. P. 1922. *The forests of India.* 4 vols. London: Bodley Head.

Stracey, P. D. 1963. *Elephant gold.* London: Weidenfeld & Nicolson.

Tennent, J. E. 1861. *Sketches of the natural history of Ceylon with narratives and anecdotes.* London: Longman, Green, Longman, and Roberts.

Wemmer, C. 1995. Gaonbura Sahib—A. J. W. Milroy of Assam. In J. C. Daniel and H. S. Datye (eds.), *A week with elephants: Proceedings of the international seminar on Asian elephants, June 1993* (pp. 483–496). Bombay: Bombay Natural History Society and New Delhi: Oxford University Press.

Whitney, C. 1905. *Jungle trails and jungle people.* London: Werner Laurie.

Williams, J. H. 1954. *Elephant Bill.* London: Rupert Hart-Davis.

Williamson, T. 1808. *Oriental field sports: Being a complete, detailed, and accurate description of the wild sports of the East.* London: William Bulmer for Edward Orme.

8 CARROTS AND STICKS, PEOPLE AND ELEPHANTS

RANK, DOMINATION, AND TRAINING
JOHN LEHNHARDT AND MARIE GALLOWAY

We contend that nonabusive training techniques can be effectively used with elephants and that a positive and respectful relationship can be established between humans and elephants in the care of humans. In support of this contention, our essay reviews the roles of dominance, rank and associated reinforcement patterns in wild-elephant societies and explores both concepts of and myths about different training modalities in human-elephant relations. We then discuss the uses of these training and management approaches to promote specific activities of elephants in the care of humans, and we address some associated ethical considerations.

What qualifies us to undertake this discussion? We have had the good fortune to work directly with, train, and care for elephants in North American zoos for the last thirty and nineteen years, respectively. This gives us an advantageous perspective on this issue that allows us to combine theoretical and ethical arguments with practical experience. Between us, we have been employed as animal keeper, museum specialist for elephants, elephant trainer, collection manager, assistant and associate curator, supervisory biologist, elephant curator, mammal curator, animal operations manager, and animal operations director. In all of these positions, which span four zoological institutions, we have had primary responsibility for the welfare of elephants and for the safety of the staff caring for those elephants. We have participated widely in elephant management through the Association of Zoos and Aquariums (AZA—formerly, American Zoo and Aquarium Association), Elephant Species Survival Plan (SSP), and Elephant Taxon Advisory Group (TAG) management and steering committees, the AZA Elephant Task Force and the Elephant Managers Association (EMA). Should we ever have an opportunity to write a book about our decades of interaction with elephants, it might be titled *Everything We Know about Life We Learned from Elephants*.

Over the past twenty years there have been significant changes in the way people perceive elephants, especially in western society, accompanied by similar changes in the way people perceive all animals and their relationship to humans. At the time we began working in the zoo business, few people questioned whether elephants should be kept in captivity or how they were kept and trained. Today, questions surrounding elephant management and training are frequent topics of conversation in AZA meetings, on animal activist websites and in elephant-housing institutions. At least one animal activist organization, People for the Ethical Treatment of Animals (PETA) even has full-time specialists monitoring captive elephants.

How have these changes come about? First, concern for animal welfare has continually increased over this time span, expressed through the rise of animal welfare organizations. Second, television, film and print publications have disseminated the groundbreaking work on elephants by field researchers such as Iain Douglas-Hamilton, Cynthia Moss, Joyce Poole, and others, work that has uncovered the incredibly rich life and complex social structure of wild elephants. The perceived similarities between human emotions and elephant experiences have made people care about elephants in a way that they care about few other animals. The growing awareness of similarities between humans and a species so obviously different from our own have led to scrutiny and questioning about how we treat these animals.

The methods used to train elephants in the care of humans now constitute one of the most contentious issues in animal management in North America. The Association of Zoos and Aquariums has debated whether to mandate a particular elephant management system, generally known as protected contact, for its member institutions. Though AZA has not decided to impose such a mandate, if it does so in the future it would essentially ban AZA-accredited zoos from employing the traditional method of elephant training and handling now called *free contact*. Free contact is used today for management of almost 16,000 captive elephants in Asian range countries, an estimated one-third of the world's total Asian elephant population (Lair 1997). Indeed, most elephants held in nonrange countries are also managed through traditional free contact methods.

Free contact is the management of elephants by humans and elephants sharing the same space, without a separating or protective barrier. Traditionally, elephant handlers in free contact use an elephant hook (a wooden or metal rod with a pointed hook on one end) to guide, or direct, the elephants' movement and, if necessary, to protect themselves from the elephant. Positive reinforcement with food is a primary training tool in free contact elephant management, but traditionally the elephant hook (or "ankus") has been used in free contact as a stimulus tool to elicit positive behavior and as a punishment tool to interrupt negative or aggressive behavior. Our essay's title mentions "sticks" to symbolize the images evoked by the use of the elephant hook. Some of these images in the popular imagination are rather dire. Our chapter will demonstrate that we believe the elephant hook is by no means the central element, nor should it ever be an abusive element in enlightened free contact training and management.

Many opponents of this method contend that managing elephants using free contact is inherently dangerous for the humans involved. Many contend that elephants under this regime can be trained to be in close contact with humans and to participate in performances only by learning through fear (heavy on the stick!), abusive training methods, and the use of force, or even torture. Prevalent among opponents of traditional free contact elephant training methods is the assumption that all free contact trainers have recourse to only one way of training elephants: through social dominance established and reinforced through physical means, including consistently inflicting or threatening pain. For instance, one animal rights activist organization has charged that traditional Asian training methods used in Thailand constitute abuse (see PETA website, www.helpthaielephants .com/).

By contrast, protected contact (heavy on the carrots!) is seen by many as being far more appropriate for the elephants and safer for the elephant managers. Protected contact is a system of elephant management in which the elephant(s) and caretaker(s) inhabit separate and exclusive spaces, either by utilizing protective barriers or by careful positioning, to minimize potential risks to the animal manager from elephant aggression. In our title, "carrots" symbolize the adaptation to elephant training of the operant conditioning training method developed by behaviorist B. F. Skinner. Positive reinforcement, with primary reinforcers such as preferred foods, is used to increase the elephant's motivation to carry out specific behaviors. Many people associate this approach only with protected contact, though in fact free contact trainers (as mentioned previously) also principally rely upon positive primary reinforcers, including favored foods, such as carrots, in their training regimes.

Can elephants be trained in close contact with trainers, even in the same unprotected space, with free contact and without abusive methods? Are human-elephant relationships possible that are not based on fear of the trainer, physical punishment, and even abusive training? One might ask how it is possible to train elephants at all, and what in their natural history indicates how one might accomplish effective training. Consideration of this natural history, combined with our own training experiences, leaves us convinced that physical dominance need not be a requisite to effective training, and that it should indeed be possible to formalize even a free contact management method of training without strong physical dominance. Below we explore both the natural history and other evidence that brings us to this conclusion.

The Relevance of Rank and Dominance

The groundbreaking work done on African elephant social structure in Lake Manyara Tanzania in the 1960s (Douglas-Hamilton 1975) and continued over the past three decades primarily by the Amboseli Elephant Research Project (Moss 2000) demonstrate a female matriarchal social hierarchy within herds. This research has shown that different individuals appear to have different social status and influence within the herds and that herds also have rank in relation to each other (see Poole and Moss, Chapter 4 in this volume, for further discussion of elephant families, bond groups, and fission-fusion societies).

Do the means by which social dominance is established in the wild help explain why—and how—dominance has been and can best be used to establish human control of captive elephant behavior? In her book, *Elephant Memories*, Moss asserts that high-ranking female elephants enforce their social dominance and maintain their rank in at least two different ways; through physical reinforcement or by establishing their leadership value to the herd and to its individual members over time in nonphysical ways. Some exercise only limited physical reinforcement of their leadership of the herd and appear to allow some flexibility in the behavior of others. Other high-ranking female elephants are much more aggressive and seem continually to reestablish their position by supplanting other elephants and pushing them around physically.

This two-pronged potential for establishment and maintenance of rank should not come as a surprise to us humans. We experience the same combination of approaches in our daily lives. At one extreme, human bullies attain rank and dominance through threat and intimidation, or by inflicting pain. Other people, however, attain a high rank by consistently acting in a manner beneficial to the group, or by using exceptional interpersonal skills. Since both elephants and humans use such similar methods to progress within the hierarchies of their respective societies, it would seem likely that social rank and dominance between the two species could be established by using the same methods.

Physical dominance has long been used by humans to attain control over elephants. A variety of methods have been used, including various combinations of restraint chutes, kraals, chains, ropes, physical deprivation, and overstimulation of the senses—the last includes such techniques as sleep deprivation, or constant exposure to people at close range. Until recently, these methods have constituted the primary approach to the human management of elephants (see also Lahiri Choudhury, Chapter 7 in this volume). However, we believe humans can also establish dominance over elephants without inflicting pain. If the human-elephant relationships are nurtured from a very early period and are maintained over time, much like the social setting of female elephants in an established wild herd, we believe it is even possible to formalize a method of training elephants that allows for free contact management yet precludes strong physical dominance. Can such a free contact management relationship also be safe for the human participants?

Social dominance appears to play a critical role in ensuring the safety of such human-elephant interaction. From our own experience as elephant managers, we can assert that nonphysical establishment of dominance can occur between humans and elephants. In many instances within our personal experiences, elephants came into human contact at between 1 and 3 years of age and received the care of a particular handler for many years in a free contact situation. These handlers often had no formal training skills but, employing empathy and care, simply provided a social context for the development of bonding and trust between handler and elephant. This has occurred most frequently where the elephant is solitary and has no other elephant companions. In this environment, the manager and other humans, can be socially dominant and safe for decades, without having to reinforce their social dominance in any significant physical way.

We once experienced an interesting example of handlers achieving social dominance over an aggressive elephant in a nonphysical manner, with a bit of help from another elephant serving as a "messenger." In 1989, the National Zoological Park in Washington, DC, where we both worked at the time, received an adult female elephant from a small municipal zoo that was closing down. This elephant had come from Thailand at about age 6 months and had been raised at that small zoo for twenty-three years, without other elephant companions. During that time, she had the same two male keepers. She had never experienced any sort of physical restraint such as being attached to a stationary barrier by ropes or chains and had learned to follow like a puppy. She had experienced no formal training or traditional establishment of human dominance through punishment. No hooks were ever used by the keepers to guide or direct her. Zoo guests were regularly brought into the enclosure in the presence of the keepers. The elephant had shown aggression only twice in her life, both times toward women whom the keepers had briefly left alone with her. After moving to the National Zoo, the elephant immediately became aggressive toward all the female keepers and could be managed only by the one male elephant keeper employed by the zoo. We developed a modified protected/free contact program for her, in which she was managed through the barriers on the male keeper's days off. We slowly built her trust and confidence in the female staff members, without using any traditional aggressive techniques. Over time, trust and positive rapport were established, and all but one of the keepers was able to work effectively with this elephant in free contact management. Thus, we had successfully used nonaggressive, nonphysical management to control this female elephant.

However, the elephant still failed to tolerate one keeper. Three times, the elephant hit this particular keeper, once knocking her down and putting her in the hospital with a bruised coccyx. We then planned for a concerted response to another recurrence of the undesirable behavior. We instructed the keeper that the next time this happened she was immediately to bring over the zoo's dominant elephant (a very large female African) and use her to displace and corner the aggressive Asian elephant. This was done the next two times that the Asian elephant was aggressive toward the keeper. It worked, and after the second cornering, the elephant showed no further aggression toward any of the keepers. From then on, all keepers managed her in a free contact situation with nonphysical control.

The history of the two elephants' relations is relevant. The African elephant had previously established her position with the new elephant by knocking her over once during the new elephant's introductory period. The new elephant appeared to have been trying to get the attention of the African elephant by spitting at her across the dividing barrier. The new elephant, raised in isolation from other elephants, didn't know how to relate normally to other elephants. The African elephant, clearly irritated by the spitting, raised her head over the barrier and pushed the new elephant hard enough to knock her completely off her feet. After this one aggressive incident, the new elephant was completely submissive to the African elephant. So, in this instance, the keeper's rank and dominance was established through association with negative reinforcement provided by a dominant elephant.

Male Elephant Social Structure and the Implications for Captive Management

Field research in Africa has shown that African elephant males live different social lives than the females. The males either are ejected from the herd (Moss 2000) by the adult females or slowly withdraw of their own initiative as they approach maturity. There are various views on the age and size that male elephants attain before leaving the herd, but one could argue that it happens about the time the males reach the same height and weight as the matriarch or their own mothers (R. Estes, personal communication). There are good reasons to postulate that this is the appropriate time for cows to cast out their sons; the cows are still large enough to oust males physically, which are not yet capable of breeding. Some researchers postulate the expulsion of males prevents inbreeding and reduces social tension caused by the male's continual need to challenge those around them physically.

The male African elephant's drive for social dominance appears to be oriented to gaining reproductive opportunities (Poole 1982, 1989a, 1989b). Following expulsion from the maternal herd, males spend as much as fifteen to twenty years attaining sufficient body size and condition to come into musth. This is a physiological condition characterized by elevated testosterone and aggression, which enhances the animal's ability to compete successfully for breeding rights with females. During this period of adolescence, bulls become familiar with their peers by sparring and testing one another's strength and thus establish their place in the male social order. Their early sparring helps establish who is stronger and more aggressive. The establishment of the male hierarchy probably helps to limit the incidence of serious fights between fully mature bulls. Thus, it appears that dominance and rank among male African elephants is established through physical force or the threat of physical force (Poole 1982, 1989a, 1989b).

The onset of musth appears to have an effect on an individual's rank within the prevailing social structure, potentially raising the musth elephant's social standing (Poole 1982, 1989a, 1989b). The competitive and combative nature of male elephants affects how they are managed and trained in the care of humans. In zoos, bulls continually challenge their environment, including their handlers, other elephants, and physical barriers. Males may become unsafe to handle directly once they reach sexual maturity (as early as 6 years old in zoos) and certainly by the time they begin exhibiting musth, which may begin occurring in the teenage years. Historically, North American zoos have kept far fewer males than females; the current ratio is about one male to every seven females (Olson 2002; Keele and Lewis 2003).

Handling mature bull elephants, or even late adolescents, in free contact is extremely hazardous for even the most talented and experienced managers. A colleague who worked for several decades breeding bull elephants in zoos equated working with bulls with jumping off a forty-story building. During the first thirty-nine stories of the descent, if asked how it was going, one could answer "so far, so good." However, a very difficult final moment is inevitable. In North America, bull elephants have a sevenfold higher probability of killing a human in their lifetime than do female elephants (Lehnhardt 1991, Table 2). This has led the AZA to develop guidelines recommending that bull elephants not be worked in free contact after age 5. This age limit was selected because captive male Asian elephants born in North America have been known to produce sperm and to exhibit signs of musth by age 6. There are no data on early onset of musth in African male elephants in the care of humans in North America, mostly because there have been very few African elephant births and hence few African male elephants available to study from an early age.

There have been a few instances of mature male elephants having been handled in North America without serious problems. King Tusk, an Asian elephant owned by Ringling Bros. and Barnum & Bailey Circus, was handled without serious incidents of aggression up to his death in 2001, at more than 50 years of age. However, these instances of manageable mature male elephants in North America are rare.

In the range states of Asian elephants, many male elephants are handled throughout their entire lives. Each of the region's different elephant-handling cultures has their own distinctive approach. Many variables appear to influence their success in handling adult males. In this discussion, we use manager survivorship as a principal measure of handling success. During the 1990s, J. Lehnhardt had the opportunity to observe the management of male elephants in three different venues demonstrating three distinct approaches: (1) timber elephant management by a government agency in Myanmar (formerly Burma); (2) management of privately owned elephants in Sri Lanka in and around the capital city of Colombo; and (3) state forest department–owned work elephant camps in Nagarahole National Park, Karnataka State, India.

In Myanmar, the timber elephants were released into the surrounding forest at night. The males were usually hobbled, with front or back feet together but were able to move about readily. They were known to have bred with locally released females and to have encountered wild bulls and cows. Managers each spent many years with their elephants. Instances of elephant attacks on managers were almost unknown in Myanmar.

In Sri Lanka, the privately owned elephants were virtually never given free forage time. There was no interface between wild elephants and the captive working elephants, and the managers had spent only a few years with their elephants. It turned out that the killing of elephant managers or of other humans by these captive elephants was common. Lehnhardt counted more than sixty human deaths among thirty elephants surveyed and nearly witnessed one death in Colombo while driving down the street with colleagues. On the street, we encountered a large animated crowd and we were told that a large mature bull elephant, by then tethered behind a temple, had just killed its mahout. We found ourselves standing next to a very nervous young man who was about to become the elephant's new mahout. He had been the apprentice mahout for fewer than two years and clearly had no interest in approaching the animal standing only a few yards away.

In Nagarahole, the primary camp Lehnhardt visited had an impressive record for mahout longevity with their elephants. Twelve years was the shortest amount of time any manager had worked with his elephant, and some had thirty years experience with their animals. The elephants were released into the forest at night and recalled to the camp twice a day for bathing, oiling with anti-fly ointment and feeding with tasty ragi balls, a native grain cooked into a thick proteinaceous dough. The camp had an adult male elephant in his forties, four older cows between 30 and 50 years of age, and two calves from these cows, with two more cows pregnant.

A remarkable incident demonstrates the extraordinarily cooperative relationship between the handlers and the elephants at Nagarahole. The camp's male elephant was a very large tusker named Gajendra, who was about 45 and had been captured only eight years previously. He had been a crop-raiding wild elephant and reportedly had killed a number of people in a local village in a series of incidents. He was blind in one eye from a gun-shot wound received during his crop-raiding days. While Lehnhardt was in Nagarahole, Gajendra's managers one day prepared to truck the elephant to a neighboring town to lead an upcoming festival. One of his assistant mahouts walked him down to the road for loading while his primary ma-hout packed belongings for the overnight trip. As Gajendra walked through the park he stopped to accept some bananas from a group of picnicking nuns. The nuns were very excited as they reached up to deliver their offer-ing of fruit directly into the big tusker's mouth.

After the treat, Gajendra was walked to a spot under a shade tree. The assistant mahouts all then left Gajendra, untethered, and went off to get the truck and back it up to that shady spot. Gajendra dozed alone in the shade for about thirty minutes. The managers then arrived with the open-bed truck and backed it up to the knoll where Gajendra was resting. When the truck stopped, Gajendra, of his own volition, stepped into the truck bed (his assistant mahouts meantime engaged nearby in an animated argu-ment) then picked up ropes piled in the truck, placing them near his legs. He then rested his tusks on the truck's cab and took another nap.

A group of about six men then began securing Gajendra to the truck with the ropes he had distributed. There appeared to be a disagreement among the group as to how the tethering should be done. Several men were arguing while climbing around and under Gajendra and tying, then rety-ing the ropes in different arrangements. Occasionally, Gajendra would open his one good eye, pick up a piece of rope and move it to a different place. It seemed he was trying politely to instruct the argumentative men about the proper placement of the ropes. The tethering was completed after about half an hour and the senior mahout arrived to oversee the departure. He was clearly dissatisfied about how the tethering had been done and be-rated the crew. He untied all the ropes and rearranged them, appearing to follow Gajendra's suggestions more closely. With the ropes now properly attached, the truck soon departed, with about twenty people sitting on, un-der, and around Gajendra on the truck bed.

Earlier that morning, Lehnhardt joined a small group of visitors allowed to go out with Gajendra's mahout to retrieve the bull elephant from his night of foraging. At dawn, the senior mahout walked about a half mile into the misty forest and climbed a tree to a height of about fifteen feet. He called several times and then listened. He apparently saw or heard what he wanted (we saw and heard nothing), then climbed back down to the base of the tree to wait. In about ten minutes we heard a light disturbance in front of us and out of the brush came Gajendra. The mahout greeted Gajendra, took hold of the bull's neck chain, which was also connected to Gajendra's foot, and then the elephant hoisted his mahout onto its back with a combined foot, head, and trunk motion by the elephant. We strolled leisurely back to camp.

This encounter with Gajendra helped us understand a lot about possible relationships between elephants and humans. The mahouts in Nagarahole could not recall any mahouts having been killed by their elephants. Wild elephants, but not any of the working elephants, had killed people. It seemed these working elephants had a near-perfect existence in the service of humans. They were bathed, oiled, and fed twice daily. They were free to roam the forest at night, to breed and socialize, and they worked only about three days a week hauling firewood from the forest for the local villages. There was also an occasional festival to lead. These animals clearly cooperated with their managers and chose to return on a daily basis, even though all were wild caught and trained only as adults. The social structure of the elephant group was probably not quite the same as that of a wild herd. Generations of calves were not present in the camps. But these elephants had a good quality of life, one that clearly presented a better alternative to being shot for crop raiding.

Even male elephants were managed effectively in this system. Unfortunately, on this short-term visit, Lehnhardt was not able to witness the captures or training that had factored into the establishment of this remarkable rapport and cooperation between the elephants and their mahouts. Presuming that the initial training was traditional, it would have employed restraint and force, combined with deprivation and overstimulation of the senses, practices that must have traumatized the elephants, at least in the short term. Significantly, if these crop-raiding elephants had not been subjected to this initial training, they would have been caught and summarily killed to remove the menace they presented to local communities. After the initial training period, however, the elephants' mahouts have gone to great lengths to cultivate and maintain a rapport with these elephants, employing remarkably humane, even somewhat pampering practices. The elephants cooperate willingly, even when given the option nightly to go back to the forest.

This example indicates that a successful, cooperative working relationship that ensures against human deaths can be maintained between humans and male elephants, even into male elephant maturity. The basic requirement for maintaining a successful relationship between humans and elephants in their care seems clear: provide for the elephant's needs. These needs include: opportunities to socialize and breed; experiencing consistency in the human/elephant relationship; and some freedom for the elephant to be independent—simply, to be an elephant. Unfortunately, these parameters, prevalent in Nagarahole Park, are not the standard in North American zoos, or even in many other situations captive elephants experience in elephant range countries.

"Carrots and Sticks": Elephant Training Modalities

We maintain that elephants can be trained, managed safely, and treated ethically in the care of humans if the appropriate requisites are in place to meet the elephants' essential needs. In free contact management, the appropriate social relationship must be established between humans and elephants and must be maintained through reinforcement that mimics the modalities present in wild-elephant social structures. The ethical question that comes to the fore stems from how the initial social dominance, which appears to be critical for a safe free contact human-elephant relationship, is established. In North America, and increasingly in Europe, the traditional physical establishment of dominance over elephants has become abhorrent to a significantly large and frequently vocal segment of the public. Full disclosure of videos showing traditional initial elephant training, whether in North America or in range states, would be met with a public outcry. A relatively recent example concerned the "Tuli elephant" controversy, regarding the capture, training, and transport from Africa of a number of young wild elephants, some destined for western zoos. It appeared to be the training methods used that caused the greatest public outcry (Patterson 1999).

In the United States, animal welfare in public institutions such as zoos and circuses is monitored by the United States Department of Agriculture, Animal and Plant Health Inspection Service (USDA-APHIS), in accordance with the Animal Welfare Act of 1966. Under this act and its subsequent amendments, several cases involving use of traditional elephant training techniques have brought USDA citations, fines, and other penalties, including suspension of USDA exhibitor's licenses, seizure of elephants, and bans from working with elephants (Farinato 2004; PETA 2006). Thus, within the United States, traditional initial training techniques are not only deemed unethical but potentially illegal. However, the majority of captive elephants today in North America and in range countries are still animals that were wild caught. In some range countries elephants are still being brought into captivity as adults. If traditional initial training methods were not available to managers charged with working with these animals—either because of ethical or regulatory reasons—would free contact training and management still be feasible?

We believe, from our collective experiences, that elephants captured at a very young age, or elephants born into a captive situation, can be trained and socialized to humans without managers resorting to abusive training. The development of training theory and the application of operant conditioning (Pryor 1984) to marine mammals in captivity have provided a model for safe, effective elephant training. One building block toward clarifying the nature of operant conditioning training of elephants is to provide clear definitions of the training terms used, to help rectify public and professional misinterpretations of elephant trainer and manager intentions, actions, and their effects. In defining and discussing these terms, we also outline the content of the free contact training process we believe is both ethically and feasibly available to elephant trainers and managers and that has been successfully used in some North American zoos.

Operant conditioning, briefly mentioned at the outset of this chapter, is learning that occurs when a response to a stimulus is reinforced. A stimulus is any object or event that causes a subject—the elephant—to respond. For example, a visual signal made by the trainer or an audible sound such as a spoken command are both stimuli. An aversive stimulus is one that the elephant wants to avoid or escape, for example, the (light) touch of an elephant hook. Operant refers to the fact that the subject of the training (in this case, the elephant) is the operator, the one doing the behavior, and—despite being also a "subject" of the training—is always actively involved in the training process. Each of the elephant's behaviors—its responses to the reinforced stimulus—are conditioned through successive approximation. "Conditioning through successive approximation" means building a change in behavior through small increments that eventually lead to the full behavior. An instance of this would be, for example, the elephant's first learning to lift its foot, then learning to lift and hold its foot in the air, and finally, after several interim steps, reliably lifting and steadily holding up the foot at the optimal height needed for a veterinarian to carry out a complete foot checkup.

A reinforcement is any action or event that increases the probability that a response will be repeated. Reinforcement of behavior can function as positive reinforcement or as negative reinforcement. The simplest way to understand the difference is to think of positive reinforcement as the addition of something the subject wants and negative reinforcement as the subtraction of something the subject does not want. Positive reinforcement can be defined as the addition of anything, which, occurring in conjunction with a behavior, tends to increase the probability of the behavior occurring again (Pryor 1984). Examples of positive reinforcement used with elephants—things the elephant will work to get again—include food (offering a carrot), praise ("good girl!"), and physical affection.

A simple behavior such as lifting the foot can be taught with positive reinforcement in several ways. One method involves "capturing" the behavior, that is, reinforcing successive approximations to the desired behavior. The keeper waits until the elephant lifts its foot even a little, then immediately reinforces that behavior by providing the elephant with a desired food item. By gradually increasing the height required of the foot before reinforcement is provided, the trainer can condition the elephant to achieve the desired height. Another method is to teach the elephant that it will receive a positive reinforcement for touching its foot to a "target." A target is a tool the elephant is trained to move toward, usually a stick with a ball attached to its end. The target can be moved or lifted until the foot rises to touch it. With either method, the keeper must be sure to offer the reinforcement at the exact moment that the desired response behavior takes place, so as to communicate clearly to the elephant that that specific behavior was the one being requested. If the reinforcement—the award—is presented either too early or too late, it will communicate to the elephant the message that the behavior they happened to be performing at the moment of the reward was in fact the desired one (Olson 2004).

Negative reinforcement is the cessation of an aversive stimulus. For instance, after an elephant is given an aversive stimulus, such as the touch of the elephant hook (the stick!), the hook is removed. To teach the elephant to lift its foot on command, the keeper touches the back of the elephant's foot with the hook. The elephant doesn't like the touch of the hook and moves its foot away—lifts the foot. The keeper immediately removes the hook (negative reinforcement) and may also give the elephant food or praise (positive reinforcement). This can also be done in successive approximations. However, the removal of the hook after the elephant has responded is what increases the probability that the elephant will repeat the behavior. To reiterate: the *removal* of the aversive stimulus (i.e., the touch of the hook) is the reinforcement and not the aversive stimulus itself. Other examples of negative reinforcement (that also occur in the wild) would be escaping a shove from another elephant's tusk (very much like the stick) or a swat from another elephant's trunk.

Given the current concern over use of the elephant hook, or ankus (also sometimes referred to as a guide), we feel we must emphasize again here that negative reinforcement is not the same as punishment. Negative reinforcement encourages the repetition of a behavior and punishment is used to extinguish a behavior. Negative reinforcement removes something in conjunction with the performance of the desired action or response. The use of an ankus (as a stimulus) to cue a behavior can be an example of negative reinforcement. The 2004 Elephant Husbandry Resource Guide provides an excellent example of negative reinforcement.

> A handler applies pressure when placing the tip of the guide (ankus) against the back of the elephant's leg to communicate to the elephant to move its leg away from the pressure. When the elephant moves its leg forward, it avoids the aversive stimulus and the behavior is negatively reinforced. This action can then be reinforced with food and praise (positive reinforcement) to further communicate to the elephant that the behavior of moving its leg was correct. (Olson 2004, 27)

Negative reinforcement is intended to have an impact on the immediate behavior—the behavior just then taking place. It allows for the desired behavior to be positively reinforced. An important distinction is that punishment cannot change the behavior it follows because the behavior has already occurred. Punishment can only affect future behavior; it is an unpleasant action that occurs immediately *after* an unwanted behavior, and it can only be intended to decrease the likelihood of the behavior recurring at a later time. Withholding a food reward when the desired behavior is not achieved is an example of a punishment. The behavior is over, and there is no chance to undo it. However, the punishment induces the subject—the elephant—to display the desired behavior the next time, which then can be reinforced.

Punishment can also be used to establish social or physical dominance. Earlier in this essay we described a zoo situation in which one elephant knocked down another. The trigger behavior (spitting) was over, but by inflicting punishment for that behavior the African elephant successfully prevented the Asian elephant from ever spitting that way again. Similarly, traditional methods of initial training of elephants employ physical punishment initially to establish dominance early in the relationship and then shift to reinforcement training to establish desired behavior patterns.

In marine mammal training, physical punishment is not an option, and early on, reinforcement of behaviors with food items, combined with nonphysical punishment (withholding the positive reinforcer, i.e., food) developed as a structured approach to training. Over the past two decades, this approach has been adapted in the development of protected contact elephant management, where the managers do not share the same space as the elephants. The consequent ability to elicit behaviors without being in the elephant's space has reduced the safety risks faced by elephant managers.

With the availability of protected contact management and improved understanding of training theory, the question becomes whether we can ethically continue to use traditional methods of free contact elephant training in North America, if they require physical dominance as a baseline for their effectiveness.

Toward an Ethical Method of Training Elephants

We have discussed in this essay how elephants live in a structured society in the wild, with social dominance and hierarchy as key components. Physical and social reinforcement of the hierarchy exist as a natural part of the social order for elephants in the wild. Humans have adapted these natural traits to manage elephants in their care. We have presented anecdotal evidence indicating that female elephants may be adaptable to human social dominance, and that male elephants in the care of humans are potentially too dangerous to trust working with humans within a nonphysical dominance hierarchy.

Can elephant training based on physical dominance continue as a viable method in the United States, especially in light of the emergence of newly prevailing ethical views and in consideration of existing legal requirements for animal welfare? If the answer is no, can free contact management continue if it is based, in part, on physical dominance?

The question must be answered separately for males and females. In the wild, male and female elephants lead different social lives. In captivity too, they require different forms of management. For male elephants, the requisites for maintaining a safe human-elephant interface cannot be met in North American zoos, once the elephant is past age 5 or 6. It makes no sense to attempt to maintain free contact management as males approach puberty: the risks are too great. For female elephants, however, it appears to us that once dominance is established, it can be maintained within the bounds of public acceptance and good welfare as well as operating within the strictures of legal responsibilities. It also appears that female elephants can be raised from birth or an early age and trained without harsh physical dominance training. Thus, female elephants could be ethically maintained in free contact management with a positive and respectful relationship between the elephants and their human caretakers.

References

Douglas-Hamilton, I., and Douglas-Hamilton, O. 1975. *Among the elephants*. New York: Viking Press.

Farinato, R. 2004. *USDA seizes the moment, orders Hawthorn to give up 16 elephants*. The Humane Society of the United States. www.hsus.org/wildlife/wildlife_news/usda _seizes_the_moment_orders_hawthorn_to_give_up_16_elephants.html.

Flynn, D., and Lehnhardt, J. 1995. Integrating the Specialized Training and Reinforcement System (STARS) into the elephant management program at the National Zoological Park. In *Proceedings of the American Zoo and Aquarium Association Northeast regional conference, Norfolk, VA, May 1995* (pp. 547–551). Bethesda, MD: American Zoo and Aquarium Association.

Keele, M., and Lewis, K. 2003. *Asian elephant (Elephas maximus) North American regional studbook update*. Portland: Oregon Zoo.

Lair, R. C. 1997. *Gone astray: The care and management of the Asian elephant in domesticity*. Bangkok: United Nations Food and Agriculture Organization Regional Office for Asia and the Pacific.

Lehnhardt, J. 1991. Elephant handling: A problem in risk management and resource allocation. In *Proceedings of the American Association of Zoological Parks and Aquariums Annual Conference 1991* (pp. 569–575). Wheeling, WV: American Association of Zoological Parks and Aquariums.

Lehnhardt, J. 1995. Working elephants in Sri Lanka and Myanmar. In *Proceedings of the 16th Elephant Managers Workshop, Port Defiance Zoo and Aquarium, Tacoma, Washington, October 10–14, 1995* (pp. 7–12). Indianapolis, IN: Elephant Managers Association.

Moss, C. J. 2000. *Elephant memories: Thirteen years in the life of an elephant family: With a new afterword*. Chicago: University of Chicago Press. (Originally published 1988.)

Olson, D. 2002. *African elephant* (Loxodonta africana) *North American regional studbook update.* Indianapolis: Indianapolis Zoo.

Olson, D., ed. 2004 *Elephant husbandry resource guide.* Silver Spring, MD: American Zoo and Aquarium Association.

Patterson, G. 1999. The Tuli elephants: Reflections of a nation. *Diversions,* August. www .garethpatterson.com/Elephant/elephant.htm.

PETA. 2006. *Tabak Royal Circus.* Norfolk, VA: People for the Ethical Treatment of Animals. www.circuses.com/fact-tabak.asp?pf=true.

Poole, J. H. 1982. Musth and male-male competition in the African elephant. PhD diss., University of Cambridge.

Poole, J. H. 1989a. Mate guarding, reproductive success and female choice in African elephants. *Animal Behaviour* 37: 842–849.

Poole, J. H. 1989b. Announcing intent: The aggressive state of musth in African elephants. *Animal Behaviour* 37: 140–152.

Pryor, K. 1984. *Don't shoot the dog: How to improve yourself and others through behavioral training.* New York: Simon and Schuster.

9 CANVAS TO CONCRETE
ELEPHANTS AND
THE CIRCUS-ZOO RELATIONSHIP
MICHAEL D. KREGER

Elephants have interacted with man throughout history. They have been used as beasts of burden in transportation, logging, and military operations. They have been exploited for their meat, hides, and ivory, and they have played a major role in religions and folklore (Gröning and Saller 1998). As the largest living land mammals, live elephants are also a source of wonder. Elephants in captivity have long been star attractions, drawing crowds to circuses and zoos. As the complexity of elephant social behavior becomes known from *in situ* and *ex situ* research, similarities to human behavior are becoming more apparent, and conservationists and animal welfare advocates have questioned the ethics of maintaining elephants in circuses and zoos.

The intent of this chapter is to provide a historical context regarding how circuses and zoos, particularly in the United States, came to acquire elephants, how circuses and zoos have collaborated in elephant management, and how they have differed. I also discuss past and present laws and regulations affecting care and use of elephants in captivity. In democratic societies, regulations often indicate a majority view on a particular issue, at least one prevailing at the time they were issued. I will not argue the morality of keeping elephants in zoos or circuses but rather current issues and likely trends in collaboration.

Elephants in Circuses

The circus, from the Latin word meaning *circle*, has thrilled people of many cultures for centuries. Along with clowns, aerialists, and others making spectacular demonstrations of human courage and skill, animal acts, including those with elephants, have historically been a mainstay of most circuses. The first modern circus, an equestrian show, was started in England in 1770 by the French and Indian War hero Philip Astley, a sergeant major in His Majesty's Light Dragoons, who had served in the American colonies (Fenner and Fenner 1970). Astley, an Englishman, started the show with trick riding on horses on a circular track. He later added acrobats, dancing dogs, a clown, and a tightrope walker to the show. In 1793, John Bill Ricketts, an Englishman employed by Astley's rival in showmanship, Charles Hughes, brought the circus to the young United States, finding a permanent home in Philadelphia (Fenner and Fenner 1970). George Washington, an expert equestrian, attended the grand opening of Ricketts's show and was so impressed that he leased his warhorse Jack to the circus for a season for $150 (Fenner and Fenner 1970).

In the United States, private traveling collections of animals had a tradition long predating the first public zoological gardens in New York's Central Park (formally opened in 1873) and Philadelphia (opened in 1874). America's first close-up look at exotic wild animals occurred with traveling menageries, which characteristically contained many animals in small cages. A few also included animal acts that featured the bravado of the "fearless" trainers more than the skill of the animals (Flint 1996; Kisling 2001). Starting in 1833, Isaac A. Van Amburgh became world famous as a lion tamer by walking into a cage of food-deprived lions, tigers, and leopards to demonstrate his "moral powers" at subduing the savage beasts, which he did by wielding a crowbar (Rothfels 2002). Van Amburgh's act made his menagerie and eventual circus financially so successful that contemporary and subsequent animal trainers often emulated him.

Solo animal acts also allowed eighteenth-century Americans to view nonnative animals. In 1796, the first elephant arrived in America, a 2–year-old female Asian elephant that Captain Jacob Crowninshield had purchased in India for $450. He sold the elephant to a United States buyer named Owen, for $10,000. It was exhibited from Pennsylvania to Rhode Island for twenty years (Kisling 2001). To calm or entice wary patrons, a 1797 poster announced, "He eats 130 weight a day, and drinks all kinds of spiritous liquors; some days he has drank [sic] 30 bottles of porter, drawing corks with his trunk. He is so tame that he travels loose, and has never made an attempt to hurt anyone" (Kisling 2001, 149).

Second to arrive was Betty, who became known as "Old Bet." A sea captain at a London auction purchased this female African elephant for $20 (Colver 1957; McClung 1993). Arriving in New York in 1804, Old Bet was sold to the sea captain's brother, farmer Hackaliah Bailey, for $1,000. Old Bet toured the East Coast moving from building to building (usually barn to barn) during the night to evade potential admission-fee shirkers. Because Bet was not trained to perform tricks, Bailey added trained animals, including a dog, a horse, and pigs, to his new menagerie to increase patronage.

Entrepreneur Nathan Howes soon leased Old Bet and developed an innovative way to cut exhibition costs. Instead of paying landlords to exhibit her in permanent buildings, Howes purchased a tent that could be pitched anywhere the elephant was touring. Bet also became known for the nature of her death. In July 1816, a farmer, possibly motivated by religious reasons, shot her to death near Alfred, Maine. "Blue laws" made it illegal to entertain on Sundays, and many people saw circus-type entertainment as frivolous and immoral, as well as a waste of poor people's money (McClung 1993; Kunhardt, Kunhardt, and Kunhardt 1995; Flint 1996). Old Bet's stuffed remains continued to tour for many years (Brown 1987).

The first elephant to perform tricks was Little Bet, "The Learned Elephant," who arrived in the United States in 1821. Little Bet could kneel, balance on alternate pairs of legs, present a foot, lie down, sit up, bow, whistle, and carry people on her trunk (Brown 1987). In an act of hyperbole typical of such showmen, Little Bet's owner advertised that her hide was so thick that a bullet could not penetrate it. On July 31, 1822, she was shot in the eye and killed by boys testing that claim. Little Bet was the last solo-touring elephant in America. After her, elephants appeared only as members of traveling menageries (Brown 1987).

Menagerie and circus promoters advertised that their attractions held moral and instructive values, probably to assuage the consciences of visitors concerned about partaking in frivolous entertainment (Flint 1996). From the 1850s until his 1891 death, Phineas Taylor (P. T.) Barnum was a master of such promotion. For example, in 1861, Barnum advertised the first hippopotamus to come to America as "the behemoth of the Scriptures" (Barnum 1889, 279), implying that a visit to see Barnum's hippo was akin to a religious experience. Owners also layered moral legitimacy onto their circuses or menageries by attracting reputable naturalists to view the animals and provide written testimonials later used for publicity (Hanson 2002). In 1884, Professor W. H. Flower, president of the London Zoological Society, attested in writing that Barnum indeed had a royal sacred white elephant from Burma (Barnum 1889). Scientific language and graphics were used in publicity material to attract "respectable" crowds.

Recognizing his market's appetite, in 1850, Barnum imported at least eight elephants captured in Ceylon. This first herd in America was also organized into America's first elephant parade, as part of a traveling exhibition called "P. T. Barnum's Asiatic Caravan, Museum, and Menagerie" (Kunhardt et al. 1995, 110). By the late 1860s, the American menagerie became part of the circus (Flint 1996). In 1870, Barnum was lured out of retirement to create the most spectacular and largest circus in the world. He merged circus with animal menagerie and American Museum as a six-tented show. To give it moral acceptance, "P. T. Barnum's Grand Traveling Museum, Menagerie, Caravan and Circus" was also called "P. T. Barnum's Great Traveling Moral Exposition of the Wonder World." Barnum advertised that, besides being a show with high moral values, it differed from other circuses in that it was clean, educational, and "ten times larger than any other show ever seen on earth" (Kunhardt et al. 1995, 224).

Circuses were very competitive since most visited the same towns. Barnum had two advantages. He had elephants and he developed the means to move his show by rail. The 1870s was the decade of the elephant wars,

P. T. Barnum's winter circus quarters in Bridgeport, Connecticut. Baby Bridgeport, the second elephant bred in captivity in the United States (*left*). Elephant
training methods (*right*).
Harper's Weekly, February 18, 1882, p. 101

with the circuses of Adam Forepaugh, Barnum, the Sells Brothers, and
Cooper and Bailey all competing to own the most elephants (Hagenbeck
1910; Kunhardt et al. 1995). German animal dealer and occasional circus
proprietor Carl Hagenbeck supplied many of these elephants (Hagenbeck
1910; Flint 1996). In 1902, Hagenbeck wrote, "In the years from 1875 to
1882 I have sold about 100 Elephants over to the U. States, of which the
greater part were sold to the Barnum & the Forepaugh shows" (Flint 1996,
106). These circuses soon after competed to claim the largest elephant and
the whitest elephant (Flint 1996). Often these claims were accompanied by
tricks such as painting the elephants white (Barnum 1889; Kunhardt et al.
1995; Flint 1996). Barnum's circuses used elephants in fairy-tale tableaus
and races, featuring more than thirty elephants. He used an elephant to
plow fields in front of his Bridgeport, Connecticut, mansion, "Iranistan,"
and, in 1882, he took pride in exhibiting Baby Bridgeport, the second elephant bred in captivity in the United States (Barnum 1889; Sheak 1922;
Kunhardt et al. 1995). The first may have been Columbia, born in 1880 at
the Bailey, Scott, & Hutchinson Circus (Sheak 1922).

Elephants in Circuses and Elephants in Zoos

Early zoos and circuses had many areas of common ground. Because there
were so few of either, they did not automatically view each other as com-

petitors. Possibly to avoid competition, zoos were often founded with the lofty goals of education and the advancement of science, thus distinguishing themselves from the collections of curiosities provided for public entertainment in traveling menageries and circuses (Hanson 2002). Circus people even advocated for establishment of zoos. In the late 1800s, John Robinson, of the Robinson Circus, promoted establishing a grand zoological garden in Cincinnati (Ehrlinger 1993). P. T. Barnum, who sometimes exchanged his animal carcasses for historical or exotic exhibit items from the Smithsonian Institution's National Museum, once wrote to Spencer F. Baird, Smithsonian's secretary, suggesting the establishment of a national zoo (Barnum 1889; Saxon 1983). In an 1886 letter to Smithsonian National Museum taxidermist William T. Hornaday, Barnum also wrote, "My partners cannot raise courage at present to start a zoological garden at Washington" (Saxon 1983, 282). Hornaday later helped establish the National Zoological Park and became the first director of the New York Zoological Park (Johnson 2001).

Nineteenth- and early-twentieth-century zoos purchased animals from circuses. Circuses had bigger animal collections and more species than most of these newly established zoos (Hanson 2002). In the 1890s, the National Zoo bought its first elephants (Asian elephants "Dunk" and "Golddust"), a lion, and a Sumatran rhinoceros from the Adam Forepaugh Circus (Horowitz 1996). Through the 1930s, the fledgling Denver Zoo received animals from the Sells-Floto Circus as well as from the newly merged Ringling Bros. and Barnum & Bailey Circus (Etter and Etter 1995). Zoos provided a home for circus animals too old to perform or too costly to maintain. Conversely, once zoos could successfully breed species or wanted to sell surplus animals, circuses were ready buyers.

Some circuses loaned animals to zoos during the circus off-season. The zoo benefited by having trained animals for exhibits, breeding, and income-generating shows and rides. In return, the loaned animals received free veterinary care, winter housing, feed, and maintenance. In the winter season of 1893–94, the Adam Forepaugh Circus loaned seventy-three animals to the National Zoological Park. The new arrivals are estimated to have brought more than 30,000 visitors to the park in a single day (Horowitz 1996). P. T. Barnum had free access to winter his animals at the Central Park Zoo (Wharton 2001).

The career of Jumbo exemplifies the intermingling of zoo and circus history in this era. Jumbo was captured as a calf in Africa in 1861, sold to a Paris zoo, and then traded to London's Royal Zoological Gardens (Brown 1987; Kunhardt et al. 1995). Jumbo was over eleven feet tall and weighed six and a half tons. Although he provided rides at the zoo for many years, eventually Jumbo's behavior became erratic and dangerous. The zoo grew anxious to sell Jumbo, and P. T. Barnum was happy to buy him, promptly claiming this was the world's largest elephant. Jumbo arrived in New York in 1882 as "The Towering Monarch of His Mighty Race, Whose Like the World Will Never See Again" (Kunhardt et al. 1995, 281). He became the hallmark of Barnum's circus. After Jumbo was killed by an unscheduled express train while crossing the tracks to his railcar on September 15, 1885, Barnum announced the animal had died pushing a younger elephant out of the way of the oncoming train (Kunhardt et al. 1995). Barnum then further capitalized on Jumbo's death and the economic value of elephants to circuses by exhibiting Jumbo's taxidermy skin (stretched to be larger than life), and skeleton as "Double Jumbo." The living Alice, billed as "Jumbo's Widow," was exhibited next to the remains (Saxon 1983, xxxiv).

Jumbo's circumstances illustrate how little was known about keeping bull elephants in early zoos and circuses. Circuses wanted bull elephants because their enormous size was such an attraction to spectators. Some circuses advertised certain cows as bulls. Zoos wanted bulls for breeding. However, neither circuses nor zoos understood the physiological condition called *musth* or comprehended why a young bull would suddenly become an unpredictably dangerous animal (Roocroft and Zoll 1994). These misunderstandings often led to killing or heavily chaining these "rogue" elephants (Lewis and Fish 1978). Although Hagenbeck successfully trained some bulls to perform, he preferred to import and sell cows because of the difficulties with adult males (Hagenbeck 1910).

Animal handling and training techniques used in zoos often were adopted from circus experiences. Some early zoo superintendents and directors had been circus animal trainers. Among these was Sol Stephan, Cincinnati Zoo superintendent from 1886 to 1937 (Ehrlinger 1993; Etter and Etter 1995). Barnum and Bailey animal trainer William Blackburne served as head keeper at the National Zoo from 1891 to 1944 (Hanson 2002). Circus people had experience in exotic animal handling, care, nutrition, breeding, and transport as well as knowledge of animal illnesses unfamiliar to many livestock veterinarians (Etter and Etter 1995).

Through the early 1900s, as the number of metropolitan zoos grew and new transportation options became available, public access to close encounters with wild animals became easier. Circuses continued to draw large crowds with performing elephants. Yet, the public no longer had to attend a circus to see an elephant; many major zoos exhibited them. Some zoo administrators and civic leaders believed a zoo was not a zoo without an elephant (Hanson 2002). Zoos also held elephant performances not unlike those seen at circuses. According to historian Elizabeth Hanson (2002), the major difference between animal performances at zoos and circuses or amusement parks was the venue, not the performance itself. Zoo animal performances were in quieter park settings rather than on dusty midways or under tents. Zoos, however, generally could not afford to purchase outright or house as many elephants as some circuses maintained.

Although both institutions incorporated elements of education and entertainment, zoos and circuses promoted themselves very differently. The circus continued to emphasize entertainment while new zoos emphasized educating visitors. Some zoos, such as the New York Zoological Park, Philadelphia Zoological Garden, and the National Zoological Park, included missions of scientific advancement and wildlife conservation in their founding charters. As the twentieth century progressed, zoos became more and more oriented toward the education, science, and conservation components of their missions. Zoos hired staff to work exclusively in these areas. Many major zoos were nonprofit organizations, with at least part of their revenue provided by municipalities (Bell 2001; Kisling 2001). By comparison, the traveling circuses received no public funding and continued to focus on profit-generating entertainment.

By the 1980s, zoo-circus collaborations began to dissolve, as zoos and circuses divided on two fundamental issues: standards for humane care of animals and willingness to portray animals in anthropomorphic and noneducational ways. Public perception has also played a role in the decrease in zoo-circus collaboration. For example, in 1998, as part of a national letter-writing campaign to stop elephant "abuse" and "exploitation" in circuses, People for the Ethical Treatment of Animals (PETA) directed thousands of letters to David Towne, the president of the American Zoo and Aquarium Association (AZA—now the Association of Zoos and Aquariums), demanding that zoos not provide elephants to circuses (Hutchins and Smith 1999). If public sentiment against elephants in circuses is accurately reflected in recent increases in anticircus litigation and legislative proposals (see the PETA website, www.circus.com; Animal Welfare Institute 2000), zoos that provide elephants to circuses could damage their reputation as institutions promoting animal welfare and conservation.

The gulf between zoos and circuses has widened even as both zoos and circuses have begun to breed animals in captivity and develop the expertise to care for them. Professional zoo societies such as AZA have in recent years developed animal care standards and philosophies expressed in their mission statements that differ from those of many circuses. Zoos have also begun to cooperate with one another on captive elephant population management and captive breeding research (Wiese 2000). At the same time, fewer circuses own elephants; instead, they contract with private elephant owners who provide elephants. With these arrangements in place, most present-day zoo-circus collaborations actually would have to be between zoos and private elephant owners.

The *AZA Standards for Elephant Management and Care*, adopted in March 2001, further limit zoo and circus collaboration (AZA 2001). These standards were developed in recognition of the unique care that elephants in captivity require to meet their behavioral, psychological, and physiological needs. The standards seek to ensure elephants are managed the same way in all accredited zoos, based on scientific findings and on the experience of AZA members. The standards cover temperature and space requirements; group composition and size; health and nutrition; reproduction; behavior management; staff organization and training; and conservation, education, and research. A circus must meet all AZA standards for elephant husbandry to qualify as a non-AZA participant in the AZA Species Survival Plan (SSP) breeding program for elephants and be able to receive elephants from zoos for breeding. Meeting these standards is difficult for circuses that spend long periods in transit, have limited space for elephants, and provide little conservation education. In addition, because circuses rely on their elephants to participate in performance tours, this may preclude them from loaning elephants to zoos for breeding purposes, although the transfer of frozen semen remains an option. Finally, circuses have relied on their own success in breeding elephants to ensure that they will always have performing elephants. Following SSP recommendations might hamper that ability by dictating to a circus which elephants should be bred, possibly breaking up a circus's own productive but genetically overrepresented breeding pair to diversify more greatly the entire captive offspring gene pool. According to Mike Keele, chair of the AZA Elephant Taxon Advisory Group / SSP, as of August 2007, no circus had applied to AZA for participant status in the SSP (M. Keele, personal communication, 2007). Today transfer of live elephants or gametes from AZA member zoos to circuses (or any nonmember institution) is nearly impossible. However, in one instance, a circus has recently loaned an elephant to an AZA-accredited zoo. If breeding occurs, the calf will belong to the zoo (M. Keele, personal communication, 2007). As in the early days of U.S. zoos, the circus benefits from the free animal care.

The AZA standards intentionally set AZA-accredited institutions apart from "roadside animal attractions," assuring the public that these standards provide a higher ethical bar for the care of elephants than do non-AZA elephant-holding institutions, including circuses. The importance of public perception of zoos and circuses has historically invited legal challenges, new regulations, media campaigns, and other activism aimed at negatively affecting attendance at these institutions. Zoo membership, public attendance, and other income-generating activities are strongly influenced by public attitude and perceptions. Animal welfare is a recurring area of public scrutiny.

Animal Welfare

Ever since Henry W. Bergh, the first president of the American Society for the Prevention of Cruelty to Animals, charged that P. T. Barnum's elephant trainers used prods and hot irons during training in the 1870s (Kunhardt 1995), the public has been suspicious about treatment of elephants in circuses. On April 13, 1883, Barnum wrote to Smithsonian Secretary Baird that he was unable to send the complete carcass of a dead elephant to the Smithsonian because, "The elephant was cut up as soon as possible to prevent Bergh's men from looking for wounds" (Saxon 1983, 233). Circuses sometimes found it difficult to provide for the elephant's well-being because of the conditions inherent in touring. For example, in the 1930s, animal trainer Clyde Beatty noted that elephants shivered severely during winter circus engagements in the northern United States, despite blankets wrapped around their backs and bellies and leather boots covering their legs (Beatty and Wilson 1941). There have been recent documented cases, however, of elephant abuse during transport. In 1997, Heather, a King Royal Circus elephant, died en route from Colorado to Texas from the heat inside her trailer, which she shared with two other elephants and eight llamas (Associated Press 1998). The surviving elephants were confiscated by police and relocated to the Rio Grande Biological Park, a zoo in Albuquerque, New Mexico.

Although modern zoo elephants usually live in relatively stable and predictable environments, circus elephants are much more frequently transported, handled, and exposed to new locations, to bright lights, to loud music, and to noise. To perform behaviors on cue, circus elephants must be well adapted to these changing circumstances. Zoo director and biologist Heini Hediger believed the circus was a better environment than the zoo to study the animal mind because circus conditions tested how well animals could adapt to novel situations (Hediger 1955/1968).

Hediger (1955/1968) and the *1998 Report into the Welfare of Circus Animals in England and Wales* (Circus Working Group 1998) both note that, unlike the zookeeper who cares for many animals during the normal workday, the circus animal caretaker works closely with a few animals around the clock. Methods of animal training in circuses are often criticized for their brutality or inhumanity (Johnson 1990; Kiley-Worthington 1990). Although Carl Hagenbeck claimed that he used positive reinforcement training (Hagenbeck 1910), many circus trainers before and since have resorted to negative conditioning using whips, chairs, and guns loaded with blanks to command the animals. Clyde Beatty wrote that American circus trainers forced, rather than coaxed, the animals to perform because American shows were often three-ringed, so every animal had to perform exactly on schedule. He observed that European circuses were slower paced, with only one performance ring and that Hagenbeck had the luxury of selecting and training only the best candidates from a seemingly unlimited animal supply (Beatty and Wilson 1941).

With elephants, a trainer must determine how to control the behavior of each member of the herd to obey commands, which allows the trainer to work safely around them. Circus and zoo elephant managers frequently remark about the need to become the dominant member of the herd (Lewis and Fish 1978; Roocroft and Zoll 1994). Elephant trainer George "Slim" Lewis asserted that fear of punishment and discomfort must be instilled in each elephant for it to respect the handler (Lewis and Fish 1978). During his 1920s–1970s career with circuses and performing zoo elephants, Lewis reported that handlers might beat the elephants before a performance. That such beatings continue routinely in circuses is a primary charge of animal protection groups and has resulted in lawsuits (Animal Welfare Institute 2000).

Changes in societal attitudes, including those of the circus animal trainer, have led to more humane management. In some instances, change is voluntary. For example, Feld Entertainment, Inc., the parent company of Ringling Bros., has produced standards and guidelines for animal care and management (Feld Entertainment 2002). The guidelines' preamble discourages punishment during animal training. In 2003, John Kirtland, then Feld's executive director of Animal Stewardship, explained that every new employee reviews the guidelines and the Animal Welfare Act as part of their training. Included in the training session is a statement that obligates animal crew or other employees to report concerns or ask questions about the care and treatment of the animals (J. Kirtland, personal communication). Contracted individuals and animal acts can be dismissed from the show on the basis of charges of mistreatment. In other instances, change is legislated. Finland, Sweden, and Denmark have banned some wild-animal acts. In the United States, many bills banning animal performances have been introduced at local, state, and federal levels. The only bills that have so far passed into law are at the county or municipality level, such as in Orange County, North Carolina, and Takoma Park, Maryland.

Circuses and zoos in the United States are regulated by the Animal Welfare Act and must be licensed by the U.S. Department of Agriculture (USDA) (Kohn 1994). The act's regulations (U.S. Code of Federal Regulations, Title 9, Chapter 1, sections 2.131, 3.125–3.142) set minimal space requirements to allow animals to make normal postural and social adjustments and include other regulations for facility design, feeding, watering, sanitation, transport, and ventilation. Animal handling and training must not cause trauma, behavioral stress, physical harm, or unnecessary discomfort. Food and water deprivation, sometimes used for veterinary, transport, or training purposes, is only permitted if the animal is receiving full dietary and nutrition requirements each day. In public exhibitions, whether at zoo or circus, barriers must assure the safety of the animals and public. Animals must have rest periods between performances. USDA has cited both zoos and circuses for violating the Animal Welfare Act in regard to elephants. The violations are usually deficiencies in veterinary care, inadequate security for the elephant and public, transportation violations, and handling that has resulted in stress or physical harm to the elephant (B. Kohn, personal communication, 2003).

In Great Britain, the Protection of Animals Act of 1911 makes it illegal to cause unnecessary suffering in animals, and the Performing Animals Act of 1925 requires anyone who trains or exhibits animals to be registered for inspection by their local authority. However, there are no articulated standards for animal care. Other laws concern safety of employees and the public, and animal transportation. New Zealand's government encourages appropriate elephant care through voluntary *Recommendations for the Welfare of Circus Animals and Information for Circus Operators.*

In 1998, the Royal Society for the Prevention of Cruelty to Animals (RSPCA) commissioned the Circus Working Group to evaluate animal welfare at circuses in England and Wales by providing recommendations to the All Party Parliamentary Group for Animal Welfare. The group, six animal protection and veterinary societies, as well as the Association of Circus Proprietors, found some animals showed indications of physical and psychological abnormalities. After reviewing circus performances, veterinary care, and other indicators of animal welfare, the group was divided between recommending a ban on having some or all nonhuman species in circuses or modifying existing legislation to include welfare standards (Circus Working Group 1998).

Despite emotionally charged arguments to the contrary, inadequate research makes it difficult to judge whether inhumane conditions confront elephants in circuses and zoos. Few peer-reviewed scientific articles in the applied ethology literature provide determinative criteria or experimentally evaluate stress, pain, or the behavioral repertoire. Studies of circus elephants have focused on elephant behavior before and after performances (Schmid 1995; Friend 1999; Friend and Parker 1999; Gruber et al. 2000). Little research has been conducted on the physiological or behavioral effects of handling, training, transport, and human-animal interaction. There is also a lack of published research on the effects of husbandry practices on zoo elephant well-being. Recently, however, the RSPCA commissioned a review of the zoo elephant literature with additional analyses of captive breeding records (Clubb and Mason 2002). Although the ensuing report's methods and conclusions have been the subject of intense debate among elephant managers (see, for example, Wiese and Willis 2004), the report is extremely critical of elephant care in European zoos and has been cited by the RSPCA in calling for an immediate phaseout of elephants in zoos.

Societal concerns, legislation, and initiatives of the animal caretakers have led to improvements in elephant welfare. This may be due as much

Circus elephants living in temporary pens or paddocks surrounded by electric wires.
Photograph courtesy of Michael Kreger, 1997

to caretakers and trainers wishing to provide the best care possible, as it is to their desire to enhance their public image. For example, many circus elephant managers are beginning to reduce the time animals spend on the traditional picket line, once 50–80% of each day (Friend 1998), chained side by side by one front and one rear foot (Schmid 1995). Instead, the elephants live in temporary pens or paddocks surrounded by electric wires. The pens permit some freedom of movement, ability to perform comfort behaviors, and social contact within a group. Another benefit of the use of pens is the reduction in stereotypic behavior, such as head bobbing (Schmid 1995; Friend and Parker 1999; Gruber et al. 2000). A criticism of circuses, however, is that elephants are forced or coaxed to perform behaviors they would not do ordinarily (Hirsch 1996; Kiley-Worthington 1990). Although critics charge that these behaviors are physically and psychologically stressful to the elephants (Johnson 1990; AWI 2000), circus elephant trainers have spoken and written about how they marvel at the physical abilities elephants can be taught to demonstrate, and how some elephants seem to "enjoy" performing (Lewis and Fish 1978; M. Kreger, personal observation). Research evaluating the critics' charges is lacking. As with zoos, circuses face occasional charges of cruelty, particularly during elephant training. Nevertheless, some elephant managers argue that humane animal training at both zoos and circuses encourages tameness, increases the human-animal

The partnership between elephants and their trainers at the circus encourages tameness, increases the human-animal bond, and is a form of environmental enrichment.
Photograph courtesy of Michael Kreger, 1997

bond, and is a form of environmental enrichment or occupational therapy that enhances the animal's well-being (Laule 1993; Roocroft and Zoll 1994; Kreger et al. 1998; Laule and Desmond 1998).

Messages

After animal welfare, a second major difference that hinders collaboration between zoos and circuses is the disparity between organizational missions. All circuses with elephants and some zoos with elephants offer animal shows. Showmanship increases visitor attendance and profits at both institutions. The zoo philosophy encouraged by professional associations such as the AZA emphasizes education. In most, but not all zoos that include elephant demonstrations, the animals do not perform "tricks" to entertain, they perform "behaviors" to educate the public about their physical traits and natural behavior, and in the demonstrations, these activities are explicitly attached to a conservation message (Hutchins and Fascione 1991). In circuses, largely similar activities are performed as tricks, used primarily to awe and entertain the public. Some critics view these as exploitative (Kiley-Worthington 1990; Hirsch 1996). The behaviors demonstrated in zoos and circuses are often the same, but the contexts are different. For example, the elephant is taught to lift its trunk over its head

In the wild, elephants may stand on their hind legs to reach tree limbs or to mate. The prolonged use of this stance in circus performances is seen by some as unnatural and exploitative but seen by others as an entertaining display of skills. *Photograph courtesy of Michael Kreger, 1997*

on command, a behavior elicited by zoo veterinarians to examine its mouth. In circuses, the same behavior is used as a salute. Other behaviors, such as headstands, are purely circus.

Both zoos and circuses often offer elephant rides. Here the similarities outweigh the differences. Both must comply with Animal Welfare Act regulations regarding care, including rest periods between rides. Both often contract with private elephant owners to provide the elephants and manage the ride activity, particularly if the ride is a seasonal attraction.

The message of the ride varies among individual institutions, not by whether the establishment is a zoo or circus. The message may be explicit, such as signage, or implicit, based on public perceptions experienced when seeing or touching a live elephant. Some zoos and circuses now place educational signs about the elephants on fences surrounding the ride, or personnel walking the elephants explain wild-elephant conservation challenges to riders. Although zoos and circuses may justify elephant rides as income-generating, educational, engendering compassion toward these animals, and a form of exercise for the elephants, critics view the rides as exploitative and a demonstration of man's dominion over wild animals (Hutchins and Fascione 1991).

Circuses based in the United States must include educational messages if they wish to travel to foreign countries with elephants. According to Anna Barry, senior biologist in the Division of Management Authority of the U.S. Fish and Wildlife Service, because Asian elephants are listed as endangered and African elephants are listed as threatened under the U.S. Endangered Species Act, circuses must supply educational information to the Fish and Wildlife Service when applying for a permit to travel to a foreign country (A. Barry, personal communication, 2003). The circus submits copies of handouts, photographs of signage, or a copy of the script used orally to inform the public of the elephant's ecological role and conservation needs.

Zoos and circuses still collaborate and share ideas on elephant management and husbandry. In the United States, circus elephant handlers have joined their zoo counterparts in the Elephant Managers Association (EMA), which provides professional guidance, conferences, and a journal for elephant managers (Grayson 2006). Through EMA, elephant trainers and managers from zoos, circuses, sanctuaries, and private enterprises collaborated with elephant veterinarians, behaviorists, and researchers to produce the *Elephant Husbandry Resource Guide* (Olson 2004). The guide provides husbandry information based on scientific research and practical experience, including that gained in traveling shows. The guide advises on topics ranging from behavior management to nutrition and transportation.

The International Elephant Foundation (IEF) is another organization with both zoo and circus ties. IEF identifies and funds *in situ* and *ex situ* elephant conservation research. Elephant-holding zoos and circuses have contributed to this fund since 1998 (Grayson 2006). The AZA Elephant Taxon Advisory Group has recently developed a partnership with IEF to review and fund projects that meet the AZA's elephant conservation objectives (Grayson 2006).

The largest circus in the United States, the Ringling Bros. and Barnum & Bailey Circus, has participated in some elephant conservation and education efforts. It joined zoo and wildlife societies in drafting and successfully lobbying the U.S. Congress to pass the Asian Elephant Conservation Act of 1997, which provides financial support for conservation programs in that species' home range. It has also provided funding to elephant conservation and rehabilitation organizations. Ringling Bros.'s show program, sold at performances, often includes sections about the biology, conservation, and training of the animals. In 1995, Feld Entertainment established the Ringling Bros. Center for Elephant Conservation (CEC) for Asian elephant breeding, conservation, study, and retirement. The CEC is a 200–acre site in central Florida. It contains paddocks for bulls, mothers, and offspring, as well as a barn with an office, laboratory, and observation room. Personnel from various zoos have consulted the CEC for information about facility design, husbandry, and calf rearing. Ringling Bros. has also assisted zoos with fundraising and provided temporary housing for Miami Metrozoo's elephants after a hurricane destroyed the zoo's elephant enclosure.

Critics have argued that the real purpose of the CEC is to ensure that the circus will always be supplied with performing elephants. However, the CEC public relations material claims that the circus has the largest and most diverse gene pool of Asian elephants outside of Asia. Perhaps it is for this reason that one of the areas of nonconsensus during the 1999 Elephant Planning Initiative for AZA accredited institutions was future collaboration between zoos and circuses (Hutchins and Smith 1999). Those who favor providing elephants or their gametes to circuses and encouraging circuses' participation in the SSP believe that the collaborations could increase the number of elephants in the captive breeding population, increase the amount of space available for SSP elephants, foster husbandry and management information exchange, and encourage circuses to participate more in elephant conservation, research, and educational programs (Hutchins and Smith 1999).

Some AZA member institutions also support the middle option, limited collaboration between zoos and circuses. This would allow zoos to work with circuses case by case. The CEC may be the only circus-managed facility of its kind, and information is already being exchanged between zoos and CEC staff. Thus Ringling Bros. may come closer than any other circus to sharing objectives with AZA-accredited zoos. If a circus demonstrates a commitment to elephant field conservation, conservation education, scientific research, and animal welfare (demonstrated through research into controversial areas such as stress physiology and behavior during transport or performances), AZA-accredited zoos may be willing to work with it. Conversely, a circus is not a zoo. Its mission of providing family entertainment by taking the show from town to town precludes it from meeting all of the standards established by AZA for maintaining elephants in a zoo. Zoos must accept this distinction and decide how important or necessary collaborations are to them. If circuses demonstrate, as they claim, that their Asian elephant gene pool is greater than that of zoos and that their elephants live longer or have a greater reproductive rate, it may be the zoos that will need to provide incentives to encourage circus participation in zoo elephant programs.

The circus industry in the United States has evolved from many large tented spectacles to a few small-tented shows and a few large high-tech arena productions. This is due in part to the growth in the number and quality of zoos, to economic or regulatory issues, and to societal concern for the treatment of wild animals in captivity. Animal management, facilities, and resources vary from circus to circus. Although circuses are primarily institutions providing entertainment for revenue, there are still opportunities for collaborations with zoos in conservation, education, and sharing of veterinary and husbandry information. This collaboration may be necessary to develop self-sustaining captive herds in North America (Wiese 2000). However, given the AZA elephant care standards, it is likely that zoos and circuses in North America will manage their populations independently, which could lead to a loss of genetic diversity, or to two nonviable captive populations over the long term. Public perception will continue to play a role in the relationship between zoos and circuses. Allegations of elephant abuse, whether in circuses or zoos, will continue to spark debate and result in litigation or regulation. Collaboration between zoos and circuses has occurred in the United States since the founding of the earliest zoos. Forms of collaboration have changed over time, particularly with respect to the exchange of live elephants and gametes. However, the exchange of information continues to benefit both institutions. If animal welfare concerns, especially, are investigated and addressed in a scientific manner, it is likely that at least some circuses will be well positioned to strengthen their relationships with zoos. This can only serve to benefit the captive elephant population, and, through targeted commitment of income generated at zoos and circuses, conservation of wild-elephant populations.

Acknowledgments

The author would like to thank Catherine Christen, Robert Hoage, Nigel Rothfels, Chris Wemmer, and an anonymous reviewer for their insightful suggestions on and improvements to this manuscript. The views expressed in this chapter are those of the author and do not represent those of the U.S. Fish and Wildlife Service.

References

Animal Welfare Institute. 2000. Performing elephants: Dying to entertain us. *Animal Welfare Institute Quarterly* 4: 17.

Associated Press. 1998. Albuquerque acquires elephants, llamas in likely circus settlement. November 26. www.amarillonet.com/stories/112698/new_llamas.shtml.

Association of Zoos and Aquariums (AZA). 2001. *AZA standards for elephant management and care.* Silver Spring, MD: AZA.

Barnum, P. T. 1889. *Struggles and triumphs; or, sixty years' recollections of P. T. Barnum, including his golden rules for money-making, illustrated and brought up to 1889, written by himself.* Buffalo, NY: Warren, Johnson.

Beatty, C., and Wilson, E. 1941. *Jungle performers.* New York: Robert M. McBride.

Bell, C. E. (ed.). 2001. *Encyclopedia of the world's zoos.* London: Fitzroy Dearborn.

Brown, R. J. 1987. The elephant comes to America. *Collectible newspapers*, April. www.historybuff.com/library/refelephant.html

Circus Working Group. 1998. *A report into the welfare of circus animals in England and Wales.* Horsham, UK: Royal Society for the Prevention of Cruelty to Animals (RSPCA).

Clubb, R., and Mason, G. 2002. *A review of the welfare of zoo elephants in Europe.* Horsham, UK: Royal Society for the Prevention of Cruelty to Animals (RSPCA).

Colver, A. 1957. *Old Bet.* New York: Alfred A. Knopf.

Ehrlinger, D. 1993. *The Cincinnati Zoo and Botanical Garden: From past to present.* Cincinnati, OH: Cincinnati Zoo and Botanical Garden.

Etter, D., and Etter, C. 1995. *The Denver Zoo: A centennial history.* Boulder: Roberts Reinhart.

Feld Entertainment, Inc. 2002. *Standards and guidelines for animal care and management* Vienna, VA: Feld Entertainment.

Fenner, M. S., and Fenner, W., (eds.). 1970. *The circus: Lure and legend.* Englewood Cliffs, NJ: Prentice-Hall.

Flint, R. W. 1996. American showmen and European dealers: Commerce in wild animals in nineteenth-century America. In R. J. Hoage and W. A. Deiss (eds.), *New worlds, new animals: From menagerie to zoological park in the nineteenth century* (pp. 97–108). Baltimore: Johns Hopkins University Press.

Friend, T. 1998. Circuses and circus elephants. In M. Bekoff, and C. A. Meaney (eds.), *Encyclopedia of animal rights and animal welfare* (pp. 107–109). Westport, CT: Greenwood Press.

Friend, T. H. 1999. Behavior of picketed circus elephants. *Applied Animal Behavior Science* 62: 73–88.

Friend, T. H., and Parker, M. L. 1999. The effect of penning versus picketing on stereotypic behavior of circus elephants. *Applied Animal Behavior Science* 64: 213–225.

Grayson, P. 2006. Pulling elephants from the brink. *AZA Communiqué.* February: 21–22.

Gröning, K., and Saller, M. 1998. *Elephants: A cultural and natural history.* Cologne: Könemann Verlagsgesellschaft.

Gruber, T. M., Friend, T. H., Gardener, J. M., Packard, J. M., Beaver, B., and Bushong, D. 2000. Variation in stereotypic behavior related to restraint in circus elephants. *Zoo Biology* 19: 209–221.

Hagenbeck, C. 1910. *Beasts and men.* Trans. and ed. H. S. R. Elliot and A. G. Thacker. London: Longmans, Green, and Co.

Hanson, E. 2002. *Animal attractions: Nature on display in American zoos.* Princeton, NJ: Princeton University Press.

Hediger, H. 1968. *The psychology and behaviour of animals in zoos and circuses.* Trans. G. Sircom. New York: Dover Press. Original work published 1955.

Hirsch, A. 1996. Circus sideshow. *The Baltimore Sun.* March 18.

Horowitz, H. L. 1996. National Zoological Park: "City refuge" or zoo? In R. J. Hoage and W. A. Deiss (eds.), *New worlds, new animals: From menagerie to zoological park in the nineteenth century* (pp. 126–135). Baltimore: Johns Hopkins University Press.

Hutchins, M., and Fascione, N. 1991. Ethical issues facing modern zoos. In *Proceedings of the American Association of Zoo Veterinarians Annual Conference 1991* (pp. 56–64). Yulee, FL: American Association of Zoo Veterinarians.

Hutchins, M., and Smith, B. 1999. *AZA elephant planning initiative: On the future of elephants in North American zoos.* Silver Spring, MD: AZA.

Johnson, S. P. 2001. Hornaday, William. In C. E. Bell (ed.), *Encyclopedia of the world's zoos* (pp. 570–571). London: Fitzroy Dearborn.

Johnson, W. 1990. *The rose-tinted menagerie.* London: Heretic Books.

Kiley-Worthington, M. 1990. *Animals in circuses and zoos: Chiron's world?* Essex, UK: Little Eco-Farms.

Kisling, V. N., Jr. 2001. Zoological gardens of the United States. In V. N. Kisling, Jr. (ed.), *Zoo and aquarium history: Ancient animal collections to zoological gardens* (pp. 147–180). New York: CRC Press.

Kohn, B. 1994. Zoo animal welfare. *Revue Scientifique et Technique* 13: 233–245.

Kreger, M. D., Hutchins, M. and Fascione, N. 1998. Context, ethics, and environmental enrichment in zoos and aquariums. In D. J. Shepherdson, J. D. Mellen, and M. Hutchins (eds.), *Second nature: Environmental enrichment for captive animals* (pp. 59–82). Washington, DC: Smithsonian Institution Press.

Kunhardt, P. B., Jr., Kunhardt, P. B., III, and Kunhardt, P. W. 1995. *P. T. Barnum: America's greatest showman.* New York: Alfred A. Knopf.

Laule, G.1993. Using training to enhance animal care and welfare. *Animal Welfare Information Center Newsletter* 4: 2, 8–9.

Laule, G., and Desmond, T. 1998. Positive reinforcement training as an enrichment strategy. In D. J. Shepherdson, J. D. Mellen, and M. Hutchins (eds.), *Second nature: Environmental enrichment for captive animals* (pp. 302–313). Washington, DC: Smithsonian Institution Press.

Lewis, G., and Fish, B. 1978. *I loved rogues: The life of an elephant tramp.* Seattle: Superior.

McClung, R. M. 1993. *Old Bet and the start of the American circus.* New York: Morrow Junior Books.

Olson, D. (ed.). 2004. *Elephant husbandry resource guide.* Indianapolis: Indianapolis Zoo.

Rothfels, N. 2002. *Savages and beasts: The birth of the modern zoo.* Baltimore: Johns Hopkins University Press.

Roocroft, A., and Zoll, D. A. 1994. *Managing elephants: An introduction to their training and management.* Ramona, CA: Fever Tree Press.

Saxon, A. H. 1983. *Selected letters of P. T. Barnum.* New York: Columbia University Press.

Schmid, J. 1995. Keeping circus elephants temporarily in paddocks: The effects on their behaviour. *Animal Welfare* 4: 87–101.

Sheak, W. H. 1922. The elephant in captivity. *Natural History.* Sept.-Oct. www.naturalhistorymag.com/.

Wharton, D. 2001. Central Park Wildlife Center. In C. E. Bell (ed.), *Encyclopedia of the world's zoos* (pp. 220–223). London: Fitzroy Dearborn.

Wiese, R. J. 2000. Asian elephants are not self-sustaining in North America. *Zoo Biology* 19: 299–309.

Wiese, R. J., and Willis, K. 2004. Calculation of longevity and life expectancy in captive elephants. *Zoo Biology* 23: 365–373.

10 WHY CIRCUSES ARE UNSUITED TO ELEPHANTS

LORI ALWARD

A cursory Internet search quickly demonstrates how hotly contested the use of elephants in circuses has become. Organizations[1] devoted to the promotion of animal welfare and the protection of animal rights argue that the use of elephants in circuses should be banned for various reasons. They contend that the elephants are caused egregious psychological and physical harm that no amount of reform can eliminate; that elephants' life spans are shortened if they are forced to live under the conditions circuses necessitate; and that circuses have contributed to the endangerment of elephant species and the destruction of elephant families. Pro-circus organizations[2] argue that elephants live longer in the circus than in the wild because they have excellent veterinary care and are not in danger from poachers or from encroachment on their habitat; that elephants are happy in the circus because they love to travel and to perform; and that although elephants were sometimes treated abusively in the past, contemporary circuses are careful to promote their elephants' physical and psychological well-being. Moreover, they argue, a circus without elephants is not a circus: people love to see the elephants and tradition demands that circuses continue to use elephants. In addition, far from contributing to the extinction of the species, these advocates hold that circuses help to preserve the species by breeding elephants, by providing habitat for elephants, and by educating humans about elephants, so that we will be motivated to work for preservation of the species.

One thing is immediately apparent when one surveys the popular arguments: no one, on either side, assumes that elephants are not owed some moral consideration. By responding to animal rights activists in the way that they do, by saying defensively that captive elephants are *not* beaten, do *not* die at a younger age than wild elephants, do *not* suffer in large numbers from tuberculosis or foot infections, and so on, pro-circus organizations and their representatives are saying that it matters that we do not beat elephants and that we do not knowingly or intentionally endanger their lives or cause them to suffer unnecessarily. In short, pro-circus folks admit that elephants are what philosophers call "morally considerable," which means that we cannot treat elephants as mere things that are owed no moral consideration, that we have some moral responsibility toward them, at least if we are in a position to affect their well-being through our actions. Thus, I too shall simply assume that elephants are morally considerable. So, rather than defending the claim that elephants are morally considerable, I shall explore what the theoretical basis of that considerability is and whether elephants' moral considerability entails that elephants should not be used in circuses.

First, I examine the two moral theories most commonly used by animal rights activists, modern utilitarianism (specifically, its animal liberation elements), and animal rights theory, as well as a relative newcomer to animal rights activism, natural law theory. I argue that none of these theories provides an adequate basis for the moral considerability of nonhuman animals. Then I consider the question of elephants in circuses in light of an alternative, called the *capabilities approach*. Capabilities approach was formulated in the context of international development and so affords a solid basis for development of public policy relating to this issue (Sen 1999; Nussbaum 2000). But as one of its principal architects, Martha Nussbaum, has shown (Sunstein and Nussbaum 2004; Nussbaum 2006a, 2006b), and as I explore here, the capabilities approach can provide a basis for the moral considerability of nonhuman animals, while avoiding some of the problems faced by utilitarianism, rights theory, and natural law theory. Finally, I discuss how capabilities approach can be applied to elephants. A happy consequence of the adoption of a capabilities approach is that we can sidestep some of the most hotly contested questions, such as whether elephants die young if kept in circuses. Furthermore, I contend, capabilities approach as applied to elephants entails that circuses should not use elephants, that circuses are unsuited to elephants. It is a separate, and perhaps more difficult question, whether we have a duty to go out of our way to promote the well-being or protect the rights of elephants by more active means, for example, by working to preserve elephant habitat. I will not address this question here because my primary concern is with elephants in circuses and clearly, if elephants are kept in circuses, their treatment and living conditions have a direct effect on the well-being of the elephants kept there.

Animal Liberation, Animal Rights, and Natural Law

Utilitarianism is the moral theory that most commonly grounds arguments about our moral responsibilities to animals, as is exemplified by the arguments about the use of elephants in circuses. In part, this is because many debates about public policy are framed as utilitarian arguments, but probably more important, with respect to debates about how we ought to treat nonhuman animals, is the success and influence of Peter Singer's *Animal Liberation* (Singer 1975), which grounds a defense of the moral considerability of other animals in classic utilitarianism. Utilitarianism is a consequentialist theory, which means that no action or policy is considered right or wrong in itself. Instead, actions or policies are right or wrong because they have good or bad consequences. The consequences that matter are the consequences to all moral agents or patients affected by the action or policy. In some classic theories, such as the utilitarianism of John Stuart Mill (1861/2001), one is required to consider the effect that one's actions would have on any human being affected by one's action or by the adoption of a particular rule. Following a suggestion of nineteenth-century philosopher Jeremy Bentham (one of Mill's mentors), Singer argues that we ought to expand the scope of the class of morally considerable beings to include all sentient beings because classic utilitarianism prescribes that we act to maximize pleasure and minimize suffering, and it is arbitrary to exclude from the scope of our moral concern animals who are capable of experiencing pleasure and pain but are not human, if our only reason for excluding them is that they are not human (Singer 1975). This seems clearly correct, and so utilitarianism is a very good candidate for grounding moral duties to nonhuman animals.

Of course, not all animals will be captured within the scope of this theory. Protozoans and most mollusks will probably be left out. Clearly elephants will fall within its scope, and so will we, which brings up the first of two primary and somewhat standard objections to animal liberation's version of utilitarianism (also see Regan 1983). Utilitarianism requires that we consider not just the consequences to elephants of using elephants in circuses, but the consequences to all sentient beings affected by the practice, some of whom are spectators at the circus; people who own or otherwise profit from the circus; people who are outraged by the use of elephants by circuses; people who are concerned with the survival of elephants in the wild; people who write articles about elephants in circuses, and so on. In other words, according to this counterargument, no matter how much harm might be done to elephants in circuses, if that harm is counterbalanced by humans' pleasure, then using elephants in circuses is morally permissible. If humans' pleasure significantly outweighs elephant suffering, then utilitarianism would not merely permit the use of elephants in circuses but would actually require that the practice be continued because the practice maximizes pleasure and minimizes pain.

This consequence of the theory should clarify why rights theorists, such as Tom Regan, have a problem with utilitarianism (Regan 1983). Rights theorists believe that some actions are not morally permissible, regardless of how much pleasure we may derive from them. They argue that we need a theory of rights that states unequivocally that certain actions are forbidden because they violate someone's rights, regardless of what benefit someone might derive if the action is performed. Therefore, for example, our right to life is protected by laws forbidding murder, and while there are acceptable legal defenses against charges of murder, such as self-defense, these acceptable defenses do not include the contention that the overall balance of utility has required that John Doe be killed.

The second objection to utilitarianism that I wish to consider is the objection that utilitarianism assigns moral value to the wrong things. What has intrinsic value and is therefore the proper object of moral deliberation is a particular type of experience, rather than the individual having the experience. Rights theorists believe that this is precisely backward: what has the highest moral value is the experiencing subject, not the experiences that the subject has. A moment's reflection will reveal that these two objections are closely related: The reason utilitarianism does not allow that there are some actions that are categorically forbidden or required is that what has the greatest moral importance according to the theory is not the experiencing subject but the subject's experience of pleasure or pain. Therefore, some individuals' experiences of pleasure, if sufficiently great in quality or quantity, can override the pain suffered by other individuals. This is a consequence of the utilitarian view that rights theorists find repugnant because it would seem to justify or even require morally reprehensible practices and actions, such as slavery, apartheid, and torture, in those cases where the overall balance of utility favors continuing the practice or performing the action (see Rawls 1971).

These classic objections to utilitarianism seem to me sufficient theoretical reason to search for a better moral theory to ground our moral obligations to elephants. In addition, there is a pragmatic reason. In calculating the overall utility of a particular practice, such as transporting elephants in closed train cars, human beings are all too likely to underestimate the pain suffered by elephants and overestimate the pleasure experienced by humans who benefit from this practice. Although it is true that human beings can distort any moral theory to justify an immoral action, it is wise, in trying to develop a theory to ground public policy, to adopt something less liable to this sort of distortion, if possible. Utilitarians claim, of course, that precise measurements of aggregate utility and disutility are possible, but this has never been persuasively demonstrated, and many moral theories offer standards which are more easily applied and whose violation is more clearly demonstrated than utilitarianism. Although possession of such standards is not a sufficient reason to adopt a moral theory, other things being equal, it is a reason to prefer one theory to another.

Animal rights theorists argue that it is a bizarre feature of animal liberation that, while it includes all sentient beings within the scope of its theory, it includes them only because those are the only beings capable of experiencing pleasure and pain and therefore of having interests, and those categories—pain, pleasure, and interest—are the morally important ones. The sentient beings themselves, it would seem, are not very significant, except insofar as they experience the morally relevant sensations. Rights theorists believe we ought to value sentient beings more highly than their sensual experiences, so that any pleasure that human beings might experience in watching an elephant perform in a circus cannot outweigh any wrongs done to the elephants kept in the circus. That is because, according to animal rights theory, the elephant has an intrinsic value that grounds the elephant's right to certain types of treatment and that right cannot be disregarded simply because to do so gives others pleasure. Different rights theorists may ground the elephants' rights differently: some may say that the elephant is a person or is person-like, while others, such as Tom Regan, might argue that the elephant is the subject-of-a-life (Regan 1983). I will focus on the latter as it is the most influential within the animal rights movement itself. The concept of a subject-of-a-life is broader in scope than that of a person, so a greater number of nonhuman animals will be included in the theory, which makes it a better theory, in the view of most activists, than other animal rights theories.

Although animal rights theory is appealing because it does value individual animals in a way that utilitarianism does not and it gives stronger grounds for forbidding certain kinds of harm to animals than utilitarianism, there are problems with the theory nonetheless. I will examine two primary criticisms. The first arises out of ecofeminist theory, which tends to be opposed to any theory deemed to be either anthropocentric or androcentric (focused on men). Animal rights theory is anthropocentric, according to ecofeminist Deborah Slicer, because it values animals and claims that they are morally considerable only insofar as they share with us certain morally salient characteristics, such as rationality or subjectivity (Slicer 1996). This is a good criticism of animal rights theory only if, by concentrating on features some animals share with us, animal rights theory neglects other animals possessing some—but not those—morally salient characteristics. In other words, the criticism is valid if some animals that ought to be accorded rights are not because we have been blinded to their intrinsic moral worth by our exclusive focus on human characteristics. Slicer's objection does seem legitimate because many animals possessing some morally salient characteristics would nevertheless be left out of the scope of animal rights theory's concept of personhood or even of its much broader concept of the subject-of-a-life. Here is how Regan defines the latter concept:

> To be the subject-of-a-life . . . involves more than merely being alive and more than merely being conscious . . . [I]ndividuals are subjects-of-a-life if they have beliefs and desires; perception, memory, and a sense of the future, including their own future; an emotional life together with feelings of pleasure and pain; preference- and welfare-interests; the ability to initiate action in pursuit of

their desires and goals; a psychophysical identity over time; and an individual welfare in the sense that their experiential life fares well or ill for them, logically independently of their being the object of anyone else's interests. Those who satisfy the subject-of-a-life criterion themselves have a distinctive kind of value—inherent value—and are not to be viewed or treated as mere receptacles. (Regan 1983, 243)

If any nonhuman animals fit these criteria, elephants are certainly among them. Nonetheless, I would still argue against using animal rights theory to ground public policy, including for elephants, because as Slicer's criticism highlights, animal rights theory's exclusion of many animal species renders it too narrow in scope. This part utilitarianism gets right: If an animal is conscious and capable of experiencing pain, we have an obligation to avoid inflicting unnecessary pain on that animal. This is a broader principle than would be justified by the subject-of-a-life criterion, which is the broadest of all the criteria that ground various rights theories. Thus, there could be many animals that are morally considerable but are not recognized as such by animal rights theory. This is a serious theoretical flaw and is sufficient reason to seek a different theory.

In addition, because we are trying not only to discover the correct moral theory to ground our moral obligations to nonhuman animals but also to develop a workable theory to ground public policy regarding elephants, it matters that not everyone will agree on which animals are persons or subjects-of-a-life. This disagreement is unavoidable, it seems, because the central moral categories of personhood and subject-of-a-life rely on subjective characteristics that are knowable, ultimately, only by the subject and that can at best only be inferred by individuals belonging to a different species. With respect to some animals, such as elephants, we feel fairly secure in our inferences, but with respect to many other species, we are less certain—and less likely to agree about—whether the members of the species are subjects-of-a-life. An animal rights theorist might argue that we should always err on the side of caution, a sentiment I agree with but which might not persuade many people. Because we are trying to establish a basis for public policy, we want a theory with broad appeal.

A possible alternative theory is *natural law theory*. This theory has not been widely used by animal rights activists but is nonetheless worth considering for several reasons. First, natural law theory is the moral theory endorsed by Matthew Scully in his 2002 book, *Dominion: The Power of Man, the Suffering of Animals, and the Call to Mercy*. This book generated a stir in the animal rights community, and so natural law theory may become more influential among individuals working to promote the interests of elephants. Second, natural law theory would not necessarily base other animals' moral considerability on characteristics that they share with us but would consider the nature of each species separately and derive the good for each type of animal from the nature of its species. Third, many legal systems, including in the United States, are grounded, at least in part, in natural law tradition. To ground our moral responsibility toward other animals on natural law might seem a promising basis for public policy. Nonetheless, I argue that Scully's version of natural law theory is problematic because it requires a teleological, or purpose-driven, view of nature and therefore ought to be rejected.

This is how Scully characterizes the key insight of natural law theory:

> All moral truth arises from the nature of things, true in themselves and in crucial respects accessible to reason. Every being has a nature, and that nature defines the ends and ultimate good for which it exists. In discerning these purposes we perceive what that being is, what it can do, what it must do to find its completion and fulfillment, and therefore what its moral interests are and how they may be advanced or hindered . . . That which advances a being onward to its natural fulfillment is good. That which frustrates or perverts its natural development is bad. (299–300)

This is a view of nature most scientists will not find congenial. Scully might think that this is irrelevant, but of course, it is not. A moral theory should not fly in the face of current scientific theory. A moral theory that is incompatible with what our best scientific theory tells us is true is a highly suspect theory.

Of course, scientists sometimes use teleological language, especially biologists or environmental scientists. It sometimes makes sense to talk about the purpose of an organ, for example, or the goal of an organism, or how well a system is functioning. Such uses of teleological language are often unavoidable and are generally regarded as benign, provided one keeps in mind that one is not assuming that nature is or has a Designer that sets ends for individual beings, based on some sort of essential nature that classes of individuals are supposed to have. But Scully's teleology is not so benign. He tells us that natural law "asserts what philosophers call a teleological view of the moral universe with a detectable structure, direction, and broad design beyond our power to alter or escape" (301). Hence, the universe has a purpose and so does everything in it. We can discover how we ought to treat individuals by discovering their ends, which will tell us what is good for those individuals and what will contribute to their flourishing. This is true of human beings and other beings as well. So we do not set our own ends; they are set for us, by a power that is independent of us.

This is an old-fashioned conception of teleology, to say the least; yet

Scully believes it is the only moral game in town. He claims that natural law "provides the only rational grounds I know of for claiming any one thing better than another, without reliance on religious belief or intuition or the constructs of theory" (301). It is understandable that Scully does not want to base his moral theory on any particular religious belief because he avowedly wishes to establish a convincing ground for public policy, and it is admirable that he does not wish to base his argument on intuition alone. Yet, it is not clear what is wrong with consulting "the constructs of theory" because natural law is a theory, or so I have always supposed. Scully disagrees, however: "Natural law is just that, a law and not a theory. If anything, it serves as a kind of anti-theory" (301). But of course, this is nonsense, as we can point to a variety of natural law theories, written over the course of several centuries. The problem with Scully's version of natural law (theory) is that he has chosen one that is incompatible with contemporary science, which is unfortunate, given that natural law theory wants to derive values from nature and so ought to rely on the findings of contemporary scientists, rather than dismissing them as irrelevant.

In conclusion, we must reject Scully's version of natural law theory, while recognizing that the theory has several virtues. Five in particular are especially apt for any theory that could effectively ground public policy regarding our treatment of elephants and other animals. First, natural law theory is not anthropocentric in the sense that Slicer worries about because our moral responsibilities to members of different species would be based on the morally salient characteristics shared by all or most members of a species. These characteristics might or might not be shared with human beings. Hence, a second virtue: All animals could be included within the scope of natural law theory. A third virtue is that the morally salient characteristics that would ground our duties to other animals would be discovered through empirical investigation and so would be verifiable by the most recent scientific research regarding various animal species. Grounding the chosen theory in science would, we can hope, forestall objections regarding subjectivity of the moral criteria and, in addition, provide a basis for widespread agreement with the empirical foundation of the theory's moral recommendations. Moreover, because our moral responsibilities to particular animals would be based on whatever morally salient characteristics are shared by members of the animals' species, our responsibilities would vary depending on the species. For example, animals that are more complex might require more freedom of movement and action, larger animals may require larger habitat, and so on. Finally, a fifth important virtue of natural law theory, shared by animal rights theory, is that the theory requires us to respect individual animals and to regard them as ends in their own right.

Ideally, we need a theory that has all of these virtues but none of the vices of the sort of natural law theory endorsed by Matthew Scully. That ideal can be met with the capabilities approach.

The Capabilities Approach

Economist Amartya Sen and philosopher Martha Nussbaum have both re-lied on the capabilities approach to discuss a number of ethical dilemmas in the field of international development. One of its best articulations is in Nussbaum's *Women and Human Development* (Nussbaum 2000), and in the following discussion, I refer primarily to that work.[3] After describing the main features of the approach, I then apply it to animals in general and elephants in particular.

Central to the capabilities approach is the concept of what it means to live a fully human life, a life worthy of a human being. Nussbaum's notion of a fully human life is similar to Karl Marx's: it grounds human dignity in the inviolability of the individual person and recognizes that the exercise of autonomy requires certain material conditions, enumerated in Nuss-baum's list of the ten central human functional capabilities (Nussbaum 2000, 78–80). The particular items on Nussbaum's capabilities list will be enumerated in the next section, when I discuss how I have applied the ca-pabilities approach to elephants. As Nussbaum explains, she compiled this list over many years, based on extensive empirical investigation, and sev-eral times modified it in light of what individuals in various cultures claimed was necessary to living a fully human life. The resulting list is, there-fore, not sacrosanct, not the necessary result of an *a priori* deduction from a particular view of human nature (Nussbaum 2000, 76).

From this brief description, we can see some similarities between a ca-pabilities approach and natural law theory: the conception of human be-ings is teleological and the method of investigation is empirical. But more important than these similarities are the welcome ways in which Nuss-baum's capabilities approach differs from Scully's natural law theory. For example, the *telos* (ultimate purpose, or goal) of human beings is not set by something independent of us, but instead is determined by us. In short, we are not being asked to swallow a teleological conception of nature that is incompatible with contemporary science. In fact, Nussbaum's capabili-ties approach does not presuppose any particular metaphysical conception of the world, and Nussbaum contends that the approach is compatible with a variety of conceptions (Nussbaum 2000, 76). Another difference from natural law theory is that the capabilities approach does not presuppose that human beings have only one overriding end but instead asserts a mul-tiplicity of ends, determined by the individual, that are not reducible to one overarching end. Hence, the list of capabilities is an irreducible plurality.

A third difference is that while natural law theory is concerned largely with what individuals do, capabilities approach is concerned with what individuals are capable of doing. In other words, natural law theory claims that we ought to act in accordance with a *telos* determined by something external to us, and to the extent that we do not, we are, in some sense, defective. In contrast, the capabilities approach asks what abilities we are capable of exercising or developing, and to the extent that we are not capable of exercising or developing particular capabilities, finds some defect in the material conditions that would otherwise make such exercise or development possible. I conclude that Nussbaum's capabilities approach is a more congenial moral theory than the sort of natural law theory espoused by Scully because Nussbaum's approach does not make the extravagant metaphysical assumptions that Scully's makes, and Nussbaum's approach is therefore compatible with contemporary science in a way that Scully's is not.

If we apply the capabilities approach to nonhuman animals, we will be required to do a great deal of investigation to discover the central capabilities of any particular species. Thus, the capabilities approach is not merely compatible with contemporary science, but it will indeed rely on the findings of scientists in the development of the moral theory. In addition, to discover the central functional capabilities of any particular type of animal, we will want to rely on data gathered by studying animals in their natural habitat. Such information will always be considered more reliable than information discovered about the behavior of members of a particular species held in zoos, laboratories, or circuses, if there is a conflict between what the former and the latter sources tell us. In other words, we can assume that animals may adapt to conditions of captivity in ways that distort their capabilities and preferences, and so I contend that we must always defer to data gathered from field studies.

The list of nonhuman animal capabilities that we develop will not be sacrosanct, any more than Nussbaum's list of human capabilities is, but can be modified in light of additional scientific discoveries. Each species will need its own list because the capabilities of various species are often distinct. There may be some overlap among the lists we develop for each species, including our own species, but clearly there will be many differences. An animal rights or animal liberation theorist might object that a capabilities approach will be cumbersome, that a simpler theory is preferable. Although it is true that simplicity is a theoretical virtue, simplicity does not trump a theory being correct. Further, although animal rights theory and utilitarianism might look simpler in the statement of the theory, they are every bit as complicated in their application. If we think about what the rights of an octopus might be or how to maximize utility for alligators and the creatures with which they share their environment, we can see that both of the two most popular theories regarding our responsibilities toward other animals could be quite complicated in their application.

We are in a position now to see why the capabilities approach retains all of the virtues of a more traditional natural law approach, while avoiding the latter's unfortunate metaphysics. I suggested at the end of this essay's first section that natural law theory has five virtues. All five are shared by Nussbaum's capabilities approach, including when we apply it to nonhuman animals. First, because each species (or genus, or family, etc.) will require its own list of capabilities, we can expect a great deal of variability among our conclusions about what we owe animals belonging to different species. Thus, the capabilities approach will not be anthropocentric in the way some ecofeminist theorists have worried that animal rights and animal liberation theories may be. Second, all animals can be included in the scope of the approach, exactly because it is constructed by looking at scientific data regarding the capabilities of different animals in order to build different lists of capabilities for different species. Indeed, one philosopher suggests it would be possible to construct lists of capabilities for plant species, as well, and so a capabilities approach might serve as the basis of a nearly complete environmental ethics. This is an intriguing suggestion, which is, of course, beyond the scope of this essay (H. Douglas, personal communication).

The third virtue the capabilities approach shares with natural law theory is that its conclusion about how we ought to treat nonhuman animals is based on extensive empirical research. It seems that Nussbaum's approach surpasses Scully's in this respect. We would have to struggle—and, I believe, ultimately fail—to make Scully's theory compatible with what scientists tell us about the abilities of various animals, whereas Nussbaum's approach requires that we amend lists of capabilities if a list turns out to be incompatible with what our best scientific research tells us about what animals can do. A fourth virtue of the capabilities approach, and, indeed, a requisite flexibility it shares with all the theories we have considered herein, is that the duties we have toward nonhuman animals would vary according to the species and particular situation of the animals. The final virtue that the capabilities approach shares with natural law theory and also with animal rights theory is that it requires respect for individual animals as ends in their own right.

Although I have not given a full defense of the capabilities approach, I have shown, I think, that the capabilities approach can be applied to nonhuman animals, including entire animal species, that the results of doing so would yield a fruitful and compelling theory of our duties to other animals, and that a capabilities approach is able to avoid some of the more obvious problems of animal rights theory, utilitarian theory, and traditional natural law theory. Therefore, it will be worth our while to see what a capabilities approach might say about our duties to elephants.

Why Circuses Are Unsuited to Elephants

Nussbaum says that the main question the capabilities approach would ask about an individual is not whether she is satisfied, nor how many resources she is able to command, but "What is this person able to do and to be?" (Nussbaum 2000, 71). Capabilities approach's main goal is to ensure that human beings are capable of living fully human lives and that capacity is defined by Nussbaum's list of central human capabilities. Thus, if we apply capabilities approach to elephants, we will not mainly be concerned with how satisfied individual elephants are or what resources are available to them. It is not that these things do not matter—in fact, having certain resources may be a necessary condition of exercising certain elephant capabilities—but it is not sufficient to show that, say, an elephant has enough to eat and is free of disease to show that we have fulfilled our responsibilities to elephants. So, instead, we will ask what individual elephants in a given situation are able to do and to be, whether they are able to live fully elephantine lives, a capacity capabilities approach—applied to elephants—would define by developing a list of central elephant functional capabilities.

The ten items on the human capabilities list are (1) life, (2) bodily health, (3) bodily integrity, (4) senses, imagination, and thought, (5) emotions, (6) practical reason, (7) affiliation, (8) other species, (9) play, and (10) control over one's environment, which is divided into two types of control, political and material (see Nussbaum 2000, 78–80). The elephant capabilities list will share some features with this human capabilities list. In fact, some human capabilities are shared by animals of all species, especially the first three: (1) life, which includes not dying prematurely; (2) bodily health, which includes reproductive health and being adequately nourished and adequately sheltered; and (3) bodily integrity, which includes being free from assault and abuse and having opportunities for sexual satisfaction and choice in matters of reproduction. The precise form of these capabilities varies from species to species, but I imagine some version of these three would show up on most lists. In any case, they would certainly be on the list of elephant capabilities. Clearly, some modification of what bodily integrity means might be necessary. It probably does not make sense to talk about elephants making choices in matters of reproduction, although it does make sense to talk about elephants choosing with whom to mate. This capability might rule out the sort of artificial insemination of elephants practiced by Ringling Bros., Disney, and some zoos, but not, perhaps, the use of contraception in some African wildlife reserves (see also Whyte and Fayrer-Hosken, Chapter 20 in this volume; Schmitt, Chapter 11 in this volume).

In contrast to the capabilities that are pertinent, some clearly would be inappropriate for inclusion in a list of elephant capabilities. The most obviously inappropriate capabilities are the capacity for practical reason, which Nussbaum defines as "Being able to form a conception of the good and to engage in critical reflection about the planning of one's life," and the capacity for political control over one's environment (Nussbaum 2000, 71). At the same time, one might want to include in the list of elephant capabilities the capacity to materially control one's environment, although the meaning of the concept would be quite different from what Nussbaum assigns to the analogous human capability, which includes the right to own real property, the right to equal employment opportunities, and freedom from unwarranted search and seizure. In contrast, the elephant's capability to control her environment might include the ability to seek out sources of food, to search or dig for water, to construct shelter, to herd with other elephants, and so on. Interestingly, how these capacities are realized will vary depending on the environment in which the elephants live, just as it does for human beings. For example, African elephants may require much more space in which to exercise this capability than Asian elephants, largely because of the different habitats in which they live (see also Mellen, Barber, and Miller, Chapter 15 in this volume; Hancocks, Chapter 13 in this volume).

I have delineated a preliminary list of nine elephant capabilities, based on Nussbaum's list of human capabilities, but modified where appropriate. I refer to the list as "preliminary" because I expect the list will be revised as other people who work with elephants comment on and discuss it. Some items may be modified and others may be added. In other words, I welcome comments on the list and suggestions for improvements because if the list is to serve as a basis for public policy regarding our treatment of elephants, then elephants' central capabilities should be described as accurately as possible.

Central Elephant Functional Capabilities

In this list, I have retained some of the original wording used by Nussbaum, as the reader will notice when my list is compared with Nussbaum's list (2000, 78–80).

1. *Life.* Being able to live to the end of an elephant life of normal length. Not dying prematurely.
2. *Bodily Health.* Being able to have good health, including reproductive health; to be adequately nourished; to live in an environment conducive to bodily health.
3. *Bodily Integrity.* Being able to move freely from place to place; having one's bodily boundaries treated as sovereign, that is, being free from assault; having opportunities for self-directed sexual satisfaction and for choice in nutriment, shelter, and other requirements for normal growth and well-being.

4.   *Senses, Thought, and Imagination.* Being able to use the senses, to imagine, and to think in a "truly elephantine" way. This will include being able to communicate over long distances, using infrasonic communication, living in an environment to which elephants have naturally adapted, and so on. By "naturally adapted," I mean something like what evolutionary biologists mean when they talk about adaptation of species. I do not refer to what might be considered a sort of "artificial adaptation," such as adaptation to conditions in zoos or circuses. Because elephants are highly intelligent animals, they can, of course, learn to adapt to many different environments, some of which could be considered highly artificial, but this adaptation to artificial environments should be considered analogous to what Nussbaum (2000) calls "'adaptive preferences.'"

5.   *Emotions.* Being able to have attachments to other animals, and perhaps to places and things; to be able to love friends and family members and to be able to grieve their absence. Not having one's emotional development blighted by overwhelming fear and anxiety or by traumatic events of abuse and neglect.

6.   *Affiliation.* Being able to live with and have social relations with others, to recognize and show concern for other elephants, to engage in various forms of normal social interaction.

7.   *Other Species.* Being able to live with concern for and in relation to other animals, plants, and the world of nature.

8.   *Play.* Being able to play.

9.   *Control over the Physical Environment.* Being able to seek and obtain food, water, and places to rest. This control must be shared with other species, of course, including our own.

The Consequences for Circuses

Although the preliminary list may be subject to revision, I shall assume that it is largely correct, in order to see what consequences these capabilities of elephants would have for circuses. Guaranteeing the first two capabilities might entail that elephants cannot be used in circuses, if animal rights activists are correct in their assertion that elephants' lives are shortened and their health impaired by being held captive in circuses. As I have mentioned before, however, spokespersons for circuses disagree with these claims, and some assert that elephants are healthier and live longer when they live in circuses than when they live in the wild. It is difficult either to verify or to disprove either side's claims because to date we possess insufficient empirical data to do so (R. Kagen, personal communication). Still, it is at least logically possible that the health and longevity of elephants could be protected better in captivity than in the wild, so let us grant the pro-circus spokespersons this point. However, this point will turn out to be irrelevant because it pertains to only two out of nine elephant capabilities, and some of the other capabilities will necessarily be thwarted when elephants are used in circuses.

We are on firmer ground with respect to the third capability, bodily integrity. For example, ensuring that elephants in circuses were able to move freely from place to place would mean that elephants could not be chained or confined to small stalls or to train cars. In short, elephants could not travel by motorized means. Even if elephants were kept in one place, it is unlikely that circuses could provide sufficient space for elephants to exercise this capacity. Moreover, even if a circus could provide sufficient space for elephants and remain in business while remaining in one place, the elephants must be made to perform on a regular schedule. Their training and their performance constitute an interference with elephants' abilities to move freely from place to place and so thwart the exercise of the capacity for bodily integrity.

In addition, some circuses are very proud that they now breed elephants using artificial insemination (see Alexander 2000). But this seems to violate bodily integrity, in a way that contraception may not. This is, to be sure, a highly controversial point. Some might argue that contraception is also a violation of bodily integrity because an elephant cannot choose contraception. In response, one might argue that if an elephant could choose, it would choose contraception, so that births were further apart, since the alternatives to contraception are claimed to be (1) sterilization, which would preclude reproduction altogether; (2) reduced health for the entire herd because too many births put a strain on resources available to the group; and (3) culling, which (in my opinion) is a violation of capabilities 1, 3, 5, and 6. In other words, we may need to adopt a "reasonable elephant standard," similar to the reasonable person standard commonly used by jurists. The reader will no doubt notice that much of what I say in this paragraph also applies to most zoos, which would mean that zoos are not suitable to elephants, either. Perhaps a handful of zoos would be capable of providing sufficient habitat to ensure that elephants have bodily integrity, but most zoos simply do not have sufficient space or resources.

The fourth capability also suggests that circuses cannot provide an environment in which elephants can exercise their capabilities. The more we learn about elephants' abilities, such as their infrasonic communication and their greeting displays, the more it becomes clear that for elephants to live a fully elephantine life, they must live among other elephants, in the wild, with as large a range as possible (regarding elephant vocalizations, see the website for the Savanna Elephant Vocalization Project, www.elephantvoices .org; see also Payne 1998). It might be contended that there are other abilities, ones we have ample evidence that elephants possess, such as the abilities to paint or to play music (see MacPherson 2001; Scigliano 2002), which require elephants to exercise their senses, imagination, and capacity for thought, and which could be developed only in captivity. Thus, the capabilities approach might permit, or even recommend, keeping some elephants in captivity because it is only in captivity that elephants can exercise their creativity in these ways. This contention does not stand, however, because elephants are not thereby exercising their abilities in what I have called a "truly elephantine" way. If there is a conflict between scientific field research data and data gathered by scientists observing animal behavior in a laboratory, the capabilities approach will defer to the conclusions of the scientists doing field research. In the same way, capabilities approach will deem the abilities exercised by elephants in the wild truer expressions of an elephant's nature than abilities exercised by elephants in captivity. In other words, in deciding what conditions are necessary for elephants to exercise their abilities for sensation, thought, and imagination in a truly elephantine way, we should take into consideration what elephants do when left to their own devices, with as little interference by us as possible, rather than focusing on what elephants do when manipulated by humans for profit or for entertainment or simply to exercise our dominance. Therefore, even though there are certain abilities elephants can exercise in circuses and that some elephant handlers claim elephants enjoy performing in the circus, the capabilities approach will conclude that circuses cannot provide the appropriate environment for elephants to exercise their abilities to use the senses, to imagine, and to think in a truly elephantine way.

The fifth capability, emotions, requires keeping family groups intact, as much as possible, and to inflict on elephants as little trauma as possible. Thus, certain practices seem to be ruled out: Hunting and culling elephants would be forbidden. In fact, this capability gives us another argument for preferring contraception to culling where elephant populations must be controlled. Spacing births through the use of contraception allows elephants to exercise their emotional capabilities in a truly elephantine way, by living within family groups, throughout their lives if female and until adulthood if male, and not suffer the trauma of being separated from family members or witnessing the slaughter of family members (for an anecdotal example regarding culling and trauma, see MacPherson 2001). Circuses, of course, cannot accommodate such large family groups of elephants and commonly engage in the practice of separating families, even separating baby elephants from their mothers at 1 or 2 years of age. Because circuses must make a profit if they are to stay in business, it seems unlikely they could ever provide the conditions necessary for elephants to exercise this capability; they simply could not keep a sufficient number of elephants nor provide them with sufficient space.

The sixth capability, affiliation, is clearly thwarted by circuses, for reasons already discussed. Circuses separate family groups and interfere with normal elephant interactions by separating elephants and keeping them confined. The sort of rich social relationships that elephants develop in the wild (see Moss 1988) are impossible in the circus. Therefore, circuses are incapable of providing the necessary conditions for flourishing of elephants' capabilities for affiliation.

Defenders of the use of elephants by circuses might claim that circuses allow elephants to interact with other species, most notably with humans, and therefore furnish the material conditions necessary for elephants to exercise the seventh capability. Although elephants do interact with humans in circuses, I would argue that interspecific interaction in the circus is relatively impoverished compared with interspecific interaction in the wild, that it is not truly elephantine interaction. Moreover, this capability also requires that elephants have contact with plant species and with the world of nature. Of course, it is possible for circuses to attempt to provide all of these things, although whether they could do so in an acceptable way seems unlikely. Nonetheless, we could grant that circuses are capable of providing the necessary conditions for elephants' exercise of their capacity to live with other species and still not conclude that circuses are suitable to elephants because they remain incapable of ensuring so many other elephant capabilities.

Circus spokespersons might think they are on fairly firm ground when it comes to the eighth capability, play. Surely, they would say, one cannot argue that elephants do not get to play in the circus because the entirety of their performances constitutes a form of play. I have no doubt some people are sincere in offering this argument, that they genuinely believe elephants are happy performing in the circus, that the elephants would choose to perform, if given a choice, and that their performance constitutes a type of play. Still, even if we were willing to grant that performance in a circus constituted a kind of play, it is an extremely impoverished sort of play compared with elephant behavior observed in the wild (on elephant play, see Denis-Huot and Denis-Huot 2003, especially the chapters "The Young," and "Drinking"). Furthermore, calling elephants' performances "play" is somewhat disingenuous. People who say this sincerely are no doubt drawing an analogy between human artists and elephant performers. After all, whose life's work is more like play than the musician or actor? What incredible freedom, compared to the work that most human beings perform! Although this may be true of human performers, the analogy to elephants is flawed. Elephants are incapable of making the choice to perform in the circus, of which they could have no concept unless forced into the circus. Furthermore, it seems unlikely that if an elephant could be made to understand the choice between living out a full life span in the wild with her family group and living separated from her family to perform every night in the circus so that her human captors could make a profit, she would choose the circus. The element of force involved in making elephants into circus performers entails that their performances are work, not play. And it is work of the worst sort: it is drudgery, slavery. Play is, of necessity, free. That some elephants might manage to enjoy some of what they do is no argument against this point. A happy slave is still a slave. Thus, circuses are incapable of providing the necessary conditions for elephants to exercise the capacity for play, which is a capability essential to elephant flourishing.

Finally, the ninth capability is control over the physical environment. How this capability is exercised in the wild depends on where elephants live, whether they must dig wells to obtain water, as some African elephants must, whether they create "rooms" in which to rest by making clearings in the forest, as some Asian elephants do, whether they must roam long distances to obtain sufficient food, or whether they can live in a comparatively small area. However particular groups of elephants at particular times exercise this capability in the wild, a moment's reflection makes it clear that no elephant can exercise this capability in the circus. To keep elephants captive in the way that circuses do, the elephants' environment must be controlled by their captors. A circus cannot allow its elephants to control their environment. Thus, this capability, too, must be thwarted if elephants are held captive in circuses.

I conclude that the capabilities approach demonstrates that circuses are unsuited to elephants and, furthermore, that circuses cannot be made suitable for this species. Even if elephants can exercise some of their abilities if held captive in the circus, they will not be able to exercise most of them. Therefore, without worrying about whether elephants are rights-bearers or about how to maximize elephant utility, we can conclude that elephants ought not be held captive in circuses.

We know that elephants have capabilities, and we know that, in many cases, elephants need a great deal of physical freedom to exercise them. Whether any zoo is capable of providing a degree of freedom sufficient for elephants to exercise their capabilities is a matter of contention (see Hancocks, Chapter 13; Mellen et al., Chapter 15; and Hutchins, Smith, and Keele, Chapter 14 in this volume), but there is no doubt that circuses are not so capable. Natural habitat zoos might be able to allow elephants to roam freely and to interact freely with other elephants and other species, but because circuses require elephants to perform, they cannot allow this and so cannot provide a proper environment for elephants.

It is interesting that most of the arguments about whether elephants should be used by circuses are disputes over whether elephants are healthy, whether they have a normal life span, and in general, whether they suffer unnecessarily. The capabilities approach as applied to elephants, while also concerned about these issues, focuses much of its argument on elephants' freedom. This is perhaps not surprising because Nussbaum's approach makes human autonomy a central value of the capabilities approach as applied to human beings. Because circuses can keep elephants only by depriving them of freedom, they can never be suited to elephants.

Notes

1. A web search using the key words "elephants in circuses" will get thousands of hits, with information about elephant welfare, survival of elephants, legislation regarding elephants in captivity, and so on. Some of the English-language websites oriented to defending the interests of animals and stopping or reforming the use of elephants in circuses include: American Society for the Prevention of Cruelty to Animals (ASPCA; www.aspca .org); Animals in Print (www.all-creatures.org/ aip); Humane Society Legislative Fund (www.fund .org); In Defense of Animals (www.idausa.org); Animal Defense League (www.animaldefense.com); People for the Ethical Treatment of Animals (www .circuses.com and www.circuswatch.com); Performing Animal Welfare Society (www.pawsweb .org); The Elephant Sanctuary (www.elephants. com); Royal Society for the Prevention of Cruelty to Animals—RSPCA (www.rspca.org); Humane Society of the United States (www.hsus.org); Born Free Foundation (www.bornfree.org); Save the Elephants (www.savetheelephants.com); Captive Animals' Protection Society (www.captiveanimals .org/elephants); and Coalette's Connection for Action (www.ccforaction.com/circuses.htm).

2. My web searches turned up fewer organizations that include among their activities the defense of the use of elephants in circuses, but they are quite professional and state their positions clearly. Most devote some effort to answering the charges of animal rights and animal welfare organizations, as well as to arguing against certain types of legislation written in opposition to the use of captive elephants for purposes of entertainment, among other things. Among the English-language organizations are Elephant Managers Association (www .elephant-managers.com); National Animal Interest Alliance (www.naiaonline.org); Outdoor Amusement Business Association (www.oaba.org); and the Ringling Bros. and Barnum & Bailey Center for Elephant Conservation (www.elephantcenter .com).

3. Although her claims develop in some different directions from the position I take here, readers may also be interested to know that Nussbaum has applied the capabilities approach to nonhuman animals in various circumstances, even commenting to some extent on treatment of circus animals. Please see *Animal Rights: Current Debates and New Directions*, a volume co-edited by Sunstein and Nussbaum (2004), and *Frontiers of Justice: Disability, Nationality, Species Membership* (Nussbaum 2006a).

References

Alexander, S. 2000. *The astonishing elephant*. New York: Random House.

Denis-Huot, C., and Denis-Huot, M. 2003. *The art of being an elephant*. New York: Barnes and Noble.

MacPherson, M. 2001 *The cowboy and his elephant: The story of a remarkable friendship*. New York: St. Martin's Griffin.

Mill, John Stuart. 1861/2001. *Utilitarianism*. G. Sher (ed.). Indianapolis: Hackett.

Moss, C. 1988. *Elephant memories: Thirteen years in the life of an elephant family*. New York: Fawcett Columbine.

Nussbaum, M. 2000. *Women and human development: The capabilities approach*. Cambridge, England: Cambridge University Press.

Nussbaum, M. C. 2006a. *Frontiers of justice: Disability, nationality, species membership*. Cambridge, MA: Harvard University Press.

Nussbaum, M. C. 2006b. The moral status of animals. *The Chronicle of Higher Education*. February 3. http://chronicle.com/temp/reprint.php?id=ydhw2skrrxq8xk3ll356hld 63d5wv47s.

Payne, K. 1998. *Silent thunder: In the presence of elephants*. New York: Penguin Books.

Rawls, J. 1971. *A theory of justice*. Cambridge, MA: Belknap Press.

Regan, T. 1983. *The case for animal rights.* Berkeley: University of California Press.

Scigliano, E. 2002. *Love, war, and circuses: The age-old relationship between elephants and humans.* Boston: Houghton Mifflin.

Scully, M. 2002. *Dominion: The power of man, the suffering of animals, and the call to mercy.* New York: St. Martin's Press.

Sen, A. 1999. *Development as freedom.* New York: Alfred A. Knopf.

Singer, P. 1975. *Animal liberation: A new ethics for our treatment of animals.* New York: Avon Books.

Slicer, D. 1996. Your daughter or your dog? A feminist assessment of the animal research issue. In K. Warren (ed.). *Ecological feminist philosophies* (pp. 97–113). Bloomington: Indiana University Press.

Sunstein, C., and Nussbaum, M. 2004. *Animal rights: Current debates and new directions.* New York: Oxford University Press.

11 VIEW FROM THE BIG TOP

WHY ELEPHANTS BELONG
IN NORTH AMERICAN CIRCUSES
DENNIS SCHMITT

The human-elephant relationship began at least 3,000 years ago. Elephants have been our partners in agriculture, industry, religion, war, and entertainment. People are drawn to elephants, and elephants have seemed to thrive in captive environments that provide social interaction and physical and mental stimulation. The assumption that elephants have been and will continue to be cared for in captivity is central to a discussion of the human-elephant relationship in the twenty-first century and beyond.

Today, while Asian elephant habitats and populations have been declining in their range states, the human population has been increasing, and human-elephant conflicts have been escalating (Lair 1997). As a result, the utopian goal of returning significant numbers of elephants from western countries to the wild is not currently realistic. The loss of rangeland habitat, combined with rapidly decreasing numbers of free-ranging elephant populations, indicates a need to establish a diversified strategy for elephant conservation, consisting of two main components: protection in the wild and propagation in captivity. Captive propagation is essential for establishing a metapopulation, for the promotion of *in situ* conservation efforts, for research, and for education.

A circus brought the first Asian elephant to North America (Keele and Dimeo-Ediger 2000; see also Kreger, Chapter 10 in this volume). Circuses in North America have been caring for and managing elephants since the late 1880s. Since then, and especially in the past twenty years, understanding elephants and their unique characteristics and needs has advanced tremendously. My own experience with circuses in North America as a veterinarian and research scientist is limited to the last seventeen years; this essay draws heavily on that experience. Modern researchers, elephant managers, and trainers have now developed considerable understanding of elephant physiology and well-being, especially in studies supported by institutions in North America, including circuses. Elephant well-being in North America continues to improve daily as a result of communication among groups responsible for elephant care.

In addition to providing entertainment to large numbers of people annually, circuses in North America have been active in promoting conservation initiatives for research and education about elephants. Today circuses

227

continue to improve elephants' well-being by adopting standards of care appropriate for elephants in circuses and traveling exhibitions (Outdoor Amusement Business Association [OABA] 2002). The continued improvement of captive elephant care is dependent on the adoption of standards by the elephant industry, including circuses and private individuals using elephants for traveling exhibitions. The activities of the Circus Unit of the OABA provide one example of the industry adopting animal care and training guidelines for performing and exhibited animals (OABA 2002).

Elephant Care and Husbandry under the Big Top

Circuses care for elephants throughout their life span, from birth through adulthood and retirement. Zoos and circuses in North America produced fifty-two viable Asian elephant offspring from 1991 to 2002 (Keele and Lewis 2003). One of these institutions, African Lion Safari, could be classified as both a zoo and a circus institution as some of the elephants have occasionally been used in circuses when the safari park is closed during the winter. Eight viable calves were born at African Lion Safari during 1991–2002. If these calves are not counted in the tally, then twenty-six calves were born to circus institutions and nineteen in zoos between 1991 and 2002. Ringling Bros. and Barnum & Bailey Center for Elephant Conservation (Ringling Bros. CEC) had five calves born from April 2001 to May 2002, exemplifying the potential for reproductive success for other elephant-breeding institutions. A number of factors contributed to the Ringling Bros. CEC's success. The staff closely monitored the females' reproductive cycles; used multiple, healthy bulls as studs; and attended to the physiological and social needs of the females. Contributing to this success was that a number of the dams were not first-time mothers and therefore were competent and experienced at giving birth and raising calves. Two of these births represented the fifth calf for their mothers. The remaining three births were the second, third, and fourth calves for their respective mothers.

The robustness of the future Asian elephant population of North America depends on the cooperation of all elephant holding institutions to maintain a viable captive population. Only thirty-five to forty elephant cows are currently producing calves in North America. This gives a sense of urgency to all who are committed to a sustainable Asian elephant population on this continent. Whether they are held by circuses or zoos, management of Asian elephants for breeding should be a cooperative effort to maximize the diversity of possible matings in North America. The zoo community's decision to work separately from the circus community is the single biggest impediment to captive reproduction of Asian elephants in North America. Managed separately, the number of experienced breeding elephants in a single pool and, hence, of possible pairings would be greatly reduced. Circuses and private institutions currently hold at least twelve male Asian elephants, which can contribute in significant ways to understanding semen physiology and genetic diversity of the captive population.

Elephants are well cared for by most circuses in North America. I am very familiar with two large circuses that have cared for more 100 elephants throughout those elephants' lifetimes. Carson and Barnes and Ringling Bros. circuses have large modern elephant facilities with outstanding, knowledgeable professional staff providing for the elephants' care. A studbook is a chronological database of individuals of a specific species or subspecies held in animal collections in a specific geographic region over a particular time period. The book contains information on sex, birth and death dates, pedigrees, and institutional transfers. An examination of the 2000 North American Asian elephant studbook reveals that the oldest Asian elephants in recent years have been circus elephants (Keele and Dimeo-Ediger 2000). The oldest Asian male was Tommy (estimated birth 1944) and the oldest female was Suzy (estimated birth 1925; Keele and Dimeo-Ediger 2000). Elephants retire from circus life and live out their lives at zoos, retirement centers, and sanctuaries (Feld 2002). Some have found new homes as companions for individually housed elephants in other institutions. These well-traveled retirees make excellent companions, as they are socially adept and integrate easily and quickly to new elephants, caretakers, and facilities.

The physical fitness and mental health of circus elephants is enhanced by the physical activity of rehearsal and performing, as well as by the novel environments encountered as they travel throughout North America. Physical fitness is difficult to evaluate in elephants. However, as I observe circus elephants in the course of my veterinary duties, few appear obese and most appear to have benefited from the physical exercise of performance, which provides the opportunity to develop muscle tone and to use the entire body. The mental health of elephants is enhanced during performances that require timing, coordination with other elephants and performers, and adjustments to novel environments. In my estimation, these activities enrich the elephants' lives and prevent boredom, a significant concern of elephant managers who maintain animals in static enclosed environments.

As a veterinarian, I am often asked to examine elephants in circuses, zoos, and sanctuaries throughout North America. Most elephant holding institutions carry out a daily routine of elephant care, including several behavioral components. I find most circus elephants are in general more accommodating of complete veterinary exams than most zoo elephants because their handlers spend more time interacting with them than most elephant keepers in zoos.

From my observations, circus elephant handlers spend more time with their elephants than do zookeepers. In addition to training, they concentrate on improving elephant care and well-being by frequent grooming and footpad inspections and continuous observation of their elephants. Circuses are active participants in several research projects benefiting captive as well as free-ranging elephants. When circus elephants need medication, veterinarians can give them injections easily, and if oral medication is indicated, circus elephants are often more compliant than most zoo elephants. Perhaps this results from the variation of their diets when they travel. In my experience, circus elephants generally do not view novel food items and environmental changes with as much suspicion as do elephants kept in zoos.

Elephant Performances and Training

Elephant performances in the circus are a series of behaviors derived from natural abilities. For instance, playful elephants in the wild sometimes stand on their heads or sit when they wallow in mud. They might roll a log over to a tree to step on so they can reach a branch with their trunk. In training elephants for performances, natural behavior is modified and these modifications are then reinforced through repetition, reward, and praise. An elephant presentation is a well-choreographed and rehearsed performance involving great coordination among elephants, trainers, human performers, and support staff, including stagehands and elephant handlers. The patience, time, and dedication that go into training elephants result in trusting relationships between the elephants and trainers. These relationships, formed over a long period of time, forge the bonds of respect that elephants and their human caretakers have for one another. An example of this involves Mark Oliver Gebel with Ringling Bros. Mark grew up with the elephants he now trains and spent his life learning from his father, Gunther Gebel-Williams, who worked with some of those same elephants for more than thirty years. The often overlooked reality is that animals weighing over 4,000 kilograms don't do anything they don't want to do. They perform because of the strong bonds of trust between themselves and their trainers.

Circuses are pioneers in developing and adopting advanced new training methods for elephants as well as other animals. Circuses participate actively in the continuous development of the body of knowledge regarding elephant behavior and training. Just as veterinarians are gaining more knowledge of elephant physiology, circuses are incorporating new training methods—ones that assimilate such new scientific knowledge—into training protocols and husbandry techniques for elephants. The consequence is continually improving elephant well-being.

Standards of Animal Care for Traveling Animal Exhibitions

Adoption of standards for elephant care and well-being is one benchmark of advances in elephant management. Today's circuses and private individuals using elephants for traveling exhibitions have adopted standards of care. In 2002, the Circus Unit of the OABA adopted recommended standards that have formalized animal care standards in the circus industry and other performance-based operations (OABA 2002). In addition, some large circuses have their own very strict animal care standards. For example, the animal care standards Ringling Bros. has instituted (see FAQ: about Animals and Ringling Bros. and Barnum & Bailey at www.feldentertainment .com/pr/pressroom.asp) meet or exceed those contained in the United States Department of Agriculture Animal Welfare Act and Animal Care Regulations (see U.S. Code 2006 Title 7; and U.S. Code 2006 Title 9) as well as the elephant care standards incorporated into the Accreditation Standards of the Association of Zoos and Aquariums (AZA—formerly, the American Zoo and Aquarium Association; see AZA 2003). Ringling Bros. has a full-time veterinary staff, veterinary technicians who travel with the animals, and on-call veterinarians in every city they visit. Ringling Bros. continually builds on its experience of over 136 years of working with elephants to do what is best for the elephants in their care. The experience of those who live with and care for their elephants twenty-four hours a day, 365 days a year is a priceless asset for the elephants in their care and for the elephant community.

Education and the Big Top

The connection between a live elephant and the public is undeniable. People remember the first time they visited a circus and saw a live elephant. How many people remember and talk about the first time they read about an elephant in a book or saw an elephant on TV?

By traveling to locations throughout North America, circuses allow the public, especially those who do not live near a zoo, to discover the unique characteristics of elephants and to interact with elephants in a positive family environment. Elephants make special appearances as ambassadors for their circus, in local media events, public "open houses," and other special activities that celebrate the circus's arrival. These activities also build awareness of all the other elephants in the world. Elephant rides provide direct elephant-human contact while exercising the elephants. Some circuses use public engagement with elephants to educate the public about elephant physiology and elephant conservation. As an example, in 2006, the Ringling Bros.'s Blue Unit (one of its three traveling circus units) focused throughout the show on the education of the public regarding Asian elephants as an endangered species and on elephant conservation as a priority.

Contributions of Circuses to
Elephant Research and Conservation

In addition to providing entertainment to large numbers of people annually, circuses in North America have been active in promoting elephant conservation and research. For example, research into the parameters necessary for safe, ethical elephant travel has improved conditions of elephants as they travel from venue to venue. Ted Friend of Texas A&M University, a respected scholar of domestic horse transportation welfare issues, has investigated the responses of elephants during transport, with funding from the U.S. Department of Agriculture (Toschano et al. 2001). Friend's study was performed in cooperation with circuses interested in documenting the well-being of their elephants, and the information has been instrumental for evaluating physiological responses of elephants during transportation in summer and winter temperature extremes. The study revealed that elephants respond to extremes in environmental temperatures during transport by using their own internal mechanisms to maintain their body temperatures within normal limits. In North America's large circuses, elephants consistently travel in well-designed and well-maintained semitrailers or railcars.

Income generated from circus elephants is being used for the benefit of all elephants. For example, the contributions of circuses have been instrumental for developing successful artificial insemination techniques for Asian elephants. In 1997, George Carden Circus International provided me with ultrasound equipment that was used to evaluate the reproductive tracts of captive elephants in North America and Europe. In 2002, Carson and Barnes Circus, through the Endangered Ark Foundation, provided matching funds for the purchase of portable ultrasound equipment that has been used in field conditions for captive elephants all over the world. In 2003, elephant veterinarians from throughout India used this equipment in workshops about using ultrasound to evaluate elephant health and reproduction. These workshops were funded by the International Elephant Foundation (IEF), through a grant to the India-based Project Elephant. Three U.S. circuses are active supporters of the IEF, along with several zoos and private elephant holding institutions.

The IEF funded a study I carried out, which focused on optimizing freezing-thawing protocols for African elephant semen. The results of that African elephant project are currently being used for the artificial insemination of some African elephants in North America. More recently, the Ringling Bros. CEC funded my two-year investigation to develop methods of freezing Asian elephant semen. All pregnancies achieved from artificial insemination in elephants, to date, have resulted from fresh-cooled semen. The efficiency and convenience of Asian elephant artificial insemination could be greatly improved with the availability of frozen-thawed semen. These projects have yielded valuable information that could help optimize the success of artificial insemination in Asian elephants. The projects provide terrific examples of how research with elephants in the care of humans can advance efforts to increase genetic diversity among all captive elephants worldwide and perhaps in the future among some isolated free-ranging populations.

Both *in situ* and *ex situ* research projects benefit the well-being of captive and free-ranging elephants—even when, as happens in range country situations, the demarcation between captive and free-ranging elephants is sometimes difficult to delineate. Member circuses have supported several IEF projects that have focused on both *in situ* and *ex situ* Asian and African elephants. These projects have provided new understanding regarding behavior, physiology, and conservation strategies. The IEF has partnered on *in situ* projects in India, Sumatra, Sri Lanka, and twenty-nine African countries. By 2006, IEF had distributed more than $870,000 for elephant conservation and research programs (J. Lehnhardt, personal communication). Some IEF funding derived from donations from circus patrons. From 2001 to 2004, for example, circus patrons attending dress rehearsals of Ringling Bros. had the opportunity to earmark their admission fee as a gift to the IEF, directly providing conservation and research dollars for elephants. Members of the OABA have donated directly to the IEF and some OABA members provide educational materials about elephants, and a pamphlet about the IEF, to circus fans attending their shows. Through these kinds of collaborations and partnerships, elephant welfare and conservation will continue to evolve.

Central to any discussion of the ethics of human-elephant relationships in the twenty-first century and beyond is the assumption that elephants have been and will continue to be cared for in captivity. One place they will continue to receive this care is in circuses. In addition to providing entertainment to many people, circuses in North America have become more active in promoting conservation of and research and education about elephants. Several circuses are directly and indirectly supporting research projects to benefit both captive elephants and free-ranging elephants, and major circuses have adopted stringent standards of elephant care. The physical and mental health of North American circus elephants is enhanced by the physical activity of rehearsal and performing as well as by the novel environments encountered as they travel throughout the continent. Circus elephants provide a hands-on, one-on-one experience that cannot be replaced by interaction through television, Internet, or movies.

References

Association of Zoos and Aquariums (AZA). 2003. *Standards for elephant management and care, adopted 21 March 2001, updated 5 May 2003.* Silver Spring, MD: AZA.

Feld, K. 2002. Address to the AZA. *Journal of the Elephant Managers Association* 13: 137–143.

Keele, M. N., and Dimeo-Ediger, N. 2000. *Asian elephant* (Elephas maximus*) North Amer-*

ican regional studbook. Portland: Oregon Zoo and American Zoo and Aquarium Association.

Keele, M., and Lewis, K. 2003. *Asian elephant (*Elephas maximus*) North American regional studbook update.* Portland: Oregon Zoo.

Lair, R. C. 1997. *Gone astray: The care and management of the Asian elephant in domesticity.* Bangkok: United Nations Food and Agriculture Organization Regional Office for Asia and the Pacific.

Outdoor Amusement Business Association. 2002. *Animal welfare and education resource materials.* Winter Park, FL: OABA (Circus Unit).

Toschano, M. J., Friend, T. H., and Nevill, C. H. 2001. Environmental conditions and body temperature of circus elephants transported during relatively high and low temperature conditions. *Journal of the Elephant Managers Association* 12: 115–149.

U.S. Code. 2006. *Animal Welfare Act of 1966. Title 7—Agriculture.* Chapter 54, transportation, sale, and handling of certain animals, sections 2131–2159. www.access .gpo.gov/uscode/title7/chapter54_.html.

U.S. Code 2006. *Animal Welfare Act of 1966. Title 9—Animals and Animal Products, Chapter 1—Animal and Plant Health Inspection Service, Department of Agriculture, Part 2—Regulations.* www.access.gpo.gov/nara/cfr/waisidx_06/9cfr2_06.html.

12 THE CHALLENGES OF MEETING THE NEEDS OF CAPTIVE ELEPHANTS

JANE GARRISON

Elephants deserve rights: to live free of pain, fear, and suffering, undisturbed in an environment that meets their needs, not the needs of those who profit unduly from their uniqueness. I have dedicated my life to defending the rights of these magnificent, sensitive beings who are the world's largest land mammals (Estes 1997) and who desperately need voices speaking out for them.

My quest to help elephants began in 1996 when an acquaintance asked me to check on some elephants that were traveling with a small circus in Southern California. What I found not only astounded me, but it changed my life (and, I hope, elephants' lives) forever. Four beautiful Asian elephants were shackled by one front leg and one back leg, unable to take even one step. It was over ninety degrees and the hot California sun was beating down on their sensitive skin (Moss 1988; Langman et al. 1996; Forthman 1998). There was no shade and no water available to them. They were chained in the sun the way someone would park a tractor when they were not using it. Three of the four elephants swayed from side to side in such unison that a passerby might actually believe the ludicrous excuse the circus gave for this behavior—that they were "dancing" (Gruber et al. 2000). The fourth elephant did not sway . . . she was too sick and weak. This elephant stood very still with her trunk lying on the hard, hot pavement. Even to my then-untrained eye, it was clear something was very wrong with her. Bones protruded through her skin and she looked as if she was about to collapse. I began a desperate attempt to get help for this obviously sick elephant. I was in constant contact with the U.S. Department of Agriculture (USDA), the agency that is supposed to enforce the minimal standards of the Federal Animal Welfare Act, the local humane societies, and the California Department of Fish and Game. I begged every agency to inspect this elephant and get her the medical attention she needed. I even made an offer to the circus for a veterinarian to examine the elephant— they declined, claiming they took great care of all their animals. Despite my pleas with the government agencies and the circus to allow this elephant to get off the road, the elephant died two months later from tuberculosis (USDA Animal and Plant Health Inspection Service [APHIS] 2003). A necropsy revealed that she was more than 1,000 pounds underweight and 80% of her lungs were destroyed. An investigation by the USDA also revealed that the owner knew the elephant was sick but shipped her across the country to perform anyway (O'Brien 1997). This was when I realized that the laws protecting elephants are extremely minimal and rarely enforced, and I decided to dedicate my life to defending the rights of elephants.

Since that time, I have been studying and defending elephants both in the wild and in captivity. I have worked on projects to help elephants in South Africa, Thailand, Mexico, India, South America, Puerto Rico, Australia, and all over the United States. Reports of my work for elephants have appeared in publications around the world, including the *Los Angeles Times*, *Wall Street Journal*, *USA Today*, and the *London Daily Mail*. I have been a frequent guest on TV and radio shows, including *The Today Show*, CNN, CBS *Evening News*, and NBC *Nightly News*. I have been a guest speaker at countless conference, events, and hearings. All of this exposure has been crucial to my mission of publicizing the plight of captive elephants and has improved my effectiveness as an advocate for elephant welfare. I have been instrumental in almost every piece of U.S. legislation that has been passed to help elephants during the past eight years, including the passage of local ordinances to prohibit the use of elephants in circuses in some counties and municipalities from California to New York.

My sincere hope is that this essay will prompt others to take a serious look at the plight of elephants in captivity and join in the quest to improve the standard of living for these amazing creatures currently kept in pitiful situations. The time has come for changes to be made in the best interest of elephants, both as species and as individuals. It is time for zoos and circuses, which profit immensely from elephants, to make the changes necessary to ensure the humane treatment of elephants before elephants are only a distant memory.

Elephants in Circuses

At present, the worst situation for elephants in captivity is their use and abuse in circuses. Of course, circuses and those organizations that have a financial interest in keeping elephants dressed in tutus and balancing on balls claim that there is nothing wrong with this situation. They claim that their animals are "family members" and that they are saving the species by using them in circuses (Ringling Bros. 2003). They even claim elephants in circuses are happier and more content than elephants in zoos or in sanctuaries and live longer than their counterparts in the wild. Nothing could be further from the truth. Circuses have a financial interest in convincing the public that there is nothing wrong with the use of elephants in circuses.

Training

The majority of elephants used in circuses in the United States were captured from the wild. Some of these elephants were the tiny victims of culls, when a country decides it has "too many" elephants and shoots entire families. A practice used many years ago during culls was to "save" the babies and sell them to zoos and circuses around the world (Clubb and Mason 2002). One can imagine the destructive consequences resulting from the distress these "saved" babies experienced when they witnessed their families being gunned down before their eyes, had ropes thrown over their small necks, and were tightly tied to the body of their dead mother (to whom they naturally huddled close to seek protection; Baer 1998).

Once captured, these babies and newly captured elephants in general go through a horrific breaking process that usually starts in their country of origin and does not end when they are shipped to circuses in the United States. This practice, which sadly I have witnessed, begins with chaining or confining a baby elephant to the point where he or she cannot move even one foot. The babies are denied food and water and sleep for many days. They are beaten constantly with bull hooks (long heavy sticks with steel points on the end), sharpened sticks, or any other "weapon" the "trainer" chooses. In some countries such as Thailand, where elephants are captured without shooting the entire family, men will take turns through the night beating the newly captured baby elephants. This is the first time the elephants learn to fear the sting of the bull hook, also sometimes referred to as an *ankus*, or *guide* (Clubb and Mason 2002). I have seen baby elephants with more than twenty abscesses on their tiny bodies following this breaking process. This breaking process is not used exclusively for wild-caught elephants (Kiley-Worthington 1989). Elephants born in captivity have also experienced this stressful process (Clubb and Mason 2002; Kane, Forthman, and Hancocks 2005). Whistle-blowers at major U.S. circuses have described in detail similar "training" sessions of elephants taking place behind the scenes. For example, Tom Ryder, who is presently suing Ringling Bros. and Barnum & Bailey Circus (Ringling Bros.) in federal district court in Washington, DC, for violating the Animal Welfare Act, has described such activities. As a representative for the animal protection community, I have personally made repeated public requests to the circus industry to allow any animal protection organization to monitor the training of captive baby elephants, twenty-four hours a day for the first few years of the baby's life. Despite circuses' claims that they train with "positive reinforcement," they have never opened their doors to us. If circuses had nothing to hide, they would want to prove this to the public. It was not a surprise when on June 3, 1999, the USDA issued a warning to Ringling Bros. for using an approach that was outdated and did not meet Animal Welfare Act standards for training baby elephants.

A recently captured baby elephant screams from pain while a trainer digs a sharpened stick into his ear.
Photograph courtesy of Jennifer Hile

This warning followed an inspection in which USDA inspectors found rope marks and burns on the legs of several baby elephants. The USDA letter to Ringling Bros. stated, "We believe there is sufficient evidence to confirm the handling of these animals caused unnecessary trauma, behavioral stress, physical harm, and discomfort to these two elephants." It is almost humorous that circuses continue to claim that they train their elephants through "positive reinforcement and food rewards"; yet, circus "trainers" are always swinging some kind of hook, whip, or stick, while they are in the ring with the elephants or transporting them.

Because of the Convention on International Trade in Endangered Species of Wild Fauna and Flora (CITES) regulations, circuses are no longer permitted to import wild-caught elephants into the United States. However, this does not eliminate the circuses' desire for baby elephants. No one can deny the adorableness of a baby elephant, and circuses want to cash in on this. Therefore, circus adjuncts such as Ringling Bros. Center for Elephant Conservation have begun aggressively breeding their elephants. They try to justify this breeding by claiming they are saving an endangered species for future generations. However, circuses are simply breeding these elephants for the circus's own financial gain and to replenish their aging population of elephants. The only way to truly save a species is to save the precious habitat in the species' homeland and work to prevent human's destruction of the species (such as poaching; Martin and Stiles 2002). Besides the hor-

A recently captured baby elephant is caged for cruel training.
Photograph courtesy of Jennifer Hile

rific training that captive-born elephants in circuses are forced to endure, these unfortunate babies are ripped away from their mothers prematurely (Eisenberg 1981; Moss 1988; Estes 1997). Elephants in the wild nurse their babies for many years. A baby elephant gains many health benefits from receiving mother's milk, and a tremendous amount of emotional bonding takes place during nursing. Circuses tear baby elephants away from their mothers at 6 months of age to begin training for the show. Considering that male elephants (when left undisturbed by humans) will remain with their mothers until they are 10 years of age and females will remain with their mothers for life, this premature separation is likely linked to low fertility rates and significant distress in both cow and calf (Taylor and Poole 1998). The one small "luxury" that should never be taken away from captive elephants is the companionship of a family member (Clubb and Mason 2002; Kane et al. 2005).

The training of elephants for circuses begins when the elephants are very young. It continues the entire time the elephant is being used in circuses. Behind the scenes, elephants in circuses are beaten, hooked, and whipped. Undercover video footage taken at the Carson and Barnes Circus winter quarters captured exactly what goes on when the audience is not present (see, for example, Clubb and Mason 2002). The footage shows the head trainer conducting a training session with several adult elephants that have been with the circus for many years, in preparation for the upcoming season. The trainer directs the other trainers in what to do to get the elephants

Elephants in circuses spend up to twenty-three hours a day chained by their legs.
Photograph courtesy of Jane Garrison

to perform. During this dialogue, the elephants are seen backing away in fear, trumpeting, and screaming. The trainer is beating the elephants with a bull hook while he states:

> Tear that foot off! Sink it in the foot! Tear it off, make 'em scream. Don't touch 'em. Hurt 'em! Hurt 'em! Don't touch 'em, make 'em scream! If you're scared to hurt 'em, don't come in the barn. When I say rip his head off, rip his f——ing foot off . . . it is important that you do it. When he starts squirming too f——ing much, both f——ing hands—BOOM—right under that chin. When he f——ks around too much you f——ing sink that hook and give it everything you got! Sink that hook into 'em . . . when you hear that screaming, then you know you got their attention. Right here in the barn. You can't do it on the road. I'm not going to touch her in front of a thousand people.

This undercover footage exemplifies how minimal are the laws in the United States that are intended to protect elephants; this circus has not been charged with violating any laws regarding animal cruelty or endangered species, and it still employs the trainer.

One would think that a federal law like the USDA's Animal Welfare Act (AWA) that requires the provision of "adequate" care and treatment in the

An elephant is chained down for the abusive training that goes on behind the scenes. *Photograph courtesy of Jane Garrison*

areas of housing, handling, sanitation, nutrition, water, veterinary care, and protection from extreme weather and temperatures would protect elephants and other animals, but it falls extremely short. Rarely is this minimal law enforced, and it is always subject to interpretation. AWA does not define what it means by "adequate" nor, for example, does it define who an "experienced attendant" might be. There are no detailed restrictions or requirements governing eligibility to apply for a USDA exhibition license. Another tremendous weakness of the AWA is understaffing. Only 100 inspectors (or two inspectors for each state) are responsible for nearly 12,000 USDA-licensed facilities falling under AWA. Understaffing virtually guarantees chronic problems with adequate monitoring and with obtaining and keeping accurate and current accounting of elephants and their health reports. In addition, no specific AWA guidelines address social, biological, environmental, or psychological needs of elephants. Only tuberculosis testing, treatment, and necropsies are detailed.

Circus Transport

The abuse of elephants in circuses does not end with behind-the-scenes training. The very nature of circuses is abusive in itself, considering the rigors of confinement, transport, and performing. Most circuses travel the majority of the year with elephants crammed in trailers or boxcars for long hours and sometimes days at a time. Elephants in circuses are forced to stand chained by their legs in their own waste in cramped, poorly ventilated trailers. Elephants are also forced to endure extreme weather conditions during transport. I have witnessed elephants transported in freezing

Heather the elephant found dead in a circus trailer.
Photograph courtesy of Animal Protection Institute / Bob Hillman

temperatures and elephants locked in trailers and boxcars while temperatures were soaring over 100 degrees. Obviously, there are times when transport proves too much for elephants. In 1997, an elephant named Heather died in New Mexico during transport with the King Royal Circus. Heather was crammed into a poorly ventilated trailer with two other elephants and eight llamas. The trailer was parked in a hotel parking lot in the hot New Mexico sun while the circus employees slept comfortably in an air-conditioned room. Albuquerque police stumbled across the trailer and noticed it was swaying from side to side. When they located the driver in the hotel and asked him to open the vehicle, they found one young African elephant named Heather lying dead next to her two companions. The police estimated the temperature in the trailer to be 120 degrees Fahrenheit. "It appears that the death of . . . elephant Heather is a direct result of gross neglect," said Michael V. Dunn, USDA's assistant secretary for marketing and regulatory programs (APHIS 1997). This case was one of the rare times when USDA actually enforced the minimal standards of the Federal Animal Welfare Act and pressed charges against the circus owners. In a highly publicized court battle, the circus industry tried to convince the court and the public that they loved and cared for their animals. However, their false claims did not hold up in court. The circus was fined $200,000 (the largest fine ever imposed on a circus) and had its USDA permit permanently revoked. However, the USDA could not prevent the circus from obtaining a license in another family member's name. Soon after the court case ended, the circus started traveling under a different name.

Confinement in Circuses

When circuses are not in transit, they set up in parking lots, large fields, or buildings such as coliseums. Setting up in a different location and city on a daily basis certainly poses a problem as to how to confine an 8,000- to 12,000-pound animal safely and humanely (Mench 1998), which is why having elephants in circuses is abusive. Logically, there is no way to give elephants access to any sufficient space in the confines of a parking lot or coliseum. Instead, circuses find the cheapest and most convenient way to restrain the elephants is to chain their legs. Elephants in circuses spend up to twenty-three hours a day chained by one front leg and one back leg. These elephants can barely take one step in either direction and most times do not have enough room to lie down. The only time the elephants are unchained is to go into center ring to "perform" ridiculous, unnatural tricks. Once their "usefulness" is done for the day, they are back on the leg chains. Here they stand exhibiting extreme stereotypic behavior as they rock back and forth in an effort to ease their mental anguish (Gruber et al. 2000). Never do these dispirited beings have a chance to walk on grass, forage, roll in sand, bath in a pond, or bask in the sun, activities essential to a full expression of an elephant's natural behavior (Baer 1998; Lindburg 1998; Poole 1998; Seidensticker and Forthman 1998). Their every movement is controlled by what is convenient and needed for the circuses profiting from their misery. Some circuses now claim to be more humane because they set up small hot-wire enclosures in each city, which give the elephants a bit of "freedom" to move. Certainly, these small temporary enclosures are better than the use of leg chains, but they are still not perfect nor could they ever be (see Gruber et al. 2000; Csuti et al. 2001). It is just not possible to humanely transport and confine an elephant to a different location each day (Clubb and Mason 2002; Kane et al. 2005).

There is only one solution; to halt the abuse of elephants in circuses requires prohibiting their use in circuses (Clubb and Mason 2002). Many cities around the country have passed laws prohibiting the use of elephants and other exotic animals in circuses, and currently several states such as Massachusetts and Rhode Island are seeking to pass similar legislation. The public is becoming enlightened to the fact that there is no way to use elephants humanely in circuses. In the future, people will look back on and equate the use of elephants in circuses with the barbaric use of animals at the Roman Coliseum. In April 2005, the *Philadelphia Daily News* reported, "The circus elephants are coming to town next week, bringing an outdated and problematic form of entertainment to all Philadelphians. Here's hoping that this is the last year such an antiquated spectacle is welcomed within our city limits."

Elephants in Zoos

Fifteen zoos, including the Detroit Zoo, have decided there is no humane way to keep elephants in captivity. These ethical zoos have closed or are phasing out their elephant exhibits and have vowed never again to have elephants in captivity. These zoos act in the best interest of the individual animal, not in the public's interest or to make money from elephant exhibitions. The zoos that refuse to close their elephant exhibits should at least try to make captivity in a zoo the best it can be.

A handful of zoos are trying their best to make captivity in a zoo the best it can be. Those zoos have converted their exhibits to protected contact, a form of elephant management that when used correctly eliminates the use of bull hooks, chains, and human domination. Elephants are provided with all available space and free access to inside and outside areas. The zoos have made a commitment not to break up companions and family members and have spoken out against abusive practices in other zoos and circuses (Douglas-Hamilton and Douglas-Hamilton 1975; Eisenberg 1981; Moss 1988; Desmond and Laule 1991; Estes 1997, 1999; Forthman 1998; Laule and Desmond 1998; Laule and Whitaker 1998; Clubb and Mason 2002; Dublin and Niskanen 2003; Kane et al. 2005). Unfortunately, only a handful of zoos have progressed in this manner. Unbelievably, a dozen or so zoos in the United States still shackle their elephants to the floor while the zoo is closed, some for up to fifteen hours per day (AZA Regents 2001). Granted, this is a tremendous improvement from 1991, when forty-eight zoos chained their elephants (Brockett et al. 1999). A zoo that still shackles, dominates, beats, and hooks its elephants should not be allowed to have elephants (Clubb and Mason 2002).

Captive Breeding

Elephants should not be bred in captivity for many reasons. Elephants are species that have such intense needs, physically, emotionally and socially, that it is impossible to satisfy their needs adequately in the confines of today's zoo or animal park (Rees 2001). Even zoos with the best intentions certainly cannot meet the needs of a behemoth species that chooses to walk up to twenty miles per day and lives in large, tight-knit family units (Sukumar 1989; Douglas-Hamilton 1998; Whyte 1993). However, because zoos view elephants as "flagship species," there exists a desire to keep a large, viable elephant population in U.S. zoos (Brown 2000). This should never be the reason to keep any species in captivity, but this is the reason behind breeding elephants in zoos. Zoos of the Association of Zoos and Aquariums (AZA—formerly American Zoo and Aquarium Association) admit they do not breed elephants with the intention of releasing the offspring into the wild (Smith and Hutchins 2000). Zoos breed elephants to restock the captive population (an intention shared with circuses; Olson and Wiese 2000; Wiese 2000). Zoos cannot keep elephants in captivity and claim that the confinement of a few individual elephants will save these species in the wild. To ensure the survival of both African and Asian elephants, their dwindling, precious habitat in Asia and Africa must be saved and killing elephants to harvest their ivory must end (Sukumar 1989; Martin and Stiles 2002). If zoos were to invest the effort and money they use for captive breeding into protection of elephants in their homelands, both Asian and African elephants would have a much greater chance of survival. It is astonishing to think what could be achieved for elephants with the several million dollars a single zoo spends on building a new enclosure for a few elephants. For example, the Los Angeles Zoo is spending almost $47 million to build a 3.5-acre elephant exhibit that will be subdivided into four enclosures. This money certainly could be better spent in Africa or Asia to secure land for free-roaming elephants and to help educate people about the dangers facing elephants. Zoos need to start making decisions based on what is best for individual animals and their species and not what is best for the zoos and their paying visitors. Little knowledge gained from holding elephants captive in zoos has yet resulted in field applications to either elephant species. Field researchers have obtained the majority of the information and knowledge of elephants' natural history that we possess today (for example, Douglas-Hamilton and Douglas-Hamilton 1975; Eisenberg 1981; Moss 1988; Sukumar 1989).

Reputable zoos interested in increasing their captive population must first try to rescue elephants in substandard conditions in circuses and unaccredited zoos. These elephants should not be overlooked because of their age or appearance. An example is the African elephant named Maggie who is housed alone at an unaccredited zoo in Alaska, where she is confined to a dark, cramped barn most of the year because of the frigid temperatures. Tragically, zoos in the United States were focused on breeding elephants and obtaining wild-caught elephants from Africa before exploring the option of rescuing this lonely elephant. Despite the lack of support from the zoo community regarding this elephant, animal advocates never gave up the fight to move Maggie to a more appropriate facility and climate. Finally, after a more than ten-year battle, it has been approved to move Maggie to a sanctuary in California. She will finally feel warmth on her skin and have more than a hundred acres to roam in the company of other rescued elephants. There was, until recently, also a lone Asian elephant named Gildah (owned by Ringling Bros.) who was kept in a tiny enclosure at a Las Vegas casino, isolated from others of her kind; she died in 2005. Again, it is a tragedy that zoos have been trying to import baby elephants from India instead of trying to rescue this lonely, dispirited elephant. Unfortunately, there are numerous elephants both within the United States and elsewhere that are in desperate need of rescue; yet, U.S. zoos turn their backs on these elephants. It is time for the AZA and the U.S. Fish and Wildlife Service to require that zoos rescue captive elephants in substandard conditions before participating in any breeding or importing of elephants.

Although I disagree with breeding elephants in captivity, I feel strongly that zoos that continue to participate in captive breeding have an ethical obligation to meet conditions that will ensure the most humane treatment possible of elephants and their offspring. If elephants are to be bred naturally, female elephants should never be transported to the bull facility (Schmid 1998). Female elephants develop such intense bonds with their companions that it violates the species' natural history to separate them even for short periods of time (Eisenberg 1981; Estes 1997, 1999; Schmid 1998; Kurt and Garaï 2001). Because males are typically housed alone in zoos, and typically live alone in the wild, the bulls should be transported to the female's facility. If a facility lacks the space to bring in a breeding bull, how will it be able to accommodate a baby elephant if it is a male? Before exploring any elephant breeding options, artificial insemination (AI) or natural breeding of elephants, zoos must first commit to meeting the following guidelines:

- Develop a policy that all female elephants born at the zoo will remain with their mothers for life (see Eisenberg 1981; Moss 1988; Estes 1997, 1999). The companionship of a family member is the one small "luxury" that can be provided in captivity. To rob an elephant of this is cruel. Separating a mother elephant and her baby is stressful, traumatic, and cruel to both the offspring and the mother.
- Zoos must commit to allow male elephants to remain with their mothers until they are between 10 and 14 years old (see Eisenberg 1981; Estes 1997, 1999; Forthman 1998; Clubb and Mason 2002; Kane et al. 2005).
- Zoos should provide elephants with at least minimal physical conditions, including, but not limited to, providing a pool in which at least two adult elephants can simultaneously fully submerge; natural substrate; adequate enrichment; and proper climate. Zoos that are located in climates that require elephants to be kept in barns for the majority of the year should not be permitted to house elephants (see Beck and Power 1988; Moss 1988; Langman et al. 1996; Seidensticker and Doherty 1996; Estes 1997; Baer 1998; Forthman 1998; Laule and Desmond 1998; Poole 1998; Rees 2000, 2002, 2003; Clubb and Mason 2002; Law and Kitchner 2002; Kane et al. 2005).

Once a zoo meets these criteria, if it seeks to explore AI, the zoo must agree that if the process of AI causes any undue stress or fear to the female elephant it will be halted immediately (see Moberg 1985a, 1985b; Baer 1998; Desmond and Laule 1998; Kane et al. 2005). The zoo must take adequate time to desensitize the elephant to the process, with no preset timetable for when the process should begin or be completed (Pryor 1984; Desmond and Laule 1994). The process should only begin once the elephant is completely comfortable with every step and should conclude when either a successful insemination occurs or the elephant is no longer agreeable to the process, whichever comes first (Line, Clarke, and Markowitz 1987; Line et al. 1989).

Rescue, AI, and natural breeding must be the *only* methods zoos use to increase their captive elephant populations. With both AI and natural breeding, zoos are faced with the dilemma of what to do if a male elephant is born. The problem is that bull elephants in zoos lead miserable lives. Most zoos that house bull elephants confine them in the smallest of the elephant enclosures because the larger enclosure usually contains several females. The bulls can become very aggressive during musth and pose a safety threat to keepers (Poole 1989). Before protected contact, many bulls in zoos could not have basic husbandry, such as foot trims, completed because of safety risks to keepers (Laule and Whitaker 1998). The challenge of housing bulls is a tremendous concern when breeding elephants because there is obviously a 50% chance a male will be born. Because of the difficulty of managing bulls in captivity and the lack of facilities available for them, some zoo professionals have discussed the option of aborting male fetuses. This is an extremely difficult subject to discuss for many reasons. Inevitably, such a discussion, for any species, will conjure up passionate opinions, as I have observed when this sensitive topic was only touched on at zoo conferences. If a zoo does not have a facility to house a male elephant, then they should not be breeding elephants under any circumstance. Zoos must commit to providing lifetime care for an unborn elephant, regardless of gender, before beginning any breeding program. Although aborting a male fetus may appear to be a humane alternative to life in a barren zoo enclosure, we must not forget the effect losing a baby has on the mother elephant (Moss 1988). Mother elephants that have lost their babies have been observed mourning them for extended periods of time. One mother elephant was observed in Africa standing over her dead baby for several days. If we are to have elephants in captivity we must limit the amount of stress and heartache they encounter, not augment it. Zoos will never abort male fetuses because of the public relations nightmare it would cause the zoo. If zoos were able to abort a male fetus without any criticism from the public, the majority of unborn male elephants would be aborted at most zoos. Certainly, the animal protection community will closely monitor the breeding of elephants in zoos and investigate whether we see a trend of male fetuses "dying" before birth. Such a trend would indicate that zoos are taking the unethical and easy way out of a situation that would raise great public outcry if healthy, unborn male elephants were being aborted because the zoos did not have enough space for them.

Import of Elephants from Range Countries

Importing free-roaming elephants from their range countries should never take place. The practice is unethical because of the cruelty involved. Most elephants captured from the wild for zoos have been babies. These babies are typically chased with high-speed vehicles or helicopters in an effort to separate them from their mothers and the rest of the herd. If separation is not "successful" (because typically the adult elephants protect the babies) darts have been used to bring down the adult elephants. In the past, darted mother elephants, upon wakening, were observed running frantically for several days in a fruitless effort to find their precious babies. The trauma and pain this causes to the mothers and the rest of the herd cannot be justified.

Considering all the knowledge that has been gained about elephants, I am appalled that in 2003 the San Diego Wild Animal Park and the Lowry Park Zoo in Tampa had eleven free-roaming elephants captured in Swaziland and imported to the two zoos. The elephants were living on over 90,000 acres, which is a grave comparison to the four-and-a-half combined acres at the two zoos. The zoos claimed the elephants were captive, merely because a fence keeps the elephants safe from poachers. The reality is that these elephants were living a life of freedom without any human domination or interference. They were not fed, bathed, artificially inseminated, hooked, or chained. These elephants were making daily choices about where and when they would walk, eat, swim, sleep, and breed. The claim that an elephant is in a captive situation merely due to the existence of a fence is spurious. The two zoos involved in this hideous plan claimed they were "saving" these elephants because of overpopulation and a culling threat, a feeble attempt to make this cruel plan more palatable to the public. The truth is that there were only thirty-six wild elephants left in Swaziland, and by capturing these eleven the zoos contributed to a one-third decrease of the wild elephant population. More than seven international animal protection organizations joined together and formed the Save Wild Elephants Coalition. This coalition fought for months in the courts to try and halt this cruel plan. The coalition even offered to translocate the elephants to other reserves within Africa if Swaziland no longer wanted these elephants. Because the zoos paid big money to "save" these elephants, the reserve owner had no interest in having the elephants relocated within Africa. This is the cruelest plan I have seen a zoo take part in. To subject juvenile wild elephants that have never been dominated by humans to such stresses of capture, transport, training, and confinement is unconscionable. These elephants will undoubtedly suffer greatly from their loss of freedom and everything that is important to them (see, for example, Poole 1998; Kurt and Garaï 2001; Clubb and Mason 2002; Kane et al. 2005). To make the plan even crueler, the San Diego Wild Animal Park sent its older, less desirable elephants off to Chicago's Lincoln Park Zoo to make room for the younger, prettier elephants. These elephants had never experienced cold and had to be confined to a barn for a large part of the year because of the freezing, windy weather in Chicago. This was despicable.

If the zoo wanted to move these older elephants to another facility, then the elephants should have been retired to a place that has comparable weather, equal or more space, and other elephants for them to socialize with, either The Elephant Sanctuary in Hohenwald, Tennessee, or the Performing Animal Welfare Society (PAWS) ARK 2000 facility in San Andreas, California (Clubb and Mason 2002). Both organizations have several thousand acres of land. At ARK 2000, elephants range on more than 100 acres, and at the Elephant Sanctuary, they have even more space. They can live a life as close to the wild as possible with room to roam and freedom to make choices. These elephants had served the zoo and the San Diego community for so many years that they deserved to be retired with respect and dignity and be transferred to the best possible facility available to them. Sadly, the San Diego Wild Animal Park did not choose this option and two of the three elephants transferred to Lincoln Park Zoo were dead within a year and a half. This truly came as no surprise to the animal protection organizations that insisted that this transfer was not in the best interest of the elephants (Mullen and Dardick 2005).

The deaths of the two elephants left the third elephant, Wankie, alone at the Lincoln Park Zoo. The zoo decided they wanted to move Wankie out of the zoo. Once again, animal protection organizations urged the zoo to do what was in the best interest of Wankie, which was to wait until the weather was warmer to move her to a sanctuary in a better climate with more room to roam. Unbelievably, the zoo made the decision to move Wankie to the Hogle Park Zoo in Utah and make the move in twenty-seven-degree temperatures. To add to this cruel plan, the zoo packed Wankie into a nonheated metal crate, just three days after she was sick with colic, for a grueling thirty-four-hour-long journey. The trip proved to be too much for Wankie, and she collapsed in the crate en route and had to be euthanized upon arrival at the zoo (see "Wankie's Final Journey," 2005; Manier and Mullen 2005).

Reports following her death found that 30% of Wankie's lung capacity had been lost due to a lung infection. It is despicable that the zoo subjected this dispirited elephant to such emotional and physical stress when there were other more humane alternatives. Again, this exemplifies zoos' making decisions that are best for the individual zoos, instead of for the individual elephants. AZA needs to take a strong position on the acquisition and disposition of elephants to help ensure that situations like this do not take place in the future. Zoos have stopped sending elephants to circuses when they no longer have use for them; now this new trend of "out with the old and in with the young" must be halted as well.

What Is the Answer for Elephants?

Taking into account the history of elephants in captivity, the unsuitable conditions they are forced to endure, the current minimal laws to protect them, cruel training methods practiced by circuses and still sanctioned by some zoos (AZA 2001b), it is time for elephants to be phased out of captivity (Clubb and Mason 2002). This is a difficult opinion for some to embrace, especially those who profit from keeping elephants in captivity. Elephants have proven themselves species that humans just cannot provide for adequately in captivity. The need for large spaces and large family groupings makes it very difficult to confine elephants humanely for our so-called entertainment, for educational purposes, or for supposed conservation benefits (Clubb and Mason 2002; Rees 2003; Kane et al. 2005).

However, now we must address what can be done immediately to help the several hundred elephants in captivity in the United States. First, all elephants should immediately be removed from circuses. This can be accomplished by passing legislation that prohibits the use of elephants in circuses or by strengthening the Federal Animal Welfare Act or Endangered Species Act. Zoos should assist legislators in passing legislation and should offer to take the elephants from circuses. Second, zoos should not import any elephants into the United States and should cease captive breeding (Clubb and Mason 2002). They should begin to phase out elephants and start focusing their resources on preserving elephants in their homelands. This can be done by preserving elephants' precious habitat and by countering the reasons that people kill elephants, including for poaching, crop raiding, and other economic incentives. Until zoos phase out elephants, some crucial guidelines should be met:

- All elephants should be maintained in protected contact (Desmond and Laule 1991; Desmond and Laule 1994; Laule and Whitaker 1998; Kane et al. 2005). This means the true form of protected contact that was designed by Gaile Laule and Tim Desmond, without the use of bull hooks, punishment, or chains. By maintaining the current captive population in protected contact, elephants will be safe against abusive keepers who still use outdated, cruel training methods and keepers will be protected from death or injury (Kane et al. 2005). This will also halt any remaining zoos from chaining their elephants overnight in the face of the extensive documentation that long-term chaining is detrimental to their physical and psychological well-being (Brockett et al. 1999).

- Family members and longtime companions should not be separated (Clubb and Mason 2002; Kane et al. 2005). If a zoo needs to transfer an elephant to another facility, it should transfer the entire herd or, at minimum, socially bonded females.
- No female elephants should be housed alone (Moss 1988; Clubb and Mason 2002; Kane et al. 2005). Zoos currently housing lone females should transfer them to a facility where they can be with others of their species.
- Elephants should be provided with daily behavioral enrichment (Poole 1998; Rees 2000; Clubb and Mason 2002; Kane et al. 2005).
- Elephants should not be confined indoors for more than a few hours each day (Clubb and Mason 2002; Kane et al. 2005). If elephants are being housed in a region where the weather is not conducive for them to be outside each day the zoo should transfer the elephants to a more appropriate climate.
- All physical punishment must be immediately halted (Pryor 1984; Kane et al. 2005). This includes the use of bull hooks, hot shots, axe handles, and other abusive devices used to create pain, fear, and dominance.

Elephants are magnificent species. They are intelligent (Hart et al. 2001) and highly social (Schulte 2000). Although numerous elephants are currently living in substandard conditions, their future does not have to be as bleak as their current situation. Humans have taken these species and exploited them for their body parts and for "entertainment." The time has come for humans to work diligently to correct the many wrongs that have been done to elephants over the years. Once humans stop these cruel actions against elephants, elephants will finally find the peace and dignity they deserve.

References

APHIS. 1997. www.aphis.usda.gov/lpa/news/1997/08/JDAVNPRT.HTM.

APHIS. 2003. *Guidelines for the control of tuberculosis in elephants 2003*. The National Tuberculosis Working Group for Zoo and Wildlife Species. www.aphis.usda.gov/ac/Elephant/TB2003.pdf.

American Zoo and Aquarium Association (AZA). 2001a. *AZA schools for zoo and aquarium personnel: Principles of elephant management*. Silver Spring, MD: AZA.

American Zoo and Aquarium Association (AZA). 2001b. *AZA standards for elephant management and care*. Silver Spring, MD: AZA.

Baer, J. 1998. A veterinary perspective of potential risk factors in environmental enrichment. In D. J. Shepherdson, J. D. Mellen, and M. Hutchins (eds.), *Second nature: Environmental enrichment for captive animals* (pp. 277–301). Washington, DC: Smithsonian Institution Press.

Beck, B. B., and Power, M. L. 1988. Correlates of sexual and maternal competence in captive gorillas. *Zoo Biology* 7: 339–350.

Brockett, R., Stoniski, T. S., Black, J., Markowitz, T., and Maple, T. 1999. Nocturnal behavior in a group of unchained female African elephants. *Zoo Biology* 18: 101–109.

Brown, J. 2000. Reproductive endocrine monitoring of elephants: An essential tool for assisting captive management. *Zoo Biology* 19: 347–367.

Clubb, R., and Mason, G. 2002. *A review of the welfare of zoo elephants in Europe.* Horsham, UK: Royal Society for the Prevention of Cruelty to Animals (RSPCA).

Csuti, B., Sargent, E. L., and Bechert, U. S. (eds.), *The elephant's foot: Prevention and care of foot conditions in captive Asian and African elephants.* Ames: Iowa State University Press.

Desmond, T., and Laule, G. 1991. Protected contact elephant training. In *Proceedings of the American Association of Zoological Parks and Aquariums Annual Conference 1991* (pp. 606–613). Wheeling, WV: American Association of Zoological Parks and Aquariums.

Desmond, T., and Laule, G. 1994. Use of positive reinforcement training in the management of species for reproduction. *Zoo Biology* 13: 471–477.

Douglas-Hamilton, I. 1998. Tracking African elephants with a global positioning system (GPS) radio collar. *Pachyderm* 25: 81–92.

Douglas-Hamilton I., and Douglas-Hamilton, O. 1975. *Among the elephants.* New York: Viking Press.

Dublin, H. T., and Niskanen, L. S. (eds.). 2003. *IUCN/SSC AfESG Guidelines for in situ translocation of the African elephant for conservation purposes.* Gland, Switzerland: IUCN-The World Conservation Union.

Eisenberg, J. F. 1981. *The mammalian radiations: An analysis of trends on evolution, adaptation, and behavior.* Chicago: University of Chicago Press.

Estes, R. D. 1997. *The behavior guide to African mammals.* Berkeley: University of California Press.

Estes, R. D. 1999. *The safari companion: A guide to watching African mammals.* White River Junction, VT: Chelsea Green.

Forthman, D. L. 1998. Toward optimal care for confined ungulates. In D. J. Shepherdson, J. D. Mellen, and M. Hutchins (eds.), *Second nature: Environmental enrichment for captive animals* (pp. 236–261). Washington, DC: Smithsonian Institution Press.

Gruber, T. M., Friend, T. H., Gardener, J. M., Packard, J. M., Beaver, B., and Bushong, D. 2000. Variation in stereotypic behavior related to restraint in circus elephants. *Zoo Biology* 19: 209–221.

Hart, B. L., Hart, L. A., McCoy, M., and Sarath, C. R. 2001. Cognitive behaviour in Asian elephants: Use and modification of branches for fly switching. *Animal Behaviour* 62: 839–847.

Kane, L., Forthman, D. L., and Hancocks, D. (eds.). 2005. *Optimal conditions for captive elephants: A report by the Coalition for Captive Elephant Well-Being.* Madison, WI: Coalition for Captive Elephant Well-Being. www.elephantcare.org/protoman.htm.

Kiley-Worthington, M. 1989. The training of circus animals. In *Animal training: A review and commentary on current practice. Proceedings of a symposium organized by Universities Federation for Animal Welfare, held at the Peterhouse Theatre, University of Cambridge, 26th–27th September 1989* (pp. 65–81). Potters Bar, Hertfordshire, England: Universities Federation for Animal Welfare.

Kurt, F., and Garaï, M. 2001. Stereotypies in captive Asian elephants: A symptom of social isolation. In H. M. Schwammer, T. J. Foose, M. Fouraker, and D. Olson (eds.), *Recent research on elephants and rhinos. Scientific progress reports: Abstracts of the International Elephant and Rhino Research Symposium, Vienna, June 7–11, 2001* (p. 20). Münster, Germany: Schüling.

Langman, V. A., Rowe, M., Forthman, D., Whittington, B., Langman, N., Roberts, T., Huston, K., Boling, C., and Maloney, D. 1996. Thermal assessment of zoological exhibits: I. Sea lion enclosure at the Audubon zoo. *Zoo Biology* 15: 403–411.

Laule, G., and Desmond, T. 1998. Positive reinforcement training as an enrichment strategy. In D. J. Shepherdson, J. D. Mellen, and M. Hutchins (eds.), *Second nature: Environmental enrichment for captive animals* (pp. 302–313). Washington, DC: Smithsonian Institution Press.

Laule, G. E., and Whitaker, M. A. 1998. The use of positive reinforcement techniques in the medical management of captive animals. In *American Association of Zoo Veterinarians and American Association of Wildlife Veterinarians 1998 Joint Conference Proceedings* (pp. 383–387). Yulee, FL.

Law, G., and Kitchner, A. 2002. Simple enrichment techniques for bears, bats and elephants: Untried and untested. *International Zoo News* 49: 4–12.

Lindburg, D. 1998. Enrichment of captive mammals through provisioning. In D. J. Shepherdson, J. D. Mellen, and M. Hutchins (eds.), *Second nature: Environmental enrichment in captive animals* (pp. 262–276). Washington, DC: Smithsonian Institution Press.

Line, S. W., Clarke, A. S., and Markowitz, H. 1987. Plasma cortisol of female rhesus monkeys in response to acute restraint. *Laboratory Primate Newsletter* 26(4): 1–4.

Line, S. W., Morgan, K. N., Markowitz, H., and Strong, S. 1989. Heart rate and activity of rhesus monkeys in response to routine events. *Laboratory Primate Newsletter* 28(2): 9–12.

Manier, J., and W. Mullen. 2005. Zoo plans to study elephant well-being. *Chicago Tribune*, May 3.

Martin, E., and Stiles, D. 2002. *The South and South East Asian ivory markets.* London: Save the Elephants.

Mench, J. 1998. Environmental enrichment and the importance of exploratory behavior. In D. J. Shepherdson, J. D. Mellen, and M. Hutchins (eds.), *Second nature: Environmental enrichment in captive animals* (pp. 30–46). Washington, DC: Smithsonian Institution Press.

Moberg, G. P. 1985a. Biological response to stress: Key to assessment of animal well-being? In G. P. Moberg (ed.), *Animal stress* (pp. 27–49). Bethesda, MD: American Physiological Society.

Moberg, G. P. 1985b. Influence of stress on reproduction: a measure of well-being. In G. P. Moberg (ed.), *Animal stress* (pp. 245–267). Bethesda, MD: American Physiological Society.

Moss, C. 1988. *Elephant memories: Thirteen years in the life of an elephant family.* New York: William Morrow.

Mullen, W., and Dardick, H. 2005. Zoo opens records on animal deaths. *Chicago Tribune*, May 19 [Metro sec.], p. 1.

O'Brien, D. 1997. Animal exhibitor Cuneo confronting new charges. *Chicago Tribune*, August 9, p. 5.

Olson, D., and Wiese, R. J. 2000. State of the North American African elephant population and projections for the future. *Zoo Biology* 19: 311–320.

Poole, J. H. 1989. Announcing intent: The aggressive state of musth in African elephants. *Animal Behaviour* 37: 140–152.

Poole, T. B. 1998. Meeting a mammal's psychological needs. In D. J. Shepherdson, J. D. Mellen, and M. Hutchins (eds.), *Second nature: Environmental enrichment in captive animals* (pp. 83–94). Washington, DC: Smithsonian Institution Press.

Pryor, K. 1984. *Don't shoot the dog: The new art of teaching and training* (rev. ed., 1999). New York: Simon and Schuster.

Rees, P. A. 2000. Are elephant enrichment studies missing the point? *International Zoo News* 47: 369–371.

Rees, P. A. 2001. Captive breeding of Asian elephants (*Elephas maximus*): The importance of producing socially competent animals. In B. B. Hosetti, and M. Venkateshwarlu (eds.), *Trends in wildlife biodiversity, conservation and management* (pp. 76–91). New Delhi: Daya Publishing House.

Rees, P. A. 2002. Asian elephants (*Elephas maximus*) dust bathe in response to an increase in environmental temperature. *Journal of Thermal Biology* 27: 353–358.

Rees, P. A. 2003. Asian elephants in zoos face global extinction: Should zoos accept the inevitable? *Oryx* 37: 20–22.

Ringling Bros. and Barnum & Bailey Circus. 2003. www.ringling.com/.

Schulte, B. A. 2000. Social structure and helping behavior in captive elephants. *Zoo Biology* 19: 447–459.

Schmid, J. 1998. Status and reproductive capacity of the Asian elephant in zoos and circuses in Europe. *International Zoo News* 45: 341–351.

Seidensticker, J., and Dougherty, J. G. 1996. Integrating animal behavior and exhibit design. In D. G. Kleiman, M. E. Allen, K. V. Thompson, and S. Lumpkin, *Wild mammals in captivity: Principles and techniques* (pp. 180–190). Chicago: University of Chicago Press.

Seidensticker J., and Forthman, D. L. 1998. Evolution, ecology, and enrichment: Basic considerations for wild mammals in zoos. In D. J. Shepherdson, J. D. Mellen, and M. Hutchins (eds.), *Second nature: Environmental enrichment in captive animals* (pp. 15–29). Washington, DC: Smithsonian Institution Press.

Shepherdson, D. 1999. Environmental enrichment for elephants: Current status and future directions. *Journal of the Elephant Managers Association* 10: 69–77.

Smith, B., and Hutchins, M. 2000. The value of captive breeding programmes to field conservation: Elephants as an example. *Pachyderm* 28: 101–109.

Sukumar, R. 1989. The Asian elephant: Ecology and management (2nd rev. ed., 1992). Cambridge: Cambridge University Press.

Taylor, V. J., and Poole, T. B. 1998. Captive breeding and infant mortality in Asian elephants: A comparison between twenty Western zoos and three Eastern elephant centers. *Zoo Biology* 17: 311–332.

Wankie's final journey [editorial]. 2005. *Chicago Tribune.* July 15, p. 28.

Whyte, I. 1993. The movement patterns of elephant in the Kruger National Park in response to culling and environmental stimuli. *Pachyderm* 16: 72–80.

Wiese, R. J. 2000. Asian elephants are not self-sustaining in North America. Zoo *Biology* 19: 299–309.

13 MOST ZOOS DO NOT DESERVE ELEPHANTS

DAVID HANCOCKS

Late in 1975, when I was employed as design coordinator for Seattle's Woodland Park Zoo, I recommended that elephants should be deleted from the zoo's master plan. Seattle had neither the climate nor the funding to meet their needs. I believed that the two elephants at the zoo, one female Asian and one female African, should be moved to a warmer place, with more space and better facilities. The public response was almost entirely abusive. Schoolchildren sent hate mail in a campaign obviously coordinated by their teachers. An angry mother called demanding to know why her child should be denied the chance to see elephants at the zoo. Seattle City Council Member Phyllis Lamphere suggested that if I raised the topic again I, rather than the elephants, would be leaving town. "I personally cannot imagine an urban zoo without elephants," she told local columnist John Hinterberger for his January 21, 1976, column in the *Seattle Post-Intelligencer.*

Removing elephants from the local zoo, no matter how miserable their existence, has traditionally been seen as ridiculous. "How can we call this place a zoo if it doesn't have an elephant?" is a general view held across the history of urban zoos worldwide, and, considering the conditions in which zoo elephants typically have been maintained, it has been surprisingly resilient. Howletts Wild Animal Park, a zoo, in Kent, England, has commented, on its website, that a decision not to keep elephants would be "high minded but difficult . . . in view of the elephants' enormous popularity with the public" (Howletts 2005).

The history of Jumbo, the big male elephant shipped in 1882 to Barnum's circus, after time at the Jardin des Plantes in Paris and at London Zoo, reveals many of the paradoxes and dilemmas of elephants in zoos. Jumbo was captured as a baby after his mother was shot dead in front of him. He was sold to various institutions, never given proper care, put on display as a monstrous curiosity, reduced to a novelty ride, moved almost constantly from place to place, and haggled over by his owners in life and after death (Hancocks 2001). Jumbo was both belittled and adored by the crowds that paid to see him. People wanted to see the massiveness of his form; yet, they saw too the superiority of their own kind. Awed by his size, they were simultaneously emboldened by their audacity in holding captive such a creature; humbled by his great hulk, they were nonetheless prideful of having human control over this giant. Jumbo satisfied the curiosity of all, the vanity of many, and the ego and greed of a few. The story of Jumbo is, sadly, not unique for a zoo elephant.

In some ways, the tenacity in wanting to keep elephants in the zoo is admittedly fathomable, even predictable. The elephant is one of the most traditional zoo animals. London Zoo curator Desmond Morris (Morris and Morris 1966) surveyed British television viewers for their Top Ten animals and, not surprisingly, the chosen list included the elephant. When Yale psychologist Stephen Kellert (1989) surveyed Americans on their most-liked animals, the elephant was the only wild exotic creature that made the list. In both polls, the preferential factors were bigness, intelligence, aesthetics, danger, and cultural and historic relationships.

Walt Whitman, in his 1891 poem *Song of Myself*, declared that, "a mouse is miracle enough to stagger sextillions of infidels" (Whitman 1982). However, for all we might admonish people to stir their imaginations and learn to see tiny creatures as marvelous, people are more readily impressed by massive size. The British poet and visionary William Blake, in 1803, wrote *Auguries* of *Innocence* as a passionate cry for a higher understanding of life and nature. He urged people "To see a World in a grain of sand, And a Heaven in a wild flower" (Blake 1917). But applying that degree of imagination is too daunting for most people, at least without chemical assistance. Most people are more excited about seeing an elephant in a zoo than looking at any frog or butterfly or iridescent beetle, no matter how wonderful such tiny creatures might be. The scale and power of an elephant is an instant and undeniable thrill. Its sheer bulk is awesome, and its strength is astonishing.

We should not be surprised, then, that zoos persist in wanting elephants. The elephant is perhaps the most impressive creature in the world's warehouse of oddities, and certainly a most appealing creature. Cynthia Moss (2000, 15) describes elephants as "intelligent, complicated, intense, tender, powerful, and funny."

Perhaps the elephant is not the strangest of all living things (consider slime mold, cuttlefish, or even the Chihuahua), but it may be the most inexplicable. Eyes of disconcerting awareness peer out from its massive head, like those of a child trapped inside a monster. So much about the elephant is contradictory. Its trunk is packed with muscles of astonishing force, yet used with startling delicacy. It is built like a truck, yet may have a delicate sensitivity for painting (as I have seen both in elephant painting demonstrations and in their finished works) and for music (à la the Thai Elephant Orchestra); professional art critics and music reviewers share this view (Mayell 2002; Scigliano 2002; Wolk 2005). It is the world's biggest land animal, yet it delights in being fully submerged in deep water. All other wild animals respect them, and adult elephants have no natural predators or enemies; nonetheless, throughout their history with humans they have been beaten many times, even to the point of collapse (Lewis and Fish 1978).

John Donne, in his 1612 philosophical poem, *The Progress of the Soul*, called it "Natures great master-peece, an Elephant, The onely harmlesse great thing" (Academy of Natural Sciences 2005). Yet, patient and accommodating as a sheepdog, and despite all the amazement and affection that it inspires, the elephant's history in zoos is replete with accounts of deprivation and hardship. In numerous instances, soldiers have been hired to shoot zoo elephants gone mad in their incarceration (Hancocks 1971; Lewis and Fish 1978). George "Slim" Lewis, in his chronicle of a career of more than thirty years training elephants recounts a very large number of savage beatings, some lasting for several hours, in order to instill fear and "respect" for the elephant trainer. In one story, about his subjugation of an elephant named Lucy: "I searched the tent for something heavier to work on her with, and found a steel automobile axle used as a tent stake. After that, we had our little argument" (Lewis and Fish 1978, 28). That sort of event would have been standard procedure with almost all zoo and circus elephants for well over a century of captive elephant management in North America. In 1992, writer Douglas Chadwick recalled what elephant trainers had told him they might do with "an elephant acting especially ornery." Paraphrasing their words, he relays that "sooner or later a handler may have to just plain beat some elephants in a last-ditch effort to establish control" (Chadwick 1992, 18–19). Even today, some zookeepers beat elephants in their care, although this does not occur as frequently as in the past.

Wretched and persistent health problems are commonplace, too. Indeed, it would be a laborious task to locate more than a few zoo elephants in excellent physical, social, behavioral, and psychological condition. Most of them stoically lead lives of loneliness, physical pain, depression, foot rot, boredom, and inactivity, occasionally interspersed for some of them with outings to the back of the barn accompanied by men with pickax handles.

A shift in public perception about elephants in zoos seems, fortunately, to be under way. In 2004, when Ron Kagan, director of Detroit Zoo, declared his belief that the elephants at that zoo should be relocated to a sanctuary with larger spaces and in a warmer climate he, too, attracted some abusive reactions, but he also received support from local politicians as well as strong encouragement from people and media around the world. Certainly, Kagan prepared the ground for his announcement much more carefully than I did for my announcement in 1975, and he more wisely made certain that he had the required political support beforehand. It is, however, revealing that reporter Hugh McDiarmid Jr. (2004), in a November 20, 2004, article in the *Detroit Free Press*, "Detroit Zoo to Free Elephants," announced that "the decision to send them away comes amid *a nationwide push to provide better care for elephants* [emphasis added]."

Another example suggests that this push for better care is not restricted to North America, as evidenced in the response by the British media to *A Review of the Welfare of Zoo Elephants in Europe* (Clubb and Mason 2002), commissioned by Britain's Royal Society for the Prevention of Cruelty to Animals (RSPCA), and prepared by Ros Clubb and Georgia Mason of the Animal Behaviour Research Group at the University of Oxford's Department of Zoology. The *Review* addressed many factors critical to the welfare of elephants in zoos, including rates of reproduction, maternal rejection, maternal infanticide, mortality of babies, foot health, weight problems, and lack of space and exercise opportunities. Many zoo professionals have rejected this report or have described it as massively flawed. An alleged mistake in one of the statistics in this 245–page document, regarding longevity of elephants in zoos compared with the wild, has generated agitation among zoo professionals. If this supposed error eventually proves actually to be a correct analysis by Clubb and Mason, one hopes zoos will then take a more patient and positive look at the *Review*'s findings. But even if it does indeed prove to embody a statistical error, that alone does not warrant the zoo profession's hostility and the unseemly haste to discredit and abandon the entire document.

The RSPCA's *Review* proposed that all breeding and importation of elephants should cease until factors responsible for poor welfare have been empirically investigated. This recommendation attracted much public attention. "Zoos Must Not Keep Elephants, Demands Report by RSPCA" was one of the more subdued headlines, in the *Independent*, one of the most literate of the British broadsheets, for an article by its environment editor Michael McCarthy on October 23, 2002. An article by Brian Sewell in the November 4, 2002, edition of the *Evening Standard*, London's most widely read tabloid, was more sensationally caustic. Headlined, "Horror Elephants Won't Forget," it referred to a photo in one of the national broadsheets showing a zoo elephant that appears to be smiling for the camera, his mouth open "as though in a guffaw," his great flat foot raised "as though to shake our hands." But the elephant performs this apparent clowning, said Sewell, only because "his two keepers . . . prod him with an ankus to urge him to perform his tricks, and . . . to make him respond, not willingly, but with obedience." He pointedly comments that a zoo is supposed to be "a place, it is so often argued, of conservation, education and research."

Sewell's sort of journalism admittedly tends to generate more heat than light, but it is noteworthy that just a few years ago such criticisms about zoos and their elephants never appeared in newspapers, even though, ironically, conditions have generally improved markedly in the past decade. Similar censures, however, are now increasingly heard. Public disquiet about keeping elephants in captivity is intensifying. The views expressed by Seattle Council Member Lamphere, who in 1976 could not imagine an urban zoo without an elephant, are in sharp contrast with the opinions expressed thirty years later by Chicago Alderman George Cardenas, who, as reported in an article by Andrew Herrman, "Pack up Elephant Exhibit, Alderman Asks Zoo," in the May 10, 2005, edition of the *Chicago Sun-Times,* called for Lincoln Park Zoo to close its elephant exhibit and ship its lone last occupant to an elephant sanctuary.

Most of the public criticism about standards of elephant care in recent years has been directed toward circuses, but zoos are increasingly now also under scrutiny. Starting in 2002, the media in Anchorage, Alaska, carried many reports and comments about an African elephant, Maggie, who had been living alone in the Alaska Zoo since 1997 without previously generating public concern. In the face of mounting public comment, Alaska Zoo in 2005 constructed a treadmill to provide exercise during the long Alaskan winters when Maggie was kept indoors. The zoo was unable to persuade the elephant to use this device. The *Anchorage Daily News* on May 16, 2006, in an article by Megan Holland, "Maggie's Treadmill Isn't Working Out Yet," quoted Alaska Zoo Director Pat Lampi's claim that the zoo had spent "more than $1 million on Maggie's living quarters so far, and that number is expected to increase as the improvements continue." Nonetheless, in May 2007, with media coverage intensifying, the zoo agreed to relocate Maggie to a facility with more space and other elephants, in a warmer climate. Both of the U.S. elephant sanctuaries offered to take Maggie and to pay all relocation costs. The zoo responded that their final decision might take until 2008. An online poll on May 30, 2007, by local NBC television station KTUU asked if the zoo should wait several months to make this decision. Seventy-six percent of the respondents said, "No"; 21%, "Yes." In September 2007, the zoo agreed to send Maggie to Performing Animal Welfare Society (PAWS) sanctuary in California. There are several astonishing aspects to this story, but most remarkable is that the media and the public gave such persistent and concentrated attention to the plight of a lone zoo elephant.

A couple of decades ago the media began to focus on wretched zoo conditions for polar bears. Many zoos have since abandoned this species, especially across Europe. Can we now expect similar results with elephants?

Different Points of View

We know too little about how to manage elephants. New information is required if zoos are to provide for their needs more accurately: to discern between what is beneficial and what is damaging, what is merely useless tradition or hearsay, and what is measured and proven. The pros and cons of existing practices need to be calibrated. Currently, standards and methods vary immensely from zoo to zoo. Many keepers are either self-taught or carry out routines and apply techniques passed from one transient keeper to another.

The prevalence of inadequate conditions for captive elephants is shameful when compared with the zealousness of zoos to own them. Despite this, it is hopeless to respond simply that zoos should never keep them because so many are determined to keep elephants in zoos under almost any circumstances, which ensures that the problems cannot be wished away. Moreover, as Cynthia Moss pointed out after spending more than thirty years studying elephants in Africa, life for elephants in the wild is frequently replete with anguish, often ended in horrible ways through human action, whereas in very good captive conditions, elephants might think they are in paradise (Moss 2002). Can we, though, create conditions to compare favorably with a secure existence in a good wild habitat?

The best beginning point for this exercise, I propose, before we even start to consider any details of management or design, would be the development of a new set of attitudes and a more sympathetic sensibility toward their captive care. At present, it seems as though a siege mentality surrounds the management of elephants in many zoos: a clandestine atmosphere prevails, with many of the dangerous hallmarks of a secret society. Training sessions are often shrouded in mystery. Even though more than half of Association of Zoos and Aquariums (AZA—formerly, American Zoo and Aquarium Association) accredited zoos have protected contact programs for their elephants, many elephants continue to be managed in an atmosphere of subjugation: the ankus remains in common use, as a tool of control and as a symbol of traditional regimes based on domination and fear. What is labeled as obedience for many zoo elephants around the world is, I fear, often only dread; that, and a sense of helplessness learned from punishments and restrictions.

In 2000, when I was acting director of Melbourne Zoo, a new manager of animal programs, Laurie Pond, began applying a fundamental change in approach to the care of the zoo's two elephants, who were suffering mental and physical anguish. Since their arrivals from the wild in 1977 and 1978, they had lived lives of almost constant stress, deprivation, and boredom. Pond introduced key changes, creating diverse opportunities for the elephants' mental stimulation and physical activity. His new approach focused on "building strong affectionate and enduring relationships between keepers and elephants" (Stroud and Kudeweh 2001, 1). The changes brought surprisingly quick results. Peter Stroud, then senior curator at Melbourne Zoo, noted, "Within a month the changes in the daily regime were clearly affecting the animals' demeanor" (P. Stroud, personal communication). These elephants appeared to be much more alert, active, and happier than before. They seemed to be much more contented and calm.

In 2002, I was invited to prepare a paper entitled "Some Zoos Do Not Deserve Elephants," for presentation at the 2003 symposium, "Never Forgetting: Elephants and Ethics," at Front Royal, Virginia. (I have modified that paper for this essay and changed the title, which is now more precisely, "Most Zoos Do Not Deserve Elephants.") To prepare for the paper, I asked Pond for his views on whether zoos deserve elephants. His response was to rephrase the question, asking why we should assume that elephants deserve zoos? He believes that elephants do not deserve the abuse routinely inflicted on them in zoos nor the inadequate attention to their needs. Pond reminded me that Barbara Woodhouse, a guru of dog keeping who had a popular worldwide television show in the 1970s, repeatedly reminded her viewers that there are "no bad dogs." He suggests the same is true for elephants. If there are "bad elephants," he says, "then we have surely made them that way." His view is that if zoos did not habitually impose restraints on them, we would find ourselves "dealing with a different animal" (L. Pond, personal communication, 2002).

A similar perspective to Pond's is manifest at two elephant facilities in North America: ARK 2000, a sanctuary operated by Pat Derby and Ed Stewart, co-founders in 1986 of PAWS in California, and The Elephant Sanctuary (TES) in Tennessee, founded by Carol Buckley and Scott Blais in 1995. Both facilities maintain several African and Asian elephants in paddocks hundreds of times larger than those found in zoos, of greatly varying topography, furnished with bodies of deep water. Derby tells me the Milwaukee Zoo and the Ringling Bros. and Barnum & Bailey Circus labeled the elephants that were shipped from them to the sanctuary as "killers" or as "dangerous." The animals in every case had their chains removed the day they arrived at the sanctuary and have "never been restrained or reprimanded since that day" (P. Derby, personal communication). Attitudes underlying the management of elephants at these sanctuaries is in contrast to the traditional zoo approach of domination and command. Derby states that "reprimands are never used in the day to day care of [our] elephants" (PAWS 1999, 56) and Buckley also notes, "at no time do we command our elephants or force them against their will" (TES 2006, question 48). Staffs at each of these sanctuaries never resort to chains, hooks, or the ankus.

The philosophies and attitudes of elephant caregivers such as Pond, Stewart, Derby, Buckley, Blais, and just a few others I know of who are associated with zoos, notably Gail Laule, Ellen Leach, and Colleen Kinzley, are unusual. Many more zoo professionals approach the management of elephants from a platform of dominance and control. In peer-review response to an earlier, similar paper I wrote on this topic, I received a four-page lecture on "training theory" that began by advising that I needed to understand that "all animals learn through a combination of reinforcement and punishment," and that "one cannot exist without the other." The ankus, I was informed, "is a tool that is used to teach, guide and direct the elephant into the proper position or to reinforce a command with a physical cue" (Anonymous 2005, personal communication). This approach seems to be considered generally acceptable and proper within the profession; the debates swirl not around the concept but about the niceties between meanings of words such as "discipline" versus "punishment." It brings to mind the spins and twists of United States officials debating the differences between describing whether people detained in such prisons as those at Cuba's Guantánamo Bay or Iraq's Abu Ghraib were subject to "interrogation" or to "torture."

Until 2006 (when one caregiver died after a sudden attack from an elephant, Winkie, who had suffered many years of beatings and other abuse at the Henry Vilas Zoo, in Madison, Wisconsin), after a combined total of thirty-one years of operation, there had been no fatalities or serious injuries to staff at either of the elephant sanctuaries in Tennessee or California, where the focus is on cooperation and free choice. Conversely, elephants killed fifteen keepers in U.S. zoos between 1976 and 1991 (Lehnhardt 1991). There were many other such deaths at zoos around the world during this period. Worldwide reports of elephants killing zookeepers appear quite regularly. It would require diligent research to obtain specific figures, but the Sydney office of the International Fund for Animal Welfare (IFAW) reports that, on average, one elephant keeper a year is killed in North American zoos and circuses alone (IFAW 2004).

Perhaps these depressing statistics are why one hears many zoo personnel routinely describe elephants as "untrustworthy" or "unpredictable." There is, however, no social evolutionary benefit to elephants for carrying such characteristics. Describing elephants in such terms is merely an excuse for our ignorance. If we could instead come to recognize that elephants are not inherently "unpredictable," and that "untrustworthy" elephants are actually the products of mismanagement, we could then see there is no sense in continuing to apply such control and domination techniques as chaining, prodding, hooking, and beating.

The problems we need to resolve for zoo elephants are many and pervasive. They include sometimes very harsh treatment but also arthritis, foot rot, loneliness, cold, stress, obesity, inadequate diets, gross inactivity, stereotypic behaviors, and, most challenging, the lack of natural long-term social structures.

Fit and Healthy

Wild elephants are characteristically active beings, usually roaming for more than two-thirds of every twenty-four-hour day (Eltringham 1982). Most zoos, by contrast, historically kept their elephants chained for at least two-thirds of each day (Taylor and Poole 1998) and sometimes, especially for male elephants, for twenty-four hours every day of the year (Lewis and Fish 1978). Less than twenty years ago a survey found that elephants were kept chained for sixteen to eighteen hours a day at a majority of U.S. zoos (Galloway 1991). Today, the AZA says chaining is an acceptable method only for temporary restraint (AZA 2003) and the European Association of Zoos and Aquaria (EAZA) nominates a maximum of three hours a day (EAZA 1997). These recommendations can be viewed as a liberal shift in policy; even so, we must hope to see a day when time spent in chains is recognized as an unbearable frustration for a big animal so adapted to such long periods of daily activity, and when chaining an elephant in a zoo will be regarded as being as inappropriate as chaining a gorilla, a giraffe, or a rhinoceros.

Much of the considerable amount of time in which wild elephants are active is spent walking. Asian family groups routinely cover up to nine kilometers a day (Reimers, Schmidt, and Kurt 2001) and twelve kilometers is normal for African herds (Wyatt and Eltringham 1974). This constant activity exercises joints and ligaments, maintains muscle tone, burns fat, and ensures good blood flow. It also contributes to mental stimulation, creating constant shifts in exposure to varying landscapes and a consequently increased richness of visual experiences and variety in visual content. Zoo elephants, by contrast, enjoy very little exercise, and little variation, if any, in their daily routines. This is very much to do with being kept in small areas and in spaces devoid of stimulus or interest. The resultant lack of exercise is a major factor in the prevalence of foot problems among zoo elephants, such as overgrown nails, split nails and soles, and abscesses (Roocroft and Oosterhuis 2001), and the situation is exacerbated when elephants must walk or stand on unyielding substrates such as concrete or asphalt. Asian elephants, especially, seem prone to foot problems in zoos; bone infections that develop from chronic diseases of the pad are a principal concern. Persistent foot problems have on occasions resulted in amputation or euthanasia for zoo elephants (Boardman et al. 2001).

The size of zoo enclosures is, of course, very much smaller than wild ranges. For many species in zoos the amount of space is not a critical problem if the enclosure offers an appropriately rich and complex environment. However, elephants are big and vigorous animals that simply must walk considerable distances each day, not just in mindless plodding circles when giving rides, but engaging in long-distance walking as both a physical and mentally stimulating exercise. Female zoo elephants are 31% to 72% heavier than their wild counterparts, probably as a result of high fat content in their diets and lack of exercise (Clubb and Mason 2002, table 10). Inactivity, moreover, not only leads to obesity but also to arthritis, joint problems, and other common disorders in zoo elephants (West 2001).

The shape of zoo enclosures and the complexity of their landscape and furnishings are therefore particularly significant design factors. Enclosure configurations can be devised to encourage exploration, and specific components within the space can stimulate activity. Multifaceted and variable (but also large) spaces offer one of the important keys to "behaviorally flexible, fit and competent [zoo] animals" (Forthman et al. 1995, 395).

Hygienic facilities for wallowing in mud and for bathing in dust or loose soil, although not necessarily easy to design or to maintain, should be considered essential. It would be best if the elephants could be allowed the exercise of making their own wallows (Roocroft and Oosterhuis 2001). Scratching posts are requisite and would be even more useful if they doubled as exercise equipment to be demolished or pushed to the ground. Variation in substrates might have psychological as well as practical benefits and in any case should provide opportunities for deep digging, maybe in areas of soft substrate regularly salted with minerals as incentive and reward. Concrete and asphalt substrates, which disallow digging, and are punishing to feet, should always be rejected.

Water is invariably given insufficient attention in enclosure design. The AZA (2003, 3) recommends that zoo elephants should have access to water while outdoors, "such as a pool, waterfall, misters/sprinklers, or wallow that provides enrichment and allows the animals to cool and/or bathe themselves." This is not sufficient. One could literally meet these guidelines by providing only a spray of mist. Eisenberg (1980) notes that water is an intensely integral part of daily life for wild African elephants. It would help if we came to regard elephants as almost essentially amphibious creatures. For an animal that can swim thirty miles (Jones 1999), that has evolved from aquatic ancestors with a long history of snorkeling (West et al. 2003), and that appears to find besotted pleasure in wholly submerging itself in water, we are merely tinkering with minimalist measurements in our current requirements. Moreover, the changing character of water can be more creatively and usefully exploited than typically happens in zoo design. Water that falls from different heights and in varying quantities; that sprays, crashes, seeps, trickles, mists, splashes, meanders, or rushes; that forms just muddy ground or shallow puddles, fast rivulets or meandering streams, bathing pools or deep lakes, all combine to provide a variety of options for different interactions. The EAZA (1997) states that zoo elephants must have daily access to a pool one meter deep. I propose instead that a pool deep enough for complete submersion, allowing elephants the pleasure of buoyancy and taking the weight off their feet, should be considered a basic requirement.

The consideration of water raises thought of its uselessness when it freezes, and the enormous difficulties of trying to cater to elephants in a cold climate. Across Europe and North America, where most zoo elephants live, it is estimated that cold wet weather or near freezing conditions confine elephants to indoor quarters (and thus greatly reduced exercise opportunities) for up to 80% of the days during five months of the year (Lindburg 1998).

Elephants are such lovable and intelligent beings they deserve to have all the things they would wish for themselves. However, even within human-imposed standards of lesser criteria, the demands of providing suitably large, complex, changeable, and challenging environments for intelligent animals of such enormous bulk and strength as elephants are beyond the capacity of most zoos. It is a problem that requires resources and commitments well beyond those currently being applied. Indeed, whether we can satisfactorily provide designed environments for the needs of *any* big, strong, intelligent animals is still open to question. Cetaceans in aquatic parks, bears and chimpanzees in zoos, and pigs in intensive farming facilities, are examples of species that routinely suffer very inadequate lives in poorly designed spaces. It is disturbing that simply providing zoo elephants with a sufficiently appropriate big space to encourage good levels of activity has so rarely been considered.

It is also probable that many zoo elephants are not provided with well-balanced diets. More research on this would be very useful. In the wild, elephants consume a very wide range of vegetation, including leaves, stems, bark, roots, twigs, herbs, flowers, and fruits (McKay 1973; Sukumar and Ramesh 1995). The typical zoo diet, however, consists basically of hay, with a smaller proportion of garden produce, some commercial concentrate feed, and vitamins and mineral supplements. Only a small number of zoo elephants have occasional access to sites for grazing and browsing (Taylor and Poole 1998). Branches and leaves are sometimes offered as a supplement, but most zoos in temperate and cool climates cannot obtain fresh browse for several months of the year. The activity involved in feeding is as important for zoo animals as is the nutrition it provides. In the wild, Asian elephants often laboriously dig out roots, and African elephants exert much time and energy tearing food from trees. Whereas wild elephants spend up to 80% of their waking hours foraging (McKay 1973), feeding activity for zoo elephants accounts for only about 35% of their time awake (Taylor and Poole 1998). The deficit of foraging opportunities is suspected to contribute to stereotypic behaviors in many species in captivity (Carlstead, Seidensticker, and Baldwin 1991); this may be especially so for elephants. Zoo elephants consume about twice as much in an hour as wild elephants because their food is readily available and requires no manipulation (Kurt and Schmid 1996).

The manner in which food is presented to zoo elephants is thus critical. Fixing fresh browse at various heights that requires stretching and hard pulling could encourage some natural feeding behaviors. Techniques to encourage other beneficial behaviors could include the securing of big posts or dead trees (which could also be dressed with securely fastened leafy branches) in such a way that they could be pushed over with exertion. Holding fixtures would need to be engineered to allow these posts and dead trees to be reset upright. Other devices could be programmed to present food at irregular intervals, or in varying locations, thus promoting exploration.

Shrubs, grasses, and, in particular, large trees play an integral part in the life of a wild elephant: They should be considered essential components of zoo exhibit design. Enclosures should be large enough to maintain grass growth over a significant percentage of the area. Sadly, allowing elephants access to live vegetation requires replacement of the plants at regular intervals. This vexatious problem demands some creative attention. Perhaps we first need to recognize that the value of the well-being of an elephant is worth the periodic investment in live vegetation for consumption, manipulation, and destruction.

Protected islands of vegetation and other environmental features located within the paddock also help to ensure the animal space is not simply an empty and undefined area. Sufficiently tall and dense vegetation in these islands will provide shade to the interior of the paddock on very warm days, promoting comfort and helping elephants to avoid sunburn. Preferably, I suggest, at least 30% of the enclosure should have dappled shade between the hours of about 11:00 a.m. and 3:00 p.m. during summer months. These vegetation islands will also define specific spaces that increase spatial cognition and enhance visual interest by alternately revealing and hiding views. Because of their bulk, elephants take a long time to warm up or to cool down (Poole and Taylor 2001). Enclosures therefore need to have open sunny areas as well as shade; the latter, in the majority of zoos, ideally provided by deciduous trees to allow winter sunshine into the enclosure. Such microenvironmental differences encourage movement and exploration.

The interrelated predicaments of foraging, mental stimulus, physical activity, and comfort for zoo elephants are so challenging, and yet so important, they should not be addressed as isolated design problems within individual zoo exhibit projects. I propose that either the World Association of Zoos and Aquariums (WAZA), or perhaps a combination of organizations such as AZA, EAZA, and the Australasian Regional Association of Zoological Parks and Aquariums (ARAZPA), form a task force of elephant ethologists, zookeepers, landscape architects, structural engineers, and environmental enrichment experts to explore novel and practical solutions that can be used as guidelines and design rules for all elephants in all zoos.

Management

Enclosure design directly affects quality of life for zoo elephants. A well-designed enclosure has to allow ways to present fresh foods that are chal-

The elephant house built at the National Zoo in Washington, DC, in the 1930's reflected the inadequate conditions long provided for most North American zoo elephants. Cramped living quarters, lack of enrichment in small outdoor yards, and long periods indoors during cold-weather months have been problems for zoo elephants even in the leading zoos of the United States, with such inadequate exhibit conditions reproduced in more inferior forms in zoos worldwide. Zoo elephants and the millions of zoo visitors throughout the world deserve much better.
Photograph, 1980, courtesy of David Hancocks

lenging and appealing for the elephants; must integrate a diversity of mud wallows, soft substrates, widely varying topography; include trees and grasses; and provide deep and shallow water bodies for splashing, paddling, and swimming, as well as other water features that entice interest and activity. Comfortable temperatures, places to relax, to lie down easily and to comfortably get back on their feet, opportunities for long walks, an abundance of different types of exercise, and all the other aspects of spacious natural habitats that provide satisfaction and interest for elephants must be assured.

In addition to introducing new design standards, some traditional aspects of zoo management must be eradicated. Zoos can provide safety from dangers that wild elephants face from human hunters: they should also ensure sanctuary from all human abuse. Elephants should not have to suffer the painful feet and teeth problems, arthritic joints, anemia, and enteritis that characterize poor zoo conditions and never should become morose and overweight through inactivity. I propose we regard it as equally plain that humans should never hit elephants, never deliberately cause them any pain, and never keep them in solitary confinement, divorced from contacts with other elephants.

It was noted earlier that elephants can be dangerous. The main factor that makes elephants so attractive to zoos, their bigness and strength, means that even in play they can inadvertently inflict great damage to frail human bodies. Further, many elephants degraded by a life of tension and punishment have inflicted deliberate injury on human trainers. A relatively recent response to all this has been the adoption of a system of management called "protected contact" (PC). Behaviorists Gail Laule and Tim Desmond, in 1989, started designing a management system for enhanced keeper safety while effectively training elephants for routine husbandry and veterinary protocols (Desmond and Laule 1991). The PC system they devised was predicated on positive reinforcement and the willing cooperation of the elephant.

Zoos in North America have moved rapidly in recent years to adopt PC as a management system for elephants, and it is now employed by more than half of all AZA-accredited zoos. Initially conceived and designed to address the two fundamental objectives of keeper safety and elephant welfare, it seems to have been accepted by many zoo managers more enthusiastically as a defense against legal liabilities than as a means of protecting elephants from potential abuse. One unfortunate aspect of this perspective is that AZA recommendations (2003) permit zoos to use the ankus even in PC. This abrogates the original criterion of a system, based on positive reinforcement only, which prohibited physical punishment, and aimed to ensure the optimal care of elephants for daily management, husbandry, and veterinary procedures (G. Laule, personal communication, 2005).

The job of elephant keeping is remarkably similar to parenting. Raising kids and caring for elephants are each enormously demanding. Success or failure in either one can set the course for a life of confusion and stress or of confidence and well-being. Although good parenting is one of the most important tasks for any human, with critical consequences for both the child and for the society he or she will live in, no formal training is required for the job, and it is rarely considered in school curricula. Most people learn to parent by doing. Failure or success relies not only on techniques, however, but also equally as much on the parent's disposition. Exactly the same, I posit, is true for elephant care. The similarities between humans and elephants, both long-lived, perceptive, intelligent, strongly social species, with many years of infancy and learned behaviors, makes them equally susceptible to acting like unpredictable thugs if they are subjected to prolonged deprivations, especially during their formative years of development.

When I was director of Woodland Park Zoo in Seattle between 1976 and 1985, like almost all zoo directors, I had no experience or training in managing elephants. Because I was prevented from relocating the elephants to

Bamboo, a teenage Asian elephant, walking the grounds at Seattle's Woodland Park Zoo in the late 1970s with her keeper, Wayne Williams, each in
distinctive headgear. An especially perceptive elephant, with an appealing
personality, her long walks around the ninety-acre zoo site included the stimulus of meeting several other species. In the late 1980s, a new administration
introduced a different philosophy. Walks around the zoo were abolished.
Chaining was introduced, for almost sixteen hours every day. Some staff
complained about use of discipline they considered harsh. By the mid-1990s,
Bamboo was a changed animal: no longer cooperative or trustworthy. In
2004, she was shipped to Tacoma's Point Defiance Zoo, then back to
Seattle in 2006. Offers by The Elephant Sanctuary in Tennessee to take her
into their care have been rejected.
Photograph courtesy of David Hancocks

a better location, I supported and encouraged the notion of only using positive encouragement and of abolishing the practice of chaining overnight.
Today, elephants at Woodland Park Zoo are managed in a PC system, but
during my tenure, as with all zoos of the time, a "free contact" system was
employed. In this environment, the establishment of an enlarged number of
keeper staff who shared my dislike for using force or hooks was encouraging. I gave approval for staff to take the elephants for walks around the
grounds, so the entire park could become their exercise yard. One Asian
elephant, Bamboo, although she was then a teenager, was the best candidate for these activities; she was sweet natured, bright, and cooperative and
interacted safely and calmly with visitors inside or outside the exhibit area.

After my 1985 departure significant changes were introduced by the new
director, David Towne. A circus-trained consultant, Allen Campbell, was

hired to prepare the elephants for a move into a new exhibit area. The keeper crew, which had included female keepers, became an all-male group. Campbell, who was killed in 1994 by a circus elephant in Honolulu, reintroduced all-night chaining for the Seattle elephants. They were chained by one front leg and one back leg from 4:00 each afternoon until 7:30 the next morning: fifteen and a half hours of unimaginable frustration. In addition, some keepers began to apply what some other staff considered "illogical" punishments, with hooks and other handling techniques introduced and used in a manner they found "harsh." Bamboo, who had been the calmest of the elephants, began attacking her new handlers: she had become yet another problem zoo elephant. More recently, staff became concerned about Bamboo's negative attitude toward a baby elephant, Hansa, born at the zoo in November 2000. Considering the depth of attachment that typifies the relationship between female and infant elephants in normal elephant society, it seems most remarkable that Bamboo reached a point where she was untrustworthy with a baby. Bamboo was removed from the group in late summer 2005 and shipped to Point Defiance Zoo, Tacoma, where she lived for nine months with other elephants also considered to be difficult individuals. This experiment failed because Bamboo remained morose and did not accept the other elephants. She was transferred back to Woodland Park Zoo in June 2006.

The story of these tragic relocations, and the deteriorated mental and emotional state that Bamboo was surely enduring, was treated by the local media as the opportunity for comic relief. Zoos repeatedly claim that their justification for keeping elephants is in large part substantiated by the respect and awe that they generate in visitors and the subsequent public support for elephant conservation. There seems to be precious little evidence for this in the real world, and it certainly does not appear to be effective with news reporters. Employing the hackneyed puns that are routinely rolled out for stories about local zoo elephants, a report by Peggy Andersen on Bamboo's first lamentable relocation in the *Seattle Post-Intelligencer* on January 14, 2005, for example, was jauntily headlined, "Cranky Elephant Packing Her Trunk, Leaving Pesky Youngster Behind."

Approached with the proper spirit, the difficult tasks of elephant keeping or of human parenting may allow an occasional mistake to be accepted as unintentional. In an atmosphere of affection and assurance a zookeeper or a parent might be forgiven by an elephant or by a child for small errors of judgment. An atmosphere of mistrust, however, causes the spirit to shrink and wither and generates hostility. A chronic state of fear is linked to distress and anxiety (Dawkins 1980; Pryor 1985), which in turn are linked to aggression (Chance 1994). It would come as no surprise, therefore, to consider that when fear is omnipresent an elephant, or a human, may exploit an opportunity to wreak fatal revenge.

I am convinced there is never any justification to hit a child. One reason is that they are so much smaller than you are. I am also certain that it is equally wrong to hit an elephant. One reason is that they are so much bigger than you are. It makes no sense to bully an animal that can so easily overpower you. That strategy may win only a Pyrrhic victory. We should imagine the day when picking up an ankus is recognized as an admission of defeat. Admittedly, I have never worked directly with elephants, and some will claim that my beliefs are naïve, but I remain adamant that the concept of punishment is alien to an elephant. This statement is violently disputed by zoo elephant managers who approach their task as a matter of training and dominance. I am loudly and frequently informed that wild elephants often punish each other.

Consider, then, how different zoo elephant management protocols might be if the observations of Joyce Poole were accepted as paramount. Poole reports from her decades of study in the field that African elephants "do not discipline their young," nor is discipline "natural in elephant society (and) therefore something that an elephant can understand" (Poole 2001). She states:

> I have no idea how this myth was started, but I have never seen calves "disciplined." Protected, comforted, cooed over, reassured, and rescued, yes, but punished, no. Elephants are raised in an incredibly positive and loving environment. If a younger elephant, or in fact anyone in the family, has wronged another in some way, much comment and discussion follows. Sounds of the wronged individual being comforted are mixed with voices of reconciliation. (Poole 2001)

Aggression has been noted as quite common between zoo elephants (Adams 1981). This may be because aggression toward zoo elephants has been quite common. Although Poole has never seen discipline or punishment administered to calves by adult wild elephants, the ankus can be applied indiscriminately in zoos even to baby elephants. At Woodland Park Zoo in 2001, Hansa, just a few months old, was allowed by her keepers to play with a rubber tub used for bathing. When the same tub was used for feeding and she tried to play with it, she was struck with the ankus. Similar mixed messages were delivered when she attempted to play with a water hose. Sometimes she was allowed to play with it, at other times she was punished. This baby elephant was observed being hit with an ankus at least eleven times in one week (Scigiliano 2002). Later, when Hansa was eighteen months old, she was beaten with an ankus more severely, for eating dirt. When news of this became public and was reported by Candace Heckman in the *Seattle Post-Intelligencer* on July 3, 2002, under the headline, "Zoo's Treatment of Elephants Too Harsh," the Zoo's deputy director Bruce

Bohmke defended the application of blows as within guidelines: "It was appropriate," he said. The newspaper noted the comment by Jane Garrison, an animal welfare campaigner, who considered that the blows were not appropriate, "Elephants in the wild are not reprimanded by their mothers for eating dirt."

Attitudes of dictatorial control lead to the view that punishment is permissible. In that spirit, beating is easily justified. Approval by administrators adds security. A willingness to hit leads sooner or later to desensitization of the antagonist, making it possible to apply increasingly harsh punishments. Although beating induces temporary submission it can also stimulate delayed aggression. Manifestation of such hostility by the elephant is then used as justification for further punishment. It is a detestable cycle, and only the humans involved can break it. Stop the punishments and you go a long way to stopping the reasons for them.

Changes in approach toward zoo elephant management will, I suggest, help to create improved elephant welfare, plus a safer working environment. Aggressive postures and an arrogant spirit, even in protected contact, combine to create stress that can eventually manifest itself in poor elephant health. Attitudes of abundant affection and high regard, however, can lead to positive results. As with so many other aspects of elephant management, there is pressing need for objective studies of the disadvantages and advantages of free and protected contact and other related methods. Changes in attitude, though, need not await such investigation, and can be applied immediately, and without risk.

Social Problems

We can expect and hope to find practical design solutions for problems of inactivity among zoo elephants. We can be optimistic that zoos will fund more research to help resolve present problems with elephant care. In addition, we may yet see changes in attitudes, but there remains an elemental problem of such magnitude that it alone makes it impossible for most zoos to keep elephants. It is such an overarching need for elephants, and one so very difficult to satisfy, that I was tempted to place it at the beginning of this essay. On reflection I decided to present other problems first; ones that, while pressing and difficult, can nonetheless be corrected by all zoos with elephants in their care.

A resolve to deal with the problems of exercise and diet, punishment and discipline, and the development of more caring attitudes, might well gain impetus in the near future. I would like to encourage those changes. The matter of satisfying elephants' social needs, however, is daunting. Lifelong social bonding in stable families is perhaps the most fundamental and irrevocable aspect that characterizes elephant life in the wild (Douglas-Hamilton 1972; Moss and Poole 1983; Sukumar 1989). To be able even to consider meeting this enormous challenge will demand levels of commitment and resources beyond anything currently applied to the care of zoo elephants.

Cynthia Moss and Joyce Poole, especially, have clearly demonstrated that adult African elephants and their young offspring in nature enjoy a close and steadfast family life with extensive gregarious connections to relatives at many social levels (Moss and Poole 1983). Most likely the same is true for Asian elephants. The astonishing sophistication of elephant communications offers a clue to the importance of their social interactions. Female African elephants, for example, can distinguish the vocalizations of at least 100 elephants from at least fourteen families (McComb et al. 2000). Female elephants spend their entire lives within the close bonds of their family. This degree of stability and the depth and complexity of the family's social interactions are completely at odds with the social conditions in most zoos.

Family units in the wild average about six to eight animals for Asian elephants, and for African ones about four to twelve individuals, in mixed groups of females and young, including prepubescent males (Douglas-Hamilton 1972; Sukumar 1989). A small number of zoos, notably Emmen Zoo in the Netherlands, Howletts Wild Animal Park in England, and Disney's Animal Kingdom in Florida, manage family groups of more than a dozen elephants, but the more typical number in most North American and European zoos is a single female Asian elephant or two female African elephants of similar ages, most of whom have never given birth (Schulte 2000; Clubb and Mason 2002, Tables 11 and 12).

It is difficult to assess the age structure for wild elephant populations, but various studies suggest that it tends to follow a bell curve (Moss 2001). A study of Sri Lanka's elephants, for example, shows a small percentage above the age of 45, curving down to about 30% under the age of 5 (Kurt 1974). The age graph of North American and European zoo populations tracks a quite different line. It shows an almost complete absence over age 45, with a bulging curve bellying out between ages 15 and 34 and a rapid depression of the curve below 15 years of age, to less than 2% (Schulte 2000; Clubb and Mason 2002, Figure 5).

The mixed age structure within elephant herds is important because elephants acquire many skills from others in their cohort, such as how to forage, where to find water, and how to care for their young. Zoo elephants may not need to learn how to find a water hole during a drought; yet, the opportunity for elephants to gain maternal skills remains crucial if zoos are to maintain social groups successfully. This opportunity, however, is almost absent in zoos.

Social bonds between females in wild family units are formed vertically, from mother to daughter to granddaughter. Most zoo elephants have been captured in the wild or taken from elephant work camps and do not share close levels of familial relatedness (Clubb and Mason 2002, Figure 2). A study of four unrelated female elephants that had been together since 1 or 2 months of age in a Sri Lankan elephant orphanage found no well-defined bonds between them (Poole et al. 1997). Nonetheless, some specific zoo females do form close associations with one another. They show much stress and agitation when their social partner is removed and a very high level of interaction when reunited after an enforced absence. These zoo relationships, however, are characteristically only between a pair of females rather than, as in the wild, among several members throughout the herd.

Very few pregnant zoo elephants have seen, or been involved in, caring for an infant. As witnessed in Africa, in the wild, all the females in the family group take a great interest in the infants that are invariably present (Lee 1987). Wild elephants give birth within the herd, receiving close attention and assistance from experienced females (Moss 2000). The exact opposite characterizes the zoo situation, where females are often removed from the group and sometimes chained by all four feet when birth is considered imminent (Schmid 1998). The lack of experienced mothers in zoos may be a significant factor in the high incidence of infant rejection and infanticide among zoo elephants (Kurt and Mar 1996). The importance of learning maternal skills is underscored in an evaluation of eight Asian zoo elephants imported from the wild at a young age, in two groups. One group, consisting of three elephants, went to a zoo that housed an old female, and remained with her in that zoo. All three elephants in this group successfully raised young. The five individuals that comprised the second group were each individually relocated to other zoos at least once. All elephants from this group at some time discarded, attacked, or killed their young (Kurt and Hartl 1995).

Relocating zoo elephants is a common occurrence (Schulte 2000). Usually the separations are permanent, although sometimes the relocations are only temporary, for breeding. In 1997, Chai, a particularly good-natured female at Seattle's Woodland Park Zoo, was shipped to Dickerson Park Zoo, in Springfield, Missouri, to be mated with a stud male. After three days on a truck, Chai arrived exhausted, entering an early estrous period, hungry, nervous, and probably disoriented. Dickerson Park staff could have chosen to assist her acclimation. Instead they set out almost immediately to establish control. They prodded her with ankuses, and, worse, in different and more sensitive contact points than the keepers used at Woodland Park. Stressed and confused she resorted to an action completely out of character for her: she swung her head at the lead keeper. The keepers decided to counter what they described as "aggressive behavior," and thus repeated the cues. Chai fell into the trap and swung her head again. It was the signal to start beating her with ax handles. The trainers had the comfort of knowing not only that this was customary procedure at their zoo but also preapproved by Woodland Park Zoo management for them to use on Chai if they deemed it necessary. One attendee reported that the pounding lasted at least an hour. A Dickerson Park Zoo spokesperson says the beating lasted "only" about five minutes in an hourlong training session (Scigliano 2002). U.S. Department of Agriculture officials investigated complaints and found that the beating Chai had endured was abusive. According to the *Seattle Post-Intelligencer* (July 3, 2002), USDA imposed a fine and instructed Dickerson Park Zoo to institute a protected contact management system immediately.

Chai did not rebel again after this incident. She did, however, show new signs of stress, and for the first time in her life began swaying and shuffling in classic stereotypic patterns. These behavioral distortions continued after her return to Seattle. Docents and staff explained to the public that they were caused by the swaying motion of the truck during the long drive from Missouri, or something she learned from another elephant at Dickerson Park, or, that old familiar zoo line, she was anticipating going inside to be fed. Visitors often simply smiled; "Oh look. She's dancing!" Their misinterpretation was never observed to be corrected by any zoo staff or volunteers (Scigliano 2002, 291).

Relocations of zoo elephants have been increasing in recent years. In the wild, separating elephants from a family unit causes enormous stress, for both young and adult animals, and can affect a young elephant for the remainder of her life (Kurt 1995; Poole 2000). In Europe, it seems babies have been separated from mothers and shipped to other zoos or circuses when only a few days old (Clubb and Mason 2002, Table 14). The AZA recommends that the minimum age for removing a female calf from its mother should be three years, the minimum age for weaning in the wild. However, considering the strength and complexity of the social bonds between mother and calf, one must question not only whether this figure is just too arbitrary but also whether separation can be acceptable in any circumstances. Trauma in early life, such as occurs with separation of young elephants from their mothers, can lead to abnormal development of the infant brain, producing long-term negative effects on mental health and on behavior, with enduring right-brain dysfunction and a predisposition to violence in adulthood (Bradshaw et al. 2005). It is an especially unnatural occurrence for female elephants and their babies to be separated. In the wild, they lose each other only through death.

The unanticipated removal of an individual from within a socially bonded group is a source of deep stress not only for elephants but also for many other intelligent social species, including primates, dolphins, and killer whales. Altered feeding and sleeping patterns after such deprivation are common among elephants, as are increased searching behaviors and vocalization interspersed by periods of energy-conserving depression. The behavioral and physiological changes can persist to such an extent they indicate not just loss but a deep sense of grief (Moss 2000; Newberry and Swanson 2001). The trauma inflicted on separated elephants is of such duration, such intensity, and such depth it would seem to be warranted, if at all, only for very exceptional circumstances. Yet in Europe about two-thirds of all captive born elephants have been separated and relocated, some of them more than once. In not one instance was the calf transported in the company of its mother (Clubb and Mason 2002, Table 14).

The commitment to maintain psychologically, emotionally, behaviorally, and physically healthy active and contented elephants in stable, multigenerational groups is imperative but is beyond the physical and financial capacity of most zoos. Certainly, I cannot imagine that any more than a very small number of zoos can provide conditions that suit elephants well. Perhaps people in the western world who wish to see elephants should expect to select only from those few zoos that can dedicate the space and resources for managing large and stable herds in spacious and complex environments in appropriate climates. In temperate climate countries, maybe the spread would look this way: one facility in Southern California and one in Florida, for North America; one perhaps in the warm climate of Spain for Europe; and one in Queensland for Australia and New Zealand.

Not being able to see elephants in the local zoo would be an inconvenience for humans, but it could mean a massive improvement for zoo elephant standards and welfare. Taking into account the relatively small amount of resources that zoos have traditionally applied to elephant needs, my sad conclusion is that not only some zoos, but, in truth, *most* zoos either will not pay for or cannot afford to create proper conditions for elephants and thus don't deserve them.

Acknowledgments
My sincere gratitude to Debra Forthman, Lisa Kane, Gail Laule, Ellen Leach, Laurie Pond, and Peter Stroud, for technical advice and for sharing their knowledge and wisdom. I greatly appreciate that Michael Hutchins kindly recommended me for this task in the first place, though there were times, as I chased obscure references and details, that I cursed him for doing that. My appreciation is also due to John Lehnhardt, Jill Mellen, and Joseph Barber for their peer reviews of my original paper. Catherine Christen and Chris Wemmer took pains gently to guide me over hurdles I had inadvertently constructed. If I have nonetheless stumbled, it is my error. Preparing the references section was, thanks to the many inconsistencies in the rules that librarians and editors have developed, almost as enjoyable as flying Qantas across the Pacific in the back of the airplane with an Australian rugby team.

References

Adams, J. 1981. *Wild elephants in captivity*. Carson, CA: Center for the Study of Elephants.

Association of Zoos and Aquariums (AZA). 2003. *Standards for elephant management and care*, Adopted 21 March 2001, updated 5 May 2003. Silver Spring, MD: AZA.

Academy of Natural Sciences (ANS). 2005. Nature's great masterpiece: The elephant. Treasures of the Ewell Sale Stewart Library. A virtual exhibition. Philadelphia: ANS. www.acnatsci.org/library/collections/elephant/.

Boardman, W. S. J., Jakob-Hoff, R., Huntress, S., Lynch, M., Reiss, A., Reite, M., and Monaghan, C. 2001. The medical and surgical management of foot abscesses in captive Asiatic elephants: Case studies. In B. Csuti, E. L. Sargent, and U. S. Bechert (eds.), *The elephant's foot: Prevention and care of foot conditions in captive Asian and African elephants* (pp. 121–126). Ames: Iowa State University Press.

Blake, W. 1917. Auguries of innocence. In D. H. S. Nicholson and A. H. E. Lee (eds.), *The Oxford book of English mystical verse* (p. 60). Oxford: Clarendon Press.

Bradshaw, G. A., Schore, A. N., Brown, J. L., Poole, J. H., and Moss, C. J. 2005. Social trauma: Early disruption of attachment can affect the physiology, behaviour, and culture of animals and humans over generations. *Nature* 433: 807.

Carlstead, K., Seidensticker, J., and Baldwin, R. 1991. Environmental enrichment for zoo bears. *Zoo Biology* 10: 3–16.

Chadwick, D. H. 1992. *The fate of the elephant*. San Francisco: Sierra Club Books.

Chance, P. 1994. *Learning and behavior*. Pacific Grove, CA: Brooks/Cole Publishing.

Clubb, R., and Mason, G. 2002. *A review of the welfare of zoo elephants in Europe*. Horsham, UK: Royal Society for the Prevention of Cruelty to Animals (RSPCA).

Dawkins, M. S. 1980. *Animal suffering: The science of animal welfare*. London: Chapman and Hall.

Desmond, T., and Laule, G. 1991. Protected contact elephant training. In *Proceedings of the American Association of Zoological Parks and Aquariums Annual Conference 1991* (pp. 606–613). Wheeling, WV: American Association of Zoological Parks and Aquariums.

Douglas-Hamilton, I. 1972. On the ecology and behaviour of the African elephant. DPhil diss., Oxford University.

European Association of Zoos and Aquaria (EAZA). 1997. *European Association of Zoos and Aquaria recommendations for elephant husbandry*. Amsterdam: EAZA.

Eisenberg, J. F. 1980. Ecology and behavior of the Asian elephant. *Elephant* 1 (suppl.): 36–55.

The Elephant Sanctuary (TES). 2006. *Frequently asked questions*. Hohenwald, TN: The Elephant Sanctuary. www.elephants.com/questions.htm, question 48.

Eltringham, S. K. 1982. *Elephants*. Dorset, England: Blandford Press.

Forthman, D. L., McManamon, R., Levi, U. A., and Bruner, G. Y. 1995. Interdisciplinary issues in the design of mammal enclosures. In E. F. Gibbon, Jr., J. Demarest, and B. Durrant (eds.), *Captive conservation of endangered species* (pp. 377–399). Albany: State University of New York Press.

Galloway, M. 1991. Update on 1990 chaining survey. In *Proceedings of the 12th International Elephant Workshop, held at Burnet Park Zoo, Syracuse, New York, October 16–19, 1991* (pp. 63–64). Syracuse, NY: The Elephant Managers Association.

Hancocks, D. 1971. *Animals and architecture*. London: Hugh Evelyn.

Hancocks, D. 2001. *A different nature: The paradoxical world of zoos and their uncertain future*. Berkeley: University of California Press.

Howletts Wild Animal Park. 2005. *Howletts Wild Animal Park* Website. www.totallywild .net/animals.php?animal=elephant/.

International Fund for Animal Welfare (IFAW). 2004. *Stop the live elephant trade: Ten reasons elephants don't belong in zoos* [brochure]. Sydney, Australia: IFAW. www.ifaw .org/ifaw/dimages/custom/2_Publications/Elephants/STOP_the_elephant_trade_ brochure_AP.pdf.

Jones, S. 1999. *Almost like a whale*. London: Doubleday.

Kellert, S. R. 1989. Perceptions of animals in America. In R. J. Hoage (ed.), *Perceptions of animals in American culture* (pp. 5–24). Washington, DC: Smithsonian Institution Press.

Kurt, F. 1974. Remarks on the social structure and ecology of the Ceylon elephant in the Yala National Park. In V. Geist and F. Walther (eds.), *Behavior of ungulates and its relation to management* (pp. 618–634). Morges, Switzerland: IUCN.

Kurt, F. 1995. The preservation of Asian elephants in human care: A comparison between the different keeping systems in South Asia and Europe. *Animal Research and Development* 41: 38–60.

Kurt, F., and Hartl, G. B. 1995. Asian elephants (*Elephas maximus*) in captivity—a challenge for zoo biological research. In U. Gansloßer, J. K. Hodges, and W. Kaumanns (eds.), *Research and captive propagation* (pp. 310–326). Fürth, Germany: Filander.

Kurt, F., and Mar, K. U. 1996. Neonatal mortality in captive Asian elephants (*Elephas maximus*). *Zeitschrift für Säugetierekunde*, 61: 155–164.

Kurt, F., and Schmid, J. A. A comparison of feeding behavior and body weight in Asian elephants (*Elephas maximus*). Paper presented at First International Symposium on physiology and ethology of wild and zoo animals, Berlin, Germany, September 18–21.

Lee, P. C. 1987. Allomothering among African elephants. *Animal Behaviour* 35: 278–291.

Lehnhardt, J. 1991. Elephant handling: A problem of risk management and resource allocation. In *Proceedings of the American Association of Zoological Parks and Aquariums Annual Conference 1991* (pp. 569–575). Wheeling, WV: American Association of Zoological Parks and Aquariums.

Lewis, G., and Fish, B. 1978. *I loved rogues: The life of an elephant tramp.* Seattle: Superior Publishing.

Lindburg, D. G. 1998. Coming in out of the cold: Animal keeping in temperate zone zoos. *Zoo Biology* 17: 51–3.

Mayell, H. 2002. Painting elephants get online gallery. *National Geographic News.* June 26. http://news.nationalgeographic.com/news/2002/06/0626_020626_elephant.html.

McComb, K., Moss, C., Sayialel, S., and Baker, L. 2000. Unusually extensive networks of vocal recognition in African elephants. *Animal Behavior*, 59: 1103–1109.

McKay, G. M. 1973. Behavior and ecology of the Asiatic elephants in southeastern Ceylon. *Smithsonian Contributions to Zoology* 125: 1–113.

Morris, R., and D. Morris. 1966. *Men and Pandas.* New York: McGraw-Hill.

Moss, C. J. 2000. *Elephant memories: Thirteen years in the life of an elephant family: With a new afterward.* Chicago: University of Chicago Press. Originally published 1988.

Moss C. J. 2001. The demography of an African elephant (*Loxodonta africana*) population in Amboseli, Kenya. *Journal of Zoology* 255: 145–156.

Moss, C. J. 2002. Presentation at Performing Animal Welfare Society banquet, May 10, Sacramento.

Moss, C. J., and J. H. Poole. 1983. Relationships and social structure in African elephants. In R. A. Hinde (ed.), *Primate social relationships: An integrated approach* (pp. 315–325). Oxford: Blackwell Scientific.

Newberry, R. C., and Swanson, J. 2001. Breaking social bonds. In L. J. Keeling and H. W. Gonyou (eds.), *Social behaviour in farm animals* (pp. 307–331). Wallingford, Oxon, UK: CABI.

Performing Animal Welfare Society (PAWS). 1999. *Everything you should know about elephants.* Galt, CA: PAWS.

Poole, J. H. 2000. Impact of capture and removal of elephants on family groups. Unpublished presentation at 11th meeting of the Conference of the Parties to the Convention on International Trade in Endangered Species (CITES-11) Nairobi, Kenya, April 9–20.

Poole, J. H. 2001. *Keynote address.* Video Presentation with Live Conference Call. Elephant Managers Association 22nd Annual Conference, Disney's Animal Kingdom, Orlando, FL. Recording OR 02 (48 minutes) in Elephant Managers Association Video Library. Indianapolis, IN: Elephant Managers Association.

Poole, T. B., Taylor, V. J., Fernando, S. B. U., Ratnasooriya, W. D., Ratnayeke, A., Lincoln, G. A., Manatunga, A. M. V. R., and McNeilly, A. S. 1997. Social behavior and breeding physiology of a group of Asian elephants. *International Zoo Yearbook* 35: 297–310.

Poole, T., and Taylor, V. J. 2001. Enriching the environments of Asian elephants: Can their behavioral needs be met in captivity? In V. J. Hare, K. E. Worley, and K. Myers (eds.), *Proceedings of the Fourth International Conference on Environmental Enrichment, Edinburgh, Scotland, 29 August-3 September 1999* (pp.160–172). San Diego, CA: Shape of Enrichment.

Pryor, K. 1984. *Don't shoot the dog: How to improve yourself and others through behavioral training*. New York: Simon and Schuster.

Reimers, M., Schmidt, S., and Kurt, F. 2001. Daily activities and home ranges of Asian elephants of the Uda Walawe National Park (Sri Lanka). In H. M. Schwammer, T. J. Foose, M. Fouraker, and D. Olson (eds.), *Research Update on Elephants and Rhinos: Proceedings of the International Elephant and Rhino Research Symposium, Vienna, June 7–11, 2001* (pp. 115–118). Münster, Germany: Schüling.

Roocroft, A., and Oosterhuis, J. 2001. Foot care for captive elephants. In B. Csuti, E. L. Sargent, and U. S. Bechert (eds.), *The elephant's foot: Prevention and care of foot conditions in captive Asian and African elephants* (pp. 21–52). Ames: Iowa State University Press.

Schmid, J. 1998. Hands off, hands on: Some aspects of keeping elephants. *International Zoo News* 45: 478–486.

Schulte, B. A. 2000. Social structure and helping behavior in captive elephants. *Zoo Biology* 19: 447–459.

Scigliano, E. 2002. *Love, war, and circuses: The age-old relationship between elephants and humans*. Boston: Houghton Mifflin.

Stroud, P., and Kudeweh, R. 2001. Redevelopment of the Melbourne Zoo elephant management program. Presented at Australasian Regional Association of Zoological Parks and Aquariums (ARAZPA) Conference. Sydney, Australia.

Sukumar, R. 1989. *The Asian elephant: Ecology and management* (2nd rev. ed., 1992). Cambridge: Cambridge University Press.

Sukumar, R., and Ramesh, R. 1995. Elephant foraging: Is browse or grass more important? In J. C. Daniel and H. S. Datye (eds.), *A week with elephants: Proceedings of the international seminar on Asian elephants, June 1993* (pp. 368–374). Bombay: Bombay Natural History Society and New Delhi: Oxford University Press.

Taylor, V. J., and Poole, T. B. 1998. Captive breeding and infant mortality in Asian elephants: A comparison between twenty Western zoos and three Eastern elephant centers. *Zoo Biology* 17: 311–332.

West, G. 2001. Occurrence and treatment of nail/foot abscesses, nail cracks, and sole abscesses in captive elephants. In B. Csuti, E. L. Sargent, and U. S. Bechert (eds.), *The elephant's foot: Prevention and care of foot conditions in captive Asian and African elephants* (pp. 93–98). Ames: Iowa State University Press.

West, J. B., Fu, Z., Gaeth, A. P., and Short, R. V. 2003. Fetal lung development in the elephants reflects the adaptations required for snorkeling in adult life. *Respiratory Physiology & Neurobiology* 138: 325–353.

Whitman, W. 1982. *Poetry and prose*. ed. J. Kaplan. New York: Library of America.

Wolk, D. 2005. Eccentric elephants. *Seattle Weekly*, Feb 2–8. www.seattleweekly.com/features/0505/050202_music_smallmouth.php.

Wyatt, J. R., and Eltringham, S. K. 1974. The daily activity of the elephant in the Rwenzori National Park, Uganda. *East African Wildlife Journal* 12: 273–289.

14 ZOOS AS RESPONSIBLE STEWARDS OF ELEPHANTS

MICHAEL HUTCHINS, BRANDIE SMITH, AND MIKE KEELE

Should elephants be kept in zoos? Contemporary zoo professionals should never take this question for granted, and it is not an easy question to answer. Meeting the physiological and psychological needs of the world's largest land mammals involves not only practical considerations, but ethical ones as well (Hutchins 2003; Hutchins et al. 2003; Veasey 2006). The Association of Zoos and Aquariums (AZA—formerly, the American Zoo and Aquarium Association) is a professional organization that accredits zoological institutions in North America. The mission of AZA accreditation is to establish, uphold, and raise the highest zoological and aquarium industry standards through self-evaluation, on-site inspection, and peer review. As part of this mission, AZA conducted an association-wide Elephant Planning Initiative in 1998–1999 to determine whether and how elephants should be held in zoos (Hutchins and Smith 1999). After much discussion and careful consideration, participants decided that modern zoos would be greatly diminished should they fail to provide their visitors with the opportunity to experience and learn about elephants. Perhaps even more importantly, it was recognized that captive elephants contribute to the conservation of their wild counterparts. But the decision came with an emphatic caveat: only zoos that can meet the complex biological and behavioral needs of elephants should have them in their collections. The current challenge lies in finding the best way to proceed.

If zoos are to be responsible stewards of captive elephants, then their primary mission is obvious: to advance elephant conservation. But this laudable goal cannot obscure the immediate husbandry and management needs that must be addressed in order to provide appropriate care for elephants in captivity. In this essay, we review the status of elephants in nature and explain how accredited zoos are becoming more integral to elephant conservation efforts. We then explore the ability of accredited zoos to care adequately and humanely for these complicated creatures. This includes a discussion of both the biological and ethical considerations involved in keeping elephants in professionally managed zoos, as well as a vision for the future of zoo elephant management.

Given the tenuous situation of elephants in the wild, the benefits associated with maintaining elephants in accredited zoos, as measured in terms of conservation, research, and education, outweigh the costs, as measured in terms of diminished animal welfare, but only if remaining husbandry and welfare challenges are effectively addressed and adequate financial and human resources are directed toward zoo elephant programs (Hutchins, Smith, and Allard 2003; Hutchins 2006; Veasey 2006). There is immediacy to this discussion. Recent studies by Wiese (2000), Olson and Wiese (2000), Faust (2005a, 2005b), and Wiese and Willis (2006) have shown that current zoo elephant populations in North America and Europe are not self-sustaining, and if nothing is done to reverse this trend, zoo populations of both Asian and African elephants will likely be extinct within the next fifty years.

The following discussion is focused exclusively on elephant programs in AZA-accredited institutions. It is difficult to generalize across regions because management and husbandry techniques, regional association infrastructure, and legal and governmental regulations affecting zoos vary so greatly.

Elephant Conservation: The Evolving Role of Zoos

Asian elephants (*Elephas maximus*) are considered "endangered" under the U.S. Endangered Species Act (ESA), are listed under Appendix I (species threatened with extinction) of the Convention on International Trade in Endangered Species of Wild Fauna and Flora (CITES), and are afforded the highest level of international protection. African elephants (*Loxodonta africana*) are listed as CITES Appendix II (trade must be controlled in order to avoid utilization incompatible with their survival) if they come from populations in Botswana, Namibia, South Africa, and Zimbabwe, and as Appendix I if they come from any other location. However, all African elephants are classified as "threatened" under the ESA, with special regulations governing trophy hunting and the ivory trade. Thus, both species are afforded a high degree of protection under U.S. law.

Hundreds of thousands of elephants have been killed for their ivory and meat (Douglas-Hamilton 1987; Chadwick 1992; Sukumar 1989). However, elephant conservation is very complex; although elephant populations are rare or declining in some areas, they are recovering or overabundant in others (Whyte, van Aarde, and Pimm 1998). Among the most significant threats to elephants' continued existence is the potential for direct conflict with humans. Elephants frequently raid agricultural areas and hundreds of people are killed annually attempting to protect their crops (Nyhus, Tilson, and Sumianto 2000; Lee and Graham 2006; Sukumar 2006).

Given this context, it is urgent that many organizations—both governmental and nongovernmental—collaborate to ensure a future for elephants and their habitats in a human-dominated world. The IUCN policy on captive breeding acknowledges that habitat protection alone may not be sufficient and that establishment of self-sustaining captive populations is needed to avoid the loss of many species, especially those at high risk in greatly reduced, highly fragmented and disturbed habitats (IUCN 1987). Reintroduction is not a current goal of the AZA's cooperative breeding program, the Elephant Taxon Advisory Group / Species Survival Plan (TAG/SSP) (Keele 1999), nor has it ever been recommended as a conservation strategy by any range country wildlife agency or international conservation organization. However, AZA-accredited zoos have greatly expanded their conservation-related activities to include a variety of programs and projects that support elephant conservation and related scientific and educational initiatives (Smith and Hutchins 1999). A recent survey of AZA zoos (AZA 2003), covering the eighteen-month period from July 2002 to December 2003, indicated that they either initiated or supported no fewer than eighty-seven elephant-related conservation and research projects.

Education

AZA-accredited zoos and aquariums invest more than $50 million annually in public education programs and are visited by more than 140 million people each year (AZA 1999). Several studies have demonstrated the power of live animals to focus people's attention and promote effective learning and attitude change (Saunders and Young 1985; Sherwood, Rallis, and Stone 1989; Stoinski et al. 2001). Zoos and aquariums offer visitors real as opposed to virtual wildlife experiences, and they cater to a diverse range of people, many of whom will never have an opportunity to view elephants in their natural habitats.

Elephants are keystone species, both in the wild and in zoos. Zoo elephant exhibits attract and hold the public's attention, making them ideal places to increase people's appreciation and understanding of elephants and the contemporary challenges to their conservation. For example, the El Paso Zoo hosts an annual elephant weekend to focus on elephant conservation and the plight of elephants in the wild. The Oregon Zoo has a popular elephant museum on its grounds dedicated to educating visitors about elephant conservation and the role elephants have played in human culture. Similarly, the Bronx Zoo's historical Zoo Center features displays that show the impact humans have had on elephant and rhinoceros populations.

AZA member zoos also conduct educational outreach programs that allow them to spread the conservation message beyond their perimeter fences. Some of these programs specifically target elephant conservation. *Suitcase for Survival*, a cooperative educational program developed by the AZA, WWF-US, and the U.S. Fish and Wildlife Service, has reached tens of thousands of schoolchildren and adults across the United States (Hardie 1987). The program discourages people from purchasing illegal wildlife products, including elephant ivory. In addition, several zoos participate in programs that train developing country biologists and increase their capacity to address local conservation needs (Wemmer et al. 1993; Jhala 1994; Cook et al. 1995; Center for the Reproduction of Endangered Species 2001; Bennett and Rabinowitz 2001; Jackson 2001; Rudran and Wemmer 2001).

Research

Captive wild animals provide scientists with research opportunities that would be impractical or even impossible in the field (Hutchins and Conway 1995; Hutchins 2001, 2003). Numerous ongoing projects focus on basic elephant life history, while others are generating data that can be immediately applied to elephant conservation in the field (Smith and Hutchins 2000).

Most of what we know about elephant reproductive biology is the result of research conducted at zoos. Research on captive elephants has provided a more detailed understanding of the physiology of both male and female reproductive systems (Jainudeen, Eisenberg, and Tilakerantne 1971; Hess, Schmidt, and Schmidt 1983; Poole et al. 1984; Cooper et al. 1990; Olsen et al. 1994; Hodges, McNeilly, and Hess 1997; Hodges 1998; Hildebrandt et al. 2006; Szdzuy et al. 2006) and chemical signals associated with estrus and musth (Rasmussen and Schulte 1998; Rasmussen, Riddle, and Krishnamurthy 2002). Furthermore, noninvasive reproductive assessment techniques developed and tested on zoo elephants, such as fecal hormone analyses, are now being used to study free-ranging animals (see Wasser et al. 1996).

Zoo biologists and their university partners continue to perfect artificial insemination (AI) and contraception techniques (Brown, Bush, et al. 1992, 1999; Fayrer-Hosken et al. 1997, 2000; Olson 1999; Brown, Goritz, et al. 2004; Brown, Olson, et al. 2004; Hildebrandt et al. 2006), both of which could have implications for managing free-ranging elephant populations. In the future, artificial insemination could be used to introduce essential genetic variability into small, isolated elephant populations. Population control is becoming increasingly necessary to ameliorate human-elephant conflicts (Pienaar 1969; Whyte et al. 1998). Although certainly not a panacea for population control, contraceptive techniques offer a potential non-lethal option for population reduction and thus can assist in park and ecosystem management (Whyte et al. 1998).

Infrasonic communication in elephants was first discovered and studied in zoo elephants (Payne, Langbauer, and Thomas 1986). This knowledge is vital for understanding how wild elephants communicate and coordinate their movements over great distances. Zoos continue to further our knowledge of elephant acoustic and chemical communication through a variety of studies (Langbauer et al. 1991; Langbauer 2000; Ortolani et al.

2000; Leong, Ortolani, Burks, et al. 2003; Leong, Ortolani, Graham, et al. 2003).

Zoo-based veterinarians and pathologists have done much to advance our knowledge of exotic animal medicine, and this has important implications for monitoring and controlling disease in free-ranging populations (Cook et al. 1995; Karesh and Cook 2001). A number of naturally occurring diseases, including salmonella (Emanuelson and Kinzley 2000, 2001), endotheliotropic elephant herpes virus (EEHV) (Richman et al. 1999; Richman, Montali, and Hayward 2000), and tuberculosis (Mikota, Larson, and Montali 2000; Mikota et al. 2001), have been studied in zoo elephants, which in turn has resulted in various detection and treatment regimes.

Technologies relevant to field conservation can be tested in zoos before being applied in nature (Hutchins and Conway 1995). For example, satellite tracking techniques, developed and tested by Wildlife Conservation Society scientists on Asian elephants in the Bronx Zoo's Wild Asia exhibit, are now being used successfully to track the movements of forest-dwelling elephants in central Africa (Nobbe 1992; Goddard 1995). Elephants do not recognize the arbitrary boundaries of national parks and reserves, and it is hoped that an improved knowledge of elephant movements can be used to help reduce human-elephant conflicts and assist in reserve design.

Zoo studies have had many other applications to field research and conservation. The efficacy of chemical repellents in keeping wild elephants away from agricultural crops has been tested in zoos and sanctuaries (Osborn and Rasmussen 1995; Rasmussen and Riddle 2004). Studies of food passage and defecation rates in captive elephants have been used to verify the accuracy of field census data based on dung counts (Barnes 2001). Growth studies, including age-size correlations in captive elephants (e.g., Wemmer and Krishnamurthy 1992), have been useful in estimating age or size in unmarked individuals and determining population age structure in free-ranging elephant populations (Western, Moss, and Georgiadis 1983; Hile, Hintz, and Erb 1997). Zoo genetic studies are also helping us to understand evolutionary patterns of adaptation, determine geographic ranges, and develop effective conservation strategies for endangered species, including elephants (Ryder 1990; Fleischer et al. 2001).

In Situ Conservation

A growing number of zoos are providing direct support for elephant field conservation and related research and educational activities. For example, the Wildlife Conservation Society has a long history of supporting elephant conservation in Africa and Asia (Goddard 1995) and is currently supporting many projects, including development of an elephant monitoring system in the Congo Basin and a study of the ecology and social organization of forest elephants in central Gabon (WCS 2003). Smithsonian's National Zoological Park has assessed the conservation status and prospects for managing critical elephant ranges in Myanmar (Leimgruber et al. 2003; Wemmer et al. 2005). The North Carolina Zoological Park has joined with WWF-Cameroon to initiate a joint study on human-elephant interactions (Usongo 2003). Similarly, the Minnesota Zoo has supported studies of elephant-human conflicts on the island of Sumatra in Indonesia (Nyhus et al. 2000).

AZA and its member institutions are the primary financial supporters of the Bushmeat Crisis Task Force, a coalition of more than thirty major wildlife conservation and animal protection organizations and zoos working to curb the illegal commercialized trade in wild animals for meat in Africa (Eves and Hutchins 2001). The bushmeat trade is the most significant threat to African wildlife today, with many species, including forest elephants, being harvested unsustainably throughout the central African region (Robinson and Bennett 2000). Bushmeat Crisis Task Force is working closely with the CITES Bushmeat Working Group, the CITES Monitoring the Illegal Killing of Elephants Program (widely known by its acronym, MIKE), African wildlife colleges, international aid agencies, U.S. and African government officials, and many other partners to explore solutions to this complex conservation challenge.

Many AZA and European zoos support the International Elephant Foundation (IEF) and several zoo biologists and administrators sit on the IEF Board. The Foundation has provided more than $1.2 million (D. Olson, personal communication) for elephant-related projects since 1998. To expand on these activities, the AZA Elephant TAG/SSP is developing a coordinated conservation action plan in cooperation with IEF that will outline important elephant research and conservation needs and prioritize them for action (AZA 2005a). This will help direct resources to selected projects of highest priority. By working together and sharing their resources— both human and financial—AZA zoos and their partners will have a much greater effect on both captive and wild elephants than they would by working alone.

Husbandry and Management

Although zoos are placing a greater emphasis on wildlife conservation, do the conservation benefits of maintaining elephants in captivity outweigh any costs to individual animals? If there are costs, what are they and how can they be minimized?

Some accredited zoos have moved toward larger, more naturalistic enclosures for elephants. However, zoo professionals realize that they can never create an exact replica of the wild, nor would they always want to. Wild elephants are subject to droughts, competition, starvation, disease, and poaching. However, unless zoos provide proper care and stimuli to meet the biological needs of their animals, captivity can be detrimental to these complex creatures (Poole 1998; Kurt and Garaï 2001). To that end, accredited zoos should constantly strive to improve husbandry and management techniques (Hutchins and Smith 1999; Hutchins 2003, 2006; Hutchins and Keele 2006; Veasey 2006). The goal is to provide captive animals with an environment that meets their physical and psychological needs, as determined by their basic biology. Animal welfare is difficult to define and measure (Mason and Mendl 1993; Veasey 2006), and, depending on the type of habitat in which they are found, wild elephants exhibit tremendous variation in their biology and behavior (Hutchins 2006). However, the professional zoo community recognizes there is much room for improvement in elephant husbandry and management techniques.

Two studies (Wiese 2000; Olson and Wiese 2000), widely reported on before their publication, brought this message home, as they concluded that despite the association's efforts, breeding success was still elusive and, if the lack of breeding success could not be reversed, North American zoos might not have elephants in the future. In addition, media reports of elephant abuse at two AZA facilities resulted in public outrage and the resignation or termination of those accountable. AZA's board of directors subsequently approved a strategic elephant planning initiative to examine a wide range of issues related to the care and management of elephants in North American zoos (Hutchins and Smith 1999). As part of this initiative, a group of experts held three meetings and presented their opinions to the AZA board and institutional directors for consideration. AZA staff also collected data on the current status of the AZA Elephant TAG/SSP and institutional facilities, interviewed elephant managers, and held discussions with zoo critics.

One of the primary directives was that the AZA Elephant TAG/SSP should develop minimum standards for elephant care and management and that all TAG/SSP-participating institutions should be required to meet those standards and to participate in the program to the fullest extent possible. In addition, they decided that, once formulated and approved, these standards should be enforced through accreditation. A subcommittee of the Elephant TAG/SSP developed the new standards with input from AZA members and assistance from AZA staff. The AZA board approved the new standards in March 2001 and AZA institutions were given a timeline for compliance.

The AZA standards cover the most essential aspects of elephant care and are among the basic requirements accredited zoos must meet to hold elephants. Although these detailed standards are a great advancement, there is a constant need to review and improve the standards as more studies are done on elephants and scientific and practical information becomes available. The standards are already in the process of being reviewed and updated. In addition, the zoo profession must be mindful to push their standards beyond the minimum and develop a program wherein elephants are allowed to thrive, not merely survive, while at the same time ensuring safe working conditions for zoo staff.

Zoo critics have claimed that captive elephants live substantially shortened lives when compared with their wild counterparts (Schmid 1998a; Clubb and Mason 2002; McKinney 2003; People for the Ethical Treatment of Animals [PETA] 2005). These generalizations were based largely on conjecture, rather than on solid scientific evidence. Wiese and Willis (2004) recently analyzed zoo elephant survival data and determined that the life spans of female elephants in North American and European zoos are comparable to those of wild elephants. Nonetheless, many zoo professionals agree with the need to provide zoo elephants with conditions that are more compatible with elephant biology. The situation is analogous to the history of gorilla management in zoos. Lowland gorillas (*Gorilla gorilla gorilla*) were once maintained in inappropriate environments and social groups. The result was shortened life spans, atypical behavior, and lack of breeding success. Improvements in their captive environments included larger and more naturalistic exhibits, environmental enrichment, improved nutrition, and appropriate and stimulating social environments, including a reduction in the practice of hand rearing. This resulted in vastly improved welfare and breeding (Beck and Power 1988; Hutchins et al. 2001). There is no reason a similar transformation cannot take place for elephants (see Hutchins 2006; Veasey 2006; Hildebrandt et al. 2006). Specific priority areas that still need to be addressed include enclosure size and design, group size and composition, environmental enrichment and exercise, management and training systems, and public and keeper safety.

Enclosure Size and Design

As the largest of all land mammals, elephants can use a great deal of space. The question of how much space is appropriate and necessary has been the subject of much conjecture and debate. The minimum indoor holding space in the AZA standards was intended to provide adequate room for animals to move about and lie down without restriction, with additional requirements for outdoor space and exercise.

The AZA will revisit and most likely increase minimum indoor and outdoor space requirements, both to provide elephants with exercise and to make room for larger group sizes. Enclosures will also need to be more complex to provide opportunities for exploration and cognitive stimulation (Mench 1998; Hutchins 2006; Veasey 2006). Though the basic spatial needs of elephants have not yet been established (Hutchins 2006), several AZA institutions, such as North Carolina Zoo, Columbus Zoo, St. Louis Zoo, San Diego Wild Animal Park, Lowry Park Zoo, and Disney's Animal Kingdom may be setting the future standard with their new or renovated facilities. These institutions provide several acres of elephant habitat supplemented with environmental enrichment programs.

One specific concern that arises because of space constraints is a method of restraint called chaining or tethering (Roocroft and Zoll 1994). Tethering has become the more popular term because some zoos are replacing metal chains in favor of synthetic materials. Some zoos use tethering to restrain elephants overnight to prevent intraspecific aggression in group-housing situations and allow subordinate animals to rest and eat in peace (Brockett et al. 1999). However, tethering also prevents normal social interaction and activities and has been shown to cause stereotypic behavior (Brockett et al. 1999; Gruber et al. 2000). Its excessive use may diminish welfare. Current AZA standards state that tethering is an acceptable method of restraint but that elephants may not be restrained for more than twelve hours. This will change as new facilities are constructed or old ones modified because AZA standards also require that all new construction and major renovations must be designed to minimize or, preferably, eliminate the need for this practice, except as a method of temporary restraint (AZA 2001). Training elephants to accept this form of restraint will still be necessary in emergency situations, including certain types of intensive veterinary care and management of aggressive behavior.

Group Size and Composition

Both African and Asian elephants live in matriarchal societies where most females remain with their natal groups and bulls disperse as they reach sexual maturity (Moss and Poole 1983; Sukumar 1989, 2003). To meet the animals' social needs, zoos should make every effort to maintain elephants in natural social groupings (Mellen and Keele 1994). However, the size and stability of elephant groups vary widely in nature, making it difficult to determine an optimal group size and composition (Sukumar 2003; Hutchins 2006). The AZA standards state that it is inappropriate to keep highly social female elephants singly and recommends that institutions should strive to hold no fewer than three adult female elephants. Further, all new exhibits and major renovations must have the capacity to hold at least three or more adult females and their young (AZA 2001).

Another relevant issue is group stability. Although most wild elephant groups are composed of related females and their offspring, captive groups are typically unrelated. In nature, female offspring generally remain with their natal group, often inheriting their mother's social status (Moss and Poole 1983), but this is not always the case (Sukumar 2003). In fact, elephants are remarkably flexible in their behavior, and unrelated females have successfully formed stable social groups in zoos (Hutchins 2006). However, there has been concern about when, or even if, female offspring should be separated from their mothers. Current standards require that offspring must remain with their mothers for a minimum of three years (AZA 2001). As elephant exhibits and female herds in them become larger, it should become increasingly possible to attain greater group stability. Optimally, zoos could manage elephants much the same way that they manage polygynous nonhuman primates. With these primates, they can mimic the natural social organization by maintaining stable family groups of females and their female offspring and rotating adult males between institutions or using artificial insemination for breeding purposes. Following this example, young male elephants would be weaned as they become sexually mature, and young females would remain with their natal group. If animals had to be moved to control population growth, then groups of females and their female offspring would be moved together. The goal is to accommodate the animals' social needs as much as possible. Behavioral assessments of zoo elephants, along the lines of Carlstead et al. (1999), will allow managers to determine whether this goal is being reached.

Environmental Enrichment and Exercise

Environmental enrichment has been defined as "an animal husbandry principle that seeks to enhance the quality of animal care by identifying and providing the environmental stimuli necessary for optimal psychological and physiological well-being." In practice, this can involve a wide range of activities "aimed at keeping captive animals occupied, increasing the range and diversity of behavioral opportunities, and providing more stimulating and responsive environments" (Shepherdson 1998, 1; 1999).

Environmental enrichment is mentioned throughout the AZA standards, and all AZA accredited institutions are now required to have a written environmental enrichment plan for their elephants and must be able to show evidence of implementation. Some progress has been made in identifying effective techniques pertaining to elephants (e.g., Stoinski et al. 2001; Law and Kitchner 2002), but elephant environmental enrichment should be an important research topic in the coming decade.

Opportunities for exercise in traditional zoo elephant exhibits have been severely limited, and new enclosure designs should take this factor into account. Obesity and all of its associated health problems have been a concern (Roocroft and Zoll 1994). Advancements in elephant nutrition and feeding practices (see Dierenfeld 1994; Hatt and Clauss 2006) will help in this regard but providing greater opportunities for exercise will be the key, as will appropriate substrates (Csuti, Sargent, and Bechert 2001; Meller, Croney, and Shepherdson 2007). Larger exhibits should provide more opportunities for movement and exploration, and environmental features such as water-filled pools, scratching posts, and mud wallows will also be important. Some institutions regularly take their elephants on walks throughout the zoo. This provides opportunities for exercise and enrichment, but if this occurs during public visiting hours, it can create public safety concerns. Some zoos, for instance, the Smithsonian's National Zoo, are considering the development of elephant "treks," fenced-in trails located around the perimeter of exhibits, or even extending out from those exhibits, that would encourage elephants to exercise regularly (T. Barthel, personal communication). This innovative design concept would provide opportunities for more extensive movement while maintaining a protective barrier between elephants and keepers.

Management System and Training

Any management system must ensure the safety and well-being of both the animals and their keepers. Elephant management systems, including training procedures, have generated considerable controversy, both within and outside the zoo community (Desmond and Laule 1993; Anderson 1997; PETA 2005). The predominant forms of elephant management are known as protected contact (PC) and free contact (FC) (Desmond and Laule 1991; Roocroft and Zoll 1994; Priest et al. 1998; Schmid 1998b; Olson 2004). AZA's Principles of Elephant Management course defines the management systems as follows: FC is the direct handling of an elephant when the keeper and elephant share the same unrestricted space. Neither the use of chains nor the posture of the elephant alters this definition. PC is the handling of an elephant when the keeper and the elephant do not share the same unrestricted space. Typically, in this system, the keeper has contact with the elephant through a protective barrier of some type. The elephant is not confined to a particular space and is usually free to leave the work area at will. However, this category also includes confined contact, the handling of an elephant through a protective barrier where the elephant is spatially confined, as, for example, when an elephant is confined in an elephant restraint device (see Schmidt 1981).

Many elephant managers believe that rather than two distinct systems there is actually a continuum of management types between no contact and FC (Olson 2004). The current AZA standards recognize a diversity of approaches to elephant management and encourage members to experiment with the dual goals of maximizing elephant health and reproduction and minimizing risk of injury to keeper staff (AZA 2001). Among AZA institutions, 81% of zoos with bull elephants use PC exclusively, and 19% use FC exclusively (M. Keele, unpublished data, 2005). Of the zoos with female elephants, 50% use PC and 44% use FC exclusively. More institutions currently use FC to manage Asian females than use PC (54% vs. 41%). Five AZA zoos use a mix of both free and protected contact for Asian females.

The primary ethical concerns regarding FC management systems involve animal behavior training methods, keeper safety, and public safety. Both PC and FC systems have adapted the same principles of positive reinforcement, negative reinforcement, and punishment in training (Desmond and Laule 1991; Peterson 1994; Hudson 1997; Priest et al. 1998). Although PC relies primarily on positive reinforcement, traditional FC management employs not only positive reinforcement but also negative reinforcement and punishment. Where a "time-out" constitutes punishment in both FC and PC, other forms of punishment are sometimes used in FC. In FC, trainers traditionally use an ankus (or "guide") to communicate desired actions to an elephant; failure to comply with a command could result in an aversive stimulus (negative reinforcement) bridged with the command to train the behavior. To eliminate an undesirable behavior, particularly those that might be aggressive or present a danger, the elephant might be punished by striking it with the ankus to cause minor physical discomfort and signaled to stop the behavior with a "no" command (Peterson 1994; Roocroft and Zoll 1994; Hudson 1997). In PC, the ankus is not necessary, though some AZA facilities have incorporated it into their system to communicate desired actions using aversive stimuli. Effective training programs should employ punishment very infrequently and only when necessary to protect keeper and animal safety (see Lehnhardt and Galloway, Chapter 8 in this volume). In FC, as well as in PC, animals will learn to comply voluntarily with commands in order to obtain food or other positive rewards without the ankus being used (Olson 2004).

Critics of FC management systems, including Desmond and Laule (1991) and Clubb and Mason (2002), maintain that FC is an antiquated management system based on dominance, physical punishment, and aversive conditioning and that it should be banned in favor of PC. Others contend that FC is not inherently abusive, that it actually allows for better care of captive elephants, that criticisms of FC are based on a misunderstanding of current training practices, and that AZA should not limit the tools available to elephant managers (see Lehnhardt and Galloway, Chapter 8). The AZA elephant standards allow the use of the ankus, or guide, traditionally used in FC (AZA 2001). The concern is not so much over the use of this tool but rather its potential misuse. The critical question is, What constitutes abuse? The AZA standards are specific about what is acceptable in elephant training and what is not, making it clear that any injury or abusive treatment will not be tolerated. Protracted and repeated use of punishment in training is of serious ethical concern, and AZA considers abusive training practices to be unacceptable (AZA 2001). Keepers now receive considerably more instruction in the appropriate use of training methods than they did just eight to ten years ago, and this has resulted in the widespread growth of effective training in zoo and aquarium animal management (Peterson 1994; Mellen and Ellis 1996; Hudson 1997; Laule and Desmond 1998).

Keeper and Public Safety

Concern over free and protected contact management systems also extends to keeper and public safety. Current AZA standards include provisions for improving the safety of elephant keepers, regardless of the management system used. For example, institutions must review their elephant management practices regularly, and elephant program managers must receive appropriate training. If an institution holds adult males or manages females in PC, they are also required to have an appropriate elephant restraint device (AZA 2001; Olson 2004).

The record of keeper safety in zoo elephant programs has not been particularly good. Since 1990, there have been 24 incidents in North America of elephants seriously injuring their keepers in accredited zoos (1.84 incidents per year); five of these events resulted in deaths (Gore, Hutchins, and Ray 2006). The most recent of these deaths occurred at the Pittsburgh Zoo in 2003. However, this was the only death in an accredited zoo in the past ten years, which suggests that improvements in elephant and keeper training may be improving safety.

Although PC management should not be considered 100 percent safe (Guerrero 1997), all but one of the above incidents (95.8%) occurred in FC management systems, that is, where keepers and elephants were sharing the same space. In the one incident that occurred in PC, the keeper violated the zoo's safety protocols. Taking into account the number of times that elephants and keepers have actually been together in FC, the risks may not seem incredibly high. However, every elephant-holding institution should consider if the risks of FC management are acceptable. Should we even have elephants in zoos if the animals cannot be maintained appropriately *and* safely? These are difficult issues that will require deep thought and considerable deliberation by the AZA leadership and individual member institutions.

FC management allows elephants opportunities for exercise outside of their exhibits and holding facilities, and some zoos have done this on a regular basis. In addition, some zoos have taken elephants to public events, such as fund-raisers or concerts. It is possible that these activities may be enriching or provide opportunities for needed physical exercise. However, zoos that take elephants outside their exhibits or holding areas need to consider carefully the risk to zoo visitors. Not many years ago an elephant escaped shortly after being walked on the grounds at an AZA member facility, and the animal narrowly missed stepping on an infant in a stroller. The legal liability and negative media associated with an elephant-related public death could be substantial, and the zoo involved in the above incident banned elephant walks immediately following the event. In the interest of public safety, the AZA board of directors amended its elephant standards in 2003, strongly discouraging the practice of walking elephants in public areas during public hours.

Zoo elephants can serve as true conservation ambassadors for their wild counterparts. The evidence presented in this chapter indicates that AZA member zoos have started the journey toward this goal. However, accredited zoos must also continue to work to improve elephant husbandry and management and do it in a safe and cost-effective manner. Development of the AZA elephant standards was a major step forward, but there is no room for complacency. In our opinion, maintaining the status quo is not an option. To breed elephants consistently and reduce associated risks to keeper staff, zoos must evolve rapidly. This means that more resources—human and financial—need to be directed to zoo elephant programs immediately. AZA institutions' willingness to move in this direction was reaffirmed at an AZA Elephant Management Strategic Planning Workshop from December 5 to December 7, 2004 (Conservation Breeding Specialist Group 2004), and at a meeting of elephant holding facility directors from January 10 to January 21, 2005, in Orlando, Florida. At the latter meeting, directors agreed that SSP participants in the breeding program would strive to hold six to twelve elephants at their facilities. Other facilities would strive to hold two to six elephants and develop the capability to hold all males. In addition, they agreed that conservation is the primary reason why elephants are in AZA facilities, that AZA standards would be regularly reviewed and updated based on new information, and that each institution needed to determine its level of commitment to the Elephant SSP, that is, whether it will be a breeding or holding institution (AZA 2005b).

It is expected that constant improvements in AZA's elephant management standards will eventually reduce the number of institutions that are able to care for elephants adequately. Instead of a few elephants at many institutions, the trend will be toward larger groups of elephants at a smaller number of quality facilities.

Admittedly, we need to learn more about elephant biology and ethology to continue to improve on the care we provide. Only by effectively enhancing the lives of captive elephants in biologically meaningful ways can we be assured that the benefits of zoo-based conservation outweigh the potential costs to individual animal welfare. To accomplish this goal, AZA must proactively formulate and implement a vision for elephant management in zoos. While some AZA institutions will have sufficient resources to fulfill this vision, others will not. The opportunity to care for, exhibit and learn from these incredible and complex animals should be considered a privilege, not a right.

Acknowledgments

The AZA Board of Directors, AZA Elephant TAG/SSP members and AZA elephant-holding institutions deserve thanks for their contributions to the AZA Elephant Planning Initiative, which provided the basis for this chapter. David Hancocks, John Lehnhardt, Chris Wemmer, Kate Christen, and other independent reviewers provided useful recommendations on an earlier version of this manuscript.

References

American Zoo and Aquarium Association (AZA). 1999. *The collective impact of America's zoos and aquariums.* Bethesda, MD: AZA.

American Zoo and Aquarium Association. 2001. *AZA standards for elephant management and care.* Silver Spring, MD: AZA.

American Zoo and Aquarium Association. 2003. *AZA annual report on conservation and science.* Silver Spring, MD: AZA.

American Zoo and Aquarium Association. 2005a. Elephants to benefit from AZA-IEF collaboration. Press release, 18 February 2005. Silver Spring, MD: AZA.

American Zoo and Aquarium Association. 2005b. Elephant directors meeting, 10–21 January 2005, Orlando, FL. Unpublished summary of meeting results. Silver Spring, MD: AZA.

Anderson, D. 1997. Protected contact issues. *Journal of the Elephant Managers Association* 8: 55–57.

Barnes, R. F. W. 2001. How reliable are dung counts for estimating elephant numbers? *African Journal of Ecology* 39: 1–9.

Beck, B. B., and Power, M. L. 1988. Correlates of sexual and maternal competence in captive gorillas. *Zoo Biology* 7: 339–350.

Bennett, L., and Rabinowitz, A. 2001. It has to be done on the ground: In-country training courses in wildlife management and research. In W. Conway, M. Hutchins, M. Souza, Y. Kapetanakos, and E. Paul (eds.), *AZA field conservation resource guide* (pp. 233–235). Atlanta: Wildlife Conservation Society and Zoo Atlanta.

Brockett, R., Stoniski, T. S., Black, J., Markowitz, T., and Maple, T. 1999. Nocturnal behavior in a group of unchained female African elephants. *Zoo Biology* 18: 101–109.

Brown, J. L., Bush, M., Monfort, S. L., and Wildt, D. E. 1992. Fertility regulation in female elephants: Applicability and potential problems of using steroidal hormones for contraception. In *Proceedings of the Elephant Contraception Symposium* (pp. 21–24). Nairobi, Kenya.

Brown, J. L., Schmidt, D. L., Bellem, A., Graham, L. H., and Lehnhardt, J. 1999. Hormone secretion in the Asian elephant (*Elephas maximus*): Characterization of ovulatory and anovulatory lutenizing hormone surges. *Biology of Reproduction* 61: 1294–1299.

Brown, J., Olson, D., Keele, M., and Freeman, E. 2004. Survey of the reproductive cyclicity status of female Asian and African elephants in North America. *Zoo Biology* 23: 309–321.

Brown, J. L., Goritz, F., Pratt-Hawkes, N., Hoermes, R., Galloway, M., Graham, L., Gray, C., et al. 2004. Successful artificial insemination of an Asian elephant at the National Zoological Park. *Zoo Biology* 23: 45–63.

Carlstead, K., Fraser, J., Bennett, C., and Kleiman, D. 1999. Black rhinoceros (*Diceros bicornis*) in U.S. zoos: II. Behavior, breeding success, and mortality in relation to housing facilities. *Zoo Biology* 18: 35–52.

Center for the Reproduction of Endangered Species. 2001. *2001 annual report.* San Diego, CA: Center for the Reproduction of Endangered Species, San Diego Zoological Society.

Chadwick, D. H. 1992. *The fate of the elephant.* San Francisco: Sierra Club Books.

Clubb, R., and Mason, G. 2002. A review of the welfare of zoo elephants in Europe. Horsham, UK: Royal Society for the Prevention of Cruelty to Animals (RSPCA).

Conservation Breeding Specialist Group. 2004. AZA elephant management strategic planning workshop. Final report. Apple Valley, MN: IUCN-The World Conservation Union/Conservation Breeding Specialist Group.

Cook, R. A., Dierenfeld, E., Karesh, W., and McNamara, T. 1995. Linking the knowledge of zoo animal health to the needs of free-ranging wildlife. In *Proceedings of the American Zoo and Aquarium Association Annual Conference 1995* (pp. 96–100). Bethesda, MD: AZA.

Cooper, K. A., Harder, J. D., Clawson, D. H., Frederick, D. L., Lodge, G. A., Peachy, H. C., Spellmire, T. J., and Winstel, D. P. 1990. Serum testosterone and musth in captive African and Asian elephants. *Zoo Biology* 9: 297–306.

Csuti, B., Sargent, E., and Bechert, U. S. 2001. *The elephant's foot: Prevention and care of foot conditions in captive Asian and African elephants.* Ames: Iowa State University Press.

Desmond, T., and Laule, G. 1991. Protected contact elephant training. In *Proceedings of the American Association of Zoological Parks and Aquariums Annual Conference 1991* (pp. 606–613). Wheeling, WV: American Association of Zoological Parks and Aquariums.

Desmond, T., and Laule, G. 1993. The politics of protected contact. In *Proceedings of the American Association of Zoological Parks and Aquariums Annual Conference 1993* (pp. 12–18). Wheeling, WV: American Association of Zoological Parks and Aquariums.

Dierenfeld, E. S. 1994. Nutrition and feeding. In S. Mikota, E. L. Sargent, and G. S. Ranglack (eds.), *Medical management of the elephant* (pp. 69–79). West Bloomfield, MI: Indira Publishing.

Douglas-Hamilton, I. 1987. African elephants: Population trends and their causes. *Oryx* 21: 11–24.

Emanuelson, K., and Kinzley, C. 2000. Salmonellosis and subsequent abortion in two African elephants. In *Proceedings of the American Association of Zoo Veterinarians and Association for Aquatic Animal Medicine Joint Annual Conference 2000* (pp. 269–274).

Emanuelson, K., and Kinzley, C. 2001. Salmonella culture and PCR results in a group of captive African elephants. In *Proceedings of the Elephant Managers Association Annual Conference 2001* (pp. 269–274).

Eves, H., and Hutchins, M. 2001. The Bushmeat Crisis Task Force: Cooperative U.S. efforts to curb the illegal commercial bushmeat trade in Africa. In W. Conway, M. Hutchins, M. Souza, Y. Kapetanakos, and E. Paul (eds.), *AZA field conservation resource guide* (pp. 181–188). Atlanta: Wildlife Conservation Society and Zoo Atlanta.

Faust, L. 2005a. Technical report on demographic analyses and modeling of the North American Asian elephant population. Unpublished report. Chicago: American Zoo and Aquarium Association Population Management Center, Lincoln Park Zoo.

Faust, L. 2005b. Technical report on demographic analyses and modeling of the North American African elephant population. Unpublished report. Chicago: American Zoo and Aquarium Population Management Center, Lincoln Park Zoo.

Fayrer-Hosken, R. A., Brooks, P., Bertschinger, H. J., Kirkpatrick, J. F., Turner, U. W., and Liu, I. K. M. 1997. Management of African elephant populations by immunocontraception. *Wildlife Society Bulletin* 25(1): 18–21.

Fayrer-Hosken, R. A., Grobler, D., van Altena, J. J., Bertschinger, H. J., and Kirkpatrick, J. F. 2000. Immuno-contraception of African elephants. *Nature* 407: 149.

Fleischer, R. C., Perry, E. A., Muralidharan, K., Stevens, E. E., and Wemmer, C. 2001. Phylogeography of the Asian elephant (*Elephas maximus*) based on mitochondrial DNA. *Evolution* 55: 1882–1892.

Goddard, D. 1995. *Saving wildlife: A Century of conservation.* New York: Wildlife Conservation Society.

Gore, M., Hutchins, M., and Ray, J. 2006. A review of injuries caused by elephants in captivity: An examination of predominant factors. *International Zoo Yearbook* 40: 51–62.

Gruber, T. M., Friend, T. H., Gardner, J. M., Packard, J. M., Beaver, B., and Bushong, D. 2000. Variation in stereotypic behavior related to restraint in circus elephants. *Zoo Biology* 19: 209–221.

Guerrero, D. 1997. Elephant management in the United States: The evolution of change. *International Zoo News* 44: 195–207.

Hardie, L. C. 1987. *Wildlife trade education kit.* Washington, DC: World Wildlife Fund and IUCN-The World Conservation Union/TRAFFIC.

Hatt, J. M., and Clauss, M. 2006. Feeding Asian and African elephants, *Elephas maximus* and *Loxodonta africana*, in captivity. *International Zoo Yearbook* 40: 88–95.

Hess, D. L., Schmidt, A. M., and Schmidt, M. J. 1983. Reproductive cycle of the Asian elephant (*Elephas maximus*) in captivity. *Biology of Reproduction.* 28: 767–773.

Hildebrandt, T. B., Goritz, F., Hermes, R., Reid, C., Dehnhard, M., and Brown, J. L. 2006. Aspects of the reproductive biology and breeding management of Asian and African elephants, *Elephas maximus* and *Loxodonta africana*. *International Zoo Yearbook* 40: 20–40.

Hile, M. E., Hintz, H. F., and Erb, H. N. 1997. Predicting body weight from body measurements in Asian elephants (*Elephas maximus*). *Journal of Zoo and Wildlife Medicine* 28: 424–427.

Hodges, J. K. 1998. Endocrinology of the ovarian cycle and pregnancy in the Asian (*Elephas maximus*) and African (*Loxodonta africana*) elephant. *Animal Reproduction Science* 53: 3–18.

Hodges, J. K., McNeilly, A. S., and Hess, D. L. 1997. Circulating hormones during pregnancy in the Asian and African elephants *Elephas maximus* and *Loxodonta africana*: A diagnostic test based on the measurement of prolactin. *International Zoo Yearbook* 26: 285–289.

Hudson, J. O. 1997. Positively common ground. *Journal of the Elephant Managers Association* 8: 18–23.

Hutchins, M. 2001. Research overview. In C. Bell (ed.), *Encyclopedia of the world's zoos* (R-Z) (Vol. 3, pp. 1076–1080). Chicago: Fitzroy-Dearborn.

Hutchins, M. 2003. Zoo and aquarium animal management and conservation: Current trends and future challenges. *International Zoo Yearbook* 38: 14–28.

Hutchins, M. 2006. Variation in nature: Its implications for zoo elephant management. *Zoo Biology* 25: 161–171.

Hutchins, M., and Conway, W. 1995. Beyond Noah's ark: The evolving role of modern zoological parks and aquariums in field conservation. *International Zoo Yearbook* 34: 117–130.

Hutchins, M., and Keele, M. 2006. Elephant importation from range countries: Ethical and practical considerations for accredited zoos. *Zoo Biology* 25: 219–233.

Hutchins, M., and Smith, B. 1999. *AZA elephant planning initiative: On the future of elephants in North American zoos.* Silver Spring, MD: American Zoo and Aquarium Association.

Hutchins, M., Smith, B., and Allard, R. 2003. In defense of zoos and aquariums: The ethical basis for keeping wild animals in captivity. *Journal of the American Veterinary Medical Association* 223: 958–966.

Hutchins, M., Smith, B., Fulk, R., Perkins, L., Reinartz, G., and Wharton, D. 2001. Rights or welfare: A response to the Great Ape Project. In B. B. Beck, T. S. Stoinski, M. Hutchins, T. L. Maple, B. Norton, E. F. Stevens, and A. Arluke (eds.), *Great apes and humans: The ethics of coexistence* (pp. 329–366). Washington, DC: Smithsonian Institution Press.

IUCN. 1987. Captive breeding. IUCN policy statement. Gland, Switzerland: IUCN-The World Conservation Union.

Jackson, W. 2001. Conservation training: A consortium of the Field Museum, Chicago Zoological Society, University of Illinois at Chicago, John G. Shedd Aquarium and the University of Chicago. In W. Conway, M. Hutchins, M. Souza, Y. Kapetanakos, and E. Paul (eds.), *AZA field conservation resource guide* (pp. 236–239). Atlanta: Wildlife Conservation Society and Zoo Atlanta.

Jainudeen, M. R., Eisenberg, J. F., and Tilakeratne, N. 1971. Oestrous cycle of the Asiatic elephant, *Elephas maximus*, in captivity. *Journal of Reproductive Fertility* 27: 321–328.

Jhala, Y. 1994. CRC wildlife conservation and management course comes to India. *CRC News*. Fall: 7.

Karesh, W. B., and Cook, R. A. 2001. Wildlife Conservation Society's field veterinary program. In W. Conway, M. Hutchins, M. Souza, Y. Kapetanakos, and E. Paul (eds.), *AZA field conservation resource guide* (pp. 49–54). Atlanta: Wildlife Conservation Society and Zoo Atlanta.

Keele, M. 1999. *AZA elephant masterplan 1997–2002*. Portland: Oregon Zoo.

Kurt, F., and Garaï, M. 2001. Stereotypies in captive Asian elephants: A symptom of social isolation. In H. M. Schwammer, T. J. Foose, M. Fouraker, and D. Olson (eds.), *Recent research on elephants and rhinos. Scientific progress reports: Abstracts of the International Elephant and Rhino Research Symposium, Vienna, June 7–11, 2001* (p. 20). Münster: Schüling.

Langbauer, W. R., Jr. 2000. Elephant communication. *Zoo Biology* 19: 425–445.

Langbauer, W. R., Jr., Payne, K. B., Charif, R. A., Rapaport, L., and Osborn, F. V. 1991. African elephants respond to distant playbacks of low frequency conspecific calls. *Journal of Experimental Biology* 157: 35–46.

Laule, G., and Desmond, T. 1998. Positive reinforcement training as an enrichment strategy. In D. J. Shepherdson, J. D. Mellen, and M. Hutchins, (eds.) *Second nature: Environmental enrichment in captive animals* (pp. 302–313). Washington, DC: Smithsonian Institution Press.

Law, G., and Kitchner, A. 2002. Simple enrichment techniques for bears, bats and elephants: Untried and untested. *International Zoo News* 49: 4–12.

Lee, P. C., and Graham, M. D. 2006. African elephants (*Loxodonta africana*) and human-elephant interactions: Implications for conservation. *International Zoo Yearbook* 40: 9–19.

Leimgruber, P., Gagnon, J. B., Wemmer, C., Kelly, D. S., Songer, M. A., and Selig, E. R. 2003. Fragmentation of Asia's remaining wildlands: Implications for Asian elephant conservation. *Animal Conservation* 6: 347–359.

Leong, K. M., Ortolani, A., Burks, K. D., Mellen, J. D., and Savage, A. 2003. Quantifying acoustic and temporal characteristics of vocalisations for a group of captive African elephants, *Loxodonta africana*. *Bioacoustics* 13: 213–232.

Leong, K., Ortolani, A., Graham, L. H., and Savage, A. 2003. The use of low frequency vocalizations in African elephant (*Loxotana africana*) reproductive strategies. *Hormones and Behavior* 43: 433–443.

Mason, G., and Mendl, M. 1993. Why there is no simple way of measuring animal welfare. *Animal Welfare* 2: 301–319.

Mench, J. 1998. Environmental enrichment and the importance of exploratory behavior. In D. J. Shepherdson, J. D. Mellen, and M. Hutchins (eds.), *Second nature: Environmental enrichment in captive animals* (pp. 30–46). Washington, DC: Smithsonian Institution Press.

McKinney, G. 2003. The ambassadorship of captive African elephant in North America: A demographic comparison of African elephant management strategies. *Animal Keeper's Forum* 30: 376–384.

Mellen, J., and Ellis, S. 1996. Animal learning and husbandry training. In D. G. Kleiman, M. E. Allen, K. V. Thompson, and S. Lumpkin (eds.), *Wild mammals in captivity: Principles and techniques* (pp. 88–99). Chicago: University of Chicago Press.

Mellen, J., and Keele, M. 1994. Social structure and behaviour. In S. Mikota, E. L. Sargent, and G. S. Ranglack (eds.), *Medical management of the elephant* (pp. 19–26). West Bloomfield, MI: Indira Publishing.

Meller, C. L., Croney, C. C., and Shepherdson, D. 2007. Effects of rubberized flooring on Asian elephant behavior in captivity. *Zoo Biology* 26: 51–61.

Mikota, S. K., Larson, R. S., and Montali, R. 2000. Tuberculosis in elephants in North America. *Zoo Biology* 19: 393–403.

Mikota, S. K., Peddie, L., Peddie, J., Isaza, R., Dunker, F., West, G., Lindsay, W., et al. 2001. Epidemiology and diagnosis of Mycobacterium tuberculosis in captive Asian elephants (*Elephas maximus*). *Journal of Zoo and Wildlife Medicine* 32: 1–16.

Moss, C. J., and Poole, J. H. 1983. Relationships and social structure of African elephants. In R. A. Hinde (ed.), *Primate social relationships: An integrated approach* (pp. 315–325). London: Blackwell Scientific.

Nobbe, G. 1992. Going into orbit. *Wildlife Conservation* 95: 62–64.

Nyhus, P. J., Tilson, R., and Sumianto. 2000. Crop raiding elephants and conservation implications at Way Kambas National Park, Sumatra. *Oryx* 34: 262–274.

Olsen, J. H., Chen, C. L., Boules, M. M., Morris, L. S., and Coville, B. R. 1994. Determination of reproductive cyclicity and pregnancy in Asian elephants (*Elephas maximus*) by rapid radioimmunoassay of serum progesterone. *Journal of Zoo and Wildlife Medicine* 25: 349–354.

Olson, D. 1999. Recipe for a successful artificial insemination. *Journal of the Elephant Managers Association* 10(1): 21–31.

Olson, D., ed. 2004. *Elephant husbandry resource guide.* Silver Spring, MD: American Zoo and Aquarium Association, Elephant Managers Association and International Elephant Foundation.

Olson, D., and Wiese, R. J. 2000. State of the North American African elephant population and projections for the future. *Zoo Biology* 19: 311–320.

Ortolani, A., Partan, S., Leong, K., Burks, K., Graham, L., Savage, A., and Mellen, J. 2000. Communication of estrus in a group of captive African elephants (*Loxodonta africana*). *Proceedings of the American Zoo and Aquarium Association Annual Conference 2000* (pp. 321–322). Silver Spring, MD: American Zoo and Aquarium Association.

Payne, K. B., Langbauer, W. R., Jr., and Thomas, E. 1986. Infrasonic calls of the Asian elephant (*Elephas maximus*). *Behavioral Ecology and Sociobiology* 18: 297–301.

Osborn, F. V., and Rasmussen, L. E. L. 1995. Evidence for the effectiveness of an oleo-resin capsicum aerosol as a repellent against wild elephants in Zimbabwe. *Pachyderm* 20: 55–64.

People for the Ethical Treatment of Animals (PETA). 2005. Elephant-free zoos. Norfolk, VA: PETA. www.savewildelephants.com/.

Peterson, J. S. 1994. Free contact elephant management at the Indianapolis Zoo. *Animal Keeper's Forum* 21: 107–110.

Pienaar, U. de V. 1969. Why elephant culling is necessary. *African Wildlife* 23: 180–194.

Poole, T. B. 1998. Meeting a mammal's psychological needs. In D. J. Shepherdson, J. D. Mellen, and M. Hutchins (eds.), *Second nature: Environmental enrichment in captive animals* (pp. 83–94). Washington, DC: Smithsonian Institution Press.

Poole, T. B., Kasman, L. H., Ramsey, E. C., and Lasley, B. L. 1984. Musth and urinary testosterone concentrations in the African elephant. *Journal of Reproduction and Fertility* 70: 255–260.

Priest, G., Antrim, J., Gilbert, J., and Hare, V. 1998. Managing multiple elephants using protected contact at San Diego's Wild Animal Park. *Soundings* 23(1): 20–24.

Rasmussen, L. E. L., and Schulte, B. A. 1998. Chemical signals in the reproduction of Asian (*Elephas maximus*) and African (*Loxodonta africana*) elephants. *Animal Reproduction Science* 53: 19–34.

Rasmussen, L. E., Riddle, H. S., and Krishnamurthy, V. 2002. Mellifluous matures to malodorous in musth. *Nature* 415: 975–976.

Rasmussen, L. E. L., and Riddle, S. W. 2004. Development and initial testing of pheromone-enhanced mechanical devices for deterring crop-raiding elephants: A positive conservation step. *Journal of the Elephant Managers Association* 15: 30–37.

Richman, L. K., Montali, R. J., Garber, R, L., Kennedy, M. A., Lehnhardt, J., Hildebrandt, T., Schmitt, D., et al. 1999. Novel endotheliotropic herpesviruses fatal for Asian and African elephants. *Science* 283: 1171–1176.

Richman, L. K., Montali, R. J., and Hayward, G. S. 2000. Review of a newly recognized disease of elephants caused by endotheliotropic herpesviruses. *Zoo Biology* 19: 383–392.

Robinson, J. G., and Bennett, E. L. (eds.). 2000. *Hunting for sustainability in tropical forests.* New York: Columbia University Press.

Roocroft, A., and Zoll, D. A. 1994. *Managing elephants:An introduction to their training and management.* Ramona, CA: Fever Tree Press.

Rudran, R., and Wemmer, C. 2001. Teaching conservation biology to developing country nationals: The National Zoo's experience. In W. Conway, M. Hutchins, M. Souza, Y. Kapetanakos, and E. Paul (eds.), *AZA field conservation resource guide* (pp. 240–244). Atlanta: Wildlife Conservation Society and Zoo Atlanta.

Ryder, O. 1990. Saving species in their habitats: the transfer of genetics research technologies. In *Proceedings of the American Association of Zoological Parks and Aquariums Annual Conference 1993* (pp. 41–45). Wheeling, WV: American Association of Zoological Parks and Aquariums.

Saunders, W., and Young, G. 1985. An experimental study of the effect of the presence or absence of living visual aids in high school biology classrooms upon attitudes toward science and biology achievement. *Journal of Research in Science Teaching* 22: 619–629.

Schmid, J. 1998a. Status and reproductive capacity of the Asian elephant in zoos and circuses in Europe. *International Zoo News* 45: 341–351.

Schmid, J. 1998b. Hands off, hands on: Some aspects of keeping elephants. *International Zoo News* 45: 478–486.

Schmidt, M. 1981. The hydraulic elephant crush at the Washington Park Zoo. *Proceedings of the American Association of Zoo Veterinarians Annual Conference 1981* (pp.100–106).

Shepherdson, D. J. 1998. Introduction: Tracing the path of environmental enrichment in zoos. In D. J. Shepherdson, J. D. Mellen, and M. Hutchins, (eds.) *Second nature: Environmental enrichment in captive animals* (pp. 1–12). Washington, DC: Smithsonian Institution Press.

Shepherdson, D. 1999. Environmental enrichment for elephants: current status and future directions. *Journal of the Elephant Managers Association* 10: 69–77.

Sherwood, K. P., Rallis, S. F., and Stone, J. 1989. Effects of live animals vs. preserved specimens on student learning. *Zoo Biology* 8: 99–104.

Smith, B., and Hutchins, M. 2000. The value of captive breeding programmes to field conservation: Elephants as an example. *Pachyderm* 28: 101–109.

Smith, B. R., and Hutchins, M. 1999. AZA elephant planning initiative. *Journal of the Elephant Managers Association* 10:150–151.

Stoinski, T. S., Ogden, J. J., Gold, K. C., and Maple, T. L. 2001. Captive apes and zoo education. In B. B. Beck, T. S. Stoinski, M. Hutchins, T. L. Maple, B. Norton, E. F. Stevens, and A. Arluke (eds.), *Great apes and humans: The ethics of coexistence* (pp. 113–132). Washington, DC: Smithsonian Institution Press.

Sukumar, R. 1989. *The Asian elephant: Ecology and management* (2nd rev. ed., 1992). Cambridge: Cambridge University Press.

Sukumar, R. 2003. *The living elephants: Evolutionary ecology, behavior and conservation.* New York: Oxford University Press.

Sukumar, R. 2006. A brief review of the status, distribution, and biology of wild Asian elephants. *Elephas maximus. International Zoo Yearbook* 40: 1–8.

Szdzuy, K., Dehnhard, M., Strauss, G., Eulenberger, K., and Hofer, H. 2006. Behavioural and endocrinological parameters of female African and Asian elephants, *Loxodonta africana* and *Elephas maximus* in the peripartal period. *International Zoo Yearbook* 40: 41–50.

Usongo, L. 2003. Preliminary results on movements of a radio-collared elephant in Lobeke National Park, south-east Cameroon. *Pachyderm* 34: 53–58.

Veasey, J. 2006. Concepts in the care and welfare of captive elephants. *International Zoo Yearbook* 40: 63–79.

Wasser, S. K., Papageorge, S., Foley, C., and Brown, J. L. 1996. Excretory fate of estradiol and progesterone in the African elephant (*Loxodonta africana*) and patterns of fecal steroid concentrations throughout the estrous cycle. *General and Comparative Endocrinology* 102: 255–262.

Wemmer, C., and Krishnamurthy, V. 1992. Methods for taking standard measurements of live domestic Asiatic elephants. In E. G. Silas, M. Krishnan Nair, and G. Nirmalan (eds.), *The Asian elephant: Ecology, biology, diseases, conservation and management* (pp. 34–37). Trichur, India: Kerala Agricultural University.

Wemmer, C., Leimgruber, P., Kelly, D., Ye Htut, and Myint Aung. 2005. Managing wild elephants in Alaungdaw Katahapa National Park and Htamanthi Wildlife Sanctuary, Myanmar. Report to U.S. Fish and Wildlife Service. Smithsonian National Zoological Park, Washington, DC.

Wemmer, C., Rudran, R., Dallmeier, F., and Wilson, D. E. 1993. Training developing country nationals is the critical ingredient to conserving global biodiversity. *BioScience* 43: 762–767.

Western D., Moss, C., and Georgiadis, N. 1983. Age estimation and population age structure of elephants from footprint dimensions. *Journal of Wildlife Management* 47: 1192–1197.

Whyte, I, van Aarde, R., and Pimm, S. L. 1998. Managing the elephants of Kruger National Park. *Animal Conservation* 1: 77–83.

Wiese, R., and Willis, K. 2004. Calculation of longevity and life expectancy in captive elephants. *Zoo Biology* 23: 365–373.

Wiese, R. J. 2000. Asian elephants are not self-sustaining in North America. *Zoo Biology* 19: 299–309.

Wiese, R., and Willis, K. 2006. Population management of zoo elephants. *International Zoo Yearbook* 40: 80–87.

Wildlife Conservation Society. 2003. *Wildlife Conservation Society annual report 2002.* New York: Wildlife Conservation Society.

15 CAN WE ASSESS THE NEEDS OF ELEPHANTS IN ZOOS? CAN WE MEET THE NEEDS OF ELEPHANTS IN ZOOS?

JILL D. MELLEN, JOSEPH C. E. BARBER, AND GARY W. MILLER

Debates about whether elephants (or polar bears or cetaceans or apes) should be kept in zoos (see, for example, Hancocks 2001; Clubb and Mason 2002) address whether these animals deserve more respect than other species, if they occupy a special place among nonhuman animals (Passariello 1999), and whether their cognitive, social, and physical needs cannot be met in zoos. Achieving full consensus on the ethics of keeping elephants in captivity is a daunting undertaking. Not surprisingly, we cannot offer a formula for achieving immediate consensus. Indeed, this essay does not debate the larger ethical issues of keeping elephants in zoos; instead, it focuses on ethical concerns for individual animals. Whether elephants should or will be housed in zoos in the future does not affect our obligation to deal as ethically as possible with those animals already in captive conditions.

Our essay presents a practical approach already in use, an approach focusing on the needs of individual animals. By presenting on a complement of animal welfare assessment tools, we hope to offer zoos a way to assess more objectively the needs of the elephants under their care and to determine whether they can meet those needs. Incrementally, we hope, assessments can help build a foundation of knowledge and ethical understanding that helps everyone achieve consensus on the ethics of keeping captive elephants in zoos. Our discussions frame some issues that pertain to the welfare of elephants in captivity and associated ethical concerns. Next we explore the "enrichment goal-setting tool" (Mellen and Sevenich MacPhee 2001) available to help zoo elephant managers assess the behavioral needs of the African or Asian elephants in their care, both at the species level and as individual animals, and hence to provide conditions that enhance welfare. Much of our discussion and commentary is derived from our own (and others') experiences at Disney's Animal Kingdom, where the tool has been used to assess and address the behavioral needs of African elephants. We discuss at length this question-and-response tool's three main components. These components consist of assessments of (1) the natural history of elephants, (2) the individual histories of a facility's own elephants, and (3) specific current exhibit constraints. We also address specific ethical considerations concerning the application of this "enrichment goal-setting tool," with reference to elephant welfare priorities for zoos, since knowing the needs of elephants, and actually providing for those needs, are two different considerations. We also describe the concept of a "welfare framework," discuss why effective coordination among a facility's animal research, training, and care programs is key to meeting the needs of elephants in captivity, and, finally, consider how institutions housing elephants can use this concept to develop effective and well-integrated animal care programs that ensure that the elephants in their care experience good welfare.

The Science of Animal Welfare: Philosophy and Practice

Attitudes toward the care of animals in captivity continually evolve, as the history of elephant care illustrates. Changing attitudes regarding the care of elephants, as with other animals, are often based on shifts in ethical concerns for these animals, sometimes in conjunction with new scientific knowledge. We know more now about the natural history of elephants than fifty years ago. Long-term field studies (see, for example, Poole 1982; Moss and Poole 1983) along with modern technology such as infrasonic sound detection (Langbauer et al. 1991; Langbauer 2000; Leong, Ortolani, Burke, et al. 2003; Leong, Ortolani, Graham, et al. 2003; McComb et al. 2003) provide insight into what it is to be an elephant in the wild. Our understanding of wild elephants has often been directly linked to studies on captive elephants, for example, ones to do with infrasonic communication (Payne, Langbauer, and Thomas 1986). With this greater knowledge, however, comes a greater responsibility to ensure that the captive conditions we provide can still meet the various needs of elephants. The science of animal welfare provides a useful approach to meeting this responsibility.

Animal welfare assessment is based on assessing how well animals cope with their environment in captivity (Broom 1996) by focusing on the specific properties of each individual animal's physical health and psychological well-being. Physical health and reproductive success have long been goals for most facilities that use animals, such as the zoo and farming communities. In the past, it was sufficient if the animals appeared physically healthy and were reproducing well. Today, most scientists agree that health, productivity, and reproductive success are not sufficient considerations (Dawkins 1976, 1980, 1990; Mason 1991; Mench 1992; Rushen and de Passillé 1992; Duncan and Fraser 1997). However, opponents of housing elephants in zoos, as well as scientists evaluating the available data, point to health concerns and poor reproductive success of zoo-housed elephants as still unresolved current welfare issues (Clubb and Mason 2002).

In the past thirty years, the focus on psychological well-being has grown

Scientist at Disney's Animal Kingdom studying social behavior and vocalizations in African elephants.
Photograph courtesy of Disney's Animal Kingdom

significantly, especially in the zoo community. Psychological well-being involves the subjective experiences of animals and is crucial to any discussion about animal welfare (Dawkins 1980, 1988, 1990; Sandøe and Simonsen 1992). An animal's psychological well-being is dependent on the degree to which it can perform highly motivated, species-appropriate behaviors (Mason 1991) and what opportunities it is given to have some control over its environment (Carlstead 1996). Animals have evolved to use particular behavioral and physical adaptations to cope with challenges in their environments. Welfare assessments of an animal's psychological well-being in captivity are important because we maintain them in environments markedly different from those in which they evolved. Captive conditions may not offer opportunities for animals to employ their behavioral or physical coping strategies or may restrict the functionality of these strategies once they are employed. In captivity, when an animal is prevented from performing highly motivated behaviors, it can become frustrated (Haskell, Coerse, and Forkman 2000). Frustration can lead to suffering. Suffering is defined by Dawkins (1980) as prolonged or intense unpleasant emotional states. These include fear, pain, isolation, frustration, and exhaustion. We are still discovering which behaviors elephants are motivated to perform, under what natural conditions they perform them, and what subjective experiences they perceive. Our understanding of elephant health and psychological well-being will continually affect our ethical decision making. Moreover, the decision whether to act on the results of animal welfare research will become an ethical choice all institutions housing elephants will someday be required to make. This will be true regardless of whether these institutions face monetary and space constraints or the need to keep visitors engaged.

Meeting the psychological needs of animals is challenging (Dawkins 1980), especially when dealing with animals such as elephants with complex cognitive abilities, social structures, and life histories. Given the present reality that elephants are kept in captivity, those institutions housing elephants need to be able to assess their elephants' physical and psychological needs to ensure they experience good welfare. The enrichment goal-setting process offers one useful tool for assessing the behavioral needs of elephants.

A Holistic Enrichment Model
for Assessing the Needs of Elephants in Zoos

Many zoos seek to enhance the quality of captive animal care by providing the environmental stimuli necessary to meet the physical and psychological needs of animals. This environmental enrichment typically involves increasing the variety and range of stimulus opportunities or choices the captive setting provides (Schwammer and Karapanou 1998; Shepherdson 1998). Jill Mellen, along with Marty Sevenich MacPhee, has proposed a holistic approach to providing enhanced environments. Mellen and Sevenich MacPhee (2001) suggest that the first step to providing effective enrichment is to assess the animals' needs, looking not only at their natural history but also their individual histories, as well as their current exhibits. Once needs have been identified, enhancements to the environment—enrichment— can be provided to meet those needs. This approach to enrichment is termed *holistic* because it focuses on all aspects of the captive environment that promote species-appropriate behaviors, provide the animals with choices, and provide the animals with control within their habitats; these are the three main goals of enrichment (American Zoo and Aquarium Association / Behavior and Husbandry Advisory Group [AZA/BAG] 1999). This approach not only promotes daily novelty by providing the animals with objects, but also enriches their environment through daily initiatives in training, keeper interactions, habitat changes, social opportunities, and diet and presentation. Research shows a holistic approach to enrichment is appropriate because many components of an animal's captive environment influence its welfare. For example, we know that in felids the level of interactions with

Young African elephant calf playing with a large ball at Disney's Animal Kingdom in an off-exhibit area. Goal of enrichment (ball) is to encourage exploratory and play behavior.
Photograph courtesy of Disney's Animal Kingdom

keepers can influence the occurrence of stereotypic behavior (Mellen, Shepherdson, and Hutchins 1998; Wielebnowski et al. 2002) and that reproductive success is influenced by social environment and the frequency and nature of interactions with keepers (Mellen 1991). We also know the quality of space appears to be equally, if not more important than the quantity of space (Wilson 1982; Chamove 1989; Mellen 1991; Crockett 1998). Even diet alterations can result in behavioral changes. As one study showed, substituting an equal weight of browse for the hay normally provided to a group of zoo elephants significantly increased feeding time and decreased inactivity (Stoinski, Daniel, and Maple 2000). Greater variety in diet has also been correlated with lower levels of pacing in felids (Mellen et al. 1998).

Since the 1990s zoos have shown considerable enthusiasm for enriching the lives of animals in captivity but have often lacked a structured approach with behavioral goals. Enrichment should be a dynamic process that encompasses a wide range of changes in husbandry practices (AZA/BAG 1999), rather than merely the addition of objects into elephant exhibits. The enrichment philosophy and the program described next offer a structured approach for developing and providing enrichment that can be integrated into daily elephant management and also provides a way to assess the elephants' behavioral needs. This process-oriented approach to developing, implementing, and evaluating enrichment has been incorporated into the Accreditation Standards of the Association of Zoos and Aquariums (formerly, American Zoo and Aquarium Association), forming the core concept of the AZA course, Managing Animal Enrichment and Training Programs. The enrichment program has six components: (1) setting goals, (2) planning, (3) implementing, (4) documenting, (5) evaluating, and (6) readjusting (Mellen and Sevenich MacPhee 2001; see also www.animalenrichment.org). The discussion on assessing zoo elephant needs focuses initially on the goal-setting component because it is fundamental to implementing effective enrichment programs.

Goal-Setting: A Key Component of Holistic Enrichment

Although the primary focus of the enrichment goal-setting tool is elephants' behavioral needs, the holistic approach taken in using this tool also addresses overall welfare needs. Setting goals helps elephant managers view enrichment in terms of an animal's entire captive milieu, by focusing on the animal's physical and social environment, the roles of human caretakers—including feeding, cleaning, and training—and the details of diet, such as food types, methods of presentation, and variety. Setting goals challenges managers to reexamine and enhance the animal's captive environment because the physical and social environment influences animal welfare, as well as reproductive success. An ideal captive environment is responsive to the animal's natural history, is guided by what we know about activity budgets in the wild, and is mediated by the effect of human caretakers and their learned skills, such as husbandry training and veterinary care.

The goal-setting tool is used to identify behavioral goals for a specific elephant enrichment program. An institution's elephant team (keepers, managers, and curators) begins by answering a series of questions about the natural history of elephants, the histories of the individual animals in their care, and the status of their current exhibits. To aid them in answering these questions, animal care staff can review the current scientific literature on elephants, refer to anecdotal accounts of elephant behavior, and draw on the experience and observations of their own elephant team members. The extent and design of the prospective enrichment program depends on the amount of information collected. The findings should be used to identity specific needs of the individual elephants and specific behavioral goals that can be promoted through enrichment activities, training, diet, or other modifications. Next, the elephant team should decide what particular behavioral goals will be set to meet the particular needs of their elephants. The process is most effective when carried out by the institution's animal care team, facilitated by key personnel, such as elephant managers and keepers, and involves veterinarians, nutritionists, and husbandry specialists.

Goal Setting: Natural History Questions

Knowing the natural history of wild elephants can help animal care staff create an enrichment plan that addresses the behavioral and physical needs of captive elephants. A complete list of questions about natural history can be found at www.animalenrichment.org. These questions help animal care staff identify behavioral goals that can be promoted by using environmental enrichment. Information about the natural history of any species can be found in scientific journals and books and from researchers (see the "References" section of this chapter). For example, keepers might ask, "What are the common comfort and self-maintenance behaviors of elephants?" The literature reveals that wallowing, tree rubbing, dusting, and even tool use (Hart et al. 2001) are common. Because these behaviors have obvious ramifications for elephants' physical health, for instance, in helping to maintain skin condition and regulate their temperature, the animal care staff would identify these comfort and self-maintenance behaviors as behavioral goals within the enrichment plan. Animal care staff would similarly inform themselves about all other key natural history questions and then add corresponding key behavioral goals to the enrichment plan.

When these behavioral goals have been identified, the team can then ask, "Which species-appropriate behaviors should and can we encourage or discourage?" For example, allowing the elephants to regulate their own temperature is an important enrichment goal because it provides them with control over their environment and is significant for their health. An exhibit with insufficient opportunities for all elephants to exercise temperature regulation at once, for instance, by using the shade of trees or bathing in pools, needs improvement. Opportunities for temperature regulation would then be added to the list of necessary enclosure design improvements, and the animal care team would begin to brainstorm about the options—more trees, a wallow, a pool, a mister, a fan, and so on.

Using natural history as the basis for assessing the needs of elephants in captivity has its limits. It is not the only guide to creating optimal captive environments. Historically, "the wild" was considered the ultimate standard for assessing the adequacy of a captive environment (see, for example, Hediger 1969), but both Shepherdson (1998) and Veasy, Waran, and Young (1996) argue that the wild is not the only, nor the best, reference for a captive environment. Often information is lacking about the behavior and activity patterns of wild populations. Further, we know that behavior in the wild is often highly variable and dependent on local environmental conditions. Animals in the wild frequently die of hunger and thirst and suffer from untreated diseases, parasite loads, and injuries. Perhaps a more realistic goal than trying to re-create the wild is trying to provide animals with choices within and control over their environment to a similar degree that they would experience in the wild. For example, the scarcity or dispersal of resources may explain why elephants in the wild forage for up to sixteen hours per day (Moss 1982). However, it is also likely that this long-duration-foraging strategy has been selected for over evolutionary time; those that did not forage for large proportions of the day ran the risk of starving. Resources are much more abundant in captivity, and the ease of access results in significantly less foraging (Shepherdson 1999). However, the act of foraging may be important to elephants, regardless of the abundance of resources, because this behavior has evolutionary significance.

Elephants in the wild face few predators other than humans. Antipredator behavior should not necessarily be promoted among elephants in captivity. However, elephants in captivity need the social structure and space to be able to respond appropriately to an aversive stimulus, whether a planned one, such as a log scent marked by a lion, or just something that the elephants perceive as aversive, such as site construction equipment left outside an exhibit. Among new elephant calves at Disney's Animal Kingdom, two natural responses to aversive stimuli have been reported. In response to the loud noise of gravel dumped from a truck, the females have formed a protective circle around a calf with their bodies. In the wild, this would typically be associated with matriarchs displaying toward predators. Another response observed at Disney's Animal Kingdom included a female initially preventing her young calf access to a pool. Presumably, the mother and other cows in range situations would also keep a young calf away from such potentially dangerous situations; in this case, as the calf matured, the cows allowed the calf access to the pool. Female social groups and sufficient space allow such expressions of natural behavior and are key features of exhibits that offer elephants control over their environment.

Goal Setting: Individual History Questions

Because welfare is the property of the individual and not the species (Broom 1996), it is important also to take account of the histories of individual elephants. Elephant managers need to address questions about individual history, to assess the characteristics of animals based on their past experiences. Three questions about individual history in the goal-setting tool segment of the holistic enrichment program follow: First, "Does the animal have medical problems?" Understanding medical problems is an important consideration so that managers may individually tailor enrichment activities. High-priority enrichment initiatives are those that promote locomotion (walking and swimming); however, for an obese elephant, these activities are even more important. High-priority features of the environment for obese elephants include pools for swimming and bathing, a large and varied habitat for walking and exercise, and a training program designed to encourage activity. Offering different food items to provide environmental novelty might be a much lower priority for an obese elephant. The needs of all individual elephants must be assessed to prioritize the enrichment features of their shared environment in captivity.

Second, "Does this animal have any behavioral problems?" Behavioral problems include stereotypic behaviors such as rocking and pacing and other forms of abnormal or undesirable behavior, such as extreme fearfulness or aggression. In many cases, these behaviors somehow derive from the presence of elements in the present or previous environment that do not meet the needs of the individual. Measures of abnormal or undesirable behaviors are useful in assessing welfare (Mason 1991; Duncan, Rushen, and Lawrence 1993). Providing novel scents to encourage environment investigation may not be appropriate for an overly fearful elephant. Similarly, elephants that habitually ingest certain objects, such as sticks and stones, may develop health problems if they are provided with inappropriate substrates. Identifying individual characteristics and needs allows the manager to plan specific enrichment initiatives for individual elephants.

The final question of the individual history section deals with other considerations: "What type of environment was the elephant reared in," and "what kind of exhibits did the elephant experience in previous institutions?" The central nervous system and behavior of animals are greatly affected by their rearing environment (Ehrlich 1959; Diamond et al. 1972; Hemsworth et al. 1986). An animal reared in an inappropriate or inadequate enclosure may continue to prefer that type of enclosure, even when subsequently housed in a naturalistic one (see Dawkins 1983). Animals' cognitive abilities and skills are affected by their early environment (Gunnarson et al. 2000). This does not mean, however, that an elephant raised in a sterile environment should not be subsequently housed in a naturalistic environment, given that naturalistic environments can better meet the needs of elephants in captivity. Rather, certain needs of an individual elephant should be considered, principally while that animal adjusts to its new environment. An elephant reared without a pool may be fearful of such a feature in a new exhibit. In that instance, desensitization training or the facilitating effect of social companions could help to reduce or eliminate the fear (see Schulte 2000). As with the natural history questions, managers should identify and prioritize behavioral goals whose realization can be encouraged through specific enrichment initiatives.

Goal Setting: Current Exhibit Questions

The third component of the goal-setting tool consists of questions about the elephant's current or future exhibit. These questions can be used in two ways. First, the elephant's current exhibit can be reviewed. For example, what is the size of the enclosure? Can the animals use all components of the exhibit? Does the physical environment contain elements of novelty?

Where and how is food provided? Answers to these questions should be compared with the behavioral goals prioritized in the natural history and individual history sections. These comparisons help identify problems with the exhibit that should be addressed to achieve those behavioral goals. For example, if the answer to the question about where and how food is currently provided is it is simply "placed in mangers in one location," this will conflict with the goal of promoting species-appropriate foraging or increased locomotion.

Second, the answers to questions about the current exhibit can be used to improve the design of new exhibits. In tandem with the natural history and individual needs already identified, answering these questions can help elephant managers create an environment that helps the elephants achieve behavioral goals. A pool, shade trees, a mud wallow, or a location for keepers to wash or spray the elephants will address the behavioral goal of regulating temperature. Foraging and locomotion can be facilitated by an enclosure of suitable size and complexity and by encouraging activity through training, enrichment initiatives, and the presence of a social group that allows individuals to be confident to explore their habitat, while avoiding negative stimuli. Ultimately, the successful achievement of the behavioral goals and the determination of whether enough space, shade, choice, and control have been provided can only be measured by assessments of animal welfare.

Welfare Assessments: Can We Meet the Needs of Elephants in Zoos?

The enrichment goal-setting tool helps us to assess elephant needs. Meeting the actual needs of elephants is the next step. Welfare assessments, their outcomes ranging from poor to good (Broom 1991), tell us whether we are effectively meeting these needs. Elephants housed in environments where their physical, physiological, or motivational needs are not met are experiencing poor welfare. If certain needs are not being met, this poor welfare is likely to translate into poor health or poor behavior. Elephants not provided with a balanced diet, for example, experience health problems (see Buckley 2001). Similarly, if an elephant is prevented from performing a highly motivated behavior, such as foraging or social interaction, it will become frustrated, as manifested by cage stereotypies and other forms of habitual or abnormal behavior. Identifying the welfare status of elephants allows us to determine the extent to which their needs are being met.

Many of these welfare assessment tools are difficult to apply in zoo environments. Moreover, no single metric can provide an overall measure of welfare (Rushen and de Passillé 1992; Fraser 1995; Dawkins 1998), and measuring many of the variables associated with welfare, such as curiosity and exploratory behavior in response to new objects (sniffing, manipulation, etc.) or normal social activities such as frequency of play in young animals (Clark, Rager, and Calpin 1997a, 1997b), can be expensive and time consuming. Combining these measures to reach a definitive conclusion is difficult and often produces only ambiguous results (Mason and Mendl 1993).

The difficulties of adequately assessing the welfare of individual animals (Mason and Mendl 1993; Brown, Wielebnowski, and Cheeran, Chapter 6 in this volume) have prompted some scientists to propose another approach. Rather than relying solely on attempts to measure welfare directly, an additional approach for zoos is to improve welfare by identifying and avoiding animal welfare problems, effectively preventing them in the first place (Fraser 1995). Broom and Johnson (2000, 169–170) state, "Where the species and the situation are well-understood, such as with laboratory animals, and in some farm practices and zoo environments, it may be possible to evaluate the causative influences instead of the consequences for the animals. Thus instead of monitoring plasma cortisol levels, it will often be more realistic to check that the animal has appropriate nutritional, physical, and social conditions."

The Welfare Framework

Identifying the conditions that enhance welfare depends on recognizing all animal care—veterinary care, nutrition, habitat/housing, enrichment, training, and research—as essential parts of a complete welfare framework. Professionals practicing in these areas of animal care—sometimes even care programs—exist to some extent in most zoos. Components of these animal care activities and programs are already assessed in some form in the AZA Accreditation Standards. The importance of enhancing and integrating such animal care can be illustrated by considering the components that make up the environmental enrichment program at a hypothetical zoo and the consequences of integrating this program with all elements of animal care.

Enrichment, Effective Processes, and Elephant Welfare: Ethical Conjunctions

Studies have shown a relationship between the use of environmental enrichment and the enhanced welfare of animals in captivity (Carlstead, Seidensticker, and Baldwin 1991; Mason 1991; Carlstead, Brown, and Seidensticker 1993; Carlstead and Shepherdson 2000; Swaisgood et al. 2001). Effective environmental enrichment—specific enrichment initiatives that positively affect behavior—increases the likelihood of improved welfare to zoo elephants. Elephants living at an institution that provides enrichment focused on the three main goals of eliciting species-appropriate behaviors, providing the animals choices within their environment, and providing them with control of their environment, have a greater "welfare potential" than elephants at an institution that provides no focused enrichment. The word *potential* is used advisedly because the elephants' actual welfare depends on how effective the enrichment initiatives prove. An enrichment initiative that is not used by any elephant is not effective at promoting the three main goals of enrichment and has no positive effect on the welfare of those elephants.

Enrichment is a process that should be part of a step-by-step program. As briefly mentioned previously in the "Holistic Enrichment Model" section and as detailed in www.animalenrichment.org, the six components for an effective enrichment program include:

- Setting Goals—based on the natural and individual histories of the elephants
- Planning—developing enrichment ideas and ensuring their safe enactment
- Implementing—scheduling the enrichment and ensuring staff are held accountable for its delivery
- Documenting—collecting information about how the animals interact with the enrichment
- Evaluating—using the information to determine how successful the enrichment is in achieving its goals
- Readjusting—making changes based on the information collected

Zoos often engage in "random acts of enrichment." Providing a ball to an elephant is a random act of enrichment if no thought has been given to what behavior the ball promotes, whether the ball is safe, the schedule on which it is provided to the elephants, or how its effectiveness is assessed. Having a process for enrichment is important because it ensures the safest, most effective enrichment initiatives are provided. The effectiveness of an enrichment program can be assessed by checking whether each of these six components is present and determining the success of each one. As noted earlier, an effective enrichment program leads to a higher welfare potential for the elephants at an institution. Elephants have a lower welfare potential in institutions lacking an enrichment program or where the enrichment program is not integrated into daily elephant management.

However, the welfare of elephants in captivity does not depend only on enrichment. Institutions having both an effective animal training program (see www.animaltraining.org) and an effective enrichment program can provide elephants with a higher welfare potential than institutions with only one of these. Similarly, institutions with effective veterinary, habitat, nutrition, enrichment, training, and research programs offer their animals an even greater welfare potential. The welfare potential of a zoo's elephants can be evaluated from a process perspective and at the institutional level by assessing whether the zoo possesses relevant and complete animal care programs and by examining the effectiveness of these programs. This process-oriented approach answers the questions about whether zoos can meet the needs of elephants from a basic management perspective. Zoos that use a "welfare framework" can objectively assess both their commitment and their ability to ensure that their elephants experience good welfare. The welfare framework helps zoo management to focus all elephant care programs on the needs of the elephants. Each program contributes to the overall care of the elephants. So, every component within each program must be effective, and personnel responsible for different programs must communicate. All the programs must have a common goal: the priority of good welfare. For most zoos, setting up effective elephant care programs will involve the organization of existing processes into an integrated framework contingent on good communication among the different programs and may involve higher personnel costs than the zoo is presently carrying.

Creating an animal care program is not the same as maintaining an effective one. All elephant facilities and programs can benefit from continual self-assessment and improvements. Using "insufficient resources" as an excuse for total inaction is not acceptable; all institutions have limited resources. Zoo managers must allocate their resources to attain the best practicable welfare of the elephants under their care. Attaining the best level of welfare feasible for that zoo by following all the steps in the enrichment process, a zoo will also realize its greatest capacity for ethical treatment of the elephants in its care. Proof of achieving "everyday ethics" of management will be zoo elephants that exhibit a range of species-appropriate behaviors and are able to exercise choices and to exercise control within their environment.

Are Zoos Doing Everything Possible to Meet the Needs of Elephants in Their Care?

Any zoo can make small changes that will improve elephant welfare, and all zoos should strive to make greater improvements over time. They can initiate an enrichment program, or they can embrace the concept of a welfare framework by ensuring that all animal care programs are connected and focused on a common goal. However, making enough changes to be able to reach a universal ideal of maximum welfare potential may not be possible, most likely for financial reasons. The goal-setting tool helps institutions only to identify the behavioral needs of the elephants, but institutions may not be able to meet those behavioral needs immediately. Zoos have to determine exactly what improvements they are actually able to provide.

Will the zoo's habitat/housing program primarily enhance the visibility of elephants to visitors, or will it accommodate the biological needs of the elephants? Among the elephants' behavioral needs, which ones are priorities for enrichment initiatives? Which aspects of enclosure design can address the needs of elephants and expectations of visitors? What are the priorities for habitat improvements that optimize the zoo's various goals? Will a zoo maintain elephants in social groups of variously aged individuals, or will it maintain pairs of cows? Some North American zoos at one time chained elephants overnight to prevent aggression between animals because they lacked facilitates to separate animals in an unrestrained condition (Galloway 1991). The idea one North American zoo recently implemented, providing a lone elephant housed inside over winter with a treadmill to promote locomotion, addressed some issues associated with having to house elephants indoors in higher-latitude zoos given inappropriate outside conditions. But it did nothing to provide this elephant with social companions, another welfare issue. In this case, unfortunately, despite the intent to afford the elephant additional exercise opportunities, outfitting her with a treadmill did not even meet the locomotion goal, as the elephant chose not to use the exercise machine (Pemberton 2006). Should zoos have elephants if it is clear that they cannot provide them with an environment that meets their needs and may never be able to do so because of limited space, personnel, or other needed elements?

To some extent, animal welfare can be scientifically assessed by monitoring variables related to physical health and psychological well-being. But these variables, such as a rise in cortisol level or diminished rocking behavior must be interpreted in light of other observations. The search for explanations for such changes and the determination of whether the animals are receiving optimal or suboptimal care must rely also on inference and deduction. When people view animals in captivity, their conclusions about the animals are always influenced to some extent by a subjectivity. The welfare of animals is assessed from our own ethical perspectives— what we personally consider to be right or wrong, good or bad. Some people see captivity itself as unethical. Other people view the expression of any abnormal behavior in elephants, such as rocking, stone swallowing, or stone sucking, as evidence that the captive environment is insufficient. Still others will maintain that this is the elephant's adaptive response to its captive environment.

Recognizing that people assess animal welfare from various ethical perspectives is important because the perspectives that predominate in a given circumstance dictate which needs of captive elephants are considered important to meet. For instance, a zoo manager may believe that providing a single female elephant with a nutritionally balanced diet, enrichment, positive reinforcement training, and a well-designed habitat is sufficient to meet her needs, even though the natural social lives of elephants are complex and lone females or very small groups of females are unusual in the wild (Poole 1994). The tools discussed here may do nothing to convince elephant managers at such an institution that the welfare of the elephant could be improved by giving the individual social partners. However, the modern-day range of ethical perspectives on captive elephant welfare is narrowed enough that it is fair to acknowledge that zoos cannot decide whether to keep elephants merely on the basis of their desire to exhibit these charismatic animals, or their wish to entertain or educate the public, or even their interest in promoting the reproductive success of elephant populations in captivity. Nowadays, their decision must also be based on their understanding of their ability to meet those particular elephants' needs. Indeed, perhaps the starkest ethical question that modern-day zoos face is whether they are equipped to house any elephants at all.

It is possible, we conclude, to assess objectively the needs of captive elephants, but the effectiveness of these assessments, and the value of the goal-setting tool, depend entirely on the zoo staff's knowledge of elephants in wild and captive settings. We also conclude that zoo managers can provide an added measure of welfare potential for their elephants by taking a process-oriented approach to examining the effectiveness and integration of animal care programs, in addition to using behavioral and physiological measures to assess welfare scientifically. Measures of welfare and welfare potential allow institutions to determine how effectively they are meeting the needs of their elephants in captivity. The practical tools and approaches presented in this essay can directly benefit many captive elephants. Our final conclusion, however, is that despite our current knowledge, and even in light of the tools offered here, meeting the behavioral and physical needs of elephants in captivity remains very challenging. Although the extent to which zoos can meet these needs should continue to be scientifically assessed, those concerned with elephants may never fully agree on whether zoos are meeting their needs.

Acknowledgments

The authors thank Marty Sevenich MacPhee, John Lehnhardt, David Hancocks, Catherine Christen, and Chris Wemmer for their comments on and suggestions for this chapter.

References

American Zoo and Aquarium Association/Behavior and Husbandry Advisory Group (AZA/BAG). 1999. Authors' unpublished notes from workshop at Disney's Animal Kingdom, of the Behavior and Husbandry Advisory Group, a scientific advisory group of the AZA, May. Lake Buena Vista, FL: Disney's Animal Kingdom.
Broom, D. M. 1991. Assessing welfare and suffering. *Behavioral Processes* 25: 117–123.
Broom, D. M. 1996. Animal welfare defined in terms of attempts to cope with the environment. *Acta Agriculturae Scandinavica Supplement* 27: 22–28.

Broom, D. M., and Johnson, K. G. 2000. *Stress and animal welfare*. London: Chapman and Hall.

Buckley, C. 2001. Captive elephant foot care: Natural habitat husbandry techniques. In B. Csuti, E. L. Sargent, and U. S. Bechert (eds.), *The elephant's foot: Prevention and care of foot conditions in captive Asian and African elephants* (pp. 53–55). Ames: Iowa State University Press.

Carlstead, K. 1996. Effects of captivity on the behavior of wild mammals. In D. G. Kleiman, M. E. Allen, K. V. Thompson, and S. Lumpkin (eds.), *Wild mammals in captivity: Principles and techniques* (pp. 317–333). Chicago: University of Chicago Press.

Carlstead, K., Brown J., and Seidensticker, J. 1993. Behavioral and adrenocortical responses to environmental changes in leopard cats (*Felis bengalensis*). *Zoo Biology* 12: 321–331.

Carlstead, K., Seidensticker, J., and Baldwin, R. 1991. Environmental enrichment for zoo bears. *Zoo Biology* 10: 3–16.

Carlstead K., and Shepherdson D. 2000. Alleviating stress in zoo animals with environmental enrichment. In G. P. Moberg and J. A. Mench (eds.), *The biology of animal stress: Basic principles and implications for animal welfare* (pp. 337–354). Wallingford, Oxon, UK: CABI.

Chamove, A. S. 1989. Environmental enrichment: A review. *Animal Technology* 40: 155–178.

Clark, J. D., Rager, D. R., and Calpin, J. P. 1997a. Animal well-being: III. An overview of assessment. *Laboratory Animal Science* (now *Comparative Medicine*) 47: 580–585.

Clark, J. D., Rager, D. R., and Calpin, J. P. 1997b. Animal well-being: IV. Specific assessment criteria. *Laboratory Animal Science* (now *Comparative Medicine*) 47: 586–597.

Clubb, R., and Mason, G. J. 2002. *A review of the welfare of zoo elephants in Europe*. Horsham, UK: Royal Society for the Prevention of Cruelty to Animals (RSPCA).

Crockett, C. 1998. Psychological well-being of captive non-human primates. In D. J. Shepherdson, J. D. Mellen, and M. Hutchins (eds.), *Second nature: Environmental enrichment for captive animals* (pp. 129–153). Washington, DC: Smithsonian Institution Press.

Dawkins, M. S. 1976. Towards an objective method of assessing welfare in domestic fowl. *Applied Animal Ethology* 2: 245–254.

Dawkins, M. S. 1980. *Animal suffering: The science of animal welfare*. London: Chapman and Hall.

Dawkins, M. S. 1983. Cage size and flooring preferences in litter-reared and cage-reared hens. *British Poultry Science* 24: 177–182.

Dawkins, M. S. 1988. Behavioral deprivation: a central problem in animal welfare. *Applied Animal Behaviour Science* 20: 209–225.

Dawkins, M. S. 1990. From an animal's point of view: Motivation, fitness, and animal welfare. *Behavioral and Brain Sciences* 13: 1–61.

Dawkins, M. S. 1998. Evolution and animal welfare. *The Quarterly Review of Biology* 73: 305–328.

Diamond, M. C., Rosenzweig, M. R., Bennett, E. L., Lindner, B., and Lyon, L. 1972. Effects of environmental enrichment and impoverishment on rat cerebral cortex. *Journal of Neurobiology* 3: 47–64.

Duncan, I. J. H., and Fraser, D. 1997. Understanding animal welfare. In M. C. Appleby and B. O. Hughes (eds.), *Animal welfare*. Wallingford, Oxon, UK: CABI.

Duncan, I. J. H., Rushen, J., and Lawrence, A. B. 1993. Conclusions and implications for animal welfare. In A. B. Lawrence and J. Rushen (eds.), *Stereotypic animal behavior: Fundamentals and applications to welfare*. Wallingford, Oxon, UK: CABI.

Ehrlich, A. 1959. The effects of past experience on the rat's response to novelty. *Canadian Journal of Psychology* 15: 15–19.

Fraser, D. 1995. Science, values and animal welfare: Exploring the "inextricable connection." *Animal Welfare* 4: 103–117.

Galloway, M. 1991. Update on 1990 chaining survey. In *Proceedings of the 12th International Elephant Workshop, held at Burnet Park Zoo, Syracuse, New York, October 16–19, 1991* (pp. 63–64). Syracuse, NY: Elephant Managers Association.

Gunnarson, S., Yngvesson, J., Keeling, L. J., and Forkman, B. 2000. Rearing without early access to perches impairs the spatial skills of laying hens. *Applied Animal Behaviour Science* 67: 217–228.

Hart, B. L., Hart, L. A., McCoy, M., and Sarath, C. R. 2001. Cognitive behavior in Asian elephants: Use and modification of branches for fly switching. *Animal Behaviour* 62: 839–847.

Haskell, M., Coerse, N., and Forkman, B. 2000. Frustration-induced aggression in the domestic hen: The effect of thwarting access to food and water on aggressive responses and subsequent approach tendencies. *Behaviour* 137: 531–546.

Hancocks, D. 2001. *A different nature: The paradoxical world of zoos and their uncertain future.* Berkeley: University of California Press.

Hediger, H. 1969. *Man and animal in the zoo: Zoo biology.* London: Routledge and Kegan Paul.

Hemsworth, P. H., Barnett, J. L., Hansen, C., and Gonyou, H. W. 1986. The influence of early contact with humans on subsequent behavioral response of pigs to humans. *Applied Animal Behaviour Science* 15: 55–63.

Langbauer, W. R., Jr. 2000. Elephant communication. *Zoo Biology* 19: 425–445.

Langbauer, W. R., Jr., Payne, K. B., Carif, R., Rapaport, L., and Osborn, F. 1991. African elephants respond to distant playback of low-frequency conspecific calls. *Journal of Experimental Biology* 157: 35–46.

Leong, K. M., Ortolani, A., Burks, K. D., Mellen, J. D., Savage, A. 2003. Quantifying acoustic and temporal characteristics of vocalisations for a group of captive African elephants *Loxodonta africana*. *Bioacoustics* 13: 213–232.

Leong, K. M., Ortolani, A., Graham, L. H. and Savage, A. 2003. The use of low-frequency vocalizations in African elephant (*Loxodonta africana*) reproductive strategies. *Hormones and Behavior* 43: 433–443.

Mason, G., and Mendl, M. 1993. Why is there no simple way of measuring animal welfare? *Animal Welfare* 2: 301–319.

Mason, G. J. 1991. Stereotypies: a critical review. *Animal Behaviour* 4: 1015–1037.

McComb, K., Reby, D., Baker, L., Moss, C., and Sayialel, S. 2003. Long-distance communication of acoustic cues to social identity in African elephants. *Animal Behaviour* 65: 317–29.

Mellen, J., and Sevenich MacPhee, M. 2001. Philosophy of environmental enrichment: Past, present, and future. *Zoo Biology* 20: 211–226.

Mellen, J. D. 1991. Factors influencing reproductive success in small captive exotic felids (*Felis* spp.): A multiple regression analysis. *Zoo Biology* 10: 95–110.

Mellen, J. D., Shepherdson, D. J., and Hutchins, M. 1998. The future of environmental enrichment. In D. J. Shepherdson, J. D. Mellen, and M. Hutchins (eds.), *Second nature: Environmental enrichment for captive animals* (pp. 329–336). Washington, DC: Smithsonian Institution Press.

Mench, J. A. 1992. Introduction: Applied ethology and poultry science. *Poultry Science* 71: 631–633.

Moss, C. J. 1982. *Portraits in the wild: Behavior studies of East African mammals.* 2nd ed. Chicago: University of Chicago Press.

Moss, C. J., and Poole, J. 1983. Relationships and social structure in African elephants. In R. A. Hinde (ed.), *Primate social relationships: An integrated approach* (pp. 315–325). Oxford: Blackwell Scientific.

Passariello, P. 1999. Me and my totem: Cross-cultural attitudes towards animals. In F. L. Dolins (ed.), *Attitudes to animals: Views in animal welfare* (pp. 12–25). Cambridge: Cambridge University Press.

Payne, K. B., Langbauer, W. R., and Thomas, E. M. 1986. Infrasonic calls of the Asian elephant (*Elephas maximus*). *Behavioral Ecology and Sociobiology* 18: 297–301.

Pemberton, Mary. 2006. Elephant not interested in using treadmill. *USA Today*. May 16.

Poole, J. H. 1982. Musth and male-male competition in the African elephant. PhD diss., University of Cambridge.

Poole, J. H. 1994. Sex differences in the behaviour of African elephants. In R. V. Sort and E. Balaban (eds.), *The differences between the sexes* (pp. 331–346). Cambridge: Cambridge University Press.

Rushen, J., and de Passillé, A. M. B. 1992. The scientific assessment of the impact of housing on animal welfare: A critical review. *Canadian Journal of Animal Science* 72: 721–743.

Sandøe, P., and Simonsen, H. P. 1992. Assessing animal welfare: Where does science end and philosophy begin? *Animal Welfare* 1: 257–267.

Schulte, B. A. 2000. Social structure and helping behavior in captive elephants. *Zoo Biology* 19: 447–459.

Schwammer, H. M., and Karapanou, E. 1998. Promoting functional behaviors in confined African elephants (*Loxodonta africana*) by increasing the physical complexity of their environment. In V. J. Hare and K. E. Worley (eds.), *Proceedings of the Third International Conference on Environmental Enrichment, October 12–17, 1997, Orlando, FL* (pp. 92–100). San Diego, CA: Shape of Enrichment, Inc.

Shepherdson, D. J. 1998. Tracing the path of environmental enrichment in zoos. In D. J. Shepherdson, J. D. Mellen, and M. Hutchins (eds.), *Second nature: Environmental enrichment for captive animals* (pp. 1–12). Washington, DC: Smithsonian Institution Press.

Shepherdson, D. J. 1999. Environmental enrichment for elephants: Current status and future directions. *Journal of the Elephant Managers Association* 10: 69–77.

Stoinski, T. S., Daniel, E., and Maple, T. L. 2000. A preliminary study of the behavioral effects of feeding enrichment on African elephants. *Zoo Biology* 19: 485–493.

Swaisgood, R., White, A. M., Zhou, X., Zhang, H., Zhang, G., Wei, R., Hare, V. J., Tepper, E. M., and Lindburg, D. G. 2001. A quantitative assessment of the efficacy of an environmental enrichment programme for giant pandas. *Animal Behaviour* 61: 447–457.

Veasy, J. S., Waran, N. K., and Young, R. J. 1996. On comparing the behavior of zoo housed animals with wild conspecifics as a welfare indicator. *Animal Welfare* 5: 13–24.

Wielebnowski, N., Fletchall, N., Carlstead, K., Busso, J., and Brown, J. 2002. Noninvasive assessment of adrenal activity associated with husbandry and behavioral factors in the North American clouded leopard population. *Zoo Biology* 21: 77–98.

Wilson, S. 1982. Environmental influences on the activity of captive apes. *Zoo Biology* 1: 201–210.

16 GIANTS IN CHAINS

HISTORY, BIOLOGY, AND PRESERVATION
OF ASIAN ELEPHANTS IN CAPTIVITY
FRED KURT, KHYNE U MAR,
AND MARION E. GARAÏ

According to most recent estimates, the total population of captive Asian elephants in the range countries numbers about 14,000 animals (Kashio 2002, table 1). The largest population is found in Myanmar (about 40% of the total population of the range countries), followed by India (26%) and Thailand (19%). Outside the range countries at least 870 Asian elephants are kept in zoos and circuses (European Elephant Group 2003). Therefore, the total captive population includes about 15,000 elephants, or 25–33% of the current total world elephant population (Kashio 2002; Sukumar 2003). Numerous reports consider captive Asian elephants to be "domesticated" animals (see Lair 1997). However, most of the elephants living in captivity have been wild caught, and a large number of elephants that have been bred in the range countries are from matings between captive females and wild males. Although captive elephants have been in close contact with humans for at least 3,500 years, they have neither been the subject of sustained captive breeding nor have they been selectively bred for particular characteristics, as has been the case with true domesticated animals, such as the horse, the sheep, or the dog (see Stevenson 2002). By assuming that elephants are domesticated, people treat them accordingly. In old-fashioned western zoos and circuses, the elephant is the only wild animal still kept more or less permanently in chains. Imagine the reaction of animal welfare organizations if suddenly zoos would keep rhinos, hippos, or tigers in chains! All people concerned with "tamed" Asian elephants should regard them for what they truly are—wild animals in captivity.

Captive elephants have an economic significance for owner and keeper/mahout. They may be an important power source in forestry applications, mainly in developing countries unable to afford tractors and forest roads. Furthermore, they have religious significance, mainly in Buddhist and Hindu countries, and social significance in nonreligious ceremonies. Captive elephants can be an attraction in urban tourism and could be useful for education (including ecotourism) concerning life sciences and conservation. Last but not least, captive Asian elephants living in jungle-based camps can still have ecological significance in selective timber harvesting and as a keystone species in the maintenance of biodiversity and ecological processes.

In many traditional elephant-keeping establishments, elephants do not enjoy "the five freedoms," which were devised by Webster (1984) regarding cattle (freedom from malnutrition, freedom from thermal and physical discomfort, freedom from injury and disease, freedom from fear and stress, and freedom to express most normal patterns of behavior). Hence, the *ex situ* maintenance of *Elephas maximus* is often accompanied by ethical dilemmas associated with traditional management practices, welfare, and conservation of the species. In our essay, we evaluate and classify the range of facilities and recommend improvements, including policy adjustments needed for improved welfare at the regional and international level.

Evolving Patterns of Elephant Management

The earliest work concerning care of captive elephants is decidedly the *Hastyayurveda* (fifth or fourth century BC). Two chapters of the *Arthashastra* of Kautiliya (ca. 300 BC–300 AD) deal with capture, keeping, care, and training of elephants. The *Matangalila* of Nilakantha, another antique source, may go back a thousand years from the present or even much earlier. Many other old Buddhist and Hindu sources also refer to the management of captive elephants (see Edgerton 1931). All these rules and regulations were (and many still are) the basis of captive elephant management in areas under the control of Buddhist and Hindu rulers in South Asia. In the nineteenth and early twentieth centuries, British civil servants such as G. H. Evans, A. J. Ferrier, A. J. W. Milroy, and G. P. Sanderson recorded many local practices, improved them, if necessary, and formulated a number of guidelines to regulate capture, care, daily food rations, working times, and working loads of timber and army elephants in India and Myanmar (see Krishnamurthy and Wemmer 1995a, 1995b; Wemmer 1995; also Lahiri Choudhury, Chapter 7 in this volume). In Myanmar, then known as Burma, state-run and private elephant establishments such as the State Timber Board (now the Myanma Timber Enterprise) had guidelines on elephant husbandry that were later prescribed as Department Standing Orders and which are still implemented. A recordkeeping system in the early days of colonial elephant management provided valuable data on the performance of individual animals. Several of these reportage methods are still routinely used.

The overall captive population, which still draws on wild populations for recruitment, is dwindling steadily due to low reproduction and high juvenile mortality but mainly due to the restriction of wild-capture operations. In Vietnam, for instance, the number of captive elephants dropped from 600 in 1980 to 165 in 2000 (Cuong, Lien, and Giao 2002). In Cambodia,

162 were left in 2000, of an estimated 300 to 600 some twenty years before (Lair 1997; Dany et al. 2002). In Sri Lanka, privately owned and temple elephants numbered 532 in 1970 and 186 in 2002 (J. Jayewardene, personal communication). In Nepal, the captive population dwindled continuously, from 328 in 1903 to 47 in 1973, then later increased to 171 in 2002 (Kharel 2002). In Indonesia, a captive population of about 400 elephants is considered stable because of regular capturing operations in recent years (Hutadjulu and Janis 2002) and even though mortality is extremely high because of miserable living conditions. Of the 117 elephants known to have been captured in Riau (Central Sumatra) over the September 2000 to March 2003 period, 109 were, by 2003, found to be either dead, lost, or expected to die (see Hammatt, Fahrimal, and Mikota 2004; also see Mikota, Hammatt, and Fahrimal, Chapter 18 in this volume). However, there are exceptions to the downward numbers. A sudden spike in the number of newborn captive elephants in northern Thailand in the early 2000s indicates that favorable socioeconomic conditions may induce a population increase in some places. The cessation of the logging work that used to discourage owners from allowing female elephants to breed (so as not to lose their working time), combined with an increasing demand for young elephants for the tourism industry, seem to be the reasons behind this uptick in births (T. Angkawanich as reported to M. Kashio, personal communication).

In Indonesia, as well as in Malaysia, keeping elephants in captivity was abandoned about 100 years ago and then reactivated in the 1980s, with the assistance of mahouts from Thailand. The present small population in Malaysia consists of thirty-six elephants, and it is said to be increasing (Daim 2002). India's captive elephant population is considered to be stable (Bist et al. 2002), while Myanmar's population seems to be decreasing slowly (Aung and Nyunt 2002). In Europe, the captive population of nearly 500 dropped to 450 in 2000, and within about twenty years was expected to decline to 120, although captive breeding has been extremely successful in recent years (Kurt 2001). In North America, the present total zoo population of about 300 will drop to approximately 10 elephants within fifty years and will be demographically extinct, unless there is continued importation or a drastic increase in birthrate (Wiese 2000).

In all range countries, considerable shifts of captive elephants from forested to urban areas are reported. For instance, India's Supreme Court imposed restriction on logging operations in the northeast and the Andaman Isles in 1996, which resulted in a great exodus of captive elephants from these regions. The number of captive elephants has increased in Kerala from about 250 in 1983 to more than 700 at present, with a strongly male-biased population structure. The city of Jaipur (Rajasthan), presently a major elephant center, has a strongly female-biased population of about 100 (Bist et al. 2002). All over Thailand, most captive elephants suddenly became unemployed after the government banned logging in 1989. Despite the evidently salubrious effect on northern Thailand birthrates, unemployment and, in due course, lack of suitable food have been the root causes of many serious problems for this country's captive elephant population. Many elephants are brought to the city of Bangkok for begging or are engaged in illegal logging (Lohanan 2002). They are often given amphetamines to speed up work. Unworkable elephants are sent to slaughterhouses, and the meat is sold (Mahasavangkul 2002). In some Thai tourist resorts, shows featuring elephant babies have become so attractive to tourists and entrepreneurs alike that a lucrative baby market has evolved, fed by illegal capture of neonates and infants in Thailand, Lao People's Democratic Republic, Vietnam and Cambodia (Kashio 2002).

A Taxonomy of *Ex Situ* Elephant Management

The multiple and diverse deployment of captive Asian elephants requires various keeping systems whose extremes can be defined as "extensive" and "intensive" (Kurt 1995; Kurt and Mar 2003). Working elephants in jungle villages of Assam and South India, as well as in Myanmar, live in extensive keeping systems. They are used as riding, carrying, and towing animals. During their resting periods they live, with hobbled front feet, in the nearby forest, where they find food and encounter their tame and wild conspecifics. In South India, depending on tradition, additional fodder (e.g., boiled rice, millet, and coconuts) is given. Many mahouts belong to tribal societies with a broad knowledge of elephants and their habitats, and few of these mahouts have either drug or alcohol problems that would impede the quality of their work with their charges. Accidents appear to be rare.

Intensive keeping systems are those in which animals are kept by temples or private owners more or less individually, fed exclusively on prepared fodder, and, at night or if idle, are shackled with long or short chains by a hind foot or by one hind and one front foot. Contact with captive conspecifics is prevented. Contact with wild conspecifics is unlikely because intensive keeping is concentrated in urban regions. Restricted movements and the practical absence of contacts with conspecifics can lead to extreme aggression toward handlers. The mahouts in charge of intensively kept elephants are often underpaid, their knowledge is fast decreasing, and many of them are addicted to alcohol and other drugs. In India and Sri Lanka, intensively kept elephants are increasingly used in religious processions, political demonstrations, or wedding ceremonies. To participate at these ceremonies, elephants are walked or transported by trucks. Sometimes accidents happen on the way. Captive elephants have also become highlights for urban tourism, as riding animals, to perform all kinds of circus tricks, or to play football or polo. In Thailand elephant shows stage scenes de-

Elephants used for ceremonies in Kerela, South India. To participate in ceremonies, elephants are walked or transported by truck.
Photograph courtesy of Khyne U Mar

picting battle attacks of war elephants, as well as noosing of wild elephants using *mela-shikar*, a traditional pole-and-noose method (see Lahiri-Choudhury, Chapter 7 in this volume, for more details on noosing methods, including mela-shikar).

The intensive keeping system has been taken over by western circuses and, in former times, also by zoological gardens. Today, the keeping system in traditional zoos as well as at the Pinnawela Elephant Orphanage in Sri Lanka can be considered as intermediate between extensive and intensive. During the day, the elephants are kept free in their paddocks, but during the night, they are shackled or otherwise kept solitary in small cages. In very modern zoos, alternative keeping systems are now evolving quickly in response to the guidelines of the *World Zoo Conservation Strategy* (International Union of Directors of Zoological Gardens / Conservation Breeding Specialist Group [IUDZG/CBSG] 1993; World Association of Zoos and Aquariums [WAZA] 2005). In these systems, elephants live in family groups. Only subadult and adult bulls are kept singly when necessary, for example, during musth. Another type of keeping system is found in the Elephant Transit Home of the Uda Walawe National Park in Sri Lanka, where young elephants are kept in a seminatural environment with minimal contacts with humans before they are brought back into the wild.

Accidents sometimes happen while traveling to ceremonies. Fortunately, this bull was not seriously hurt but had minor bruises.
Photograph courtesy of Khyne U Mar

The economic, religious, and educational significance of captive elephants does not track precisely with the divisions between intensive and extensive keeping systems. Such significance is relatively high for elephants kept for cultural functions and for those in modern zoos, timber and tourist camps, or the Pinnawela Elephant Orphanage but is relatively low for elephants kept in the Transit Home, in urban tourist centers, in circuses, or in traditional zoos (Kurt and Mar 2003). Environmental conditions and fitness of elephants are better in forest camps and modern zoos than in establishments with intensive keeping, such as circuses or traditional zoos. Monotonous food, consisting of the leaves, branches, and stems of one to three plant species, is common fare for intensively kept elephants. Elephants living in jungle-based extensive systems enjoy a much more diverse diet. In modern zoos, as well as in the Pinnawela Elephant Orphanage, the food is considered to be at an intermediate level of diversity (Kurt and Mar 2003).

The daily life of elephants kept in extensive systems includes, besides work, the search for food, skin care, and socializing. Stereotypies are absent or rare. Their daily species-specific activity is relatively high. Elephants kept in intensive systems are often idle. Stereotypies occur regularly. The variety of daily species-specific activity is low. Obesity and foot problems

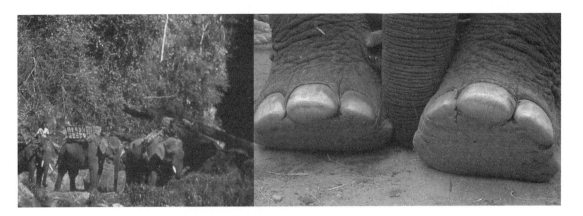

Extensively kept working elephants amid natural foraging site (*left*); naturally trimmed footpad of timber working elephant (*right*).
Photograph courtesy of Khyne U Mar

When elephants walk on flat hard surfaces without receiving adequate nail and foot care, the footpad thickens unduly.
Photograph courtesy of Khyne U Mar

are common. The same applies to elephants kept in traditional zoos and circuses. However, diverse daily activities are found in the lives of the elephants of modern zoos and the Pinnawela Elephant Orphanage. Limited social behavior is a consequence of management systems that either do not

Intensively kept temple elephants spending their time on natural soil do not suffer from overgrown footpads, but their sedentary lifestyle does lead to prolongation of toenails.
Photograph courtesy of Khyne U Mar

allow intraspecific interactions or where they are rarely possible because of the limited social structure of the captive group's members. In most intensively kept populations, the age-sex structure is limited, for example, mainly subadult and adult bulls for festivals in Kerala; or older females in circuses; or infants and juveniles in Thai tourist resorts. Groups in traditional zoos consist mainly of middle-aged or old females and rarely include bulls or younger members. But the populations in the Pinnawela Elephant Orphanage, as well as those in jungle-based camps, include various age and sex classes. In summary, welfare is low in intensive keeping systems as well as in traditional zoos and the Elephant Transit Home, but welfare is significantly higher in modern zoos, extensive keeping systems, and the Pinnawela Elephant Orphanage.

Reproduction

In the forest camps of India's Tamil Nadu and Karnataka states, captive females had an annual fecundity rate of 0.16 per adult female between 1969 and 1989. Given the relatively high survival rate at young ages, this population could grow strongly with the higher fecundity rate observed in recent years (Sukumar 2003). In Myanmar, the annual fecundity rate of timber elephants ranks below 0.1 per adult female, and their large population is declining slowly under the prevailing natality (average 3.1% of the population) and mortality (3.3% of the population) rates (Aung and Nyunt 2002). Improved health care and nutrition, especially for pregnant and nursing females, could help managers to achieve a stable population size, as seemingly existed in timber camps in colonial Burma at the beginning of the twentieth century (see Williams 1950). Relatively successful breeding occurs also in other jungle-based establishments: of 198 privately owned extensively kept elephants in Assam, only 56 (28%) were captured from the wild (Ashraf et al. 2003). In the Jaldapara Wildlife Sanctuary (115 km²) of West Bengal, 67 captive elephants of the Indian Forest Department are breeding so successfully that a further increase in population would affect the resources and the need for effective birth control is becoming a major issue (J. T. Mathew, personal communication with F. Kurt).

In jungle camps, up to 90% of the extensively kept females reproduce (Sukumar 2003). Of the 57 females of the national parks and reserves of Nepal, 16 (28%) gave birth between 1979 and 2000 (Kharel 2002). This figure is still higher than in European zoos, where in the ten years between 1992 and 2002, only 42 (19.9%) of 211 adult females reproduced (European Elephant Group 2003). However, this low fecundity rate is also due to lack of suitable males. In the Elephant Training Centers of Sumatra, where the animals are kept intensively, captive propagation occurs, but neonate mortality seems to be 50% or more (Reilly and Sutkatmoko 2002).

There is no self-sustainable captive population of *E. maximus* at present. Nevertheless, there are smaller populations with a high potential for reproduction, such as at the Pinnawela Elephant Orphanage, at modern zoos, and in intensive keeping systems. In intensive keeping systems of Kerala, India, and Sri Lanka, reproduction is practically absent. In the city of Jaipur, India, only two births have been recorded during the past 100 years. Most owners of intensively kept elephants seem not to be interested in breeding them because of the economic losses incurred during the last months of pregnancy and the two-year lactation period. In addition, many intensively kept elephants are neither physiologically nor psychologically able to reproduce because of serious retardations in body growth (Kurt and Kumarasinghe 1998) and absence of a species-specific socialization process (Garaï 2002a).

Jungle camps are potentially important for *E. maximus* conservation not only because of the relatively high elephant reproduction rate but also because many of their captive-born babies have been fathered by wild bulls. Thus, many jungle camps function as gene traps. Their populations harbor a relatively diverse gene pool, a conservation asset. In European zoos, the situation is the opposite. As of 2003, for the 236 females, there were only 16 actively breeding bulls, and in several cases, fathers had already reproduced with daughters (European Elephant Group 2003).

Policy and Practice

Conservation agencies aim to conserve biodiversity, emphasizing maintaining threatened species, such as the Asian elephant, and safeguarding genetic diversity and ecological processes. All captive Asian elephants are important for conservation, if one considers that they account for 25–33% of the entire dwindling world population of this species. This fact was first strongly affirmed in 1993 by a coalition of modern zoos. They formulated the *World Zoo Conservation Strategy* and started to take responsibility for *ex situ* conservation of certain animal species, including the Asian elephant (IUDZG/CBSG 1993; WAZA 2005). Accordingly, these zoos expressed the desire to keep captive elephants in ecological and social conditions that are as natural as possible, as reflected in a 2002 publication from the Federation of Zoological Gardens of Great Britain and Ireland:

> Elephants must only be kept in zoos as part of an overriding conservation mission so that they are in actively managed breeding programmes. . . . Their presence must enable progressive educational activities and demonstrate links with field conservation projects and benign scientific research, leading to continuous improvements in breeding and welfare standards. Zoos must exercise a duty of care so that standards of husbandry practices, housing, health and welfare management are humane and appropriate to the intelligence, social behavior, longevity and size of elephants. (Stevenson 2002, vii–viii)

India's Central Zoo Authority prescribed standards and norms for all zoos in its 1992 "Recognition of Zoo Rules," published that year in India's *Official Gazette*. However, no such standards and norms exist for the care of hundreds of privately owned elephants in India, as well as in the other range countries. In recent years, some Indian publications have enumerated these care imperatives, including Bist's draft "Standards and Norms for Elephant Owners" (1996) and the *Kerala Captive Elephants (Management and Maintenance) Rules, 2003* (Government of Kerala 2003). If implemented, these guidelines would ensure improved living conditions for captive range country elephants. However, few of these elephants live in a social and ecological environment or in physical and psychological conditions conducive to their species-specific behavior, and the importance of facilitating breeding has hardly been recognized.

In recent years, national and international animal welfare organizations, including Elephant Care International, Elephant Family, Friends of Elephants, International Elephant Foundation, International Fund for Animal Welfare, Kerala Elephant Welfare Association, Royal Society for Prevention of Cruelty to Animals, Save the Elephant Foundation, Universities Federation for Animal Welfare, have variously campaigned, conducted training, and provided resources to improve living conditions of captive elephants in several range countries, including India, Indonesia (Sumatra), Sri Lanka, and Thailand. These organizations' activities and their assuredly important stationary and mobile elephant clinics mainly address the fitness—health and body condition—of individual elephants, rarely touching on the socioecological considerations needed to maintain near-natural conditions.

In many densely populated countries of South Asia, relatively large numbers of wild elephants have survived because many people venerate the elephant as a symbol associated with Buddha and with a number of Hindu gods: "Elephants kept in temples or participating in cultural or religious festivities can reinforce this sentiment of sacredness among the people" (Sukumar 2003, 400). Nevertheless, ethical dilemmas remain: Elephants used for festivities live in chains and social isolation.

Upgrading Human Knowledge of Elephants

Conservation-oriented elephant keeping must be based on the ecological and behavioral demands of the species under natural conditions. In our view, most people concerned with captive elephants have little knowledge of wild elephants. Accordingly, they run the risk of considering the abnormal as normal. Weaving is considered by ethologists as a stereotypy, hence as an abnormal behavior. Many of those who own or manage captive elephants consider it normal that their elephants weave. Accordingly, they maintain that weaving is requisite for blood circulation or that this stereotypy is also found in wild elephants, which has never been confirmed by field researchers.

That captive elephants are domesticated and can be made to reproduce like farmed cattle, horses, or rabbits is erroneous because they are socially highly organized wild animals. Mating, parturition, and raising elephant offspring depend on female choice, social environment, and life history. In European zoos, all of 138 successful matings occurred only when the bull was taller than the female (Kurt and Garaï 2007). Shifting females from their home places to a bull kept in a different establishment often fails because the translocated female must first establish her social rank in the new group (Garaï 1992). When brought back to her home place she must reestablish her rank, a process that often induces the fetus to be resorbed or aborted (Kurt and Mar 1996). In Nepal, where females are often translocated after mating, nine out of twenty-five neonates (36%) were either stillborn, too weak to survive, or were killed by the mother (Kharel 2002; Kurt and Garaï 2007). However, in the Pinnawela Elephant Orphanage, where mothers and offspring grow up in more or less close contact, eight orphaned and three captive-born females gave birth twenty-two times between 1983 and 2003, and all neonates survived the first years. Only one mother, Sharmi, did not accept her offspring. Sharmi had a very low rank in the Pinnawela herd, and reportedly was subsequently sent to the Dehiwala (Sri Lanka) zoo for mating.

Life history is important for determining elephant reproductive success. Adult female wild elephants encourage offspring three years and older to care for neonates and infants by keeping them close to the group, by protecting them during sleep and bath, and by offering them prepared food. This facilitates the duties of mothers and enables maturing females to learn vital maternal skills before becoming mothers. The importance of an undisturbed genesis of maternal behavior in captive Asian elephants can be shown from recent experiences in European zoos. Of sixty-three reproducing females, twenty-nine were weaned from their mothers early in their life (i.e., at the latest in their third year), and thirty-four were separated in their fourth year or not at all. On reaching maturity, the early weaned group gave birth to fifty-one offspring. Only 33.3% of these offspring were accepted by the mothers, and 67.7% were stillborn, killed, or not accepted. The second group, the thirty-four elephants weaned late or not at all, gave birth to sixty-one offspring; 80.3% of these sixty-one newly born were accepted by their mothers and only 19.7% were stillborn, killed, or not accepted. Accordingly, the European Endangered Species Programme for elephants in member institutions of the European Association of Zoos and Aquaria demands "that matriarchal family units are developed and kept together . . . to keep female offspring within their family group during their life" (Dorresteyn 2001, 14).

In the range countries, all captive-born elephants are weaned between 1 and 4 years of age because the process of taming and training is easier, and less dangerous, for humans in the absence of the elephant mothers. Methods to tame weaned offspring or newly caught elephants consist mainly of depriving them of movement for several days or weeks by means of ropes (Assam, parts of Karnataka), crushes (Thailand), or cradles (parts of Myanmar). This breaking-in process is combined with beating and other tortures, deprivation of food, and water, as well as overstimulation with drums and fire.

Three methods are used in the taming of elephants in Myanmar; the one-sided or "half" crush, the two-sided crush, and the cradle, or "sling," method. Normally, crushes are used for captives measuring 2.1 meters or taller and cradles for those measuring below 2.1 meters in height. If a half crush is used, trainee elephants suffer less injury than if a two-sided crush is used, but the trainers are put at greater risk of being injured. Specially trained hunting elephants, called *koonki* elephants, are useful in this operation as they help restrain the new trainee by pushing it alongside the crush. In a two-sided crush, the risk of trainers' being injured by the elephant is lower than with the one-sided crush, but the elephant can be subjected to bruises and concussion as the result of struggling and aggression experienced during the first few days of capture. Two-sided crushes are normally used for aggressive and intractable elephants. They are also useful for general veterinary inspection of tamed elephants. The trainers employ food and water as a reward during the breaking operation. Until it becomes thoroughly obedient, usually after about three days, the calf is not allowed to see any other elephants. Later on, trained elephants are brought alongside the crush and fed and handled in full view of the captive. As soon as it has learned to be obedient, the trainee is taken out of the crush, but its movements are confined by a breast band (cradle) or by being tied to a tree. Thereafter its progress is rapid. Breaking procedures are normally conducted at night, to avoid heat strokes and unnecessary loss of energy.

To improve the management of captive elephants, it is important to find humane methods for taming captive-born and captured elephants. The same applies for captive bulls during the period of musth, when they still suffer from tight fettering, beating, and several other forms of torture. Not all traditional methods are designed to attain control over elephants by physical dominance. The oldest-known taming methods in South India work fundamentally differently. Here the elephants to be tamed are first kept singly in kraals of 4 to 6 meters wide and long. The trainer stays outside, so this is a protected contact method. The taming and basic training is done by means of positive reinforcement (Krishnamurthy 1992; Kurt 1992).

Improvement of Facilities

In western zoos, the present trend is toward large enclosures, which include a pool, wallows, and scratching trees. Bulls older than 10 years are no longer chained, and today only 25% of European zoos holding elephants chain females and young animals during nighttime. Chains are being replaced by box stalls, and direct contact between keeper and elephant is giving way to the system of "indirect contact" or "no contact" (European Elephant Group 2003). However, these trends do not apply to most zoos in the range countries, where elephants are almost permanently chained, except when brought to their daily bath and used for shows or riding. An improvement of the keeping conditions in these locations is urgently needed.

Keeping elephants in chains is cheap but is accompanied by serious accidents (see Haufellner, Schilfarth, and Schweiger 2002). In elephants, chaining also causes foot problems, such as uneven wear of toenails and soles and bedsores, or pressure sores. More or less permanent chaining leads to notorious weaving. This stereotypy can be considered as a symptom of social isolation that finally leads to abnormal social behavior, including infanticide (Kurt and Garaï 2002). Social isolation in more or less permanently chained elephants is further enforced by the fact that they are lined up according to sex and size. However, the members of a group of wild elephants organize themselves in a typical spatial pattern in which neonates and infants stay close to their mothers or allomothers—females who give maternal care to the offspring of other mothers—and juvenile females stay close to neonates and infants (Kurt and Garaï 2007). Most members of the Pinnawela herd try to follow this pattern when not chained (Garaï 2002a). From the point of view of animal welfare and conservation, the number of intensively held elephants must be kept as low as possible, and it makes no sense to increase their populations with animals from well-breeding populations such as jungle villages or the Pinnawela Elephant Orphanage. It is important for the future of elephant conservation that many of the range country facilities in urban areas give way to more adequate keeping systems. Permanent chaining should be replaced with keeping idle elephants singly in box stalls or group-wise in paddocks. This has been done by numerous western circuses with the successful cessation of aggressive behavior and increased tractability (F. Knie, personal communication).

About 60% of captive elephants in range countries are kept extensively, that is, they still live in social groups close to or in the jungle (see Kashio 2002; Kurt and Garaï 2007). However, new roles must be found to avoid an unwanted exodus of elephants into urban areas. This is a real possibility whenever logging bans are imposed in range countries where logging had been the main use of elephant power. With their extreme cross-country mobility, elephants are unique mounts for antipoaching patrols, researchers, and ecotourists and are used, mainly in India and Nepal, as irreplaceable partners in the management and study of wildlife. Since antiquity, specially trained hunting elephants, so-called koonkies, have carried elephant catchers into the herds of wild elephants for mela-shikar, or selective noosing. Koonkies still play an important role in taming, training, and translocating wild elephants in India, Malaysia, Myanmar, and Thailand.

In Sri Lanka, a relatively large wild population of about 3,500 elephants are subject to considerable human-elephant conflicts. Here, extensively kept elephants could be used, as in other South Asian countries, for patrolling in protected areas and for protecting men and crops from elephant raids. Today, ecotourism on elephant-back is developing strongly in Sri Lanka and is already established in Thailand. Those former working elephants not suitable for tourism could be used for diverse conservation work. Thailand harbors 100 national parks and other nature reserves. If only four elephants are assigned to patrol in each of them, 400 elephants could survive in a near natural habitat (Salwala 2002). Conservation of captive Asian elephants in jungle camps also means employment of local, often tribal, elephant people. The importance of their traditional knowledge of elephants and of the biodiversity of the natural habitat can be usefully employed for conservation and applied research.

Alternative Keeping Systems

About 2,000 years ago, the *Arthashastra* proposed the establishment of a *Mrgavana*, a fenced park for tamed elephants and other wild animals. This idea should be adopted again. Large areas of at least a quarter of a kilometer could be fenced and furnished with adequate sources of water and shady places to become a home for orphaned, surplus, dangerous, or otherwise problematic elephants. Elephants kept in such elephant parks have to be given supplementary foods and regularly checked by veterinarians. Apart from this, contacts between men and these elephants should be reduced to a minimum. In Spain, a large park is being established for surplus zoo-born Asian elephant bulls. South Africa has much experience with keeping elephants in fenced range areas, and it would be worthwhile for Asia-based elephant managers to learn about this from South African experts (Garaï 2002b). The Pinnawela Elephant Orphanage may become a good example of such an elephant park after renovation. We believe that elephant parks would be a preferred solution for maintaining elephants belonging to evolutionarily significant units, like the elephants from Borneo (Fernando et al. 2003) and Sumatra (Fleischer et al. 2001), instead of trying to keep them intensively, using unskilled mahouts.

Along with the Pinnawela Elephant Orphanage in Sri Lanka, several other more or less similar establishments in Thailand, Assam, or Malaysia also provide care for orphaned elephants. According to our studies in Pinnawela, orphans without social contacts with older females tended to show retarded body growth, social isolation, and susceptibility to several diseases (see Kurt 2001; Garaï 2002a). Sufficient allomothers should be present in all orphanages. It is not difficult to find them, because many adult females, even those that have never reproduced before, are altruistic and therefore capable of nursing orphans and are willing to do so. Allomothers should also be used to raise and train orphans in transit homes because it is known that translocated juvenile orphans of African elephants that are not socially integrated with other elephants may show social pathologies later in life, for instance, bulls killing rhinos or females killing newborns (Garaï and Carr 2001; Garaï 2002b). As is the case with transit homes, elephant parks and orphanages should be placed at the peripheries of protected areas. Here they could either attract visitors and, in due course, free natural ecosystems from unwanted human disturbances, or under ideal conditions, the elephants they contain could eventually go back to the wild. An experiment in this kind of arrangement is currently under way in Sri Lanka's Uda Walawe National Park (Mohamed 2002).

Well-maintained elephant parks and transit homes allow elephants to be kept in groups and free of chains, which allows elephants to express their natural behavior and provides them with exercise and socialization opportunities. Elephants kept in such a way convey to the visitor a better understanding of the species and its requirements than "dancing" elephants or elephants that are forced to ride bicycles, paint on canvas, play football, or perform on the guitar or mouth organ, activities that misinform and even deceive the visitor.

The pressures on a number of wild populations by human-elephant conflicts, poaching of young elephants, and other disturbances by man are the likely reasons that will cause a continuous inflow of wild Asian elephants into range country captive populations in the years to come. Governmental and nongovernmental conservation agencies have to fight for modern elephant keeping facilities with high welfare and management standards but also consider conservation, such as captive propagation of genetic and behavioral diversity, and conservation education. This goal is best reached by maintenance of jungle-based establishments. Where selective logging has been stopped, local people and captive elephants working in these jungle stations should be employed in research, ecotourism, and education. Every effort must be made to maintain traditional knowledge of medical plants, as well as indigenous capturing and management methods.

Elephant facilities in zoological gardens must be improved so that they reach the standards defined by the *World Zoo Conservation Strategy*. For elephants living in intensive keeping systems elephant welfare must be improved and, if necessary, the size of these populations must be restricted or reduced. Government and nongovernment organizations should encourage the formation of regional databases and studbooks to monitor the demography of captive elephants. However, this can only be achieved after a proper registration of captive elephants, owners, and elephant holding facilities has been established.

Another important step is the continuing education of range officers, managers, and representatives of animal welfare organizations regarding scientific knowledge about elephants. The life of elephants in their natural habitat must be the model for conservation-minded keeping systems. Wherever necessary, mahout training schools should be established as they have been in Thailand, and vocational training for mahouts' families should be set up to supplement the mahout's income, for instance, through integrated farming, weaving traditional fabrics, and so on. A database of species-specific ecological and behavioral characteristics must be established to help managers change unsuitable keeping systems and improve living conditions for captive elephants. Furthermore, an international veterinary consultant group must be set up to give free advice and medical care in the range countries. Finally, alternative keeping systems, such as elephant parks, orphanages, and transit homes must be promoted, where the animals can be kept in near-natural conditions and, if suitable, where release into their natural habitat may become possible.

References

Ashraf, N. V. K, Choudhury, B., Mainkar, K., and Barman, R. 2002. *Elephant health camp.* Report. New Delhi: Wildlife Trust of India.

Bist, S. S. 1996. Standards and norms for elephant owners: Draft for comments. *Zoo's Print* 11 (6): 52–54.

Bist, S. S., Cheeran, J. V., Choudhury, S., Barua, P., and Misra, M. K. 2002. The domesticated Asian elephant in India. In I. Baker and M. Kashio (eds.), *Giants on our hands: Proceedings of the International Workshop on the domesticated Asian elephant, Bangkok, Thailand, 5–10 February 2001* (pp. 129–148). Bangkok: United Nations Food and Agriculture Organization Regional Office for Asia and the Pacific.

Cuong, T. V., Lien, T. T., and Giao, P. M. 2002. The present status and management of domesticated Asian elephants in Viet Nam. In I. Baker, and M. Kashio (eds.), *Giants on our hands: Proceedings of the International Workshop on the domesticated Asian elephant, Bangkok, Thailand, 5–10 February 2001* (pp. 111–128). Bangkok: United Nations Food and Agriculture Organization Regional Office for Asia and the Pacific.

Daim, M. S. 2002. The care and management of domesticated elephants in Malaysia. In I. Baker and M. Kashio (eds.), *Giants on our hands: Proceedings of the International Workshop on the domesticated Asian elephant, Bangkok, Thailand, 5–10 February 2001* (pp. 149–156). Bangkok: United Nations Food and Agriculture Organization Regional Office for Asia and the Pacific.

Dany, C., Weiler, H., Tong, K., and Han, S. 2002. The status, distribution and management of the domesticated elephants in Cambodia. In I. Baker and M. Kashio (eds.), *Giants on our hands: Proceedings of the International Workshop on the domesticated Asian elephant, Bangkok, Thailand, 5–10 February 2001* (pp. 179–188). Bangkok: United Nations Food and Agriculture Organization Regional Office for Asia and the Pacific.

Dorresteyn, A. 2001. Forward planning and EEP management for elephants in EAZA institutions. In H. M. Schwammer and S. de Vries (eds.), *Beiträge zur Elefantenhaltung in Europa* (pp. 13–16). Münster, Germany: Schüling.

Edgerton, F. (trans.). 1931. *The elephant-lore of the Hindus, the elephant-sport (Matanga-Lila) of Nilakantha.* New Haven, CT: Yale University Press.

European Elephant Group. 2003. *Elefanten in zoos und safariparks Europa.* Münster, Germany: Schüling.

Fernando, P., Vidya, T. N. C., Payne, J., Stuewe, M., Davison, G., Raymond, J. A., Andau, P., Bosi, E., Kilbourn, A., and Melnick, D. J. 2003. DNA analysis indicates that Asian elephants are native to Borneo and are therefore a high priority for conservation. *Public Library of Science (PLoS) Biology* 1 (1): e6.

Fleischer, R. C., Perry, E. A., Muralidharan, K., Stevens, E. E., and Wemmer, C. M. 2001. Phylogeography of the Asian elephant (*Elephas maximus*) based on mitochondrial DNA. *Evolution* 55: 1882–1892.

Garaï, M. E. 1992. Special relationships between female Asian elephants (*Elephas maximus*) in zoological gardens. *Ethology* 90: 187–205.

Garaï, M. E. 2002a. Social behaviour of the elephants at Pinnawela Elephant Orphanage, Sri Lanka. In H. M. Schwammer, T. J. Foose, M. Fouraker, and D. Olson (eds.), *A research update on elephants and rhinos: Proceedings of the International Elephant and*

Rhino Research Symposium, Vienna, June 7–11, 2001 (pp. 32–40). Münster, Germany: Schüling.

Garaï, M. E. 2002b. *Managing African elephants: Guidelines for introduction and management of African elephants on game reserves.* 2nd rev. ed. Vaalwater, South Africa: Elephant Managers and Owners Association.

Garaï, M. E., and Carr, R. D. 2001. Unsuccessful introduction of adult elephant bulls to confined areas in South Africa. *Pachyderm* 31: 52–57.

Government of Kerala, India. 2003. *Kerala captive elephants (management and maintenance) rules, 2003.* Kerala, India: Government of Kerala.

Hammatt, H., Fahrimal, Y., and Mikota, S. 2004. Implications of new data for Sumatran elephants in captivity: Time for change. In J. Jayewardene (ed.), *Endangered elephants, past present and future: Proceedings of the Symposium on human elephant relationships and conflicts, Sri Lanka, September 2003* (pp 61–64). Colombo: Biodiversity and Elephant Conservation Trust.

Haufellner, A., Schilfarth, J., and Schweiger, G. 2002. Haltungsbedingte Unfälle mit Elefanten in Zoos. *Elefanten in Zoo und Circus* 2: 3–18.

Hutadjulu, B., and Janis, R. 2002. The care and management of domesticated elephants in Sumatra, Indonesia. In I. Baker and M. Kashio (eds.), *Giants on our hands: Proceedings of the International Workshop on the domesticated Asian elephant, Bangkok, Thailand, 5–10 February 2001* (pp. 59–66). Bangkok: United Nations Food and Agriculture Organization Regional Office for Asia and the Pacific.

International Union of Directors of Zoological Gardens / Conservation Breeding Specialist Group (IUDZG/CBSG). 1993. *Executive Summary, The World Zoo Conservation Strategy: The role of the zoos and aquaria of the world in global conservation.* Brookfield, IL: Chicago Zoological Society.

Kashio, M. 2002. Summary of the international workshop on the domesticated elephant. In I. Baker and M. Kashio (eds.), *Giants on our hands: Proceedings of the International Workshop on the domesticated Asian elephant, Bangkok, Thailand, 5–10 February 2001* (pp. 17–22). Bangkok: United Nations Food and Agriculture Organization Regional Office for Asia and the Pacific.

Kharel, F. R. 2002. The challenge of managing domesticated elephants in Nepal. In I. Baker and M. Kashio (eds.), *Giants on our hands: Proceedings of the International Workshop on the domesticated Asian elephant, Bangkok, Thailand, 5–10 February 2001* (pp. 103–110). Bangkok: United Nations Food and Agriculture Organization Regional Office for Asia and the Pacific.

Krishnamurthy, V. 1992. Care and management of elephant calves in captivity. In E. G. Silas, K. Nair, and M. Nirmalan (eds.), *The Asian elephant* (pp. 82–85). Trichur, India: Lumiere Printing Work.

Krishnamurthy, V., and Wemmer, C. 1995a. Timber elephant management in the Madras Presidency of India (1844–1947). In J. C. Daniel and H. S. Datye (eds.), *A week with elephants: Proceedings of the international seminar on Asian elephants, June 1993* (pp. 456–472). Bombay: Bombay Natural History Society and New Delhi: Oxford University Press.

Krishnamurthy, V., and Wemmer, C. 1995b. Veterinary care of Asian timber elephants in India: Historical accounts and current observations. *Zoo Biology* 14: 123–133.

Kurt, F. 1992. *Das Elefantenbuch.* Hamburg, Germany: Rasch & Röhring.

Kurt, F. 1995. The preservation of Asian elephants in human care: A comparison between the different keeping systems in South Asia and Europe. *Animal Research and Development* 41: 38–60.

Kurt, F. (ed.). 2001. *Elefant in Menschenhand: Forschungsberichte aus Sri Lanka.* Fürth, Germany: Filander.

Kurt, F., and Garaï, M. E. 2002. Stereotypies in captive Asian elephants: A symptom of social isolation. In H. M. Schwammer, T. J. Foose, M. Fouraker, and D. Olson (eds.), *A research update on elephants and rhinos: Proceedings of the International Elephant and Rhino Research Symposium, Vienna, June 7–11, 2001* (pp. 57–63). Münster, Germany: Schüling.

Kurt, F., and Garaï, M. E. 2007. *The Asian elephant in captivity: A field study*. New Delhi: Cambridge University Press and Foundation Books.

Kurt, F., and Kumarasinghe, J. C. 1998. Remarks on body growth and phenotypes in Asian elephant *Elephas maximus. Acta Theriologica* 43 Supplement 5: 135–153.

Kurt, F., and Mar, K. U. 1996. Neonate mortality in captive Asian elephants (*Elephas maximus*). *Zeitschrift für Säugetierkunde* 61: 155–164.

Kurt, F., and Mar, K. U. 2003. Guidelines for the management of captive Asian elephants and the possible role of IUCN/SSC Asian Elephant Specialist Group. *Gajah* 22: 30–42.

Lair, R. C. 1997. *Gone astray: The care and management of the Asian elephant in domesticity.* Bangkok: United Nations Food and Agriculture Organization Regional Office for Asia and the Pacific.

Lohanan, R. 2002. The elephant situation in Thailand and a plea for co-operation. In I. Baker and M. Kashio (eds.), *Giants on our hands: Proceedings of the International Workshop on the domesticated Asian elephant, Bangkok, Thailand, 5–10 February 2001* (pp. 231–238). Bangkok: United Nations Food and Agriculture Organization Regional Office for Asia and the Pacific.

Mahasavangkul, S. 2002. Domestic elephant status and management in Thailand. In H. M. Schwammer, T. J. Foose, M. Fouraker, and D. Olson (eds.), *Research update on elephants and rhinos: Proceedings of the International Elephant and Rhino Research Symposium, Vienna, June 7–11, 2001* (pp. 71–82). Münster, Germany: Schüling.

Mohamed, M. R. 2002. Behaviour of juvenile elephants released from the Elephant Transit Home to the Uda Walawe National Park. In *Elephant Update 6, Newsletter of the Biodiversity and Elephant Conservation Trust, Sri Lanka.*

Reilly, J., and Sukatmoko, P. 2002. The elephant training center at Way Kambas National Park, Sumatra. *Gajah* 21: 1–40.

Salwala, S. 2002. The role of private organizations in elephant conservation. In I. Baker and M. Kashio (eds.), *Giants on our hands: Proceedings of the International Workshop on the domesticated Asian elephant, Bangkok, Thailand, 5–10 February 2001* (pp. 223–226). Bangkok: United Nations Food and Agriculture Organization Regional Office for Asia and the Pacific.

Stevenson, M. F. 2002. *Management guidelines for the welfare of zoo animals: Elephants* Loxodonta africana *and* Elephas maximus. London: Federation of Zoological Gardens of Great Britain and Ireland.

Sukumar, R. 2003. *The living elephants.* New York: Oxford University Press.

Tun Aung and Thoung Nyunt. 2002. The care and management of domesticated elephants in Myanmar. In I. Baker and M. Kashio (eds.), *Giants on our hands: Proceedings of the International Workshop on the domesticated Asian elephant, Bangkok, Thailand, 5–10 February 2001* (pp. 89–102). Bangkok: United Nations Food and Agriculture Organization Regional Office for Asia and the Pacific.

Webster, J. 1984. *Calf husbandry, health and welfare.* London: Collins.

Wemmer, C. 1995. Gaonbura Sahib—A. J. W. Milroy of Assam. In J. C. Daniel and H. S. Datye (eds.), *A week with elephants: Proceedings of the international seminar on Asian elephants, June 1993* (pp. 483–496). Bombay: Bombay Natural History Society and New Delhi: Oxford University Press.

Wiese, R. J., 2000. Asian elephants are not self-sustaining in North America. *Zoo Biology* 19: 299–309.

Williams, G. H. 1950. *Elephant Bill.* Garden City, New York: Doubleday & Co.

World Association of Zoos and Aquariums (WAZA). 2005. *Building a future for wildlife: The World Zoo and Aquarium Conservation Strategy.* Bern, Switzerland: WAZA.

PART III

ELEPHANTS AND PEOPLE IN NATURE

THE ETHICS OF CONFLICTS
AND ACCOMMODATIONS

17 RESTORING INTERDEPENDENCE BETWEEN PEOPLE AND ELEPHANTS

A SRI LANKAN CASE STUDY

LALITH SENEVIRATNE AND GREG D. ROSSEL

Until contemporary times, elephants and people enjoyed a long tradition of interdependence in Sri Lanka. Elephants figured prominently in work, warfare, culture, religion, and pageantry throughout Sri Lanka's history. If not for the help of elephants, the country's civilization might not have reached the heights it attained. A thousand years ago, elephants hauled millions of tons of brick and stone, then pounded into place the foundations of this island country's vast cities, irrigation canals, and dams. They also assisted in the defense of those cities from invaders arriving from across the seas (Parker 1909). Historical agricultural practices and landscape planning also allowed for a balanced coexistence of people and elephants. Sri Lanka's religious and sociocultural traditions have long supported a remarkably benevolent attitude toward wildlife, with the elephant placed at the pinnacle of esteem. The world's first recorded wildlife sanctuary, the Mahameuna Garden in Anuradhapura, established in the third century BC, bears testimony to the level of this regard for wild animals (Parker 1909). The preeminent position of the elephant meant that people treated it with respect and care in captivity and provided for its sustenance in the wild by preserving adequate habitat.

The advent of colonialism 500 years ago precipitated changes in the social and ecological fabric of the nation, adversely affecting these links of people-elephant interdependence. A master-servant attitude slowly unraveled the old social and economic structures of Sri Lanka's communities, and with them the relatively balanced coexistence of people and wildlife. Elephant habitat became, effectively, the playing fields in which colonial masters proudly practiced their most ostentatious sport—the hunting of elephants in large numbers. Famous sportsmen of the early nineteenth century, like Major Thomas William Rogers are credited with having shot more than 1,400 elephants in just a few years (Cannon and Davis 1995). During these centuries, vast tracts of land were cleared, and cash crop plantations of tea and rubber were planted, which decreased available elephant habitat. The elephant also lost its place as a workhorse, owing to the advent of mechanization and firearms.

By the early twentieth century, modern medical advances, unfortunately not matched by general economic improvement, led to a human population explosion, putting continued pressure on available forestland, leading to increased people-elephant conflict. When independence arrived in 1948, the country's population evidently had lost the art of self-governance and the land lacked a true national spirit. Postindependence politics only aggravated these problems, as some took advantage of the "one-man-one-vote" system of government thrust on a population lacking a sense of common responsibility. Although remarkably, more than 20% of the country's natural landscapes now enjoy some sort of habitat protection under the auspices of either Sri Lanka's Wildlife Conservation or Forest Conservation Department, the fragmented nature of these protected areas means that elephant ranges often significantly overlap with populated areas.

Crop raiding by elephants is the main reason for conflict. Cultivated crops represent a food resource superior to what is found in the remaining patchwork of conservation areas. Male elephants are prone to high-risk, high-gain strategies, such as raiding crops. Farmers attempt to protect their crops by chasing the elephants away, shouting, building fires, and lighting firecrackers. When these efforts fail, they resort to poison, snares, trap guns, and shotguns, leading to elephant deaths and injuries. Elephants respond to the villagers' protective measures with aggression. Previously wounded elephants tend to become rogues, going after the farmers with deadly intent. These conflicts result in both elephants and people being harassed, traumatized, and killed.

Thus, over the span of only a few centuries, Sri Lanka's once revered elephants have been transformed into little more than pests, albeit unusual pests, still adorned with some measure of ornamental and cultural value within both elite and popular Sri Lankan traditions. The country's remaining elephants, estimated at between 2,000 and 4,000 (Fernando 2000), are undeniably embattled, pitted against rural Sri Lanka's human populations in a contest for resources and livelihood. Yet, in parallel, those humans, too, are embattled. Uneven demographic, economic, and social development has left Sri Lanka's burgeoning rural populations stranded, with few short-term options and consequently fewer long-term prospects. The tragic effects of the December 2004 Indian Ocean tsunami only worsened conditions for many of Sri Lanka's poorest citizens. Any ethical issues to do with treatment of Sri Lanka's elephants are inherently paralleled with ethical issues regarding the quality of life for the country's human populations, especially its rural poor.

The parallels are transparent. Elephants today remain a significant component of Sri Lanka's greater cultural identity and heritage. That cultural identity must be restored, perhaps most urgently in impoverished rural regions, as part of Sri Lanka's recovery from its colonial trauma. Elephants, in their own right, deserve protection and long-term conservation in a cultural landscape. Sri Lanka's rural villagers, also in their own right, deserve opportunities for bettering their lives and their children's future. The most ethical solutions for these twinned problems will be approaches that help build opportunities for better human livelihoods while also mitigating pressures on elephants. Stated from the reverse perspective, optimal ethical treatment of Sri Lanka's elephants entails a parallel consideration for the welfare of Sri Lanka's villagers. Such ethics-based solutions are worth carrying out on a small scale, village by village, refining approaches as indicated by initial pilot experiences. With time, these innovations can be replicated on a larger scale around the country.

Reflecting on the situation, a group of like-minded engineers decided to help address these pressing ethical issues by working on restoring the lost interdependence between people and elephants. Our group is endeavoring to develop innovative solutions directed at mitigating people-elephant conflicts and promoting communities' greater appreciation of elephants. In the process, we have been reminded of the futility of seeking solutions by backtracking in history, as reflected in the words of English novelist L. P. Hartley (1953, 1): "The past is a foreign country; they do things differently there." We have informally named our activities the Elephant Detection Project. Early on we realized we would be unsuccessful unless the community spirit lost over the past several generations could be restored so people would begin to realize the value of taking ownership of the issues confronting them.

To develop solutions, we set up a field station in the southeastern Sri Lankan village of Pokunutenna, bordering Uda Walawe National Park (Seneviratne and Rossel 2001). We chose to address two urgent community and environmental imperatives, both suited to our skills and interests: the need to combat crop raiding by elephants and the need for rural electrification. The intertwined solutions to these problems—developing a nonintrusive crop protection system and promoting rural electrification with renewable energy—represent good community rallying points. Both solutions rely on simple, low-cost, and replicable applications of engineering technology. These same technological approaches can result in battery-powered elephant alarms and in local electrification plants that give villagers a chance to read at night. Employing these similar approaches, both to help villagers to set up elephant detection systems and to help provide them with access to electrification, our informal engineers' group not only facilitated ethical treatment of problem elephants, something very important to our own conservation-minded interests, but also promoted rural economic and social development. Both positive outcomes are of equal ethical importance in the quest to restore rural human-elephant interdependence.

Protection from Crop Raiding:
Nonintrusive Detection and Alarm Systems

Our aim was to develop an effective, independent, low-cost, and nonintrusive alarm and deterrent system to protect individual villages and their respective croplands without restricting elephants' freedom of movement into large, yet unprotected, natural areas. Our hope was that local communities would adopt any useful new approaches we could demonstrate to them. An effective crop protection system could help improve future prospects for survival of Sri Lanka's elephants, in part because it will simultaneously improve community economies. With their crops secured from elephants, local communities might even start thinking of the economic advantages elephant neighbors could offer, such as new avenues of income based on ecotourism. Such new economic opportunities could further motivate communities to value and protect elephants.

Starting in May 1999, we began conceptualizing, developing, and field-testing several designs for detection and alarm systems. Among the approaches we have explored with varying success are seismic detection, laser and optical detection, tripwire detection, and infrasound detection. Further details about each will be discussed.

Our design for a seismic detection system was based on U.S. Army surplus miniature seismic detectors. We modified these to enable detection of elephant movement and then buried the detectors along the peripheries of crop fields. They transmit a radio pulse when elephants approach, which triggers an alarm at a central radio receiver, alerting the villagers, who can then rush down to the field and chase away the elephants before they begin to forage on crops. This high-tech system has the disadvantage of bearing a high acquisition and operating cost. An unexpected additional drawback of these seismic detectors has been elephants' uncanny ability to sense them even when buried in the ground or hidden under rocks. Elephants remove them regularly and destroy them by trampling. Because of these drawbacks we decided to defer further investigation of this form of seismic detection. However, we are exploring other lower-cost techniques also associated with elephant weight and movement. A promising example may be the use of buried fiber optic cable, which could take advantage of changes in light transmission characteristics caused by the weight of walking elephants.

We have experimented with a series of laser beams and later with simple optical light beams. Laser and optical beams cost less than conventional seismic detectors, but several practical implementation obstacles cropped up in the field tests that were not apparent in laboratory tests. These include the lack of a simple villager-friendly method for aligning these lasers and light beams, and the high cost of providing a weather- and windproof mounting system, necessary in the field to ensure stability. These laser and optial detection systems require further research to improve their practicality.

Tripwire detection was the simplest in concept and turned out to be the most promising. This system uses a trip line to signal intrusion. We use fishing line, of up to 250 meters in length to create tripwire lines. We insert the fishing line through eyelets fixed at a height of 1.5 meters, on wooden stakes spaced 3 meters apart. One end of the fishing line is connected to the pole of a weatherproof electrical switch, which is mounted on a tree. An electrical cable extends from the switch to a simple flashlight-battery-powered signaling device, often a bicycle horn, located in the house of the villager who lives closest to the field. The switch trips the moment an elephant pulls the fishing line by crossing it, and the resulting electrical connection sets off the signaling device in the villager's house. The fishing line is rigged to slip off the switch pole after it is pulled.

To prepare this set-up, the scrub has to be cleared and a footpath constructed along the path of the fishing line. This allows the villager to get to any point along the line within minutes of the elephant intrusion alarm sounding, so he can chase the elephant away before it begins to feed. The cost of a 250–meter tripwire set-up, including 60 meters of electrical cable and a bicycle horn, is less than US$20. To protect distances in excess of this length, several sections of line can be used, each one connected to a separate alarm. The wooden stakes used for suspending the line are readily available from the forest, and their use causes hardly any environmental impact. The maintenance cost is marginal, as the line rarely breaks; instead it gets dislodged from the switch as it is pulled. After being dislodged, the line can easily be reconnected to the switch, providing repeated protection.

From our observations, villages with frequent people-elephant conflicts typically consist of about fifty houses, together sharing mixed rice paddy and *chena* (rotating slash and burn) farmland of around 15 hectares. Elephants usually enter the farmland from somewhat predictable crossing points bordering the forest. Normally, the boundaries needing protection vary from about 300 to 1,500 meters and thus require from one to eight sets of alarm systems. The total cost to the community is about US$160, an affordable amount, considering the benefit. From our few years of operational experience in a few communities across the country, the results are positive and encouraging. In these communities, the villagers are now well trained in setting up the system and versatile at maintaining it on their own. They have taken ownership responsibility of the system because they see value in it. They are willing to adopt this new technology and the conflict-diminishing approach it engenders not only because it is more ethical in terms of elephant welfare, but also because its low-cost design and its simplicity are ethical with respect to the welfare of the community. That is because, unlike potentially more complex and more expensive systems, both the demands and payoffs of this detection system were developed with close regard to the local community's own needs and resources. We are now in the process of introducing the system to more communities across the country to spread awareness of it and evaluate its effect from a countrywide perspective.

Elephants use infrasound, the sound produced at frequencies below the range audible to people, to communicate over long distances (Payne, Langbauer, and Thomas 1986). We are developing a system to detect automatically these very low frequency vocalizations of intruding elephants, as an anti-crop-raiding early warning system. Our system consists of the sensor module, a data acquisition system, and a software module. Currently, we are in trial and refinement at the prototype stage, having developed both the hardware and software. Software development has required recording and analyzing many hours of elephant calls. We are refining this software, particularly to install algorithms that will automate recognition of specific calls, such as juvenile calls. Prototype field testing has already begun; once this is successfully completed, the device must be produced as an affordable unit. To keep costs low, we have used a single-board computer (a stripped-down version of a personal computer) as the processing device. We have established a formal partnership with the Tiergarten Schönbrunn (Vienna Zoo), to work on the further development and testing phases of this infrasound project and other elephant conservation projects. We also have an informal affiliation with ElephantVoices, allowing us to benefit from their extensive research in Kenya on infrasound in African elephants.

Automated Deterrence System

A successful detection system provides a crucial alert to the villager. Then, the villager must still actually chase and deter the intruding elephants. We thought of going a step further and developing an automated deterrence system. Elephants are fearful of fire, bright lights, and loud noise. With this in mind, we developed a solar-powered light and siren system, which randomly sequences lights, illuminating different cropland sections at twenty- to sixty-second intervals. When the tripwire detects intruding elephants, the lights and sirens throughout the entire cropland parcel are activated for a two-minute period, then switch to the sequencing setting. The 12–volt direct current (DC) amber-color strobe lights flash 130 times per minute. Amber was chosen after experimentation with several colors, ranging from red to yellow, showed it to be the most consistently threatening to elephants. The elephants avoid a 60–meter radius of the flashing amber strobe light. In fact, we observed that if the lights are switched on after elephants have intruded, they immediately move away. We also used 12–volt DC sirens to generate loud noises, but the elephants do not seem to react to the siren, so the effectiveness of this sound deterrent is questionable.

The main drawback to the deployment of this deterrence system is its cost. A typical setup, consisting of the solar panel, battery, 15 strobe lights, and 4 sirens costs US$500, unaffordable for most communities. If the deterrence system is to provide an ethical and feasible option, it needs further development aimed at lowering its costs. In addition, we must ascertain whether the elephants are likely to habituate to the whole system and ignore it over time.

Rural Electrification with Renewable Energy

In most developing countries, including Sri Lanka, rural electrification has accelerated over recent years, not by design but as a result of ad hoc extensions of national grids (Seneviratne and Rossel 2004). The investment required to generate electrical capacity to serve these predominantly inefficient and unprofitable extensions happens in an equally unplanned manner, with the result that utilities services are plunged into debt. This contributes to the regular power shortages and power outages that deeply affect the lives of so many in the developing world. Cash-strapped utilities can no longer afford to provide uninterrupted power to the profitable and economically crucial industrial and urban sectors, let alone continue grid expansion to rural areas. Sadly, we may safely assume that the poorest 20% of households in the developing world will not receive regular grid power in the foreseeable future. In the case of Sri Lanka, typical of the developing world, over half of the country's rural households remain without electricity. These households represent the least economically prosperous and most remote segments of the country's population.

Yet reliable and affordable power is vital to uplift rural living standards, which, in turn, will have a positive economic effect on the whole country. The average rural household in Sri Lanka spends about 400 rupees per month (US$4) on kerosene for bottle lamps and lanterns and on disposable flashlight batteries. This amount, a staggering 25% of each household's monthly cash income, could instead finance a renewable energy system. Long-term improvements in rural areas would require only a generator system producing enough power for a few light bulbs and an outlet for small appliances in a village's households, plus, ideally, sufficient reserve power for the village to sustain a community-wide economic enterprise. In addition, using renewable energy sources will eliminate the grave safety risks, such as explosion, associated with kerosene bottle lamps, and provide higher-intensity light, so that village children can practice reading without fearing intrusion by wild animals.

Village-level or multiple-village cooperative power generation from renewable energy sources such as solar, wind, and biomass can provide sustainable alternative power independent of the national grid. These power enterprises can also provide ideas, and a framework, for conducting other socioeconomic ventures, as well as the revenue to finance them, thus further improving local economies. Electrification is an important, easily identifiable rallying point to unite communities and impel other community initiatives, one reason our engineers' group has been working with local communities to implement local electrification. As localities perceive the benefits of collective effort and the positive outcomes of local group initiatives, they are likely to apply such effort to many other endeavors, possibly with significant conservation effects.

In Sri Lanka, most of the poorest villages are also buffer zone communities, adjacent to conservation areas and biodiversity hotspots critical for protection of the global environment. If a village-level cooperative society can successfully administer its own local grid, its members can begin planning other locally-based entrepreneurial activities, even including buffer-zone ecotourism ventures. In fact, implementation of local renewable electrification projects offers a general model for how to approach the sustainable development of buffer zone communities bordering national parks and protected areas. We are confident that empowering villagers through local electrification projects could directly benefit the ethical treatment and conservation of elephants by sustainably bettering the lives and livelihoods of the people who share the elephants' living space. Facilitating villager-based renewable-energy electrification can translate to better conditions for elephant and wildlife conservation in these buffer zone communities. Our research and project development over the past several years indicates that multivillage-level activities may have even greater potential for long-term success than single-village enterprises.

The village of Pokunutenna, site of the solar and wind power station we have developed, is typical of Sri Lanka's economically underprivileged buffer zone villages. As discussed earlier, we have conducted extensive elephant detection system experiments in this village, which borders the northeast corner of Uda Walawe National Park, an important elephant refuge. The village consists of about thirty households, spread over about 2 square kilometers. The nearest grid power line is approximately 10 kilometers away. Before this project, the villagers depended only on kerosene bottle lamps and lanterns for lighting. In 2001, we installed a set of solar panels and a supplementary wind generator (Seneviratne and Rossel 2004). The power thus generated is stored, through a charge controller, in sealed rechargeable deep discharge batteries. The stored DC power is converted to alternating current (AC) by means of an inverter. The AC power is then distributed through a low-cost underground cable network to clusters of homes and street lamps. Each cluster has a distribution box containing one fuse per household or per street lamp. This distribution scheme provides up to 50 watts of power per household.

Future expansion of this local grid system would allow for formation of small-scale electricity using microenterprises that could address the community's specific needs, providing paid employment for their operators along with other long-term economic benefits for the entire village. One microenterprise already operating in Pokunutenna is a battery-charging center for rechargeable flashlight batteries. Other potential enterprises include a communications center, an agricultural products processing center, a health center, and an educational center. To galvanize the village and give it the ownership of the system, we initiated the formation of a village cooperative society, with its own elected officials, a general committee, and governing constitution and bylaws. Power users pay a monthly tariff to the cooperative to help finance the power system.

Our work on biomass-based electrification started with a single-village project in Endagalayaya village, adjacent to Pokunutenna. Villagers agreed to provide their own "dendro-thermal" energy materials to feed a biomass electricity generator, each villager growing up to a quarter hectare of fuel wood yearly, allowing for enough electricity production to power the village households and at least one community enterprise. Very quickly we determined that biomass offered significant financial advantages over solar or wind power in this locality, principally because the villagers can grow their own renewable fuel source—mostly gliricidia trees (*Gliricidia sepium*), which they coppice rather than cutting down (see also Kapadia 2002). We have also started experimenting with growing some native tree species for this purpose. An additional advantage is that biomass-based electrification lends itself easily to the development of a robust project design framework that ensures coordinated and sustainable implementation and operation of multivillage systems, thus avoiding some of the inertia to which new stand-alone village cooperative systems are prone.

To capitalize on these advantages, we have worked with a group of neighboring villages, setting up a for-profit rural electrification company ("Flowing Currents") in association with a not-for-profit organization ("Aspira"). These two units together compose our electricity-generating social enterprise. The nonprofit company provides a combination of reforestation assistance, electricity tariff subsidy, and subsidy of certain real expenses like house wiring, enabling the for-profit to install biomass village electrification systems in a scalable and sustainable manner. The nonprofit partners with grant-giving foundations, while the for-profit partners with financial institutions to allow postpilot expansion. In this way, we are testing, and hope we can prove, our hypothesis that a market exists for renewable energy among the rural poor in Sri Lanka and elsewhere. In this scenario, villagers pay back development costs over time, when they can afford to do so.

This approach should enable scaling up from the two pilot biomass village systems already operating since 2006 to twenty-five systems, reaching a minimum of 875 low-income households in the next few years, and many more in the five years after that. Because this biomass-based electrification relies on coppicing of plantation trees, providing this local electrical power automatically takes place hand-in-hand with environmental improvements, raising villagers' awareness of the benefits of environmental conservation.

To date, this biomass electrification endeavor has been the fastest-progressing project of all those outlined in this essay. That is in part because its social enterprise framework enhances its appeal to investors. We believe that similar social enterprise funding should prove equally appealing to potential investors in directly elephant-related projects. Applying a similar for-profit plus nonprofit-combined business and philanthropy model to such activities as further development and implementation of early detection systems or elephant exclusion fences would directly benefit both elephant conservation and ethical treatment of local human communities in Sri Lanka.

Free-ranging elephants are subject to continuous harassment and unethical treatment resulting from the ever-increasing demands of human populations. Our experiences in mitigating the conflicts between people and elephants have taught us that supporting straightforward yet innovative community initiatives offers an effective means to uplift rural communities, both psychologically and tangibly, by rallying community spirit and simultaneously providing new economic and social opportunities. The ultimate consequence is that people will better value their elephant neighbors and therefore treat them with respect and even protect them. Such solutions are easier to formulate and implement than outsiders (nonvillagers) commonly believe. Furthermore, they have an added advantage. Local solutions will not require government to take on the kind of debt that has resulted from financing traditional mega-development projects.

The multivillage approach we have been exploring in the biomass project may demonstrate the best level for these activities. The admittedly restricted, but almost entirely positive, effects of our solutions offer a marked contrast to the huge footprint of mega-projects, with their mostly negative effects on the environment and on elephants and their failure to deliver on local people's reasonable expectations for material progress. In contrast, our multivillage project level and for profit/nonprofit framework allows private investors to provide financing that the participating villagers should actually be able to repay. Villagers unified by experiencing affordable local electrification are likely to be able to convince similar investors that they also care about collectively implementing humane elephant exclusion practices. Step by local step, people-elephant interdependence can be regained in Sri Lanka, with positive ethical outcomes for all.

Acknowledgments

We thank the villagers of Pokunutenna and Endagalayaya in Sri Lanka for their unstinted enthusiasm and support, and the U.S. Fish and Wildlife Service for a grant under the Asian Elephant Conservation Fund for development of elephant detector systems.

References

Cannon, T., and Davis, P. 1995. *Aliya: Stories of the elephants of Sri Lanka.* Ferntree Gully, Australia: Airavata Press.

Fernando, P. 2000. Elephants in Sri Lanka: Past, present, and future. *Loris* 22: 38–44.

Hartley, L. P. 1953. *The go-between.* London: H. Hamilton.

Kapadia, K. 2002. Home-grown power plants: The case for wood-based energy systems in Sri Lanka. *Refocus* 3 (6): 34–39.

Parker, H. 1909. *Ancient Ceylon: An account of the aborigines and of part of the early civilization.* London: Luzac & Co.

Payne, K., Langbauer, W. R., Jr., and Thomas, E. 1986. Infrasonic calls of the Asian elephant (*Elephas maximus*). *Behavioral Ecology and Sociobiology* 18: 297–301.

Seneviratne, L., and Rossel, G. 2001. Living next door to a national park. *Sri Lanka Nature* 3 (2): 51–56.

Seneviratne, L., and Rossel, G. 2004. Lighting up a village: Community RE generating systems for the developing world. *Refocus* 5 (1): 26–28.

18 SUMATRAN ELEPHANTS IN CRISIS
TIME FOR CHANGE
SUSAN K. MIKOTA,
HANK HAMMATT, AND YUDHA FAHRIMAL

No animal evokes stronger emotions than the elephant, and no topic related to elephants evokes stronger opinions than keeping elephants in captivity. At the March 2003 workshop, "Never Forgetting: Elephants and Ethics," professionals from diverse disciplines with expertise in elephant management convened to discuss this topic. Relevant to this essay are their consensus agreements that (1) captivity should not be considered a substitute for the protection of elephant populations in their natural habitat and (2) all elephants should be treated ethically.

Most elephants now in captivity were once wild. Of the estimated 60,000 Asian elephants worldwide, 16,000 are captive (Asian Nature Conservation Foundation 2006). In this chapter, we discuss the specific situation of Sumatran elephants in Elephant Training Centers (ETCs) operated by the government of Indonesia. These ETCs were established in the mid-1980s in an attempt to alleviate human-elephant conflict and now hold 300–400 captive elephants. We maintain that (1) the ETCs have failed to accomplish their stated purpose, (2) the methods used to capture elephants are inhumane and result in unnecessary injuries and deaths, and (3) the current rate of removal is not sustainable and will ultimately result in the demise of wild elephants on Sumatra.

Background

Indonesia is an archipelago of about 17,500 islands straddling the equator between Australia and Asia and covering 1.3 percent of the earth's surface. Indonesia is the world's fourth most populous nation (after China, India, and the United States). Sumatra, the westernmost island, was the recipient of transmigrants (resettled families) in a Dutch colonial program intended to relieve crowding on Java. This program continued throughout most of the twentieth century and contributed to environmental pressures.

A burgeoning human population, rampant illegal logging, and conversion of land to palm oil plantations threaten Indonesia's remaining forests, especially on Sumatra. Humans and elephants struggle for survival, and when there is conflict, losses occur on both sides. Escalating human-elephant conflict led the government of Indonesia to seek solutions in the early 1980s. Various strategies of deterrents, translocation, and culling have been ap-

plied elsewhere. However, Indonesia chose the capture of wild elephants as its primary approach. The rationale for capture and placement into ETCs was to "domesticate" and train elephants for agriculture, forestry, and tourism, anticipated to become revenue-generating ventures.

The Sumatran Elephant

The Sumatran elephant (*Elephas maximus sumatranus*), found only on Sumatra, is a subspecies of the Asian elephant (*E. maximus*), which is listed as "endangered" in the *2006 IUCN Red List of Threatened Species* (IUCN 2006) and is included in Appendix I of the Convention on International Trade in Endangered Species of Wild Fauna and Flora (CITES; UNEP-WCMC 2007). Analysis of DNA has confirmed the evolutionary separateness of the Sumatra elephant and supports its recognition as an evolutionary significant unit, meaning it is substantially reproductively isolated from conspecific populations and represents an important component in the evolutionary legacy of the species (Fleisher et al. 2001; Fernando, Vidya, Payne, et al. 2003; Fernando, Vidya, Ng, et al. 2003).

Although counting elephants would seem simple, in reality, they are difficult to census in dense tropical forests. Questionnaire surveys conducted in the mid-1980s (Blouch and Haryanto 1984; Blouch and Simbolon 1985, both cited in IUCN/SSC-AESC 1990) suggested a total wild population of 2,800–4,800 Sumatran elephants. More than one-third of these occurred in Riau province in east-central Sumatra, nearly one-half in the four

(Facing page) Top row: Deforestation and conversion. Little lowland rain forest remains on Sumatra. Illegal logging and conversion to palm oil plantations has decimated former elephant habitat, so human-elephant conflict is on the increase, resulting in scenes like those below. *Second row*: Leg injuries. Improper application of chains can result in superficial or deep injuries. Applied too tightly, chains embed into tissue, causing painful lacerations. This elephant (on right) lost all the skin on the lower 18 centimeters of her legs; the entire circumference of each leg. The injury was so deep that tendons were exposed and severed. These unnecessary injuries can be prevented by using chains of appropriate strength and by covering chains with used fire hose or heavy plastic tubing (inexpensive and readily available locally). These elephants are all dead. *Third row*: Neck wounds. The kah, a device made of wood and wire, used on newly captured adult elephants to restrict movement. The elephant is tethered to a tree by a chain that extends from the kah. Any head or neck movement causes deep lacerations. These wounds often become maggot infested. Once the kah is removed, wounds will heal with proper care, but such injuries are unnecessary. The use of this archaic device should be discontinued. This elephant died. *Bottom row*: Training and results. New captures are often trained within days or weeks of capture, even if injured. Harsh training techniques result in additional stress and many elephants lose condition during this process. Ana (left-most elephant in left-hand photo) captured in August 2001, was under treatment for capture-related injuries, and our advice that training be postponed until the injuries healed was ignored. Ana died in October 2001. Training for Ema (in right-hand photo) was temporarily discontinued when her condition deteriorated. Ema "escaped" from the Minas ETC twice. We do not know whether she is still alive.
Photographs courtesy of Hank Hammatt

southern provinces of Lampung, South Sumatra, Jambi, and Bengkulu, and the remainder primarily in the northern province of Aceh (Tilson et al. 1994). A 1990 action plan identified forty-four discrete populations. Three of these were extinct by 1990 (Santiapillai and Ramono 1993; Tilson et al. 1994).

In 1993, an elephant Population and Habitat Viability Analysis (PHVA) was conducted in Indonesia. The PHVA brought together biologists, scientists, and conservationists from Indonesia, Sri Lanka, Thailand, Malaysia, India, Australia, New Zealand, Ireland, the United Kingdom, and the United States, and government officials and policy makers from Indonesia. PHVAs are administered by the Conservation Breeding Specialist Group (a part of the IUCN, or World Conservation Union) and their purpose is to help experts develop recovery plans for threatened species and habitats. Population demography, genetics, and ecology data are compiled and entered into a computer program. Biological data (age at first reproduction, reproduction and mortality rates, inbreeding depression, etc.) are combined with environmental variables (e.g., potential catastrophes such as drought, disease epidemics, or removal by man) to predict the survival of animal populations.

In the 1993 simulation, with a growth rate close to zero (deaths equal births), no inbreeding depression, and no removal, the chance of surviving 100 years for populations of 10, 25, and 50 elephants was 65%, 95%, and greater than 99%, respectively. Removing elephants had a dramatic effect even under scenarios with a low harvest (defined as the removal of four elephants—one adult female, one juvenile female, one juvenile male, and one adult male every four years for twenty-five years). The probability of survival to 100 years decreased to 10% for a population of fifty elephants and to 1% for a population of twenty-five elephants. Thus, the PHVA predicted that the harvest of even one elephant per year for twenty years would likely result in the extinction of any population of less than fifty elephants. An initial population of 100 had a greater potential for survival (>99%), although the herd would be reduced to about half of its original size (Tilson et al. 1994).

The prospects for survival became even more dubious with a high harvest rate (defined as one elephant per year but continuing for fifty years). Assuming a population growth rate of 1% but with a high harvest rate, a population of twenty-five had a 0% chance of survival, and a population of fifty had only a 3% chance. Larger populations fared better (100 elephants had a 98% chance of survival), but overall herd numbers still decreased. Even assuming a growth rate of 2%, the survival of populations with fewer than fifty elephants was still less than 5% (Sukumar and Santiapillai 1993; Tilson et al. 1994). The conclusion in 1993 was that the survival of habitat capable of supporting major populations was critical to the survival of Sumatran elephants. The PHVA further cautioned that if elephants were removed from the wild, the "annual off take" must be sustainable and "unless there is substantial improvement in the veterinary care of the elephants, and sufficient financial and trained manpower resources are available . . . increased capture of elephants cannot be justified" (Sukumar and Santiapillai 1993, 60).

At the time of the PHVA, habitat fragmentation had increased the number of separate populations on Sumatra to forty-seven. Nine of these populations were composed of fewer than twenty-five elephants and considered nonviable (Tilson et al. 1994), and indeed, three of these were extinct by 1997 (Lair 1997). As of 2002, only three of twelve populations in Lampung Province had survived (these twelve were among the forty-four populations identified by the 1990 action plan). Two of these had healthy populations (of about 500 and 180 based on dung-count surveys), but the third was smaller and probably nonviable (Hedges et al. 2005). In 2002, the Indonesian government stated (based on park officials' "estimates") that a maximum of 3,354 elephants remained in the wild, with over 50% residing in unprotected areas (Susmianto 2002). In 2007, WWF reported that the population had declined further, to between 2,400 and 2,800 ("Indonesia Seeks Plan to Save Rare Tigers, Elephants," 2007).

Capturing "Problem" Elephants

The government of Indonesia began capturing "problem" elephants in the 1980s. Although historically, captive elephants had been kept in Indonesia, such traditions had not persisted into the twentieth century. Thai mahouts and *koonki* elephants (elephants used in capture operations) were therefore brought over to train the new Indonesian *pawangs* ("mahouts" or "elephant handlers"). The first ETC was established in 1986 at Way Kambas in Lampung Province in southern Sumatra. As of 2004, there were six major ETCs. It was reported that 520 elephants were captured between 1986 and the end of 1995 (Lair 1997; Suprayogi, Sugardijito, and Lilley 2002; Susmianto 2002). Incredibly, this is the exact figure cited at the 1993 PHVA as the recommended maximum holding capacity of the ETCs (Tilson et al. 1994). Despite that recommendation, a five-year plan (1996–2001) to increase the captive population to 900 was formulated by the Directorate of Forest Protection and Nature Conservation, within the Indonesian Ministry of Forestry (Lair 1997; Susmianto 2002).

The Capture Process and Its Effects

In 2002, two of the authors (S. Mikota and H. Hammatt) observed the capture process in Riau Province. The basic procedure is the same throughout Sumatra. The capture team consists of six pawangs and two koonki elephants. Two pawangs carry dart guns and the others have spears. They track the wild elephant and then dart it with xylazine, a sedative. Once sedated, they apply chain hobbles to the front and back legs and place a *kah* (a neck yoke made of wood and wire) around its neck. The sedated elephant is chained to the koonki and walked or dragged out of the forest to a truck. The capture team typically withholds food and water for the first few days to weaken the elephant and make it easier to control. It may take several days to move a newly captured elephant out of dense forest.

During capture elephants sustain injuries, including deep leg wounds from unprotected chains, penetrating neck wounds from the kah, and abscesses from unclean darts. When Mikota and Hammatt were present the team followed the recommendations to discontinue use of the kah and to cover chains with fire hose to reduce leg injuries. The method of administering the darts, however, is deeply ingrained and our suggestions for improvement were not accepted. A number of the Thai-trained Indonesians still work for the forestry department. Younger pawangs learn from their peers. There is no formal training program. The capture equipment and supplies were brought into Indonesia ca. 1984 and had not been upgraded or replenished. It is, unfortunately, the darting techniques and subsequent infections that result in death.

The dose of the capture drug (xylazine) is very high compared to that used in India and Malaysia. The same high dose is often used regardless of the size of the elephant. The use of high loads (the load propels the dart from the gun) and high charges (the charge is inside the dart and moves a plunger to expel the drug) at close ranges results in unnecessary trauma. This, combined with a lack of basic hygiene in the preparation of darts, needles, and syringes and the absence of any postcapture wound treatment, sets the stage for infection.

Externally the dart abscesses may appear small and localized, but beneath the skin the infection may be extensive. Among darted elephants that we examined, we consistently detected very high white blood cell counts, indicative of more generalized infections. Death results from septic shock when bacteria or bacterial toxins are shed into the bloodstream. This may occur within days or weeks of capture, or the infection may spread slowly between the muscle layers, causing a debilitating infection and death from septic shock months later. Veterinarians refer to this type of infection as *necrotizing fasciitis*, a term that literally means inflammation and death of fascia, the fibrous tissue that permeates muscles and muscle groups. Necrotizing fasciitis also occurs in humans (often as the result of traumatic injury) and is an emergency situation, requiring invasive surgery and intensive care. Brogan and Nizet (1997) reported a mortality rate of 76% in humans in the absence of early intervention. Proper hygiene, prompt wound treatment, and the administration of prophylactic antibiotics in the field are needed to prevent this condition in elephants. The addition of a trained veterinarian to the capture team could help address these issues.

The numbers of elephants that have been captured or that have died as a result of capture is uncertain (Lilley and Saleh 1998). In the 1990s, officials at Way Kambas stated that only two elephants died during the capture and training of 600 wild elephants, a figure most professionals experienced in capture operations would not regard as credible (Lair 1997). We experienced difficulty in obtaining accurate information from Indonesian officials, a phenomenon that has also been observed by others (Reilly and Sukatmoko 1999; International Elephant Foundation 2001). As explained by one Indonesian, the content of official reports is strongly influenced by a culture in which mistakes and failures are thought to be shameful (Suprayogi et al. 2002). Thus cultural norms offer an explanation but not a justification.

Although Suprayogi et al. (2002) reported that the government discontinued capturing wild elephants in 1999, apparently it was government funding for these operations that actually ceased (Susmianto 2002). Captures were ongoing in Riau Province in 2000 and have continued. In 2004, seventeen elephants were captured and six more in early 2005 (N. Foead WWF-Indonesia, Jakarta, Indonesia, personal communication, May 2005). Funding sources include local palm oil plantations and villages affected by crop-raiding elephants. Our records indicate there were 117 known captures in Riau Province between September 2000 and March 2003. Seventy-four of these elephants were placed in the Riau ETC and forty-three were translocated. Of the seventy-four new captures taken to the Riau ETC, sixty-three died by September 2003. We anticipated three more deaths because of observed injuries and documented infections, and four captured elephants were lost in the forest. (Elephants that are reported as "lost in the forest" have either escaped or have been intentionally released, often because they are thin or ill.) By September 2003, we expected only four elephants (6%) of the seventy-four captured in this interval to survive. At best, if the four lost elephants were found alive and the three injured elephants recovered, the survival rate for elephants captured between September 2000 and March 2003 and taken to the Riau ETC would be only 15% (eleven of seventy-four). In addition, thirty-nine of the forty-three translocated elephants were likely to die, based on the expected survival rate of 6% among the seventy-four known (Riau ETC-placed) cases, combined with the fact that infections were documented in thirteen of the thirteen translocated elephants examined by the authors. Thus, of the 117 elephants captured over the September 2000 to March 2003 period, 109 elephants (93%) were either dead, lost, or expected to die (Hammatt and Fahrimal 2003).

It is speculation whether these appalling death rates are unique to Riau or whether they are indicative of a historic and pervasive problem. However, we surmise from the data collected the absurdity of the claim that only two deaths occurred in the course of capturing and breaking of over 600 elephants (Lair 1997).

The selection of elephants for capture is an issue of great concern. Crop raiders are typically single adult bulls or adult females leading family groups. However, most elephants in the ETCs are in their teens or younger. ETC rosters provided by the government to the International Elephant Foundation (IEF) for use during their 2000–2001 ETC visits listed capture dates for 262 of the elephants examined, and ages for 319 of them. Although age estimations are made by different ETC personnel and a standard technique is lacking, the numbers are nonetheless revealing. These government-provided data show that the median age at capture is 6 years ($n = 262$ elephants) and the median age of elephants in the centers visited was 12 years ($n = 319$). Thus, at the time of capture, half were under 6 years of age and half were over 6 (Hammatt and Fahrimal 2003). One would expect a higher proportion of older elephants among the captives, since captures were initiated in 1984 and purportedly limited after 1999.

We infer from the data that younger elephants are preferentially captured or that older elephants do not survive. Indeed, some pawangs revealed to the IEF teams that younger elephants are captured because it is easier to do so and that adult elephants do not often survive the transition to captivity. Although government officials told the IEF teams that captures were no longer taking place, the teams observed recently captured elephants at several ETCs. Based on conversations with pawangs, and on observations of independent conservationists, IEF suggested that the actual rate of survival of elephants going through the wild-to-captive transition in the ETCs may be less than 50% (International Elephant Foundation 2001). At least in Riau, we have documented that the percentage is much worse. Retraining in immobilization techniques is urgently needed. Our plans for a Sumatra-wide elephant immobilization and translocation workshop were under way and funds secured, when Mikota and Hammatt were compelled to leave Indonesia following harassment by Indonesian authorities in July 2003.

Elephant Training Center Problems

The ETCs are plagued by a multitude of health care, husbandry, and management issues. (Krishnamurthy 1992; Lewis 1998; Lilley and Saleh 1998; Stremme 1998; Mikota et al. 2000a, 2000b; Lewis and Yusuf 2001). Parasites, wounds, and nutrition are the primary health issues of elephants that have survived capture and training. The inadequacy of the diet was first noted in 1992 by a respected elephant veterinarian from India who recommended the addition of a grain ration (Krishnamurthy 1992). This recommendation was not adopted. Elephants are typically taken into the forest to feed during the day. If they have a caring pawang, they will be moved to a new location once they have browsed an area. Pawangs lead their elephants back to camp in the evening, and while supplemental food is usually supplied, it consists only of banana stems or coconut leaves. Although these elephants are not stressed by heavy workloads, Suprayogi et al. (2002, 187) have suggested that they are stressed by lack of activity, inadequate access to food, and "spending most days in drag chains and hobble."

Based on comparative data available for twenty-six captive elephants we observed and evaluated at the Riau Province ETC between 2001 and 2003, 58% either lost weight or grew at less than 40% of the rate documented for other Asian elephants (Sukumar 1989). Fifty percent of the Riau elephants tested in 2000 (eighteen of thirty-six) and 40% of those tested in 2003 (eighteen of forty-three) were anemic (H. Hammatt and S. Mikota, unpublished data). This condition likely has a nutritional basis, possibly confounded by intestinal or blood parasite problems. The widely held belief that Sumatran elephants are smaller than their mainland counterparts may be untrue, particularly if this was based on observations of ETC elephants. Indeed, measurements of five consecutive newly captured adult wild elephants in Riau were comparable to those of mainland Asian elephants. Further research in this area is needed.

Veterinary supplies and equipment at the centers are inadequate, and there is no veterinary training in elephant care. The stark living conditions and remote locations of many ETCs have disadvantages, and not all camps have veterinarians. Pawangs are poorly paid and typically receive only half of their salary, with the balance due paid when government funds are dispersed near the end of the fiscal year. There are few incentives to do a good job and it is difficult, if not impossible, to fire staff that perform poorly.

Despite the plethora of reports, material assistance from the international community was lacking until recently. Mikota and Hammatt worked for three weeks in early 2000 to survey and improve elephant health care conditions at the Riau ETC. The IEF, partnering with Fauna & Flora International (FFI) and supported by a grant from the Asian Elephant Conservation Fund of the U.S. Fish and Wildlife Service, initiated a project in 2000 to improve the management and care of captive Sumatran elephants. Working under the auspices of WWF-Indonesia, Mikota and Hammat moved to Sumatra in 2001 and initiated a project, which Fahrimal later joined, to improve health care at the Riau ETC and establish a Sumatran Elephant Resource Center. The Center, headed by an Indonesian senior veterinary officer and an Indonesian field veterinarian, was to have provided supplies and veterinary training to all the ETCs in Sumatra.

The level of care received by an individual elephant is largely at the discretion of the pawang. Many pawangs treat their elephants with kindness and some genuinely love their elephants. A few are abusive. Ironically, almost all seem unmoved by the suffering of newly captured elephants. Such apathy may stem from lessons allegedly taught by the Thai mahouts (namely, that newly captured elephants must feel pain to prevent their escape). It may also be that feelings of concern are linked only to the individual elephant that guarantees employment.

The government has taken little action, although the problems of captive Sumatran elephants have been clearly identified in reports previously cited. As noted by the International Elephant Foundation, "Until the political and financial situation in the country is improved, wild elephants and elephants in the ECCs need help from the international community, which must not only provide immediate relief, but can also be a voice for conservation needs of the elephants" (IEF 2001; the ETCs are sometimes referred to as Elephant Conservation Centers or ECCs). Economics are an impediment, but the reluctance to institute recommended cost-free improvements remains inexplicable. Management changes, such as directing pawangs to move elephants to new feeding areas more frequently, would provide better nutrition for the elephants. If properly executed, such changes might even decrease the need for and expense of supplemental feed. Similarly, insisting that pawangs bathe their elephants daily would minimize external parasites.

The ETCs remain underresourced and a financial burden for the government. The demand for trained elephants has been below expectations. In 1995, the Ministry of Forestry requested that fourteen logging companies use at least one pair of trained elephants per 10,000–20,000 hectares (25,000–50,000 acres). The request went unheeded as the companies preferred the use of mechanical equipment and thought the care of elephants too difficult (Hutadjulu and Janis 2002). In 1995, the government spent about Rp1854 million ($800,000) to maintain captive elephants. Between 50% and 55% of the annual national budget for elephant conservation has been used to operate the ETCs since fiscal year 1997–98 (Hutadjulu and Janis 2002). The shortcomings of the ETCs are acknowledged by Indonesian nationals, conservation organizations, and government officials (Santiapillai and Ramono 1993; Hutadjulu and Janis 2002; Sitompul et al. 2002; Hedges et al. 2006).

Underlying Problems

Social and political problems affect the elephants of Indonesia. Lack of political will, inadequate laws, powerful incentives favoring logging and forest conversion, and inadequate conservation incentives have been cited by the government as the root causes of declining elephant populations (Susmianto 2002). Additional factors include a burgeoning human population, corruption, and consumer demand in western countries for Indonesian products.

In 2003, the human population in Indonesia exceeded 230 million. One hundred sixteen humans per square kilometer (300 per square mile) competed with elephants and other wildlife for space and food. By comparison, in the United States, there were thirty humans per square kilometer (78 per square mile; Population Reference Bureau 2003). Indonesia had the highest rate of deforestation in the world, with a loss of 3.8 million hectares a year ("Deforestation in Indonesia Worst in the World, NGO Says," 2003). Sumatra lost 60% of its remaining lowland forest between 1990 and 2002 (Glastra 2003). According to one source, illegal operations account for 50–70% of all logging in the country (compiled data reported in Glastra 2003). Indonesia's Environment minister stated in January 2004 that at least 75% of the logging was illegal (Paddock 2004). The two pulp mills in Riau (Riau Andalan Pulp and Paper and Indah Kiat Pulp and Paper) are among the largest in the world. These mills consumed half of Indonesia's and twice Riau's legal wood supply in 2000 (Glastra 2003). Illegal logging is driven by high mill debts, insufficient legal timber, and ineffective law enforcement.

Consumer demand provides an impetus to deforestation, and western consumers share the responsibility for the loss of Indonesia's forests and elephants. Timber, pulp, paper, and palm oil are Indonesia's major exports, with paper products reaching end markets primarily in Europe. Indonesian products also reach the United States, and increased production of robusta coffee has been linked to deforestation in Indonesia (O'Brien and Kinnaird 2003).

Economic instability and decentralization of the central government have affected many aspects of life in Indonesia. The absence of effective law enforcement and the diversion of land-use authority to provincial governments (except for national parks and reserves) together with the allocation of timber rights to powerful conglomerates and politico-business families have been disastrous for the environment. Honest government officials who attempt to intervene may face serious personal threats. The central government has been unable (or unwilling) to alter this situation and illegal logging has become semi-legal (Jepson et al. 2001).

Beginning in 1997, fires emanating from Indonesia covered much of Southeast Asia in smoke. These fires, set by individuals and even more often by plantation and timber companies, diminished human health and threatened wild orangutan populations. Despite protests from neighboring countries, the fires continue each year. If Indonesia's forests are destroyed, unchecked flooding will result and an ominous future awaits both humans and elephants. Flooding has already resulted in human deaths. Hundreds were left dead or homeless in 2003 from floods, attributed by both conservation organizations and government officials, including President Megawati Sukarnoputri, to uncontrolled deforestation (Brummitt 2003; "Rampant

Deforestation Blamed for Langkat Flash Flood" 2003; Unidjaja and Gunawan 2003).

Underlying the aforementioned problems is widespread corruption, known in Indonesia as "KKN" (collusion, corruption, and nepotism). Indonesia received a score of 1.9 (on a scale where 0 = highly corrupt and 10 = highly clean) on the Transparency International Corruption Perceptions Index (Transparency International 2003). Corruption correlates directly with the loss of biodiversity. A recent analysis demonstrated that corruption surpasses poverty or human population pressure in explaining decreases in elephant populations (Smith et al. 2003).

Poaching elephants for ivory is a growing problem for Asian elephants. Incidents of poaching have occurred in Sumatra. Mikota and Hammatt observed that tusks were routinely trimmed on newly captured bulls in Sumatra, even those to be translocated, and that ivory was also removed from all elephants that died, even the tushes from females. Where is that ivory? The lack of transparency by the Indonesian government on this matter is disturbing.

Alternatives to Capture and the Future of the ETCs

Human-elephant conflict takes its toll on both sides. Tragedies associated with the loss of homes, crops, or human lives cannot be discounted. In self-defense or in revenge, villagers or plantation owners may take matters into their own hands, as they did in 1996, when twelve elephants were poisoned and buried in Riau (Bangun 1996, cited in Lair 1997) and in 2002 when seventeen elephants died of suspected poisoning in North Sumatra ("Death of 17 Elephants in Tapanuli South Still a Mystery," 2002). These are just examples, and many incidents are unreported. Farmers view elephants as crop-raiding pests, according to Hutadjulu and Janis (2002), but a survey of attitudes in Aceh (Jepson et al. 2002) showed that people in that province (even in conflict areas) like elephants and support their conservation. Human-elephant conflict can result in public protest, negative media coverage, and hostile actions toward conservation organizations (Susmianto 2002). In one case, angry villagers held demonstrations and kept officials hostage (in orangutan cages!) until they agreed to capture problem elephants (W. Azmi and B. Suprayogi FFI, personal communications, 2002).

What are the alternatives? Certainly habitat preservation is crucial. WWF-Indonesia has secured a portion of the last large, remaining lowland forest in Sumatra (Tesso Nilo) as a national park to be managed as an elephant conservation area. A conflict mitigation plan has been implemented to use "flying squad teams" consisting of four elephants and eight mahouts designed to reduce wild elephant raids by driving wild elephants back to their habitat (N. Foead, personal communication).

Other conflict mitigation techniques can be tried. Although many farmers simply expect the government to capture and remove "problem" elephants, some have taken a more proactive approach to protect their crops. Crop protection teams that use watchtowers as lookout posts and inexpensive items such as fire torches, noisemakers, and bright lights are being tested in Lampung (Sitompul et al. 2002; Hedges et al. 2006). Other crop protection methods include clearing forest boundaries so that elephants can be seen, planting buffer zones of deterrent crops, and choosing alternative crops (Sitompul et al. 2002).

Indonesia adopted their system of capture and training for productive uses as a solution to conflict. We have shown that the ETCs have failed to achieve their intended purposes. There is no evidence that capturing elephants in Sumatra has reduced human-elephant conflict. In fact, conflict has escalated. Forestry officials reported sixteen elephant attacks from 1998 through 2002, but in the first five months of 2003, there were forty-eight, with at least three human fatalities (Paddock 2004). Capturing elephants for placement into ETCs is not a viable long-term solution. Under the current system in Indonesia, elephants may be severely injured during capture or training and often endure great suffering. Many die. The projected income from use of elephants is negligible and not likely to improve substantially.

At several meetings in Indonesia in 2000–2003, Indonesian officials expressed a desire to institute captive breeding programs in the ETCs. The rationale has been that they need to breed elephants to be able to "loan" them to foreign zoos. However, such loans are unlikely to have significant benefit and may actually contribute to further elephant captures and deaths (Hedges et al. 2006). A maximum number of elephants have been designated for each ETC, and ETCs strive to maintain this number because it determines support funds from the central government. We consider it abhorrently unethical to consider breeding captive Sumatran elephants. We have demonstrated that the removal of each elephant from an ETC (as a loan to a western zoo, for example) will directly result in the deaths of eight to ten wild elephants, as the ETCs will capture additional elephants to maintain their unofficial "quota." We have also substantiated the inadequacy of existing diets and the poor growth rates of captive Sumatran elephants.

From a conservation perspective, captive elephants are the "living dead." The IUCN does not believe release of captive African elephants to the wild can contribute meaningfully to the conservation of the species (Dublin and Niskanen 2003) and likely this holds for Asian elephants. Wild populations of Sumatran elephants are clearly threatened. From a welfare perspective, certain issues beg to be addressed. Elephants that must suffer the interruption of their lives at the hand of man deserve, at the very least, humane treatment. Archaic capture techniques that cause painful wounds are inexcusable. The IEF and FFI strongly recommended in 2001 that wild elephant captures should cease immediately (IEF 2001). A moratorium on capture requested by the authors and WWF-Indonesia, with support from FFI and the Wildlife Conservation Society (WCS) was granted in 2001 but never enforced. And so the captures continue.

Although translocations offer an alternative to long-term captivity, these must be meticulously planned and executed and should include follow-up monitoring of released elephants. Guidelines for the translocation of African elephants have been compiled by a task force jointly convened by the IUCN Species Survival Commission's African Elephant and Reintroduction Specialist Groups. The purpose is to provide informed advice to governments, nongovernmental organizations, and others wishing to reintroduce or supplement elephant populations. The guidelines arose because of a lack of guidance that often resulted in poorly planned translocations with adverse consequences to both elephants and humans. The guidelines address topics such as precapture monitoring, environmental and ecological impact, staffing and budgeting, social, behavioral and veterinary considerations, security, implementation (capture, transport and release), and postrelease monitoring (Dublin and Niskanen 2003; also, see Whyte and Fayrer-Hosken, Chapter 20 in this volume). Similar guidelines are needed for Asian elephants.

Elephant translocations have been carried out in Africa, India, and Malaysia, but translocation is neither simple nor a panacea. Translocated elephants may attempt to return to their home ranges or may cause problems in new areas. Such was the case in 2002 in Riau when translocated elephants subsequently destroyed several houses near the release site, angering villagers. The issue of suitable release sites remains problematic in Indonesia, as adequate surveys have not been conducted to identify such areas, and even protected habitats are under constant siege by illegal logging perpetuated by ineffective law enforcement. Even habitat set aside for captive elephants is at risk. The Riau ETC, established in Sebanga in 1991 on 5,000 hectares (12,350 acres), was reduced to 220 hectares (543 acres) by 2001 due to illegal logging (Susmianto 2002). At the least, translocation operations require skilled veterinarians and humane handling. Sumatran elephants with life-threatening injuries have been translocated, an intolerable practice.

In the interim, capture techniques and conditions at the centers must be improved. There must be increased willingness to accept help and technology currently not available in Indonesia. Indonesians have only been keeping elephants in captivity since 1986—they have much to learn from their Asian counterparts and also from the West.

Culling has been used to reduce populations within the confines of South Africa's rigorously managed parks. The practice is highly controversial and has met with public protest. The Indonesian government aborted its plans to cull Sumatran elephants in 1992 and 1994 as a result of media reports and public outcry (Lair 1997). Santiapillai and Ramono have stated that "capturing elephants is more humane than shooting them as pests" (as quoted in Lair 1997, 79). Considering the despairingly low survival rates and the slow, lingering deaths we have witnessed, culling might have been more humane. However, we do not view culling as a solution for Sumatra, as there is no reason to believe that it would not foster even greater abuses.

And what of the ETCs? At best, with improved husbandry and care, the ETCs could provide a long-term care facility for displaced elephants. But, what if conditions remain unchanged? Nazir Foead of WWF-Indonesia (personal communication) suggests that the ETCs do meet a need, and that it is the management that is the problem, not the policy.

In conclusion, Sumatra's ETCs do not represent an effective use of limited conservation resources, which would be better allocated to habitat preservation and alternative conflict mitigation methods. The current rate of removal of wild elephants from Riau Province (purported to have one of the largest number of elephants remaining on the island) is clearly not sustainable, based on the 1993 PHVA analysis. Foreign zoos should not take elephants from Sumatra. To do so will promote additional captures.

Views toward elephants and nature are a product of basic beliefs and are strongly influenced by religion and economics. If we consider ourselves to be a part of nature, we will strive to preserve that which supports us. If we consider that we are above nature, we have two choices. We can be stewards of the earth's resources or we can simply use them up. Do we have the knowledge and good will to use nature and at the same time preserve it?

Christian and Muslim religious leaders in Indonesia have called for a moral movement to end corruption (Nugroho and Unidjaja 2003). A 2003 World Bank report also holds hope for the world's religions to rally to protect nature: "Ultimately, the environmental crisis is a crisis of the mind," state Palmer and Finlay. "If the information of the environmentalists needed a framework of values and beliefs to make it useful, then where better to turn for allies than to the original multinationals, the largest international groupings and networks of people? Why not turn to the major religions of the world?" (Palmer and Finlay 2003, 16).

Until we adjust our values to reflect what is important to life (clean air and water, human rights, the survival of species) and reallocate our resources accordingly, we are only putting a Band-Aid on environmental problems that need major surgery. Population and land-use planning should consider and respect the other species with whom we share our planet and on whom our own lives depend. The general attitude of most people (and governments) favors short-term gain. If the majority of people felt that we must preserve things for our children's children, this would drive societal decisions in a very different direction in every country. West-erners must reduce consumption and become more knowledgeable about the effects that their chosen purchases have on the other side of the world. Caring consumers can be a powerful lobby.

Ultimately, Indonesia must decide whether Sumatran elephants are to survive, and if so, decisive action must be taken quickly. Before the gov-ernment can solve the problems in the ETCs, they must admit the prob-lems, and stop denying and underreporting deaths of elephants. In the race for diminishing resources, elephants are losing. Surely the floods, fires, and human deaths are a wakeup call. Similar events led to Thailand's logging ban in 1989. Can Indonesia's chainsaws be stopped in time? Political will, strong leadership, public education, and international pressure are needed. Environmental anarchy must end. Neither culling nor capture would be necessary if the government set aside and protected habitat for its elephants. It is Sumatra's Elephant Training Centers and not Sumatra's elephants that should be managed into extinction.

Indonesian government officials consistently state that resources are lacking to care for captive elephants properly. This rationale of poor re-sources (for patrols) is also used to whisk away complaints about illegal log-ging and destruction of national parks and preserves.

Recent news from Indonesia suggests that the situation is not improv-ing. On September 4, 2007, the *Jakarta Post* reported that the National Police headquarters had named over 350 suspects, including dozens of Forestry ministry officials and police officers, in illegal logging cases. "In Riau province alone, there are a total of 189 suspects of illegal logging—and among them are forestry officials" ("Officials Suspected of Illegal Log-ging," 2007). On September 5, 2007, a special team to combat illegal logging in Riau was announced by the president of Indonesia. Twelve nongovern-mental environmental organizations expressed opposition to the team, with the executive director of the Indonesian Forum for Environment pointing to two members of the special team, the minister of Forestry and the governor of Riau, as allegedly involved in illegal logging activities (Harahap 2007). On September 7, Indonesia pleaded that more money was needed from foreign governments to stop deforestation (Thompson 2007). Indonesian officials always suggest the necessity for foreign financial sup-port to meet such needs.

We suggest that the reality is something quite different. There is no for-eign organization or government that can solve the issues of corruption and illegal logging in Indonesia. That must happen from within. In the absence of political will, no amount of monetary support will save the habitat or the Sumatran elephants. The people and leadership must demand and se-cure the assets of their nation. A consortium of nongovernmental organ-izations led by a WWF–World Bank Alliance on Illegal Logging estimated that each year US$1.5 billion of government revenue is lost because of illegal logging (WWF-World Bank Alliance on Illegal Logging 2005; see also WWF-World Bank Global Forest Alliance 2005). Indeed, the Indonesian Ministry of Forestry stated in May of 2005 that uncontrolled logging practices have caused the state to suffer losses amounting to US$19.1 billion over the past five years ("Illegal Logging Causes State Losses of Rp 180t[rillion]," 2005). Thus, if the government reduces illegal logging losses by even a small percentage, recovered funds could be sufficient to protect the forests that benefit elephants and people. As seen so dramatically in recent years, forests and mangrove are essential for the protection of the people from floods and tsunamis. The environment does matter.

Unless these larger issues of corruption and illegal logging are addressed in Indonesia, Sumatra's elephants will be doomed. Must we accept this grim reality? Those who struggle in the trenches to preserve nature are often caught up in a system of poorly funded efforts that are directed at helping various immediate causes (elephant conservation, for example) without addressing the global issues (such as overpopulation and overconsumption). It is easy to despair, but despair is not a solution. We can all play a role and we must. Western consumers should choose related purchases wisely and advocate socioeconomic changes in Indonesia that will facilitate stewardship of this country's unparalleled biodiversity. Given Indonesia's tumultuous history and current challenges, it will not be an easy path. But, "even if the rope breaks nine times, we must splice it back together a tenth time. In this way, even if ultimately we do fail, at least there will be no feelings of regret" (Dalai Lama 1999, 129).

Acknowledgments
The advice and support provided by Nazir Foead (WWF-Indonesia), Simon Hedges (WCS), Steve Osofsky (WWF-US), and Martin Tyson (WCS) during the authors' work on Sumatra is gratefully acknowledged. Financial support was provided by a Guggenheim Fellowship, Nancy Abraham, IEF, and many zoos and individuals.

References
Asian Nature Conservation Foundation. 2006. Status and Distribution http://www.asian-nature.org/status.php.
Brogan, T. V., and Nizet, V. 1997. A clinical approach to differentiating necrotizing fasciitis from simple cellulites. *Infections in Medicine* September: 734–738.
Brummitt, C. 2003. Logging may have caused Indonesia floods. *Associated Press*. November 4.
Dalai Lama. 1999. *Ethics for the new millennium.* New York: Riverhead Books.
Death of 17 elephants in Tapanuli South still a mystery. 2002. *Analisa Daily.* July 5.
Deforestation in Indonesia worst in the world, NGO says. 2003. *Jakarta Post.* October 29.
Dublin, H. T., and Niskanen, L. S. (eds.). 2003. The African Elephant Specialist Group in collaboration with the Re-introduction and Veterinary Specialist Groups 2003.

IUCN/SSC AfESG Guidelines for the in situ *translocation of the African elephant for conservation purposes.* Gland, Switzerland; Cambridge, UK: IUCN.

Fernando, P., Vidya, T. N. C., Ng, L. S. G., Schikler, P., and Melnick, D. J. 2003. Conservation genetic analysis of the Asian elephant. In J. Jayewardene (ed.), *Endangered elephants, past present and future: Proceedings of the symposium on human elephant relationships and conflicts, Sri Lanka, September 2003* (pp. 40–43). Colombo: Biodiversity and Elephant Conservation Trust.

Fernando, P., Vidya, T. N. C., Payne, J., Stuewe, M., Davison, G., Raymond, J. A., Andau, P., Bosi, E., Kilbourn, A., and Melnick, D. J 2003. DNA analysis indicates that Asian elephants are native to Borneo and are therefore a high priority for conservation. *Public Library of Science (PLoS) Biology* 1: 110–115.

Fleisher, R. C., Perry, E. A., Muralidharan, K., Stevens, E. E., and Wemmer, C. M. 2001. Phylogeography of the Asian elephant (*Elephas maximus*) based on mitochondrial DNA. *Evolution* 55: 1882–1892.

Glastra, R. 2003. *Elephant forests on sale: Rain forest loss in the Sumatran Tesso Nilo region and the role of European banks and markets.* Frankfurt: WWF Germany. www.wwf .se/source.php?id=1015046.

Hammatt, H., and Fahrimal, Y. 2003. Implications of new data for Sumatran elephants in captivity: Time for change. In J. Jayewardene (ed.), *Endangered elephants, past present and future: Proceedings of the symposium on human elephant relationships and conflicts, Sri Lanka, September 2003* (pp. 61–64). Colombo: Biodiversity and Elephant Conservation Trust.

Harahap, R. 2007. Activists reject team to tackle Riau logging. *Jakarta Post.* September 8.

Hedges, S., Tyson, M. J., Sitompul, A. F., and Hammatt, H. 2006. Why inter-country loans will not help Sumatra's elephants. *Zoo Biology* 25: 235–246.

Hedges, S., Tyson, M. J., Sitompul, A. F., Kinnaird, M. F., Gunaryadi, D., and Baco, A. 2005. Distribution, status, and conservation needs of Asian elephants (*Elephas maximus*) in Lampung Province, Sumatra, Indonesia. *Biological Conservation* 124: 35–48.

Hutadjulu, B., and Janis, R. 2002. The care and management of domesticated elephants in Sumatra, Indonesia. In I. Baker, and M. Kashio (eds.), *Giants on our hands: Proceedings of the International Workshop on the domesticated Asian elephant, Bangkok, Thailand, 5–10 February 2001* (pp. 59–66). Bangkok: United Nations Food and Agriculture Organization Regional Office for Asia and the Pacific.

Illegal logging causes state losses of Rp 180t. 2005. *Jakarta Post.* May 18.

Indonesia seeks plan to save rare tigers, elephants. 2007. *Reuters India.* August 29. http://in .reuters.com.

International Elephant Foundation (IEF). 2001. *Support for the improved health and health care management of captive populations of Sumatran Asian elephants.* Final report to the United States Fish and Wildlife Service. Azle, TX: IEF.

IUCN-The World Conservation Union (IUCN). 2006. *2006 IUCN Red List of Threatened Species.* Gland, Switzerland: IUCN. www.redlist.org.

IUCN-The World Conservation Union/Species Survival Commission-Asian Elephant Specialist Group (IUCN / SSC-AESC). 1990. *The Asian elephant: An action plan for its conservation.* Comp. C. Santiapillai and P. Jackson. Gland, Switzerland: IUCN.

Jepson, P., Jarvie, J. K., MacKinnon, K., and Monk, K. A. 2001. The end for Indonesia's lowland forests? *Science* 292: 859.

Jepson, P., Nando, T., Hambal, M., and Canney, S. 2002. *Some results of the first survey of public attitudes towards elephants in Nanggroe Aceh Darrusalam Province, Indonesia conducted during March 2002.* SECP-FFI Technical Memorandum No. 5. Banda Aceh, Indonesia: Sumatran Elephant Conservation Program—Fauna & Flora International.

Krishnamurthy, V. 1992. Recommendations for improving the management of captive elephants in Way Kambas National Park, Lampung, Sumatra, Indonesia. *Gajah* 9: 4–13.

Lair, R. C. 1997. *Gone astray: The care and management of the Asian elephant in domesticity.* Bangkok: United Nations Food and Agriculture Organization Regional Office for Asia and the Pacific.

Lewis, J. 1998. A Veterinary Assessment of Sumatran Elephant Training Centers, a report on the Visit of Dr. John Lewis of International Zoo Veterinary Group, on behalf of Fauna & Flora International, to the Sumatran Elephant Training centers at Lhokseumawe, Sebanga, and Way Kambas, Sumatra 29.04.98–13.05.98. Internal report. London: Flora & Fauna International.

Lewis, J., and Yusuf, I. 2001. A Second Veterinary Assessment of Sumatran Elephant Conservation Centres. Notes on the visits by Dr John Lewis (veterinary advisor to FFI, UK) & Dr Irwandi Yusef (veterinary advisor to the Sumatran Elephant Conservation Programme, Aceh, Indonesia) to the Bentayan, Lahatm, and Seblat ECC's, 13–17 March 2001. London: Fauna & Flora International.

Lilley, R. P. H., and Saleh, C. 1998. *Captive Elephants in Crisis.* WWF report on a survey of Elephant Training Centres in Sumatra, Indonesia, November 9–20. Submitted to WWF Asian Elephant Action Planning Workshop, Vietnam, December 1–6, 1998.

O'Brien, T. G., and Kinnaird, M. F. 2003. Caffeine and conservation. *Science* 300: 587.

Mikota, S. K., Hammatt, H., Azmi, W., and Manullang, B. 2000a. Medical evaluation of captive elephants, Sebanga-Duri Elephant Conservation Center, Riau Province, Sumatra, Indonesia. Unpublished report.

Mikota, S. K., Hammatt, H., Azmi, W., and Manullang, B. 2000b. Report of Sumatran Elephant Conservation Workshop, Bogor, Java, April 2000. Unpublished report.

Nugroho, I. D., and Unidjaja, F. D. 2003. Indonesian law toothless against corruption. *Jakarta Post.* November 17.

Paddock, R. C. 2004. Unkindest cuts scar Indonesia. *Los Angeles Times.* January 2.

Palmer, M., and Finlay, V. 2003 *Faith in conservation: New approaches to religions and the environment.* Washington, DC: World Bank. www.worldbank.org.

Population Reference Bureau. 2003. *World population data sheet.* www.prb.org.

Rampant deforestation blamed for Langkat flash flood. 2003. *Jakarta Post.* November 5.

Reilly, J., and Sukatmoko, P. 1999. The Elephant Training Centre at Way Kambas National Park, Sumatra: A review of the centre's operations and recommendations for the future. Department of Biological Sciences, Manchester Metropolitan University.

Santiapillai, C., and Ramono, W. 1993. Why do elephants raid crops in Sumatra? *Gajah* 11: 55–58.

Sitompul, A. F., Hedges, S., Tyson, M. J., O'Brien, T. G., and Santoso, J. 2002. Human-elephant conflict around two national parks in Indonesia, and conservation implications. Abstract of paper given at the Society for Conservation Biology's Annual Meeting in Canterbury, UK, July 14–19, 2002.

Smith, R. J., Muir, R. D. J., Walpole, M. J., Balmford, A., and Leader-Williams, N. 2003. Governance and the loss of biodiversity. *Nature* 426: 67–70.

Stremme, C. 1998. Significant veterinary problems caused by the training methods utilized by elephant training centers (ETC) in Sumatra, Indonesia. 29.04.98–13.05.98. Internal report. London: Fauna & Flora International.

Sukumar, R. 1989. *The Asian elephant: Ecology and management* (2nd rev. ed., 1992). Cambridge: Cambridge University Press.

Sukumar, R. 2001. The Asian Elephant Specialist Group, activities for the inter-sessional period. *Species*, Spring 2001: 21.

Sukumar, R., and Santiapillai, C. 1993. Asian elephant in Sumatra: Population and habitat viability analysis. *Gajah* 11: 59–63.

Suprayogi, B., Sugardjito, J., and Lilley, R. P. H. 2002. *Management of Sumatran elephants in Indonesia: Problems and challenges.* In I. Baker and M. Kashio (eds.), *Giants on our hands: Proceedings of the International Workshop on the domesticated Asian elephant, Bangkok, Thailand, 5–10 February 2001* (pp. 183–194). Bangkok: United Nations Food and Agriculture Organization Regional Office for Asia and the Pacific.

Susmianto, A. 2002. Action plan for elephant conservation, country chapter for Indonesia. Presented at the June 3–4, 2002, Sumatran Elephant Workshop, Palembang, Sumatra, Indonesia.

Thompson, G. 2007. Indonesia says more money needed to stop deforestation. *Australian Broadcasting Corporation News*. September 7. http://abc.net.au/news/stories/2007 09/07/2027346.htm?section=world.

Tilson, R., Soemarna, K., Ramono, W., Sukumar, R., Seal, U., Traylor-Holzer, K., and Santiapillai, C. 1994. Asian Elephant in Sumatra: Population and habitat viability analysis report. Apple Valley, MN: IUCN/SSC Captive Breeding Specialist Group.

Transparency International. 2003. *Annual report*. Berlin: Transparency International. www.transparency.org.

Unidjaja, F. D. and Gumawan, D. 2003. Megawati joins the chorus blaming forest destruction. *Jakarta Post*. November 6.

United Nations Monitoring Programme-World Conservation Monitoring Centre (UNEP-WCMC). 2007. *UNEP-WCMC Species Database: CITES-Listed Species*. Cambridge: UK: UNEP-WCMC. www.unep-wcmc.org/species/index.htm.

Wiese, R. 1997. Demographic analysis of the Asian elephant population in North America. In *Proceedings of the 18th Annual Elephant Managers Association Workshop, Fort Worth Zoological Park, November 1–4, 1997* (p. 57). Fort Worth, TX: Elephant Managers Association.

WWF-World Bank Alliance on Illegal Logging. 2005. *Joint statement presented to the Indonesian Minister for Forestry by NGOs led by the WWF-World Bank Alliance on Illegal Logging*. March 24. Washington, DC: WWF-World Bank Alliance on Illegal Logging. www.illegal-logging.info/item_single.php?item=news&item_id=770& approach_id=15.

WWF-World Bank Global Forest Alliance. 2005. *Annual report*. Washington, DC: WWF-World Bank Global Forest Alliance. www.worldwildlife.org/alliance.

19 HUMAN-ELEPHANT CONFLICTS IN AFRICA

WHO HAS THE RIGHT OF WAY?

WINNIE KIIRU

Human-elephant conflict is any human-elephant interaction that results in negative effects on human social, economic, and cultural life, on elephant conservation or on the environment (Dublin, McShane, and Newby 1997). Human-elephant conflict occurs when elephants raid human settlements, feed and trample on food crops, and destroy farm installations, such as water pipes and reservoirs, food stores, fences, and houses (Kiiru 1995a; Tchamba 1995). Elephants occasionally kill or injure people who attempt to protect their property from such depredations. In areas of extreme conflict, elephants prevent people from undertaking their normal daily activities, prevent children from attending school and cause family members to spend nights in the fields guarding crops. In Africa, villagers protect themselves using traditional methods, which range from simple scare techniques to dangerous snaring, which sometimes result in deaths and injuries to elephants (Hoare 2000a; Sitati 2003).

Africa has experienced unprecedented growth in human population, from an estimated 100 million at the turn of the twentieth century to about 500 million today, at the outset of the twenty-first century. Demand for land for the increasing numbers of humans resulted in the compression of space available for elephants (Laws 1970; Western and Lindsay 1984; Barnes et al. 1999). Most of the traditional elephant migration routes have been curtailed, and now elephants are confined in protected areas or isolated refugia outside protected area networks (Cumming, du Toit, and Stuart 1990).

In this chapter, I offer an account of human-elephant conflict in Africa, with particular focus on Kenya. I highlight the various methods used to mitigate the conflict in different areas of the continent and then discuss at some length conflict mitigation approaches used in this eastern African country. I also discuss what is known about the nature, intensity, and extent of human-elephant conflict, accumulated from scientific research in various African countries. I conclude with thoughts on the way forward, with regard to government policy and the roles of scientific research and of rural communities in promoting coexistence between people and elephants.

Some Considerations about Human-Elephant Conflict in Africa
West Africa

Human-elephant conflict is not a new phenomenon in West Africa. In Burkina Faso, for example, villagers blamed elephants for damaging their property around the Nazinga Ranch in the rainy season, while during the dry season, they reported elephants knocking down millet granaries (Damiba and Ables 1993).

A study at Banyang-Mbo Wildlife Sanctuary in Cameroon noted that elephant damage to crops was seasonal. It occurred mainly during the rainy season from August to October, with damage concentrated on particular fields and villages surrounded by natural vegetation; higher incidents of elephant crop raiding were recorded where farms were located closer to the sanctuary edge. Besides elephants, other predators, such as cane rats, domestic goats, and grasshoppers also damaged crops. However, elephants were known to cause damage to the greatest number of different crop varieties (Nchanji 2001).

In Benoue Wildlife Conservation Area, North Cameroon, conflicts between humans and elephants can be severe. Elephants are reported to have devastated several crops, causing great losses to the local communities. Areas closest to the park experienced the most severe crop loss. In the arid north, elephants from Waza and Kalamaloue National Parks raided surrounding communities. An increasing trend of crop raiding by elephants was experienced in Waza and Kalamaloue National Park from the 1980s to the mid-2000s. The conflict started to escalate in the 1980s, when an estimated 10 hectares of crops were destroyed at one time, and two elephants were killed in retaliation (Tchamba 1995). Traditionally, farmers used various strategies to deter elephants, including fencing, poaching, illegal encroachment of farms into the conservation area, and scaring problem elephants. Apparently, none of these strategies were effective (Tchamba 1995). In the Waza-Logone region, one researcher reported an even wider range of traditional methods used to deter elephants (Tchamba 1996). These methods included collective prayers and magical practices; beating drums and empty barrels, burning sheep dung, lighting fires around croplands, and throwing sticks and stones at elephants.

Elephant crop raiding is a major problem in Gabon. Elephant hunting was banned there in 1981. Subsequently, perhaps consequently, increasing cases of crop raiding by elephants were recorded, prompting Gabon's government to authorize "control shooting" to help ameliorate the conflict. This was not successful; in fact, the "control shooting" was seldom implemented. Hunters require high-powered firearms to shoot elephants, but these weapons were not readily available, and the government was unable to honor payments due to the hunters (Lahm 1996). In Gabon, traditional deterrents used against elephants include lighting fires or lamps at plantation perimeters, beating on metallic surfaces, hanging bottles and tin cans on cables or vines as noisemakers, making scarecrows, and abandoning the location. Ghana is not an exception with regard to human-elephant conflict (Barnes, Azika, and Asamoah-Boateng 1995). A large influx of elephants from forest fragments in Ghana's Kakum National Park destroys crops in surrounding farms just before harvest time (Barnes 1997).

Central Africa

In the Dzanga-Sangha Forest Reserve area of Central African Republic, logging companies built access roads into the forest, allowing human immigration into previously inaccessible areas. The immigrants subsequently established diamond mining and agricultural activity in the reserve. The disturbances pushed elephants to the periphery of the forest, increasing the human-elephant interface. Elephants abandoned these portions of forest reserve when they were completely destroyed, moving into the remaining forest blocks. Meanwhile, elephant conservation projects became well established in the area, and the elephants developed a heightened sense of security due to increased protection. Elephants gained the courage to venture into surrounding human settlements, thus escalating crop damage, especially of cassava and coffee trees. Some of the mitigation measures carried out against the elephants in this area include culling, which was tried in 1992 but failed, shooting of blank cartridges, and snaring (Turkalo and Kamiss 1999).

Several traditional deterrent methods were used effectively in the Dzanga-Sangha Reserve area. Concentrating crop fields along the main road has proved a successful deterrent, because the elephants were evidently frightened away by roadside activity. Burning Chinese bamboo near the fields, with its resulting explosions, also scared away the elephants, and fencing off areas with ropes bearing various noisemaking objects has also worked (Turkalo and Kamiss 1999).

Southern Africa: Zimbabwe

In southern Africa, conflict has been reported in all seven range states: South Africa, Botswana, Namibia, Zimbabwe, Mozambique, Angola, and Zambia. In the communal lands of the Zambezi valley in Zimbabwe, migrants moved into areas previously devoid of human activity due to prevalence of tsetse flies, carriers of trypanosomiasis, also called African sleeping sickness. The new settlements were established on marginal agricultural land. Attempts to grow maize, millet, and sorghum, the main food crops in the area, were often fruitless, and conflict with elephants compounded the subsequent food production crisis. Through CAMPFIRE (Communal Areas Management Program for Indigenous Resources), communities inhabiting these marginal lands were encouraged to zone their holdings into grazing areas, settlement areas, and wildlife areas, to enhance conflict management (Child 1995; Murphree 2001). For a period, the CAMPFIRE model worked, by delivering benefits to rural people at both communal and household levels and reducing the demand for new farmlands.

When the benefit flow to the household level was reduced significantly in the 1990s because of weaknesses in the CAMPFIRE model, these rural communities reverted to agriculture on areas zoned for wildlife, thus escalating the human-elephant conflict (Bond 2001). Such conflict was mainly characterized by elephants raiding maize, sorghum, millet, and cotton crops. The use of capsicum oleoresin as an elephant repellent was pioneered in these areas in an attempt to address the severe crop-raiding problems (Osborn 2002). Sport hunting of problem elephants was also used to address conflict in these areas, with variable success (Taylor 1993).

Human-Elephant Conflict in Kenya

In Kenya, severe human-elephant conflict was first reported in 1992 (Mnene 1992; Ngure 1992, 1995; Kangwana 1993, 1995; Kiiru 1995a, 1995b; Sitati 2002, 2003). Kenya's elephant herds were just recovering from severe poaching of the 1970s and 1980s, which had severely reduced their numbers, from an estimated 167,000 in 1974 to about 20,000 in 1987 (Douglas-Hamilton 1987). The remaining elephants had retreated to protected areas where poaching was relatively less severe (Poole 1996). After the ivory ban of 1989 and the subsequent establishment of the Kenya Wildlife Service (KWS), elephant poaching in Kenya became minimal. Elephants slowly started to venture outside the protected areas, and Kenya Wildlife Service field officers started reporting cases of elephants raiding entire maize fields. Soon the press was highlighting the problem. The KWS Elephant Program, charged with managing elephants countrywide, was inundated with calls to control raiding elephants, and researchers were deployed to the field to investigate the problem. Human-elephant conflict was evident in varying degrees of intensity across Kenya's elephant range. The conflict was particularly severe in areas where high-potential agricultural land and prime elephant range were in close proximity (Mnene 1992; Ngure 1992; Kiiru 1995b).

Elephants made regular forays into neighboring farms in the central highlands of Kenya, particularly along the boundaries of Mt. Kenya Forest and the forests of the Aberdares ranges. The problem was exacerbated by the presence of human settlements inside the dense forest. The government continued to excise large areas of forest for settlement, resulting in islands of farmland surrounded by dense forest. The remaining forest provided a perfect refuge for elephants during the day. Elephants usually raided farms at night, and when mature crops became available the raids became most severe (Kiiru 1995a). Farmers were forced to spend nights in the fields guarding their crops. They used flashlights, fire, and loud noises to scare the elephants away. Success rates were variable, and farmers often lost large areas of cropland to elephants.

Kenya Wildlife Service rangers were sometimes called on to assist in scaring away elephants. Sadly, there were hardly enough rangers to deal with the problem. Elephants often invaded during the growing season and destroyed everything. Kenya Wildlife Service rangers patrolled large areas on foot, often arriving after the elephants had left a trail of destruction. When rangers arrived in time, they were hesitant to shoot the offending elephants. They had all witnessed the slaughter of the 1980s. They were acutely aware of the losses the elephant population had suffered. Shooting elephants was not going to be easy for these men. Many of their counterparts had lost their lives in bush wars, protecting elephants from poachers. So, they resorted to scaring elephants using explosives and by shooting in the air (Kiiru 1995a; Poole 1996).

In the rangelands of northern Kenya, particularly in the Laikipia and Samburu areas, the conflict was equally intense (Thouless 1993, 1994). The problem was compounded by the existence of small farms adjacent to large ranches. Unsuspecting peasants, hoping to escape the population pressure in the highlands, purchased barren plots of land in Laikipia District. An estimated 3,500 elephants ranged between the big ranches, some of which were especially "elephant friendly," and the few existing forest areas. Elephants made frequent forays into the small farm plots at night, retreating into the ranches or surrounding forest fragments during the day (Kiiru 1995a).

In the arid and semiarid parts of the country, occupied largely by pastoralist communities, the advance of agriculture through irrigation schemes and large-scale wheat farms, facilitated by international aid agencies, added to the complexity of the human-elephant conflict scenario. For the seminomadic Samburu, Rendille and Boran people, insecurity in this region, fueled by a bandit economy that thrived on gun running and cattle rustling, made settlement an attractive option. Unfortunately, many irrigation and farming schemes were set up without reference to the migratory nature of elephants and were located on elephant migratory routes or prime dry-season watering points. These tribes who traditionally lived in harmony with elephants could no longer tolerate them, and the stage was set for serious human-elephant conflict. Downstream, river flow was severely affected by water offtake for irrigation. Mud wallows traditionally used by elephants also dried up (Kiiru 1995a).

Flawed government policies encouraged human settlements around the world-famous Maasai Mara National Reserve and Amboseli National Park (Kangwana 1995; Sitati 2003). Migrant communities leased or bought land from the Maasai for commercial agriculture. Large areas of forest were cleared, giving way to wheat farms around Narok, adjacent to the Mara. The Kimana irrigation scheme was established around Amboseli. Elephants venturing into these areas were bound to locate the farms. Conflict was thus inevitable. Near watering holes, elephants killed people as they returned home from drinking sprees. Schools were closed because school attendance became impossible due to the presence of elephants in forest fragments adjacent to human settlements (Sitati 2003).

Approaches to Resolving Human-Elephant Conflict in Kenya

KWS has experimented with a large number of conflict resolution techniques to complement the efforts of local communities. Problem animal control teams were set up in 1993. Members of these rapid response teams were trained and given authority to shoot offending elephants and then were deployed to conflict hotspots around the country. They learned "humane" shooting techniques designed to eliminate elephants quickly and "painlessly." The regular KWS rangers continued to use scare techniques such as shooting in the air and "thunder" flashes. This kept the elephants away, but only for short periods. KWS encouraged villagers to dig deep trenches or moats around their plots to keep the elephants out. In some areas, elephants filled the moats by digging mounds of soil with their tusks and dumping the soil into the trench. The moats required constant surveillance and maintenance.

As the conflict intensified, it became necessary to find more effective methods of reducing the levels of damage. Farmers were suffering huge losses without recourse to compensation. The Kenyan government abolished the wildlife damage compensation scheme in 1989. The scheme had been crippled by fraudulent claims, and the exchequer was unable to raise enough money to pay for all claims. The government continued to pay compensation for human deaths and injuries. The amounts paid were very low, about US$500 compensation for death and US$200 for injury. The bureaucracy involved in processing those claims made it impossible for illiterate peasants and other potential claimants to receive their due (Kiiru 1995a).

The KWS installed about 1,000 kilometers of electric fencing in different parts of Kenya with varying degrees of success. Elephants constantly challenged the electric fences. Matriarchs were sometimes seen pushing the live wires apart with their tusks and hurling young ones through the gap. Electric fences were only effective when properly maintained, a big challenge for communities and KWS, especially because communities often were too busy tending crops to clear undergrowth by fences, as required (Thouless 1993; Thouless and Sakwa 1995).

Studies on elephant fertility control initiated in Kenya have explored immunocontraception techniques, which work by causing the elephant's immune system to produce antibodies that surround the eggs, thus preventing fertilization (Poole 1993). The challenges have included identifying target females, administering the treatment, and monitoring the subjects for long periods to study the effect. Experiments could only be carried out in confined populations of known elephant herds. Despite progress, more investigation will be required before elephant fertility control becomes available as a tool for elephant management.

In severe cases of conflict, the KWS used capture and translocation, or elephant drives, to remove problem elephants. Both methods proved very expensive and required meticulous planning. Translocation is manpower-dependent, requiring large numbers of people to execute. Recipient areas must be carefully chosen to avoid transferring problems from the source area. Elephant welfare is an important consideration; both capture and translocation subject elephants to high stress levels, which could lead to death. Elephants have highly developed social structures and the separation of family members, which often occurs during translocation, is a traumatic experience (Moss 1988; Njumbi et al. 1996; Omondi et al. 2002).

The KWS has encouraged communities living in close proximity to elephants to establish sanctuaries. In Shimba Hills, near the Kenyan coast, long negotiations resulted in the formation of the Golini Mwaluganje Elephant Sanctuary. To enable elephants to cross from the Shimba Hills National Reserve to the Mwaluganje Forest, local residents vacated an area across the Pemba River valley. That corridor and the Shimba Hills National Park were fenced to minimize human-elephant conflict around the reserve. Similar community-based sanctuaries, such as Il Ngwesi, Shompole, and Kimana Elephant Sanctuaries, have developed into viable projects providing benefits to both the local communities and the elephants. The economic incentives accruing from those projects have helped change the attitudes of the people toward elephants while providing important income for rural development. The local land-use zoning that accompanies the projects helps to minimize human-wildlife conflict (Kiiru 1995b).

The Kenyan experience illustrates the challenges of managing human-elephant conflict in many parts of Africa. Wildlife departments, many of them with meager financial and human resources, face the daily challenge of protecting people's lives and property while trying to conserve healthy elephant populations. The Kenyan case illustrates that there are many different approaches to conflict management but each measure presents specific challenges and, often, a combination of methods must be used to resolve human-elephant conflict.

Understanding Human-Elephant Conflict:
The Status of Scientific Research

Across Africa the lack of scientific information has hampered elephant managers in their attempts to resolve conflicts. That paucity has only begun to be addressed in the past few decades. Managers have long suspected that, in many areas, the conflict scenario has been exaggerated and highly politicized. To wildlife managers, the level of damage caused by elephants has often seemed much less severe than that caused by other agricultural pests. Managers have been under pressure to shoot elephants to placate local people who claim they are able to identify habitual crop raiders and that destruction of these elephants would arrest the problem. The social impact of crop raiding on communities has not been well understood. The relationship between land-use changes and the intensity of human-elephant conflict urgently requires study (Hoare 2000a, 2000b).

Managers have been ill equipped to collect systematic records of conflict situations at the local level. They are forced to depend on reports made to the stations by local people. The distance from villages to the reporting stations often determines the number of incidents reported, with more reports made by inhabitants of villages closer to the reporting station. Invariably, unsatisfactory levels of intervention by wildlife authorities in response to reports have discouraged villagers from providing regular information on conflict (Kiiru 1995b).

On the national scale, planning was impeded by lack of information on the extent of the problem in a given country. Georeferenced data on conflict sites and land-use types were lacking. The wildlife authorities were aware of seasonal fluctuation in the intensity of conflict, but lacked objective information on seasonal patterns. Data on the migration of elephants were inconsistent, and this made it difficult for managers to envisage upcoming crop-raiding patterns. Links between relevant government departments were weak, and those departments ignored the relationships between land-use transformation and elephant crop-raiding or other human-elephant conflicts (Kiiru 1995b).

Research scientists around Africa have made attempts to study human-elephant conflict and to provide quantitative and qualitative information on the nature, extent, and intensity of conflict across the elephant range. Although scientists have not exhaustively studied many of the areas previously discussed, they have acquired valuable information since the 1990s.

Elephants and Other Agricultural Pests

An evaluation of different levels of agricultural damage has shown that elephants rank relatively low in the list of damage caused by agricultural pests (Naughton, Rose, and Treves 1999). In many areas, particularly where agricultural land borders forest or protected areas, a number of wildlife species are reported as pests. Economic analyses done by Naughton et al. (1999) showed that on average, other pests, including baboons and other primates, birds, rodents, and insects caused more agricultural loss than elephants. However, elephants are perceived as the greatest threat. This is attributed to various factors, including the large size of elephants, the unpredictable nature of their raids, and the high levels of damage caused per raid. The social disruption occasioned by elephant presence in an area further contributes to this perception.

In areas of high human population density, raiding activity by elephants is usually confined to a few plots close to the elephant refuge area, and the costs of the damage are borne by a small section of the community. Many of the other pests stage regular and well-distributed raids. Elephants tend to be perceived as significant pests at the local level, whereas on a national scale the level of damage is demonstrably insignificant (Sitati 2003).

The media and local communities often exaggerate perceived elephant problems. The "elephant problem" in most of Africa is compounded by the additional socioeconomic costs of living with elephants. These costs include restriction on people's movements; sleepless nights due to the need to guard property and the consequent increase in the prevalence of diseases such as malaria, related to exposure to mosquito bites experienced during guard nights; reduced school attendance; poor employment opportunities; and competition for water between elephants and people and their livestock. Many of these costs are difficult to quantify but contribute significantly to the negative attitude communities have toward elephants (Sitati 2002, 2003).

Seasonality and Intensity of Crop Raiding

Recent studies have shown that elephant crop raiding in the savanna ecosystems demonstrates a seasonal peak, usually corresponding to the late wet season (Sitati 2003). The majority of crop-raiding incidents involve predation on maturing food crops. In the rain forests of West Africa, crop raiding almost always occurs during the wet season, suggesting that elephants seek to minimize the risks associated with raiding. A similar strategy could also explain why elephants living amid dense human populations concentrate their crop-raiding forays on agricultural plots close to the forest boundary (Kiiru 1995b; Kofi Sam 1999; Turkalo and Kamiss 1999).

Scientists have proposed several hypotheses to explain the growing intensity of crop raiding. The transformation of elephant range into agricultural land results in loss of elephant habitat and increases the probability of contact between elephants and human settlements. This suggests a direct association between the amount of land transformed by agriculture and the level of problem-animal activity. The existence of reasonably secure elephant refuge areas, either protected areas or forest fragments, near human settlements increases the intensity of conflict. Some studies track how rainfall patterns also influence intensity of crop raiding. Higher rainfall increases the biomass and yield of dry-land crops, leading to an increase in elephant crop raids (Hoare 1995; Hoare and du Toit 1999).

Habitual Crop Raiding

A study carried out in the semiarid lands of Zimbabwe suggested that the nature of elephant crop raiding is related to the behavioral ecology of male elephants (Hoare 1999; Hoare and du Toit 1999). Crop raiding by elephants represents opportunistic feeding forays by a segment of the male elephant population. During this Zimbabwe study, male elephants were found to occupy the peripheries of the protected areas. They entered human settlements in daytime, penetrating deeply into areas of dense settlement, and exhibiting high levels of tolerance to disturbance. This evidence supported earlier observations of the existence of habitual crop raiders and fence breakers in Kenya's Laikipia district and the communal lands of Zimbabwe (Thouless 1993, 1996; Thouless and Sakwa 1995; Hoare 1999).

A similar study of Asian elephants showed the raiding frequencies and economic effect of males on crops to be more than five times greater than that of females (Sukumar and Gadgil 1988). To explain this, Sukumar (1991) proposed the concept of a male strategy of risk-taking that maximizes reproductive success through better nutrition. He suggested that strategies for managing problem animals should focus on male elephants. In a follow-up survey for Africa, assessing evidence for the existence of habitual problem elephants among savanna and forest elephants, Hoare (1995, 1999, 2000a) concluded that Sukumar's hypothesis could not be strongly supported by the evidence available across twenty-two sites in eight different African countries. One of the challenges facing that study, Hoare noted, was the difficulty of recognizing individual animals during nocturnal raids, particularly in forests. Hoare noted the danger (to the elephant) of management authorities categorizing male elephants as "repeat offenders," which could result in their rapid elimination. An even worse consequence was the promotion of the "habitual raider" theory by local communities, resulting in undue pressure on management authorities to eliminate male elephants.

Hoare's study proposed caution when explaining the activities of problem animals. He suggested that a segment of any elephant population is involved in problem incidents. At any one time, raids are conducted by a variable number of individuals belonging to that cadre, whose members are difficult to identify. It is likely that even when the culprits are known, removing them from the population by translocation or shooting will result in fairly rapid cadre replenishment with other members of the larger population (Hoare 1995, 1999, 2000a).

Conflict mitigation

Research into new methods of deterring elephants from crop fields continues. The use of chili (*Capsicum* spp.) as an elephant repellent was pioneered in the communal lands of Zimbabwe and has been applied successfully to deter elephants from crop fields (Osborn 2002; Osborn and Parker 2002; Sitati and Walpole 2006). Elephants have been effectively repelled from fields by burning a mixture of capsicum powder and elephant dung. Ropes pasted with a mixture of grease and chili powder have also been used to reduce the incursions of elephants into fields. In Kenya, ropes pasted with a mixture of used engine oil, chili powder, and tobacco powder have been used as effective deterrents in Transmara and Loitokitok, near Amboseli National Park. Chili is a fast-growing crop offering high economic returns to farmers. Studies in Zimbabwe have shown that the promotion of chili as a cash crop not only provides farmers with an elephant deterrent but also reduces their demand for expansive farmlands due to its high yields per unit area. Additionally, the palatability of chili to other crop-raiding animals, such as greater kudu (*Tragelaphus strepsiceros*), bush pigs (*Potamochoerus larvatus* [*P. porcus* in Parker and Osborn 2006]), and chacma baboons (*Papio cynocephalus ursinus*), is low compared to that of crops like maize, sorghum, and cotton (Parker and Osborn 2006).

The Way Forward

Vast populations of Africa's elephants are found outside national parks and protected area networks. Even those inside the national parks depend on the surrounding rangelands for dispersal, but in these rangelands, the elephant habitats are disappearing as a result of deforestation and agricultural expansion. As the land available for dispersal disappears, the compression of elephants into protected areas will have a serious impact on the conservation of any and all biodiversity in those areas. Hope lies in communities that have coexisted with elephants throughout history (Kiiru 2004).

In a workshop held in 2001 in the Maasai Mara Game Reserve, Kenya, to discuss conflict and conservation in Maasai Mara, the senior warden of the game reserve said:

> Historically, Maasai communities have coexisted with wildlife in this ecosystem fairly harmoniously. It is no coincidence that the richest wildlife areas in the country are within Maasai and related Samburu areas. However, as communities become more sedentary and change their lifestyles and as human populations increase,

> there is an inevitable increase in conflict with wildlife over access to resources. Wherever wildlife and people coexist, there will be some form of competition and conflict. The challenge is to manage that conflict and to reduce it. It is unlikely that conflict can be totally eradicated, but it needs to be controlled at a level that local people can tolerate and at the same time people need to see the benefit from wildlife to offset those costs of conflict. (Sitati 2002)

Local residents expressed similar sentiments during a forum held in the Amboseli area by the Maasai Resource Coalition (Dapash 2002). The people appreciate that wildlife is an important part of their lives. They realize that conflict cannot be totally eradicated but can be reduced to tolerable levels. The Maasai appreciate that the world around them has changed and that commercial agriculture has changed the land-use in rangelands. The survival of elephants in Africa depends on the ability of rural people to appreciate the importance of coexistence and to create space for them.

Kenya, like many African countries, lacks a comprehensive land-use policy. Most African countries inherited flawed land-use practices as part of the colonial legacy. After independence there was an urgent need to allocate land to the masses, as the struggle for independence was a struggle for land. Little effort was expended on formalizing the land-use process. This legacy has had many negative impacts on the conservation of natural resources. Africa's exponential human population growth and the commensurate demand for space have further complicated patterns of land use. This competition for space is the greatest threat to conservation in Africa today. National-level land-use planning that recognizes the importance of creating protected areas with adequate dispersal zones, wildlife corridors, and buffer zones remains the most progressive way of resolving human-elephant conflicts. In many African countries, however, this complicated and potentially politically volatile crisis may take a long time to resolve.

In the absence of national land-use planning, the management of human-elephant conflict can only be approached through area-specific initiatives. These range from highly specialized, expensive methods, such as electric fencing, translocation, and fertility control, to local and inexpensive practical methods, such as collective guarding of crops and use of capsicum repellents. Centralized systems of conflict management incorporate many problems, including lack of resources for wildlife departments and managerial weaknesses. Enhancing the capacities of wildlife departments to manage human-elephant conflict by provisioning them with necessary equipment and training is important. But building the capacities of rural farmers to manage conflict resolution and empowering them to be the frontline guards in these efforts is imperative. This calls for the development of low-cost, practical, and effective conflict mitigation tools to be deployed within communities.

The role of further scientific research, providing data on the complexities of human-elephant conflict, cannot be overstated. Information gaps in the scientific knowledge available to wildlife managers continue to prove a major hindrance to the strategic management of human-elephant conflict and the development of effective mitigation tools.

African governments must be encouraged to develop a policy environment that supports elephant conservation. Of principal importance is development and implementation of land-use policies that recognize wildlife management as a feasible and economically viable land-use option. Opportunities still exist to designate new conservation areas in Africa but realizing such opportunities will depend on the capacity to enhance the value of wildlife conservation and reduce the costs borne by rural communities. Many African communities appreciate the intrinsic value of wildlife conservation. Building on this ethic by creating an enabling environment for elephant conservation while minimizing conflict is the way forward.

References

Barnes, R. F. W. 1997. A proposed solution to the Kakum elephant problem. In B. Bailey (ed.), *Facing the storm: Five years of research in and around Kakum National Park, Ghana. Research Colloquium Report* (pp. 15–18). Washington, DC: Conservation International.

Barnes, R. F. W., Azika, S., and Asamoah-Boateng, B. 1995. Timber, cocoa, and crop raiding elephants: A preliminary study from Southern Ghana. *Pachyderm* 19: 33–38.

Barnes, R. F. W., Craig, G. C., Dublin, H. T., Overton, G., Simons, W., and Thouless, C. R. 1999. *African elephant database 1998*. IUCN/SSC African Elephant Specialist Group. Gland, Switzerland and Cambridge, UK: IUCN.

Bond, I. 2001. CAMPFIRE and the incentives for institutional change. In D. Hulme and M. Murphree (eds.), *African wildlife and livelihoods: The promise and performance of community conservation* (pp. 227–243). Oxford: James Currey.

Child, G. 1995. *Wildlife and people: The Zimbabwean success*. Harare: Wisdom Press.

Cumming, D. H. M., du Toit, R. F., and Stuart, S. N. 1990. *African elephants and rhinos: Status survey and conservation action plan*. IUCN/ SSC African Elephant and Rhino Specialist Group. Gland, Switzerland: IUCN.

Damiba, T. E., and Ables, E. D. 1993. Promising future of an elephant population: A case study in Burkina Faso, West Africa. *Oryx* 27: 97–103.

Dapash, M. O. 2002. Coexisting in Kenya: The human-elephant conflict. *Animal Welfare Institute Quarterly* 51 (1): 10–12.

Douglas-Hamilton, I. 1987. African elephants: Population trends and their causes. *Oryx* 21: 11–24.

Dublin, H. T., McShane, T. O., and Newby, J. 1997. *Conserving Africa's elephants: Current issues and priorities for actions*. Gland, Switzerland: WWF.

Hoare, R. E. 1995. Options for the control of elephants in conflict with people. *Pachyderm* 19: 54–63.

Hoare, R. E. 1999. Determinants of human-elephant conflict in a land-use mosaic. *Journal of Applied Ecology* 36: 689–700.

Hoare, R. E. 2000a. African elephants and humans in conflict: The outlook for coexistence. *Oryx* 34: 34–38.

Hoare, R. E. 2000b. Projects of the human elephant conflict taskforce: Results and recommendations. *Pachyderm* 28: 73–77.

Hoare, R. E., and du Toit, J. T. 1999. Coexistence between people and elephants in African savannas. *Conservation Biology* 13: 633–639.

Kangwana, K. F. 1993. Elephants and Maasai: conflict and conservation in Amboseli, Kenya. PhD diss., University of Cambridge.

Kangwana, K. F. 1995. Human-elephant conflict: The challenge ahead. *Pachyderm* 19: 11–14.

Kiiru, W. 1995a. The current status of human-elephant conflict in Kenya. *Pachyderm* 19: 15–19.

Kiiru, W. 1995b. Human-elephant interaction around the Shimba Hills National Reserve, Kenya. MSc thesis, University of Zimbabwe.

Kiiru, W. 2004. Human-elephant conflict mitigation measures in Kenya. In J. Jayewardene (ed.), *Endangered elephants, past present and future: Proceedings of the symposium on human elephant relationships and conflicts, Sri Lanka, September 2003* (pp. 156–159). Colombo: Biodiversity and Elephant Conservation Trust.

Kofi Sam, M. 1999. *The distribution of elephants in relation to crop damages around Bia Conservation Area during the 1999 rainy season.* IUCN/SSC African Elephant Specialist Group Report. Nairobi: IUCN African Elephant Specialist Group.

Lahm, S. A. 1996. A nationwide survey of crop raiding by elephants and other species in Gabon. *Pachyderm* 21: 69–77.

Laws, R. M. 1970. Elephants as agents of habitat and landscape change in East Africa. *Oikos* 21: 1–15.

Mnene, R. 1992. *Interactions between people and wildlife around Shimba Hills National Reserve and Maluganji Forest.* Nairobi: Kenya Wildlife Service Rural Services Design Project Report.

Moss, C. 1988. *Elephant Memories: Thirteen years in the life of an elephant family.* New York: William Morrow and Company.

Murphree, M. 2001. Community, council and client: A case study in ecotourism development from Mahenye, Zimbabwe. In D. Hulme and M. Murphree (eds.), *African wildlife and livelihoods: The promise and perfomance of community conservation* (pp. 177–194). Oxford: James Currey.

Naughton, L., Rose, R., and Treves, A. 1999. *The social dimensions of human-elephant conflict in Africa: a literature review and case studies from Uganda and Cameroon.* Report to the African Elephant Specialist Group, Human-Elephant Conflict Taskforce, IUCN/SSC. Gland, Switzerland: IUCN.

Nchanji, C. 2001. Crop damage around northern Banyang-Mbo Wildlife Sanctuary southwestern Cameroon. Unpublished report of the Cameroon Biodiversity Programme/ Banyang-Mbo Wildlife Sanctuary Project, Yaounde, and the Wildlife Conservation Society, New York.

Ngure, N. 1992. History and present status of human-elephants conflicts in the Mwatate-Bura area, Kenya. MSc thesis, University of Nairobi.

Ngure, N. 1995. People-elephant conflict management in Tsavo, Kenya. *Pachyderm* 19: 20–25.

Njumbi, S., Waithaka, J. M., Gachago, S. W., Sakwa, J. S., Mwathe, K. M., Mungai, P., Mulama, M. S., Mutinda, H. S., Omondi, P., and Litoroh, M. 1996. Translocation of elephants: The Kenyan experience. *Pachyderm* 22: 61–65.

Omondi, P., Wambwa, E., Gakuya, G., Bitok, E., Ndeere, D., Manyibe, T., Ogola, P., and Kanyingi, J. 2002. Recent translocation of elephant family units from Sweet waters Rhino sanctuary to Meru National Park. *Pachyderm* 32: 39–48.

Osborn, F. V. 2002. Capsicum oleoresin as an elephant repellent: Field trials in the communal lands of Zimbabwe. *Journal of Wildlife Management* 66: 674–677.

Osborn, F. V., and Parker, G. E. 2002. A community-based system to reduce crop damage by elephants in the communal lands of Zimbabwe. *Pachyderm* 33: 32–38.

Osborn, F. V., and Parker, G. E. 2003. Towards an integrated approach for reducing the conflict between elephants and people: A review of current research. *Oryx* 37: 80–84.

Parker, G. E., and Osborn, F. V. 2006. Investigating the potential for chilli *capsicum* spp. to reduce human-wildlife conflict in Zimbabwe. *Oryx* 40: 343–346.

Poole, J. H. 1993. Kenya's initiatives in elephant fertility regulations and population control techniques. *Pachyderm* 16: 62–65.

Poole, J. H. 1996. *Coming of age with elephants: A memoir.* New York: Hyperion Press.

Sitati, N. 2002. Human elephant conflicts in Trans Mara district, Kenya. In M. J. Walpole, G. G. Karanja, N. S. Wasilwa, and N. Leader-Williams (eds.), Wildlife and People: Conflict and Conservation in Maasai Mara, Kenya. Unpublished proceedings of a workshop series organized by the Durrell Institute of Conservation and Ecology, University of Kent, England.

Sitati, N. 2003. Human-elephant conflict in TransMara district adjacent to Masai Mara National Reserve. PhD diss., University of Kent, Canterbury, UK.

Sitati, N. W., and Walpole, M. J. 2006. Assessing farmer-based measures for mitigating human-elephant conflict in Transmara District, Kenya. *Oryx* 40: 279–286.

Sukumar, R. 1991. The management of large mammals in relation to male strategies and conflict with people. *Biological Conservation* 55: 93–102.

Sukumar, R., and Gadgil, M. 1988. Male-female differences in foraging on crops by Asian elephants. *Animal Behaviour* 36: 1233–1235.

Taylor, R. D. 1993. Elephant management in Nyaminyami District, Zimbabwe: Turning a liability into an asset. *Pachyderm* 17: 19–29.

Taylor, R. D. 2000. *A review of problem elephant policies and management options in southern Africa.* A report to the African Elephant Specialist Group, Human-Elephant Conflict Taskforce, IUCN/SSC. Gland, Switzerland: IUCN.

Tchamba, M. N. 1995. The problem elephants of Kaele: A challenge to elephant conservation in northern Cameroon. *Pachyderm* 19: 26–32.

Tchamba, M. N. 1996. History and present status of the human/elephant conflict in the Waza-Logone region, Cameroon, West Africa. *Biological Conservation* 75: 35–41.

Thouless, C. R. 1993. *The Laikipia elephant project, final report.* Nairobi: Kenya Wildlife Service and WWF Eastern Africa Regional Office.

Thouless, C. R. 1994. Conflict between humans and elephants in northern Kenya. *Oryx* 28: 119–127.

Thouless, C. R. 1996. Home ranges and social organization of elephants in northern Kenya. *African Journal of Ecology* 34: 284–297.

Thouless, C. R., and Sakwa, J. 1995. Shocking elephants: Fences and crop raiders in Laikipia district, Kenya. *Biological Conservation* 72: 99–107.

Turkalo, A., and Kamiss, A. 1999. *Elephant crop raiding in the Dzanga-Sangha Reserve, Central African Republic.* IUCN African Elephant Specialist Group Report. Gland, Switzerland: IUCN.

Western, D., and Lindsay, W. K. 1984. Seasonal herd dynamics of a savannah elephant population. *African Journal of Ecology* 22: 229–244.

20 PLAYING ELEPHANT GOD
ETHICS OF MANAGING WILD AFRICAN ELEPHANT POPULATIONS
IAN WHYTE AND RICHARD FAYRER-HOSKEN

In contrast to the situation in many other parts of Africa, the elephant populations of the southern African region have been unaffected by the excessive illegal killing (poaching) and population declines experienced elsewhere and consequently have burgeoned over the past 30 years. Nearly 60% of Africa's total estimated elephant population of 461,000 occurs in southern Africa (Blanc et al. 2003). Botswana's population now stands at an estimated 120,000, Zimbabwe's at 80,500 (Blanc et al. 2003), while in South Africa, the population has risen from a few dozen in the early 1900s (Hall-Martin 1992) to over 14,000. Many populations in these countries are considered to warrant some form of limitation as they now pose threats to these countries' biodiversity. The question is how to limit these populations? What methods can be used and what are the ethical issues relevant to each?

Elephants are wonderful animals with unique qualities, and people empathize with many of these characteristics. They are long-lived, intelligent, have a strong sense of family, and show some perception of death through their fascination with carcasses of other elephants. This empathy with elephants generates strong emotional feelings in people that complicates rational decision making about their management. Wonderful as they are, when they are confined in national parks or wildlife preserves (hereafter referred to as "reserves") elephants are capable of detrimental effects on some species in particular and on overall biodiversity. This negative effect is especially apparent in African ecosystems in which elephants occur at high densities, leading directly to the fundamental dilemma facing managers of reserves in which elephants are found—what should be done about high densities of elephants? Managers are faced with this primary question: Is the reserve you are managing to be designated as an elephant sanctuary or is biodiversity conservation a higher priority? It may be entirely valid to decide in favor of an elephant sanctuary. But where there is consensus that controlling elephant numbers, for the sake of biodiversity, is the management priority, managers must decide just when to intervene. This decision needs to be linked with some sort of biodiversity monitoring program that can indicate when elephant impacts are beginning to exceed levels considered acceptable. The Kruger National Park (KNP) has developed such a system (Whyte et al. 1999) that also recognizes that flux is normal and desirable, and that differing densities of elephants (including very high and very low) will contribute to the maintenance of maximum biodiversity and/or ecosystem heterogeneity. This policy no longer recognizes the concept of carrying capacity but looks instead at elephant impacts. It subscribes to modern disequilibrium theory (e.g., Gillson and Lindsay 2003) but also recognizes that losses of species from a reserve or ecosystem are contrary to a policy of maintaining biodiversity. Space limitations preclude any discussion on these issues here, but while we accept that others may hold different opinions, managers are still faced with the dilemma of losses of species.

If elephant control to promote biodiversity conservation is considered necessary, a further decision must be taken on the method of control—translocation, contraception, or culling. The technology of translocation has now been developed to the point that intact families and even the largest of adult bulls can be moved. Research into contraception as a potential control technology has progressed well. Culling methods have also been refined to a point where trauma is minimized. These three options (with variations) are the only ones that are available to managers for controlling populations of any species, but what are the ethical issues inherent in each of them when used to control elephant numbers? Is any one method more acceptable than others?

To some people, the answers to elephant population control questions might seem straightforward, but the ethical issues surrounding each of these control options, and the particular circumstances inherent in each individual control situation, inevitably complicate any decision. Is translocation better than contraception? Can either of these two options be implemented? If not, is culling an acceptable option? If the answer to these questions is "no," can we just allow nature "to run its course"? The problem with this is that in most reserves these days nature can no longer run its natural course. Most reserves (even the largest ones) are increasingly being reduced to islands in a sea of anthropogenic land uses. Natural movements and migrations of most species can no longer take place.

In African reserves where elephant populations are adequately protected, their densities will increase and their impacts on the environment will increase accordingly. Species lost from a system may not be replaceable. The "elephant problem" is in fact a human problem—there are not too many elephants, there are too many people, and these people are now taking up space that was historically elephant range. But the outcomes of conservation efforts depend on decisions made by people, and where elephants are concerned, consensus is rare. In this emotionally charged arena, such decisions should be rational and ethical. But whose ethics should apply? To a rural African with little access to protein, an average westerner, or an animal rights person, ethical elephant management will mean very different things.

We intend to present a balanced view of the advantages and disadvantages of the available African elephant control options and the ethical considerations posed by each. We also briefly consider the advantages, disadvantages, and ethical considerations attached to another choice sometimes taken by managers, namely, the "nonmanagement option." Ultimately, the ethical viewpoints of individual readers will determine their own personal views, but it is hoped that these insights will be of some assistance in clarifying people's thinking.

Natural History of African Elephants

Some of the dilemmas associated with management of African elephants stem from the nature of the animals themselves, and it is necessary to understand elephants and their society to understand the dilemmas. African elephants are intelligent animals that live in a structured, family-oriented society in which individuals (particularly females) have strong permanent bonds with related animals (Moss 1988). Males tend to leave their mothers at around 14 years of age and then show little social affinity toward others, but females stay with their mothers as long as they are both alive. This results in matriarchal groups of up to forty-five or more individuals that may span four generations of related individuals (Poole, Kahumbu, and Whyte, forthcoming). These groups are broadly pyramidal in shape, with the matriarch at the head, the young adult females in the center, and many juveniles and calves forming a wide base. Growing up in such a family has many advantages to calves. It offers security, with many older aunts and sisters offering care and supervision. There are numerous role models for teaching and learning, and many siblings for playmates. The adaptive value of the system improves the survival of calves (Moss 1988). Matriarchs are the repositories of social knowledge in this society (McComb et al. 2001), and as leaders of these groups, they have a crucial role to play.

When matriarchs die, the group tends to split, each daughter forming a new matriarchal group of her own. These new groups maintain close social contact with each other, and collectively are known as a "bond group" (Moss 1988). Bond groups and families vary in their degree of cohesion, but it is known that such elephants can have social bonds with up to twenty-five other families representing up to 175 other adult females (McComb et al. 2001). Sometimes families, or even subgroups of families, may split off temporarily from bond groups, while at other times, groups may join to form large gatherings. At such times, mixing of families may occur with juveniles "visiting" others across family groups. It is this complex social structure and behavior that sets elephants apart from most other animals and complicates ethical management.

The Fundamental Ethical Dilemma

Wherever African elephant populations become contained and adequately protected from illegal killing in reserves, their numbers can increase at rates of up to 7% per year (Calef 1988; Woodd 1999; Whyte 2001a). At this rate, a population doubles every ten years. At higher population densities, growth rates slow down, usually as a result of increasing interbirth intervals. This interval varies from population to population, depending on food availability and other factors, and may range from 2.9 to 9.1 years (Laws, Parker, and Johnstone 1975).

At low elephant densities, grasslands tend to revert to woodland (Dublin, Sinclair, and McGlade 1990; Dublin 1995), but at high densities, elephants reduce the diversity of habitats (Western and Gichohi 1989; Moolman and Cowling 1994; Herremans 1995; Cumming et al. 1997; Johnson 1998; Johnson, Cowling, and Phillipson 1999; Lombard et al. 2001; Mosugelo et al. 2002). High elephant densities change woodland to grassland (Leuthold 1977; Bourlière and Hadley 1983) and may also eliminate some plant species (e.g., Leuthold 1977; Page 1999). Structural changes to vegetation caused by elephants in Amboseli National Park (Kenya) have even resulted in the extirpation of both lesser kudu and bushbuck (Western and Gichohi 1989) and more recently also gerenuk and giraffe (D. Western, personal communication). There is evidence to suggest that in extreme cases of elephant utilization, even local climate patterns may be affected (Altmann et al. 2002). Even in KNP, where elephant densities were kept low for twenty-seven years, impacts have been significant (see references quoted in Whyte et al. 2003), including losses of Marula trees (*Sclerocarya birrea*) from certain landscapes (Jacobs and Biggs 2002). On the African continent, where so little protected habitat remains, there is often a need to conserve animal and plant communities that are more localized and thus more vulnerable to extinction than elephants (Hoeft and Hoeft 1995). This justification explicitly recognizes that restoring nature to its natural state or states on the original scale is now impossible and that reestablishing a species once lost to a system may also be impossible, particularly while elephants in high densities remain part of that system.

This gives rise to the fundamental elephant management dilemma: Because it is ultimately not possible to prevent losses of biodiversity in reserves where elephant densities are not controlled, should the biodiversity losses be accepted and the reserve be managed as an elephant sanctuary? Or should elephant numbers be limited and the reserve be managed for the conservation of biodiversity?

The choice is a policy decision that *must* be made if reserve management is to be rational. A manager cannot simply ignore increasing elephant numbers and hope that other species will not be lost in the long term. The problem will not resolve itself in favor of all species. In some cases, it may be completely justifiable to decide that elephants take priority over biodiversity. This was the case with the Addo Elephant National Park in South Africa, proclaimed to protect the last few elephants that occurred there (see Moolman and Cowling 1994; Johnson 1998; Johnson et al. 1999). But this decision must be taken with the full awareness that as elephant densities increase, some other species will inevitably be lost, habitats will be transformed, and ultimately starvation will limit further elephant population growth.

Either way the decision is taken, it will trouble the consciences of those involved. If the choice is for an elephant sanctuary, there may be no going back once extensive habitat damage has occurred, and responsibility for losses of species will have to be accepted. When plant or animal species have been lost, it may be possible (through the management of the elephant population) to maintain the remaining biodiversity, but the restoration of the system to its former richer state of biodiversity (Hoeft and Hoeft 1995) and function may not be possible. However, if biodiversity conservation is the management strategy decided on, the responsibility of limiting elephant numbers will impose an equally heavy burden. In large conservation areas, this will usually mean the culling of excess individuals. The unavoidable ethical dilemma will be in weighing up the sacrifice of individual elephants against the sacrifice of other species. Which is ethically more acceptable?

Methods of Elephant Population Management

Assuming that a reserve *is* to be managed for biodiversity objectives, what elephant management options are available to managers? The limitation of any species' population growth, be they mosquitoes or elephants, can only be effected by managers in three fundamental ways. These are

- By moving excess individuals elsewhere (translocations)
- By preventing recruitment (contraception and/or sterilization)
- By killing excess individuals (culling and/or hunting).

The first two of these options are nonlethal and therefore, from an ethical point of view, may seem more acceptable than killing. In fact all of these options come with their own sets of both practical and ethical considerations. Elephant managers who have already made the difficult choice of seeking to limit elephant population in favor of overall reserve biodiversity immediately face the next dilemma-ridden task of assessing each option's advantages, disadvantages, and related ethical considerations to determine the one best suited to their own circumstances.

Translocation

The techniques for translocating adult elephant bulls and intact family units were only pioneered in 1994. Specialized equipment was needed and had to be designed and developed before such operations could be implemented. But with access to such equipment and expertise, the translocation of elephants can be routinely conducted in a reasonably humane way (Dublin and Niskanen 2003).

Advantages, Disadvantages, and Ethical Considerations of Translocation

Translocation may appear to be the best management choice for population control. It is non-lethal, and generally, families can be translocated as intact units. It has the additional advantage of allowing the establishment of other elephant populations elsewhere while at the same time preserving the genetic material from that specific gene pool. In culls the genetic material is lost. Furthermore, if there is a legal and controlled market for free ranging elephants, translocations can be financially rewarding to the reserve.

Translocations entail a few problems and disadvantages. There is a risk that the group of elephants selected for translocation may not be an intact unit or may include members from other families. To circumvent this problem, the African Elephant Specialist Group (Dublin and Niskanen 2003) advocates that a study be conducted before the translocation to determine the relationships between the individual elephants. This is to ensure that intact families are captured for translocation. This is known as a pretranslocation study, or premonitoring. In some environments where the terrain and vegetation allow it, and where the scale of the operation is not too large, it has been possible to conduct such pretranslocation exercises (Omondi et al. 2002). But for larger-scale operations, this may prove unfeasible.

In southern Africa today, saturation point has almost been reached. Most reserves wishing to acquire elephants already have them. Reserve managers understand the consequences of too many elephants and do not wish to increase their populations any further. Even the 10,813 km² Limpopo National Park (LNP), recently established in Mozambique on the eastern boundary of the KNP, is unwilling to accept more than the 125 translocated elephants it has received so far. Additionally, there are still many people living in LNP who have not been adequately consulted and the management authorities are hesitant to proceed with more translocations.

The KNP elephant population is censused annually using a method that attempts to count every individual (see Whyte [2001a] and Joubert [1983] for methods). This is the most intensive census of an elephant population anywhere in Africa and has been conducted since 1967. In 2003, the population estimate was 11,700 (Whyte 2004). The estimated rate of increase of the population is 6.8% per year (Whyte 2001a) and the stabilization of the population would thus require the annual removal of around 6.8% (±800 animals). Thus local supply exceeds demand by a large margin, and translocations will not answer the management needs of the KNP population, nor any other large elephant population in Africa.

A further consideration is that while translocations may provide some relief to the donor population, the elephant problem is being transferred elsewhere, and it will only be a matter of time before similar biodiversity dilemmas rear their heads in the recipient area. Once all potential recipients have sufficient elephants, translocation will cease to be a management option.

So what are the ethical questions? Is it ethical to uproot animals that show very distinct fidelities to a home range and place them in an unknown environment? Initially, this must be highly stressful, particularly to the matriarchs who have responsibility for leading the group. The whereabouts of food and water resources are unknown, and the boundaries and potential dangers are also unknown.

A second consideration: is it ethical to proceed with the capture and translocation of a group of elephants when you cannot be sure of whether the captured group is an intact family? If terrain and tractability of the elephants do not allow premonitoring, there will always be uncertainty about the structure of the group, and some family members may be left behind. How does this affect the family? And how does it affect those left behind? A translocation operation is very similar to a cull in terms of its effect on the donor population. Both operations use helicopters to round up the elephants, and heavy recovery vehicles on the ground. To the elephants that remain behind, family members have simply disappeared. How does this affect them?

The "welfare costs" of either of these options are as yet unclear. Trauma may be experienced by translocated elephants, but they do settle, and the traumas will diminish in time. Fragmented families will at least still be with known (probably closely related) families from their bond group and so such traumas may be minimal. Although we still do not know the welfare costs of translocation, at least the animals have not had to be culled.

Contraception

Contraception is the second nonlethal option and has been the focus of considerable research. Dobson (1993, 1994) and Poole (1992, 1993) gave earlier consideration to effects of contraception on the dynamics of elephant populations. South African National Parks has encouraged further research into contraception with the hope that it may one day provide another ethical and practical option. Two methods of contraception have been investigated thus far—hormonal control using subcutaneous estradiol-17-β implants, and the immunocontraception technique, using porcine zona pellucida (pZP) antigen vaccinations.

Current contraception techniques are expensive and logistically difficult to apply in large free-ranging populations. Computer modeling has shown that to stabilize an elephant population approximately 75% of all breeding females must be under treatment at any one time (Whyte et al. 1998; van Aarde et al. 1999). In KNP, with a current population of more than 12,500 elephants, this would involve about 4,500 females, all fitted with radio collars to allow locating them from a helicopter when follow-up treatments were due. Collars are expensive, and their batteries run down. The additional trauma potentially imposed by recapture to replace batteries raises another ethical consideration. The pZP vaccine, a random "blanket" treatment that keeps 90% of a wild population's breeding females inoculated at all times, potentially could obviate this particular ethical concern.

Two Contraception Methods:
Advantages, Disadvantages, and Ethical Considerations

Any contraception technique presents an inherent ethical dilemma, in that family size will be reduced. Family size and structure constitute an integral part of elephants' social lives. Limiting the number of offspring a female can have will mean that fewer daughters will be born to her, which will change the structure of such families from a broad-based pyramid with many young adults and juveniles into a much narrower one with fewer siblings and role models. Large families offer learning and security advantages that are reduced in small families, and in smaller families with fewer aunts and older sisters, it can be expected that calf survival may be compromised (Moss 1988). It is not known whether affected families in bond groups might band together as surrogate family members under such circumstances. If so, this might alleviate some of the disadvantages of smaller families.

Another consideration is that in a long-lived species such as the elephant, contraception will not reduce the population in the short term. Preventing conceptions in all females will only stabilize the population, and this will only occur once all the pregnant females have calved. Gestation time in elephants is twenty-two months, and so for a period of about two years after all females have been treated, the population will still increase. A decline in the population will only occur in response to natural mortalities. This will likely be a very slow process.

The estradiol-17-β project was initiated in KNP in 1996 but was quickly terminated on humane grounds because of side effects (Whyte and Grobler 1998). The rationale was to use estradiol-17-β to prevent ovulations through slow, sustained-release subcutaneous implants. The sole advantage of this method was that it was shown to be effective in preventing conceptions. The disadvantages and difficult ethical considerations regarding using estradiol-17-β implants are many.

Females under this treatment are induced into a sustained state of "false estrus." The high levels of estrogen are metabolized and secreted in the urine in significant quantities of pheromones detectable by bulls. The pheromonic signal to the bulls is that these cows are in estrus, when in fact they are not. In KNP, this resulted in bulls harassing cows, causing separation from their families and even from their small calves (Whyte and Grobler 1998). Three of the ten calves whose mothers had been treated died during the research period. Over the same time interval, none of the twenty calves involved in the pZP project died.

Two and three years after termination of the estradiol-17-β project, ultrasound examination of the elephants' ovaries showed them still to be completely inactive, which suggested that they had been permanently affected (Fayrer-Hosken et al. 2001). The "permanent" inactivity of ovaries for older females with established families might be acceptable, but for younger females, sterility is a dubious option. This would deprive them of the natural processes of gestation, parturition, and raising a baby, and it would deprive them of the advantages of large families described previously. Another problem with permanent sterility of a large proportion of a population is the impaired potential to recover their numbers should a catastrophe, such as epidemic disease or large-scale poaching, reduce the population.

In addition, as with many hormonal contraceptives, there are significant medical side effects, especially if the administration is sustained. Known side effects of estradiol-17-β include cystic ovaries; edema (swelling) of the vulva with prolapsed vagina or rectum; aplastic anemia (bone marrow failure); open cervix leading to a pyometra (a severe uterine infection); cystic hyperplasia (cellular proliferation with fluid-filled vesicles) of the endometrium and cervix; hypertension; myocardial infarction (heart attack); endometrial or mammary carcinoadenomas (cancers); and hepatic adenoma (benign liver tumor) (Fayrer-Hosken et al. 2001).

With estradiol-17-β, implants must be replaced every six months and they cannot be delivered remotely, so each replacement requires full immobilization of the animal and a surgical procedure. Helicopters and veterinarians must be used and therefore costs for treating large populations of elephants would be prohibitive. Another issue is that the meat of these treated elephants would contain high levels of estradiol that might adversely affect any predatory or scavenging animals consuming the meat (Whyte and Grobler 1998). Finally, family size, structures, and group dynamics will be altered, with all of the unknown consequences that these changes may have for elephant social groups. With estradiol-17-β, the ethical and health considerations far outweigh any contraceptive advantage and so this method is not considered a humane option in wild elephants.

Immunocontraception through pZP vaccination of adult elephant females has been shown to contracept them successfully (Fayrer-Hosken et al. 2000; Fayrer-Hosken et al. 2001). The pZP antigens are purified from pig oocytes. In the vaccine, the pZP glycoproteins (pZP1, pZP3α, and pZP3α) are combined with an adjuvant, or safe immunostimulant, to produce quality humoral response. The vaccine stimulates the animal's immune system to produce antibodies that bind to the elephant cow's maturing follicular oocytes, which prevents the binding of sperm cells and fertilization of the oocyte.

Counted among the advantages of using pZP is that this method has no known somatic or behavioral consequences (Whyte and Grobler 1998; Barber and Fayrer-Hosken 2000; Fayrer-Hosken et al. 2000). The vaccine has been shown to be 80% effective when free-roaming elephants are darted remotely. Remote delivery of the vaccine does not require surgical procedures and can be accomplished by trained technical staff. This obviates the need for full-time deployment of veterinarians.

After initial inoculation, only a single annual booster is needed to maintain the contraceptive effect of pZP. Furthermore, the vaccine is 100% reversible once the treatment has been terminated. The vaccine is also safe when administered to pregnant animals and has no known effects on the neonate (Fayrer-Hosken et al. 2000; Delsink et al. 2002; Turner et al. 2002). Also, as the vaccine has no hormonal component, there will be none of the side effects associated with hormonal contraceptives.

An immunocontraception program in wild mares showed that their ovarian activity was reduced after five years of annual treatments (Kirkpatrick et al. 1997); they cycled less regularly and had a reduced luteal phase. Both the survival rates and general body condition of long-term immunocontracepted mares improved markedly (Kirkpatrick and Turner 2002), due to the absence of lactational and gestational stress. The reduced ovarian activity was reversible; three years after cessation of pZP treatments, the mares started cycling again.

There are some disadvantages to using pZP, as well as important ethical considerations. Currently, this technology requires at least two or three initial inoculations at three-week intervals to elevate antibody levels to the point that they will provide the required contraceptive effect. This lasts for twelve months, and annual boosters are necessary thereafter. This makes field delivery very labor- and cost-intensive and severely limits its use in large populations. The practicalities of large-scale vaccine application, when booster shots are also required, have not yet been sufficiently worked out. To do so would require population modeling to calculate the proportionate numbers of females that would need to be vaccinated at each treatment. However, a single-administration, multiple-release pZP vaccine (Turner et al. 2002) has recently been developed and shown to be effective in horses. This provides the initial three vaccinations in a single administration. This one-shot vaccine has been shown to deliver adequate antibody titers in elephants (H. Bertschinger, personal communication) but has as yet not been field tested. If effective in elephants, this would greatly facilitate the practicalities of field delivery. The lower cost and logistical benefits might render the technique suitable for use in larger populations.

As with the estradiol technique, family size, structures, and group dynamics will be altered with the use of pZP, with all of the yet unknown consequences that these changes may have for elephants, including social problems, reduced calf survival rates, and so on. As treated females do not conceive, the mating frequency for these females will increase. Under normal circumstances, females will come into estrus and mate only once in every four years, if conception occurs with that mating. This frequency increases to around once every fifteen weeks when conceptions do not occur. When contraception is in effect, how will frequent estrus and mating affect a female's behavior and her family's circumstances? And how will it affect males' behavior? It is postulated that this may also have some health effects as, in captive elephants, multiple estrous cycles without breaks due to pregnancy have resulted in increased incidences of leiomyomas (uterine fibroids) and cysts (Montali et al. 1997; Montali et al. 1998). Although the increased risks of leiomyomas and cysts are cause for concern, there is as yet no evidence to suggest that these may cause discomfort or contribute to an increased risk of mortality. A properly implemented contraception program that reduces conception frequency but allows a female an occasional calf may considerably reduce the probability and risks of uterine pathologies.

The choice of which females to vaccinate would also need careful consideration. Young or maiden females should not be treated. They should be allowed to remain fertile while associating with older, more experienced females who play a key role in the natural learning processes associated with gestation, parturition, and acquisition of maternal skills.

One problem with current pZP contraception technology is that it requires that animals under treatment must be located for every booster shot. This may preclude the technology's use in large populations. Still it has been successfully applied in a smaller population (Delsink et al. 2002). In a large reserve, it might yet be possible to regionalize an elephant management program, using immunocontraception only in certain manageable parts of the range. As only older females would be immunocontracepted, aerial selection by size might be a realistic possibility. Also, the immunocontraceptive has no effects on pregnant females, so all older females could be integrated into the program, obviating the necessity for radio-collaring.

Once primed, treated elephants would only need an annual booster. The recently developed, single-administration, multiple-release pZP vaccine (Turner et al. 2002), would further facilitate this approach.

Culling

Culling is an extremely emotive subject, particularly when it is applied to elephants. Although to some, culling elephants may be anathema, it is usually seen by reserve managers concerned with the maintenance of biodiversity as a necessary evil, to be used only in situations where the other two nonlethal options (translocation and contraception) cannot be implemented. Two basic methods of culling have been used, using rifles to shoot elephants with bullets or using the drug scoline (succinylcholine).

Advantages, Disadvantages, and Ethical Considerations of Culling

In countries such as Uganda (Laws and Parker 1968) and Rhodesia—now Zimbabwe—(Dunham 1988; Thomson 2003), elephants were shot from the ground by highly trained rangers using semiautomatic rifles. Up to fifty elephants were often culled within a few minutes using this technique. In South Africa, for personnel safety, the culling of elephants in KNP was always conducted from a helicopter, and, for this same reason, the elephants were initially culled using scoline. This compound is a neuromuscular blocking agent that paralyzes the animal, rendering it harmless once it is recumbent. An additional advantage of using scoline is that it metabolizes into compounds that are present in all mammals, and the meat can therefore be used for human and animal consumption.

However, research in KNP showed that using scoline to kill elephants is inhumane (Hattingh et al. 1990). In animals such as African buffalo (*Syncerus caffer*), scoline acts quickly, as all of the body's muscles are affected simultaneously and death is rapid (Hattingh et al. 1990). In elephants, however, the locomotory muscles are immobilized initially, rendering the animal recumbent and then the respiration is arrested. The heart muscle continues to function and the animal, still conscious, eventually dies of asphyxiation. The use of scoline was therefore terminated. The use of analgesic drugs such as etorphine would not be deemed inhumane by these standards, but such drugs would render the meat unusable either for human or animal consumption. Subsequently, all elephants culled have been shot from helicopters. Helicopters are used because, despite all efforts to the contrary, an animal that is shot by rifle may be recumbent, but only be stunned and not dead. Such an animal presents a danger to people working on the ground, as it may suddenly get to its feet in a wounded condition. Protecting personnel is one ethical consideration; achieving the most humane cull possible is of course another paramount ethical consideration. To preclude merely stunning or wounding, all such culling is done using rifles of an adequate caliber. In all cases, the shots are aimed to penetrate the brain immediately at the base of the skull.

The advantages of culling include the fact that large populations can be reduced in size effectively (except where a population may be so large that even this option is not logistically possible). Also, the processed by-products of the cull, such as meat, hides, and ivory, can be sold, making the method financially viable, providing conservation activities (such as antipoaching work) with much-needed funding. However, anthropocentric use of the by-products brings the ethics of culling into question. If culling is practiced purely for ecological reasons, it can perhaps be ethically justifiable, particularly if contraception and translocations are not feasible. But culling purely for economic, sustainable harvest motives may be ethically unacceptable.

Among the disadvantages of culling is a certain amount of disturbance experienced by related elephant groups that may be nearby at the time of the cull. The degree of disturbance will depend on how close they are and will be worse when a larger group needs to be split. However, this disturbance is almost identical in nature to a translocation operation—the only difference being the sound of gunshots. In addition, there is always a concern (as with a translocation operation) that not all members of a family will be selected for culling, which would leave some members behind without their families.

Two major considerations with culling are, who will do the work and how will the carcasses be processed? While KNP has the necessary staff, equipment, and processing facility, very few other institutions do. It is one thing to shoot elephants, but to process them thereafter is another matter. They cannot be sent to local abattoirs that would not have the facilities to handle elephants. Public and animal health requirements also complicate the processing if the meat is to be used for human consumption.

Another disadvantage of a cull is the negative media publicity and public reaction that always accompanies such an operation. Although these pressures usually come from a well-meaning press and public, they can be severe and will contribute to the unpleasantness of the task. A manager will not only have the killing of the elephants on his or her conscience, but may also be faced with a negative and hostile media and public. Involved staff members need to be prepared for these eventualities.

As with translocations, the ethical considerations related to culling include the effects on the bond-group / family members left behind, who experience both the short-term disturbances of the operation and the longer-term effects of losing their culled relatives. In culls from smaller populations from which only few animals must be removed, premonitoring may ameliorate this problem, but where large-scale culls must be conducted, it will probably prove logistically unfeasible to premonitor for all groups. The possible consequence of not thoroughly premonitoring, namely, the failure to cull some family members, remains one of the significant ethical concerns about culling.

More generally, the major ethical question surrounding culling can be simply put. Is it ethical to kill elephants? Although many believe it is unethical, it must be clearly understood that killing individual elephants must be weighed against the losses of other species that will occur if elephant numbers are not limited. Is it ethical to allow species to be lost from a system when preventing this loss is possible?

The Nonmanagement Option

Just as culling elephants is a management decision, opting for nonmanagement must also be considered a management decision. It should be a conscious decision based on the priorities set for the particular reserve and must be taken with the full recognition of the consequences of that decision. Just as managers are held accountable for a decision to cull elephants, so too should they be held accountable for a decision that leads to losses of other species.

Here we offer a simplified, and generalized, description of the consequences of the nonmanagement option. When elephant densities are low, their effect on the environment will also be low, and as their numbers increase, their impact increases. Initially, impacts will consist of declines in favored food plants, but eventually these species may be entirely lost. Switches to other food plants will result in their loss as well, and eventually woodlands will be transformed into grasslands. This then results in the loss of vertebrate and invertebrate species that depend on woodland (Western and Gichohi 1989). Ultimately, elephant numbers will also be affected through starvation and possibly disease when food resources are depleted, as in a drought. The changes brought about by this process will scarcely be reversible. A drastic reduction in elephant numbers may allow the reestablishment of some species, but it is unlikely that such a system could be restored to its former state of species richness and structural diversity. As the sea of humanity widens around these islands of conservation, the possibility of recolonization by most species (other than birds and flying insects) will become increasingly remote.

Advantages, Disadvantages, and Ethical Considerations of the Nonmanagement Option

The main advantage of the "do-nothing" option is that elephants do not have to be managed; in large reserves, as we have seen, such management would usually mean culling. The disadvantages of this option are that ultimately there will be habitat degradation and losses of some other species, as well as elephants suffering from starvation. Two basic ethical dilemmas lie at the core of the elephant management problem. (1) Is it ethical to allow losses of other species from a local environment when it would have been possible to prevent them? (2) Is it ethical to allow elephants to modify a local environment so that their populations, and more particularly those of other species, will probably be threatened by starvation and possible extirpation? These are anthropogenic problems—there are not too many elephants, but too many people—and people must seek ways of addressing them. Perhaps what is still lacking in elephant management is a "higher-order ethic," similar to Aldo Leopold's "land ethic" (Leopold 1949) in which a true respect for ecosystems affirms the right to continued existence of all biotic and abiotic elements in a state as close as is possible to the natural. But for now, consensus on any of the ethical dilemmas posed in this chapter will scarcely be possible.

Countering elephant population growth cannot occur without ethical dilemmas. The dilemmas are complex. There are no easy choices, and there is no universal, ethically "correct" solution to the problem. For every reserve, the circumstances will be different. For each reserve, all management options and their probable consequences must be evaluated to determine what is sufficiently practical and ethically most acceptable. Do we want to limit elephant numbers? If not, can we accept the consequences for the habitat, the other species occupying it, and ultimately for the elephants themselves? If we wish to maintain management focus on biodiversity, which method shall we use to limit elephant numbers? Is translocation an option? Is there space and a demand elsewhere for the elephants? If not, is contraception an option? Do we have the means and the funding to provide immunocontraception? Who will do the work of immunocontraception and its follow-up? If neither of these options is feasible, should we cull? Who will do the work of culling, and who will process the carcasses?

In smaller populations, nonlethal options (translocations and immunocontraception) would probably be viewed as those most ethically acceptable and may offer a solution for some reserves. This would depend on the numbers of elephants involved and the resources available. But where large numbers of elephants need to be removed, culling is likely to be the only option. Contraception alone will not reduce the population and because of limitations on space and demand for elephants, neither will translocations. For managers, the decision to cull or contracept will be controversial and deeply troubling and must be weighed against the consequences of allowing the elephant population to grow unchecked. To "do nothing" may also be a valid management option, but its enactment also must be carefully considered, as losses of biodiversity, such as the extirpation of bushbuck (*Tragelaphus scriptus*) and lesser kudu (*T. imberbis*) from Kenya's Amboseli National Park (see Western and Gichohi 1989) is a possible outcome. This also will be a deeply troubling decision. Ultimately, killing individual elephants must be weighed against the degradation of biodiversity and, eventually, the loss of species. These are each topics for considerable debate.

Although wildlife managers must be held accountable for decisions to cull elephants, so too must they be accountable for the consequences of a decision not to manage elephants. Equally, pressure groups whose successful opposition to rational elephant management actions results in biodiversity declines should be held co-accountable to the public for any inactions and their consequences.

Individual attitudes to these dilemmas will be shaped, at least in part, by whether the person is involved in the management decision-making process. It is fairly easy to have an opinion on the relative ethics on each of the issues if you are not directly involved. But when the long-term conservation outcome depends on your decisions, your view is likely to be somewhat modified. Those who make decisions about elephant management are held fully accountable by society at large, and even the most stubborn alternative ethical viewpoints probably will have to cede to managers' understanding of the real-world constraints.

Finally, all national parks in the African savanna, provided the elephant population is effectively protected, will face these ethical dilemmas; if not now, then at some time in the future. Braving these dilemmas constitutes an unavoidable component of the management of these wonderful animals.

References
Altmann, J., Alberts, S. C., Altmann, S. A., and Roy, S. B. 2002. Dramatic change in local climate patterns in the Amboseli basin, Kenya. *African Journal of Ecology* 40: 248–251.
Barber, M. R., and Fayrer-Hosken, R. A. 2000. Evaluation of somatic and reproductive immunotoxic effects of the porcine zona pellucida vaccination. *Journal of Experimental Zoology* 286: 641–646.
Blanc, J. J., Thouless, C. R., Hart, J. A., Dublin, H. T., Douglas-Hamilton, I., Craig, C. G., and Barnes, R. F. W. 2003. *African Elephant Status Report 2002: An update from the African Elephant Database*. IUCN/SSC African Elephant Specialist Group. Gland, Switzerland; Cambridge, UK: IUCN.
Bourlière, F., and Hadley, M. 1983. Present-day savannas: an overview. In F. Bourlière (ed.), *Ecosystems of the world, Volume 13: Tropical Savannas* (pp. 1–17). Amsterdam: Elsevier Scientific Publishing Company.
Calef, G. W. 1988. Maximum rate of increase in the African elephant. *African Journal of Ecology* 26: 323–327.
Cumming, D. H. M., Fenton, M. B., Rautenbach, I. L., Taylor, R. D., Cumming, G. S., Cumming, M. S., Dunlop, J. M., et al. 1997. Elephants, woodlands and biodiversity in southern Africa. *South African Journal of Science* 93: 231–236.

Delsink, A. K., van Altena, J. J., Kirkpatrick, J., Grobler, D., and Fayrer-Hosken, R. A. 2002. Field applications of immunocontraception in African elephants (*Loxodonta africana*). In J. F. Kirkpatrick, B. L. Lasley, W. R. Allen, and C. Doberska (eds.), *Fertility control in Wildlife. Reproduction Supplement* 60: 117–124. Nottingham, UK: Nottingham University Press.

Dobson, A. P. 1993. Effect of fertility control on elephant population dynamics. *Journal of Reproduction and Fertility* 90: 293–298.

Dobson, A. P. 1994. Effect of fertility control on elephant population dynamics. In C. S. Bambra (ed.), *Proceedings of the 2nd international NCRR (National Centre for Research in Reproduction) conference on advances in reproductive research in man and animals, held in Nairobi, Kenya, 3–9 May 1992* (pp. 293–298). Nairobi: Institute of Primate Research and National Museums of Kenya.

Dublin, H. T. 1995. Vegetation dynamics in the Serengeti-Mara ecosystem: The role of elephants, fire and other factors. In A. R. E. Sinclair and P. Arcese (eds.), *Serengeti II: Dynamics, management and conservation of an ecosystem.* Chicago: University of Chicago Press.

Dublin, H. T., and Niskanen, L. S. (eds.). 2003. The African Elephant Specialist Group in collaboration with the Re-introduction and Veterinary Specialist Groups 2003. *IUCN/SSC AfESG Guidelines for the* in situ *translocation of the African elephant for conservation purposes.* Gland, Switzerland; Cambridge, UK: IUCN.

Dublin, H. T., Sinclair, A. R. E., and McGlade, J. 1990. Elephants and fire as causes of multiple stable states in the Serengeti-Mara woodlands. *Journal of Animal Ecology* 59: 1147–1164.

Dunham, K. M. 1988. Demographic changes in the Zambezi Valley elephants (*Loxodonta africana*). *Communications from the Mammal Society* 56: 382–388.

Fayrer-Hosken, R. A., Grobler, D., van Altena, J. J., Bertschinger, H. J., and Kirkpatrick, J. F. 2000. Immuno-contraception of African elephants. *Nature* 407: 149.

Fayrer-Hosken, R. A., Grobler, D., van Altena, J. J., Kirkpatrick, J. F, Bertschinger, H. J., and Hofmeyr, M. 2001. Conservation of the African elephant using contraceptives. In *Proceedings of the Society for Theriogenology 2001* (pp. 37–42). Nashville, TN: Society for Theriogenology.

Gillson, L., and Lindsay, K. 2003. Ivory and ecology—changing perspectives on elephant management and the international trade in ivory. *Environmental Science & Policy* 6: 411–419.

Hall-Martin, A. J. 1992. Distribution and status of the African elephant (*Loxodonta africana*) in South Africa, 1652–1992. *Koedoe* 35: 65–88.

Hattingh, J., Jitts, N. I., Ganhao, M. F., and de Vos, V. 1990. The responses of elephant and buffalo to succinylmonocholine. *South African Journal of Science* 86: 546.

Herremans, M. 1995. Effects of woodland modification by African elephant *Loxodonta africana* on bird diversity in northern Botswana. *Ecography* 18: 440–454.

Hoeft, R., and Hoeft, M. 1995. The differential effects of elephants on rain forest communities in the Shimba Hills, Kenya. *Biological Conservation* 73: 67–79.

Jacobs, O. S., and Biggs, R. 2002. The impact of the African elephant on marula trees in Kruger National Park. *South African Journal of Wildlife Research* 32: 13–22.

Johnson, C. F. 1998. Vulnerability, irreplaceability and reserve selection of the elephant-impacted flora of the Addo Elephant National Park, Eastern Cape, South Africa. MSc thesis, Rhodes University.

Johnson, C. F., Cowling, R. M., and Phillipson, P. B. 1999. The flora of the Addo Elephant National Park, South Africa: Are threatened species vulnerable to elephant damage? *Biodiversity and Conservation* 8: 1447–1456.

Joubert, S. C. J. 1983. A monitoring program for an extensive national park. In R. N. Owen-Smith (ed.), *Management of large mammals in African conservation areas* (pp. 201–212). Pretoria: Sigma Press.

Kirkpatrick, J. F., and Turner, A. 2002. Reversibility of action and safety during pregnancy of immunization against porcine zona pellucida in wild mares (*Equus caballus*). In

J. F. Kirkpatrick, B. L. Lasley, W. R. Allen, and C. Doberska (eds.), *Fertility control in Wildlife. Reproduction Supplement* 60: 197–202. Nottingham, UK: Nottingham.

Kirkpatrick, J. F., Turner, J. W., Liu, I. K. M., Fayrer-Hosken, R. A., and Rutberg, A. T. 1997. Case studies in wildlife immuno-contraception: Wild and feral equids and white-tailed deer. *Reproduction, Fertility and Development* 9: 105–110.University Press.

Laws, R. M., and Parker, I. S. C. 1968. Recent studies on elephant populations in East Africa. *Symposia of the Zoological Society of London* 21: 319–359.

Laws, R. M., Parker, I. S. C., and Johnstone, R. C. B. 1975. *Elephants and their habitats.* Oxford: Clarendon Press.

Leopold, A. 1949. *A Sand County almanac and sketches here and there.* New York: Oxford University Press.

Leuthold, W. 1977. Changes in tree populations of Tsavo East National Park, Kenya. *East African Wildlife Journal* 15: 61–69.

Lombard, A. T., Johnson, C. F., Cowling, R. M., and Pressey, R. L. 2001. Protecting plants from elephants: Botanical reserve scenarios within the Addo Elephant National Park, South Africa. *Biological Conservation* 102: 191–203.

McComb, K., Moss, C. J., Durant, S. M. Baker, L., and Sayialel, S. 2001. Matriarchs as repositories of social knowledge in African elephants. *Science* 292: 491–494.

Montali, R. J., Hildebrandt, T. B., Göritz, F., Hermes, R., Ippen, R., and Ramsay, E. 1997. Ultrasonography and pathology of genital tract leiomyomas in captive Asian elephants: Implications for reproductive soundness. *Verhandlungsbericht des Internationalen Symposiums über die Erkrankungen der Zoo-und Wildtiere* 253–258.

Montali, R. J., Hildebrandt, T. B., Göritz, F., Hermes, R., Porter, K., and Tsibris, J. 1998. High prevalence of uterine leiomyomas in captive Asian elephants and their implications for reproductive soundness. *Proceedings of the Third International Elephant Research Symposium*, Springfield, MO.

Moolman, H. J., and Cowling, R. M. 1994. The impact of elephant and goat grazing on the endemic flora of South African succulent thicket. *Biological Conservation* 68: 53–61.

Moss, C. J. 1988. *Elephant memories: Thirteen years in the life of an elephant family.* London: Elm Tree Books.

Mosugelo, D. K., Moe, S. R., Ringrose, S., and Nellemen, C. 2002. Vegetation changes during a 36–year period in northern Chobe National Park, Botswana. *African Journal of Ecology* 40: 232–240.

Omondi, P., Wambwa, E., Gakuya, F., Bitok, E., Ndeere, D., Manyibe, T., Ogola, P., and Kanyingi, J. 2002. Recent translocation of elephant family units from Sweetwaters Rhino Sanctuary to Meru National Park, Kenya. *Pachyderm* 32: 39–48.

Page, B. R. 1999. Detecting extirpation and changes in abundance in woody species: A case study from the N. E. Tuli Block, Botswana. Poster Presentation at the Workshop on Long Term Ecological Monitoring in southern Africa, Skukuza, South Africa, August 17–18, 1999.

Poole, J. H. 1992. Logistical and ethical considerations in the management of elephant populations through fertility control. Paper presented at the NCRR Elephant Reproduction Symposium, May 1992, Nairobi, Kenya.

Poole, J. H. 1993. Kenya's initiatives in elephant fertility regulation and population control techniques. *Pachyderm* 16: 62–65.

Poole, J. H., Kahumbu, P., and Whyte, I. J. Forthcoming. *Loxodonta africana.* In J. Kingdon, D. Happold, and T. Butynski (eds.), *The Mammals of Africa, Vol. 1.* London: Academic Press.

Thomson, R. 2003. *A game warden's report.* Hartebeestpoort, South Africa: Magron Publishers.

Turner J. W., Jr., Liu, I. K., Flanagan, D. R., Bynum, K. S., and Rutberg, A. T. 2002. Porcine zona pellucida (pZP) immuno-contraception of wild horses (Equus caballus) in Nevada: A 10–year study. In J. F. Kirkpatrick, B. L. Lasley, W. R. Allen, and C. Doberska (eds.), *Fertility control in Wildlife. Reproduction Supplement* 60: 177–186. Nottingham, UK: Nottingham University Press.

van Aarde, R., Whyte, I. J., and Pimm, S. 1999. Culling and the dynamics of the Kruger National Park elephant population. *Animal Conservation* 2: 287–294.

Western, D., and Gichohi, H. 1989. Segregation effects and the impoverishment of savanna parks: the case for ecosystem viability analysis. *African Journal of Ecology* 31: 269–281.

Whyte, I. J. 2001a. *Conservation management of the Kruger National Park elephant population.* PhD diss., University of Pretoria.

Whyte, I. J. 2001b. Headaches and heartaches—the elephant management dilemma. In D. Schmidtz and E. Willot (eds.), *Environmental Ethics: What really matters, what really works* (pp. 293–305). New York: Oxford University Press.

Whyte, I. J. 2004. Census results for elephant and buffalo in the Kruger National Park in 2003. Scientific Report 02/2004. Pretoria: South African National Parks.

Whyte, I. J., Biggs, H. C., Gaylard, A., and Braack, L. E. O. 1999. A new policy for the management of the Kruger National Park's elephant population. *Koedoe* 42: 111–132.

Whyte, I. J., and Grobler, D.G. 1998. Elephant contraception research in the Kruger National Park. *Pachyderm* 25: 45–52.

Whyte, I. J., van Aarde, R., and Pimm, S. L. 1998. Managing the elephants of Kruger National Park. *Animal Conservation* 1: 77–83.

Whyte, I. J., van Aarde, R. J., and Pimm, S. 2003. Kruger National Park's elephant population: its size and consequences for ecosystem heterogeneity. In J. T. du Toit, H. C. Biggs, and K. H. Rogers (eds.), *The Kruger experience: Ecology and management of savanna heterogeneity* (pp. 332–348). Washington, DC: Island Press.

Woodd, A. M. 1999. A demographic model to predict future growth of the Addo elephant population. *Koedoe* 42: 97–100.

21 TOWARD AN ETHIC OF INTIMACY
TOURING AND TROPHY HUNTING FOR ELEPHANTS IN AFRICA
REBECCA HARDIN

Elephants' interactions with people are particularly revealing of power and status relations between humans. These human relationships, often gendered or racialized in nature, and often reflective of postcolonial politics, can determine elephants' fates. I set forth some historical, ethnographic, and ethical ideas not only about the relationships of elephants to humans but also the roles—real and imagined—elephants play in human histories and cultures. In particular, I explore through brief historical commentary two familiar and economically significant modes of human interaction with elephants, safari hunting, and tourism in Africa. I then present in more ethnographic depth some highlights from a case study of elephant conservation-through-tourism in a protected area of southwestern Central African Republic (CAR). My review of historical relations of domination and appropriation of elephants disaggregates European attitudes and actors from the colonial era and also integrates an analysis of symbolic and embodied practices in a contemporary site. This approach enables a more nuanced analysis of conservation efforts from the colonial era, moving beyond the common but sterile categories of "western" or "African" in thinking about human-elephant relationships. Such rethinking reveals conceptual, as well as philosophical and political alternatives. It enables what I call a practical ethic of intimacy, which respects and values various sustained interactions with elephants.

Ecotourism—tourism activity integrated with wildlife conservation—is a relatively new but rapidly growing industry. Its development in relation to another widely practiced form of nature tourism, "trophy," or "safari," hunting," is underdocumented. I use trophy hunting or safari hunting in reference to leisure hunting practices for the collector of animal trophies (such as antlers, or the whole head); these practices are, for the most part, distinct from subsistence hunting, or hunting for management purposes. I recount several key historical elements of the practices of ecotouring and of trophy hunting for elephants in Africa and particularly the CAR. I argue that the salient categories for conflicting philosophies on touring, trophy hunting, and the treatment of elephants are not those of western versus non-western. Rather, within and across these categories exist regionally specific relationships between humans and wildlife within Europe, and within

Africa. In their encounters with one another under colonial and post-colonial circumstances, these traditions have given rise to distinct kinds of touring and trophy hunting industries, reflecting different philosophies or attitudes.

Today's activities of elephant hunting and elephant tourism not only draw from distinct philosophies but also connect to divergent political agendas for the scientific and technical management of contemporary wild elephant populations. Whereas trophy hunting evidences more of an ethic of domination, tourism works through an ethic of appropriation—terms about which I shall have much more to say. This chapter looks at the cultural and ethical frameworks within which elephant trophy hunting and elephant touring have developed. The two activities also share some attributes, including their history of having become global industries during the era of colonial encounters and their propensity to make an abstraction of elephants as symbols of power, wildness, or wealth. Such attitudes now circulate beyond geographical boundaries, creating new political and practical challenges to the protection of elephants from ever-expanding and intensifying capitalist processes of consumption.

Considering the deep historical relationships between people and elephants means considering how symbolic lines between animals and people have been constituted across different times and places, and how this relates both local and trans-local economic exchange and integration to social differentiation among groups of people. Such questions are far from merely academic. Differences within Europe, or within Africa, between types of hunters (subsistence, trophy, or otherwise), between hunters and animal rights activists, between scientists and managers, and between wealthy and impoverished communities, will determine the future of elephants and many other species alongside them. Taking a historical view of these different economic, cultural, and ethical stances has the advantage of allowing new philosophical categories to emerge in our thinking about elephants and about wildlife in general. This expansion of philosophical possibilities occurs even as new forms of engagement with elephants emerge and coexist with more traditional forms.

I suggest the notion of an ethic of intimacy—in general, a state of intertwined or interdependent lives. Intimacy is a concept that can include persistently uneven power relations and elements of suffering that exist alongside caring, compassion, and respect in human relationships, and even in those relationships that cross species boundaries. As a base for a system of ethical thinking, it also has normative elements: it privileges the knowledge and experience of those who have lived in close contact with elephants, be they ecologists, horticulturalists, or safari hunters.[1]

Engaging an ethic of intimacy advances the concerns and contributions of those who otherwise must struggle to wrest recognition of their authority from the increasingly market-driven global discussion of the value(s) of elephants. This is not to suggest that intimacy can exist only outside of markets. However, many dominant, marketable visions of human-elephant interaction, on television, in ecotourism brochures, or at safari club conventions, romanticize relationships between individual humans and individual elephants. There is a particular and growing fascination with watching celebrities feed, bathe, or medically treat elephants. A brief discussion of recent film and television and film productions to follow illustrates this point. Such performed moments, delightful for popular consumption, are anathema to the sorts of intimacies I describe as central to this ethical framework.

There are many humans living and working in close contact with elephants, whose lives and stories seem nearly invisible in the popular imaginary. For instance, field ecologists' conservation-oriented publications are largely confined to a specialized professional audience. Although they are not necessarily socially marginalized to the same extent as small-scale farmers living in elephant home ranges, neither group is easily heard in global marketplaces (Naess 1995). Wealthy ecotourists and trophy hunters, however, are conferred instant authority by the attentions of global markets and the conventions of contemporary reality television. An ethic of intimacy values and encourages us to search for and recognize the knowledge and experience of all these entities and recognizes their connections to one another despite distinct or even opposed practices of knowing elephants.

The notion of intimacy thus takes into account the varying contexts (historical, geographic, economic, and cultural) that have shaped connections between humans and elephants (Bird Rose 1999). Yet this notion does not preclude consideration of new forms of connection across these species, and attendant responsibilities. A full explication of the relationship between intimacy and responsibility is beyond the scope of this chapter. However, I do embrace the idea that there are more "organic" (inherited, subsistence-related, or landscape-based) and more "contractual" (professional, intellectual, political, or commercial) forms of intimacy with elephants, each giving rise to distinct but equally valuable and profound forms of responsibility. These potentially overlapping forms are crucial for the future (Jonas 1990). Finally, the idea of intimacy binds together the ethics of humans and wildlife, not by making wildlife commensurate with humans, but by recognizing the intense and persistent interdependence between the two, at both material and symbolic levels. It also allows us to escape from stale debates over intrinsic versus economic value of elephants.

In analyzing my own field research results and other relevant publications, I both draw from and diverge from philosopher Peter Singer's influential work, which presents sentience and personhood as morally relevant differences qualifying some animals to stand within a rights framework.

Singer is firm in his conviction that such advocacy of animal rights is not anathema to humanism. In fact, he argues, it springs from the abstract universalism of Enlightenment thinking (Singer 1999). He also claims this intellectual participation in a "western way of thinking that dominates our society" (1999, 156) and enables one most effectively to reach as many people as possible. Since Singer's initial work on the topic of animal welfare, related approaches have proliferated, either refining a rights-based framework, advocating less universalistic and more contextual compassion, or both.

These approaches have been concomitant with debates about the well-being of individual animals, or groups of animals, vis-à-vis a broader systemic sense of ecological or environmental ethics (Rolston 1994). As a social/cultural anthropologist, I cannot foreclose on the validity and possible concomitant power of alternative philosophical approaches, however, especially those emerging from the traditions of intimacy with elephants that are often marginalized or subjugated by dominant ways of thinking. Investigating today's dominant representations of human-elephant interaction calls first for a careful historical look at the history of human efforts to dominate elephants through hunting and display of their bodies.

Histories of Hunting for Trophies

Distinct regional hunting traditions within Europe had profound effects on hunting and wildlife management policies in former European colonies. As we shall see, ritualized trophy, or sport, hunting by individuals in northern Europe contrasts with the collective hunting of more Mediterranean regions. The former, increasingly powerful within Europe from the 1500s through the early 1900s, created clear parallels between patrician social mores and increasing class stratification at home in Europe, and the etiquette or economy of hunting and game management in different parts of Africa where Europeans were striving to assert their dominance—both over one another, and over Africans (Beinart 1989; Neumann 1996). These multiple rivalries are important to remember, to aid in thinking through the political impasses of today's increasingly polarized confrontations between, for instance, animal rights activists and hunters. Today, these same distinct European histories of hunting also contribute to varying traditions of animal rights and wilderness protection across European regions. This, in turn, helps shape conflicting visions for wildlife management in elephant habitat such as the forests of the Congo basin.

French anthropologist Bertrand Hell (1994) notes that the Germanic regions of northern Europe, as well as several Central European countries, share a tradition that dates to the Middle Ages or earlier, of hunting as selective harvesting. Socially speaking, this northern European management regime entails vast territories managed with minute attention to the details of trophy-bearing species, under completely private territorial control by the elite. From Alsace to Austria, Hell argues, the fundamental traits of this system are the same: limited numbers of land owners, legal provisions about minimal surface for hunting parcels (at least 200 hectares), and purchase of small private forests where "round up" hunting methods for mass prey, often involving large groups of hunters, are forbidden.

France and Mediterranean Europe, however, tend toward more collective "gathering" or round up hunts, described by residents of this region as ancestral. Such practices are based on convictions that stock replenishes itself and must be controlled to prevent a menace to agricultural production. Here, hunters reject strict management; elite and nonelite hunters have varying arrangements to share forests for hunting, and the only important distinction is between wild and domestic, or cultivated, spaces.

The Vosges Mountains, in the historically disputed territory of Alsace Lorraine, exemplify this divide: in the Alsatian watershed, descending the crest of the Vosges eastward toward Germany, the German occupiers enforced the law of 1881, letting purchasers have vast hunting grounds for individual hunting only in search of trophies. Toward France, on the western slopes, there have long been village-based hunting organizations that carry out less-restricted hunting in teams, with complex meat-sharing practices. In 1981, densities of hunters in relation to overall population on each side of the Vosges demonstrated the continued relevance of these distinctions, as they reflected broader regional realities within Europe. Less than 1% of the population were hunters on the eastern side, versus between 2% and 6% on the western side. The former corresponds to the sort of figures one finds in Germany; the latter is much closer to those percentages found in general throughout Italy, Spain, and Greece.[2]

This corresponds to cultural and political differences such as the limit between Germanic and Romance languages, or the dispersed modes of political organization among the onetime Northern Tribes of Europe versus the Mediterranean traditions of commercial and fortified city-based Republicanism. Such distinctions have persisted ever since the initial, fragile unification of Western Europe under the Carolingians, an empire led by avid huntsmen who drew from northern hunting traditions and spread them southward, without, however, completely displacing alternative regional traditions.

Such differences continue today and are constantly reconfigured. They explain some of the contemporary distinctions between animal rights and wilderness protection activities across European regions. In the Anglo-Saxon northern European regions such as Germany and the United Kingdom, Protestantism and Victorian social mores combined to create philosophical perspectives such as John Locke's sympathy toward animals, Charles Darwin's observations of our "closeness" to them, and Jeremy Bentham's claims that animals' "sensitivity" makes them deserving of some rights. France, however, better exemplifies an Aristotelian legacy of dominion over animals, transmitted via Zeno and the Stoics to Augustine, then to Catholics more broadly via Thomas Aquinas (Plender 2001).

Present-day intellectual and philosophical divergences regarding the treatment of animals can be seen at various scales, but nearly all have intellectual histories that are not merely about "the West" but rather connect to different local and regional human relationships to animals, broadly, and to wildlife, specifically. Within the United States today, such differences are at play both regionally and nationally, with important religious and scientific overtones. Some thinkers (often from the right side of the political and social spectrum; see Scully 2002) advocate merciful "dominion" over animals from the God-given distance between them and humans; others (often from the left side of the U.S. political and social spectrum—think of Jane Goodall and other primatologists) are concerned with preservation of animals who are our "closest relatives" in an evolutionary sense.

These differences, past and present, are also linked to different wildlife management paradigms throughout the developing world and offer us a way to think about the connections between political, economic, and ethical issues. Scholar and translator Martin Thom (1990) notes how rivalries between France and Germany after the Franco-Prussian War, when the race to colonize Africa mounted to its final pitch in the late 1800s, politicized the aforementioned regional differences within medieval Europe anew. Evidence from colonial archives in France and Germany supports this vision of continually reinvented cultural differences; when the ivory trade was at its peak at that century's turn, the management of elephants became a major conflict between France and Germany as they vied for control over the forests of today's Cameroon and CAR, and Germans deplored French hunting management strategies as unscientific and rapacious—in Africa, as at home (Hardin 2000). The Germans lost out entirely by the time of World War I, but their competition with France was a major factor in French efforts to assert colonial control in this resource-rich equatorial African region.

The Dzanga Sangha Reserve area is located in areas historically contested by colonizing French and German forces, and there are running tensions even now between French and German economic interests. Mostly, French loggers draw from an ethic of domination that dovetails with a vision of the forest as a multiuse forest concession and has long proven compatible with French trophy hunting operations and even culling elephants to protect crops. However, a suite of U.S.- and German-led forest conservation projects favors the appropriation and protection of the forest through ecotourism and research. They advocate ceding hunting concession and quota allocations to conservation agencies. Each side has its alliances in regional

Schematized map showing the location on the African continent of the trinational Sangha River Conservation Region, including the Dzanga Sangha Dense Forest Reserve in southwestern Central African Republic.
Map by Rebecca Hardin; based on a hand-drawn sketch by Stephanie Rupp

and national political structures, and this has often produced paralysis in the long-range planning for the area (Hardin 2001).

In this face-off, "western" management practices are not pitted against African cultural understandings of elephants. Rather, various African traditions of human and animal relations mingle with the vestiges of distinct European ones, producing heated debates today. These European and African regional traditions have been intersecting, overlapping, and shaping human interactions with wildlife since European colonization of Africa began, with several implications for the ethics of elephant hunting, then and now. I have drawn out some of the distinct cultural and historical elements of competing philosophies of elephant management today (hunting as "wise use" integrated with logging, versus conservation through touristic support for "protected areas"). Let me now turn to analysis of the colonial history that these apparently opposed views share, and to the transformation of such ancient European differences into nationalist discourses and transnational market systems suffused with images of white male dominance over the natural world.

Kenneth Cameron, historian of the safari, traces the transformation of the journeys called "*safir*" (a word in both Urdu and Arabic) by early Indian and Arab traders, and tells the story of a handful of British men who went about "wrenching the industry away from Indian control; redefining it" (1990, 46). Earlier safaris, originally coordinated by Hindu and Moslem traders for inland travel from the East African coast, sometimes even included Hindu women traders. They operated quite differently than their British colonial counterparts with regard to the organization of porters, camps, and payment. Their camp layouts often constituted circles around sleeping women or key expedition members. The safaris run by British men out of their offices in Mombasa and Nairobi emphasized the militaristic aspects of the expeditions, inherited from the Arab slave-trading caravan camps that would later inspire the starkly divided and gridlike "Native" and "European" camp styles of Henry Morton Stanley and others.

Robert Baden-Powell served in the British colonial armed forces, where he both relied on the wilderness skills of African soldiers and trackers and transmitted to them European traditions of military organization and disciplined hierarchy. He fused these forms of knowledge into a single cultural activity, founding the Boy Scouts, and thus popularizing a powerful combination of virility, wilderness, and warfare (Mackenzie 1987). Such dominant and male-dominated cultural practices all too often belie their own remarkably multicultural roots and become emblems of national culture, as the charisma and influence of Theodore Roosevelt's outdoorsman persona illustrates.

Cameron notes: "Nothing can be done easily now about the gender prejudices of safari or about its racialism; they are part of the excess baggage that we drag into the 21st century" (1990, 191). Yet, his work is a careful and conscious departure from the mainstream literature about safari hunting's "pioneers." For instance, this literature, largely published by specialized presses such as Rowland Ward and Safari Press (which focus on works by, for, and about big game hunters) praises women trophy hunters and explorers for their ability to emulate the military discipline of their husbands or fathers, while remaining "feminine and sensitive" (Dyer 1996, 81). In contrast, Cameron describes in detail one seventeenth-century traveler of Africa, Mary Hall Celia Fiennes, who departed from then dominant military-style camp practices, to "sleep in the middle of her camp, her tent or netting surrounded by African companions for protection, rather than set back from their sleeping space" (1990, 69). Unlike the celebrity speaking tours enjoyed by many return travelers from the colonies, when she returned to London, "the Royal Geographical Society had her make only one speech on her accomplishment—for an audience of children" (1990, 70).

Adult audiences in Europe craved violent tales of virile daring against wild animals. This relates to the popular vision of nature as female—profoundly seductive, prolific, and fertile but also potentially dangerous (Ortner 1974). Such complex economies of images are found repeatedly in the writing and thinking of explorers and hunters, from early European voyages (Raffles 2002), through colonial expansion (Spurr 1993), to today's conservationist discourse (Sawyer and Agrawal 1997). They even characterize the present era of packaged hunts and tours, perhaps especially with respect to Africa. The Swanepoel and Scandrol safaris' publicity material for a 1997 hunting tour in Zambia proclaims: "Africa with her mystery, her freedom, her untrammeled spaces, and her barbarism had become my mistress." Such images from today's travel brochures echo those of publicity materials from the colonial era. They remind us of how the natural world can be conflated with those social worlds that are being subjugated in a context of expanding empire, thereby reinforcing political processes of appropriation and domination of both culture and nature.

Dyer's work on "lady" hunters reminds us that those engaged in such domination need not be, and have not historically been, only men. Cameron, however, cautions us that those, often women, who attempted to create more balanced encounters across social and geographical boundaries, were themselves frequently marginalized by the dynamics of intertwined celebrity and expert knowledge formation perpetuated through social clubs and scientific societies of that era.

Cameron and Dyer concur, however, that "[U.S.] President Theodore Roosevelt . . . spent a year in Africa, hunting extensively, and was the first client to make really full use of the best of the new 'white hunters.' His safari . . . did a great deal to popularize the modern safari" (Dyer 1996, 46). Cameron notes: "Through the enormous publicity that his presence generated—countless newspaper articles, a feature film, magazines, his own books—he shared the experience with the world. Thereafter, 'safari' was an institution, very shortly a fashion" (1990, 49). The hunting safari of eastern and southern Africa became "The Hunt," a crucial part of the fabric of British colonialism, as well as a global commodity (Mackenzie 1987).

Far from mere fantasies, such understandings of nature as a feminine force to be tamed and converted to productive use informed the political economy of science and government in the colonial era. Bolstered by the nationalist charisma of Teddy Roosevelt and others, these notions helped justify the exclusion of many women, especially those seeking autonomous and original experiences overseas, from the sources of funding and support that their male counterparts enjoyed. However, perhaps the lack of resources was, in part, responsible for the very different and more "local" ex-

Photographs and text about the Central African Republic from the collection of materials on tourism and trophy hunting displayed at France's 1932 colonial fair. *Images reproduced by the Archives Nationales d'Outre Mer, Aix en Provence, France, Carton Agence France Outre Mer 357, Dossier Chasse / Tourisme*

periences of travel and collaborative work one can discover from the study of female travelers during the late nineteenth and early twentieth centuries. This was not only true in Africa; historian Pamela Henson chronicles the challenges overcome by early botanist Agnes Chase: "Denied access to institutionally supported fieldwork, Chase broke through the barriers and established more informal and egalitarian ties with Latin American scientists than many of her male colleagues ever had" (Henson 2002, 598).

Responding to social mechanisms of exclusion that functioned along both racial and gender lines, such "ties" and "breakthroughs" altered, yet also reflected, many of the more nationalistic and imperialistic rivalries described in the next section. Such informal and egalitarian ties are the "seeds" of the circumstantial but often consciously crafted intimacies I describe across cultural and species boundaries. The traces of female scientists and travelers, like those of local experts and guides, are harder to find in the historical record than are the exploits of Teddy Roosevelt or Ernest Hemingway. Yet, understanding the genesis of these informal and egalitarian ties may be crucial to the eventual growth of ethically innovative and politically effective systems for sustainable human-nature interactions. For a moment, though, we can step back from the cultural forms in question, and reflect on some of the political and economic ones that are at stake in the management of elephant populations in equatorial Africa.

Legacies of Rivalry, Not Rationality

In " French equatorial Africa," a colonial term for several territories administered together in the western Congo basin, the early hunting industry received less international attention and opened up to a worldwide clientele much later than in the British colonies. France and Britain previously competed regarding plantation production output and concessions for trade in animal products. However, their rivalry intensified with the advent of tourism and leisure use of colonial territories. The French government generated documents about colonial tourism development that often asked, "Can we not do as well as the English have?"[3]

The interior forests and savannas of French colonial Africa did not develop an expedition-related industry such as the British colonies of southern and eastern Africa. There, as Cameron describes, dynamic commercial and political elites from multiple cultural traditions within and beyond Africa drove the industry's development. In the French Congo, stark confrontations occurred between the Africans from further north who served as guards and militia for the French, and those multiple groups from within the Congo basin, who relied to different degrees on hunting for their traditional economies of foraging, fishing, or farming. Each of these subsistence groups had distinct relationships to elephants, though many shared the element of ritual passages to manhood through the hunting of elephants. Many such groups were more or less forcibly integrated into work collecting ivory and other forest products for more global trade during the 1800s (Coquery-Vidrovitch 1972). Some, however, evidenced enormous ingenuity and brinksmanship in temporarily transforming their societies, and their relationships to elephants, to stockpile and control ivory in light of those trading opportunities (Harms 1981).

Trophy hunting and related tourism in French equatorial African regions have remained, to the present, a "*chasse gardée*." This French phrase evokes heavily protected or managed hunts that are thus not true contests to determine who is most fit or skilled. This hunting metaphor for political cronyism, ironically, captures well the actual politics of hunting in today's former French colonies. Permits and access rights in today's Cameroon, Congo (Brazzaville), and Central African Republic unfold, to a large extent, through intimate conversations among members of several intersecting elites. Since these territories became colonies, concessions within them have been renewed from year to year, unlike in parts of formerly British southern Africa where concession management is more long term.

The process is thus effectively subject neither to policy control nor market forces that would encourage sound management practices for maintaining wildlife populations over the medium to long term. Overhunting in a given season is almost encouraged by such factors, and the year-round presence of hunting professionals is rare, leaving animals vulnerable to poachers during the off-season, despite the relative protection conferred by hunting operations during part of the year (Fay 1995; Wilkie 1998). Poaching, of course, is a tricky, contestable term, a broader discussion of which is regrettably beyond the scope of this chapter. The growth of the arms trade and increasing civil conflict in much of sub-Saharan Africa have rendered the distinction between poachers and other hunters even more difficult but have also made the threat of illegal elephant hunting increasingly dire.

Former British colonies have led the African continent in trophy hunting's transformation from an adventure open to only a select few into a much broader practice, indeed in the transformation of travel into a mass industry. The advent of automobile technology after the turn of the twentieth century led to a major shift away from what had long been only an elite European tradition of pilgrimage and holiday travel to particular seaside, mountaintop, or cultural and religious sites (Urry 1995; Nash 1996). Railroads and road networks became crucial components of broader tourism development. Tourist buses began to define a worldwide genre of group tourism, the "motor tour." In colonial territories, expeditions sponsored by companies such as automobile manufacturer Citroën pitted their new machines against the rigorous terrain of tropical regions, generating publicity for their products as well as interest in the sights and experiences of these "exotic" locales. In the eastern and southern African savannas, buses or vans became crucial for wildlife viewing at a safe distance. They remain so today, as what was once a colonial industry has morphed into a more global one that is largely focused on bringing non-Africans to the continent to explore (Bandy 1996).

Equatorial Africa's advantage in this rapidly shifting competitive economy of leisure use has always been that it offers unique and adventuresome opportunities "off the beaten path" of touring and hunting. The disadvantages are many, however. Rational management, based on assessments of actual elephant population and migration dynamics, is elusive at best. As during the colonial era, management today is mostly mired in the ways that members of elites use wildlife as a resource in their economic and political deal making. Recently established protected areas, and the ecotourism practices and village-based trophy hunting tours they are pioneering, do represent a challenge to these historically rooted systems, and to the gender and race biases built into them (Roulet 2004). However, such newly decentralized practices come with their own contradictions and challenges, as we shall see in a closer look at a Central African forest reserve.

The Dzanga Clearing: Today's Ecotourism Encounters

> The elephants, having at last found in these reserves a tranquility to which they were no longer accustomed, are now quite numerous.
>
> (Anonymous 1942, 7)

The Dzanga Sangha Dense Forest Reserve in Southwestern Central African Republic, established in 1991, is managed by WWF Germany, WWF-U.S., and the Deutsche Gesellschaft für Technische Zusammenarbeit (GTZ, the

German government's agency for international development). It has one principal tourist attraction: a clearing of approximately one by three kilometers called Dzanga, where natural mineral and salt deposits in the soil attract elephants to drink, dig with their tusks, and bathe. The forest surrounding Dzanga is laced with clearings of various sizes, maintained by the activity of forest elephants (*Loxodonta cyclotis*), who fell trees and use clearings for mud baths and other purposes. Other herbivores such as forest buffalo (*Caffir nanus*), bongo (*Tracephalegus euryceros*), sitatunga (*Tragelaphus spekei*), several species of duiker (*Cephalophus*), giant forest hogs (*Hylocherus meinertzhageni*), red river hogs (*Potamochoerus porcus*), and western lowland gorillas (*Gorilla gorilla gorilla*) also use the clearings for sunning, bathing, and grazing on tender regrowth vegetation. Dzanga is by far the largest clearing of this kind in the immediate region, and on any given morning or evening one can see from ten to over one hundred elephants along with other animals basking and frolicking.

To promote relatively safe observation, the reserve administration has constructed a platform, perched in treetops at an entrance to the clearing. To get to the platform, one must drive on an old logging road fifteen kilometers from the nearest town, then walk approximately one kilometer through the forest to the clearing, crossing the Modoubou stream, where water is at shin level during the dry season, and waist level in the rainy season. This excursion is most often spoken of as "seeing the elephants" and is the one activity accomplished by almost every tourist who visits the reserve, as well as by local residents on occasional organized day trips.

Once at the clearing, visitors may move about, but most often stay on the observation platform, watching and photographing the goings-on beneath and around them, while listening to grey parrots and hornbills in the surrounding trees. Most of the conversations on the platform are variations on a theme and familiar to the guides. Questions and remarks from visitors usually address the varied colors of elephants' coats (because of their use of mud in various stages of dryness to keep insects from them); aspects of their morphology (the enormity of genitals, the size and shape of tusks, the hilarious movements of the young when trying to use their trunks); or their social behavior (group size, gestures of communication with trunks and ears, etc.).

Almost all visitor groups notice the research platform of Andrea Turkalo in a nearby tree. Working in collaboration with the New York–based Wildlife Conservation Society, she has been counting, identifying, and observing these elephants since 1991. Her research camp and its personnel (largely local hunters and gatherers known as BaAka, who have experience tracking elephants) have in many ways become part of the clearing. Certainly its presence has deterred poaching in the area, as has the nearby temporary tourist camp.

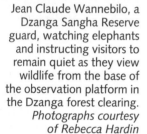

Jean Claude Wannebilo, a
Dzanga Sangha Reserve
guard, watching elephants
and instructing visitors to
remain quiet as they view
wildlife from the base of
the observation platform in
the Dzanga forest clearing.
*Photographs courtesy
of Rebecca Hardin*

The forest around the Dzanga clearing appears to have the highest density of elephants in the region. Many tourists I interviewed had encounters with elephants on the way to or from the platform; many return to camp terrified but exhilarated. I have seen diplomatic corps ambassadors and their drivers emerge from the stream, splattered with mud, clasping one another's hands with mixed laughter and tears about the elephant that tried to charge them. The trip to the clearing, particularly when interrupted by elephants blocking the road, spraying water with their trunks, or trumpeting at the stream crossing, is a sort of ritual, where normal social orders are upset or reversed, only to be strengthened or reestablished afterward (Turner 1969).

Elements of fear, wonder, connection, observation, and communication collide in this short walk to the clearing. Most visitors return moved, either simply by what they have seen or by the rush of feelings that comes from the real awe and terror that an unexpected encounter with an elephant—let alone with forty of fifty of them—can inspire. All of this is framed by what may be an equally unexpected bonding experience with humans one hardly knows. Facing down an elephant is no small feat, but trackers and guides alike at Dzanga have learned alternate routes back to the parking lot to avoid angering the animals. They are familiar with where to look for elephants and how to assess their impatience.

One French military commander stationed nearby who brought his troops as tourists in 1995 described it thus:

> We ran into an elephant in the water, where we were up to our knees . . . one of us had to move. The guide checked to see if it was a male or a female, since apparently it is the females that charge, usually; the trackers lined up and started hitting the water with their machetes. They told us to follow, but it took half an hour to get rid of the elephant. . . . My men were armed, but with no ammunition. . . . I could have given them ammo, but for what? I asked the guide and he told me particularly NOT to fire into the air, but I think that is what I would have done [if things had not gone well]. . . . Well, since we're supposed to rely on the guide, we did . . . and he was good. . . . We even bought him drinks later, at that bar in the center of town.[4]

Certainly, there is room for improvement in this sort of elephant-human encounter, for much is left to sheer circumstance and to the clarity of communication across barriers of species, language, culture, race, and social hierarchy. Military tourists are particularly tricky for local guides, for they seem least inclined to follow guides' advice in such situations; perhaps this also reflects guides' ambivalence about the French military—memories of colonial conflicts endure and inform today's interactions. Yet, even this difficult example of the tourist experience at the Dzanga clearing demonstrates the power that makes it so popular: the power to suspend the sorts of social hierarchies I have described elsewhere in this chapter as characterizing the history of trophy hunting and ecotourism.

Many things happen in the moments of confrontation with elephants, and in their aftermath, first and foremost the valorization of local forms of knowledge, and the fleeting yet profound empowerment of BaAka, sometimes known as Pygmies, who, back in a village or town setting, may be seen by their other African neighbors or by non-African visitors as little more than mendicants. People are also, in these moments, sharing and exchanging feelings, memories, and material goods that create bonds across social categories that would otherwise divide them. A local guide who might otherwise be unceremoniously tipped and thanked is taken out for drinks and entertained as a local hero after an exciting elephant encounter. The value of such interactions is difficult to quantify and hard to relate to more systemic change but seems to this author undeniably important.

Yet social hierarchies in the broader community are reinforced in some disturbing ways by the new economic activity of elephant tourism. Revenues from this activity are not accruing to the long-term residents of the region, such as BaAka. Rather, those who make the most money are neighboring Africans who arrive from other countries or larger towns to establish businesses or partake in the ever-growing "bushmeat" trade, which develops in relation to such dynamics of economic boom and bust (Barnes 2002). It is difficult to weigh such risks to elephants against the long-term benefits of facilitating human contact with and understanding of these animals (and, through that, with other humans). Such contact can foster new intimacies that might be mobilized for their protection. Economic and ecological monitoring processes are not yet in place to make such calculations with confidence. Until they are, conservative policies that limit tourism and trophy hunting appear most appropriate and would likely increase the lure of the area over the medium to long term.

To summarize, the activities I have described occur in a zone where trophy hunting and logging managed by French interests compete with conservation efforts managed by U.S. and German interests. The current situation thus mimics the earlier struggles between Germany and France for colonial control of the area, echoes the larger colonial rivalry between Britain and France over distinct models for economic development, and even calls to mind the distinctions between more Anglo-Saxon and Mediterranean modes of hunting and managing animal populations. Obviously, none of these past rivalries have been transmitted unchanged through time. Perhaps, in part because of the tradition of elite trophy hunting in northern Europe, the environmentalist and animal rights movements in Germany and England enjoy broad and powerful popular support today relative to France, for example, where movements for hunters' rights are seen as more "grassroots" and linked to the welfare of farmers, rather than to fox hunting nobility (as in England).

Back in CAR, we see perceptible institutional effects of such deeply historical and cultural legacies, aligning animal protection with the building blocks of civil society at regional and local levels and aligning hunting revenues with the maintenance of state-related infrastructures. Throughout the late 1990s, nonhunting tourism revenues were associated with the CAR Ministry of Environment and Tourism, whose technical counsel came from Germany. These revenues flowed through international conservation nongovernmental organizations (NGOs) to support the local conservation NGOs they have spawned. The CAR's revenues from trophy hunting and logging fees, however, were associated with the Ministry of Environment and Forests, whose technical counsel came from France. These monies flowed through the local mayor's office to support offices, guesthouses, and related government buildings in the area.

For elephants the result of such competing tactics for human management of the environment is ever more varied and intense pressure on their populations and habitats. Such competing visions have made rational, sustainable management of the Dzanga Sangha Reserve nearly impossible. Rather it has become a hub of economic activities, attractive to in-migrants.

As both African and expatriate populations increase, the circuits connecting the area's resources to consumers of elephants, as meat and as central symbols of the tourism and hunting industries also increase. Finding alternatives to such structures of rivalry should thus seem clearly linked to the fate of those few remaining elephants that still live in wild populations. The search for alternatives, however, comes with a prerequisite for further analysis along ethical lines.

Culture, Ethics, and Elephants

In this chapter, I have considered some of the roots of domination and appropriation of elephants by humans who have not shared their lives with them in any sustained way. I next discuss the broader dynamics these interactions reflect, specifically, appropriation in ecotourism markets and domination in trophy hunting. At the same time, I hope to stay close to some of the historical and ethnographic cases previously discussed. They have given rise to many of my conceptual points, in what Lynn (2002, 314) calls an "ecology of theoretical insights and empirical cases." Like Lynn, I seek a practical ethic, one that responds to recent debates in ethics and philosophy literatures that contrast "abstract, impartial, absolute, universal perspectives versus concrete, local, historically specific, contextual perspectives" (Warren 1999, 131).

Appropriation

Even in broad treatises on the topic of environmental ethics writ large, elephants have a tendency to appear as crucial symbols:

> Ecological ethics—what is that all about?
> . . . It is about the mother elephant who tries in vain to protect her young before a danger for which nature did not prepare her, before ivory poachers with high technology weaponry bent on murder within an ever shrinking elephant habitat. . . . It is about humans who think that they are the crowning glory of creation and the lords of all creatures but behave in the world like an elephant in a china shop—though anyone whom an elephant ever stroked with the velvety lip of her trunk can testify that this metaphor is all wrong. (Kohak 2000, 1)

Philosopher Erazim Kohak is certainly not alone in perceiving elephants as "gentle giants." His consideration of them as a mascot for the environment, as the ultimate symbol of man's violence against nature's "gentle glory," is effective. It illustrates the possibilities for humans, as ethical actors, to identify with and protect even those species whose existence presents us with some conflict or conundrum, often because they are valuable to us and thus worth consuming. His perspective suggests that con-

trolling our own technological abilities and economic desires is a prerequisite for being less noxious as actors within complex ecological systems.
One is even left with the impression that he has experienced some intimacy
with elephants—or at least the "velvety lip" of one elephant's trunk.

Yet there are two problems with such a perspective. First, it does not
take into account the complexity of human-animal relationships, where
a cross-species caress of the kind he describes is valued precisely because
of the very real possibility of a violent interaction. Elephants, like many
wild animals, can intimidate and harm people, even kill them and devas-
tate their property. That such behavior by elephants is most often in response to human incursions into "ever shrinking elephant habitat" does
not alter the social fact that humans fear elephants. That fear, inspired by
the elephant's sheer size and fed through generations of tales recounted,
can create respect and reinforce a certain "safe" distance between humans
and such massive animals. It can also, however, degenerate into anger, especially when the pattern of human incursion entails competition for
space and resources.

During my work in the Dzanga Sangha Reserve, I regularly had elephants
in my "backyard" at night. With some discomfort, I often deferred a late-
night trip to the outhouse to avoid confronting them. My house backed up
on a zone newly cleared and planted with food crops to feed the growing
population arriving for work in a nearby sawmill. In the mornings, I often
saw and heard angry stories from women whose fields had been ransacked
by elephants having a midnight snack at their expense. I saw such women
march to the Reserve Director's office, crushed cornstalks and damaged
cassava branches in their arms, to demand restitution. More than a mere
headache for reserve management, the problem has spurred experiments
with electric fences and other approaches to defend family food plots
from the elephants. My firsthand account of human-elephant conflict reflects situations repeated throughout elephant range countries, as mani-
fold sources attest (see, for instance, Kiiru's Chapter 19 in this volume).
Such strained negotiations, neither violent and victimizing of elephants,
nor gentle and loving of them, nevertheless constitute a component of environmental ethics as it must confront clear land and wildlife management
challenges.

A second problem with appropriating elephants as a symbol for mobi-
lizing support in environmental causes is that it renders them as passive
victims, awaiting the protection of, and from, all-powerful human beings.
This is a crucial contrast with many historically rooted, local negotiations
with elephants and introduces interactions predicated on our abilities ut-
terly to destroy them or heroically to save them. Often missing from these
new arrangements for interacting with elephants, however, is a more nuanced biological and ecological understanding of elephants as agents,

who are also capable of adapting, deciding on, defending, and abandoning their own territories.

One area of difficulty in my framework is distinguishing between new forms of intimacy that are personally engaged in with elephants as agents and those performances of intimacy that are more patronizing, or profiteering—that simply appropriate elephants without either recognizing or engaging them as agents. Such performances often rely on stock narratives about adventure and exploration that are evocative of the colonial era attitudes previously analyzed. They are thus tailored to popular demands for storylines that replicate basic cultural assumptions of men as apart from or opposed to nature and women as connected to or conflated with it. This often involves either the destruction of nature (by powerful male figures) or salvation and redemption of nature (by artistic and benevolent female figures), rather than more complex and less predictably gendered interactions over time. A 1996 film about Indian elephants, featuring Goldie Hawn, while in many ways movingly sincere and personally engaged, reflects such market preferences for watching lithe celebrities be "stroked by the velvety lip" of elephants' trunks.

The trailers for the film, also released as a Public Broadcasting Service Home Video, read: "Goldie Hawn loves everything about India, especially the elephants. On a previous trip she found her special elephant, now she is going back to find her again. She spends time with . . . an Asian elephant advocate with his own pet elephant Tara." During the film, a charmingly reluctant and nervous Hawn overcomes her fear of elephants to bathe them, ride them, and generally become their advocate. Advocate against what? The primary factor threatening elephants, we are told, is Indian population growth. This leads to situations of "conflict with poor villagers who have no control over where they live." This voice-over accompanies footage of screaming male villagers with lit torches chasing a confused and frightened elephant through the night (for more on conflict between people and wild elephants, see, for example, chapters by Sukumar [Chapter 2], Kiiru [Chapter 19], and Seneviratne and Rossel [Chapter 18] in this volume).

Hawn's charmingly faltering intimacy with elephants in the film is obviously no match for the real intimacies lived by pet elephant Tara's European owner and Indian handlers, who also appear briefly in sequences where Hawn is actually interacting with elephants. It raises tricky questions beyond the scope of this chapter about the distinction between historically and contextually rich forms of intimacy and more performative or strategic ones. Hawn's own performance is aimed at spurring better advocacy for and public awareness about these elephants among an American viewing public. In this sense, it is like the appearances of *Animal Planet* star Jeff Corwin, who can be seen on camera (and thereafter on Internet sites such as YouTube and MySpace) not only bathing with an elephant, rolling his body over and around hers, but also caressing her face and trunk. Because of screenwriters and voice-over techniques, in a key sequence, the viewer sees him communicate telepathically with the elephant, as she reads his thoughts, and we hear her comments on his haircut, his lifework saving animals, and so on.

The unintended results of such a performances, and the market preferences they reinforce, have more difficult implications. First, they create a fantastic notion of instant, complete, and even slightly erotically charged intimacy across species boundaries that belies the long-term, often grueling work of caring for and building trust and communicative capacity across such divides. Second, they play into what one scholar of wildlife management terms "a popular racial stereotype of 'primitive' Africans [or, in this case, Indians] as part of the natural landscape" (Neumann 1996, 125).

Geographer Rod Neumann goes on to document how, in the history of protected areas in Africa, sometimes a "native" presence in the parks was tolerated. Such was the case, according to colonial archives, in Parc National Albert, one of the first and largest protected areas created in what is today the Democratic Republic of Congo, formerly Zaire. There, local hunters and gatherers were regarded as part of the fauna and therefore "left undisturbed" (Neumann 1996, 125). There is continuity from such colonial ideas to the integrated conservation and development regimes of the Central African Republic that I previously described.

These recent projects prominently feature people like the BaAka as being in need of protection but do not often refer to them in their lists of "stakeholders," which do include categories such as farmers, logging company officials, and immigrants who work in the logging or diamond-digging industries (Giles-Vernick 2002). An adequate consideration of how distinct groups of Africans have forged various relationships to elephants is beyond the scope of this chapter. However, this relegation of elephants, and of those African hunters and gatherers who have historically known them most intimately, to a "protected" status, with no option to participate actively in establishing or revising the spatial boundaries of protected areas, makes elephants the ideal object for western managers and movie stars to manipulate in their common search for star-standing, sponsorship, and support.

As powerful symbols, elephants make sense as mascots for political or intellectual environmental movements. However, we must work hard to understand the ironies inherent in these associations. There are great, worldwide pressures on elephant habitat. Yet, for historical reasons, locals may not see how their own agricultural expansion is a problem for elephants. Instead, they may perceive elephants as powerful individually, and, as groups, capable of changing or expanding their territory. The torch-wielding Indian villagers in the PBS Hawn film, or my farming and foraging neighbors at Dzanga Sangha, may see themselves as fighting with elephants about the limits of human versus elephant space. A certain amount of respect is implied in this conflict, which may reflect centuries of intimate negotiation with elephants. Those engaging in an ethic of intimacy might seek better to understand their conflicts and cohabitations, recognizing and reinforcing elements that can assist with resolving today's challenges to elephants' continued existence in the wild.

The Indian villagers of the Hawn film, like the protected BaAka in colonial and contemporary African conservation settings, are at the fulcrum of increasingly global practices of environmental protection that are rooted in the colonial moment. These approaches to protection often reflect the male-dominated cultural norms and compelling narratives of adventure and appropriation that are the legacy of colonial culture (Haraway 1989). The colonial era's gender and class biases often still shape the politics of elephant protection. Even within progressive conservation practices, there remain many biases in key market sectors—fund-raising for protection efforts and ecotourism sales, for example. The prevailing logics still seek the symbolic (and, often, real) appropriation of elephants according to outdated colonial models of great white adventurers with civilizing missions.

Especially ironic is that many western conservationists have trouble seeing the likenesses across the various levels and types of human consumption that conspire to make elephants scarcer despite measured conservation and management victories. This is what leads Kohak (2000, 9) to claim the "innocent greed of the affluent" as the biggest ecological time bomb threatening elephants and other wildlife. Kohak's work reminds us that understanding how some are "appropriating" elephants by using their wealth is crucial to finding ethical ways forward. Those who donate money to conservation, or go on photographic safaris themselves, often are no more able to see their own daily choices (expansion into second homes or third cars) as harmful to elephants than are the villagers whose expansion into new agricultural lands brings them into direct and intimate conflict with elephants.

Domination

Advocates of hunters' rights have been quick to notice these ironies of conservationist and protectionist practice. Indeed, they are used in efforts to create new communities across hunting categories. Many wealthy leisure hunters began their hunting careers doing varmint control or subsistence hunting on family farms and nurture a sense of kindred spirits with poor hunters, as opposed to those less emotional links they might create with wealthy naturalists and conservationists. Similarly, many self-identified "native hunters" have commented on their experience of being pushed aside by wealthy conservationist interests that do not acknowledge the legitimacy and importance of the kill in human interactions with natural systems:

> For Native American people . . . the moral universe includes all animals and plants. Every living thing has basic rights and should be treated with respect, regardless of appearance, personality, or perceived relationship with humans. But within indigenous traditions there is also a deep and lucid awareness that taking plant and animal life is how we survive each day. What matters is that we conduct ourselves respectfully toward every organism, consciously recognizing and honoring this dependence. (Nelson 1992, 30)

Such an argument simply may not extend to the high-prestige stakes of trophy hunting. Yet many trophy hunters have some sort of subsistence hunting in their past, and they see hunting as a nearly primordial need and right. For some, hunting is analogous to sexual instincts in humans (Causey 1989). Few would argue that sex should only occur for reproduction—for it is also an important pleasure in itself, and a crucial building block in human intimacy. Likewise, hunting no longer need occur for subsistence only but is an important pleasure in itself, and a crucial practice for maintaining certain elements of human intimacy with landscapes and animal populations.

This view turns upon an ethical acceptance of death as natural, primordial, and significant to life. Others disagree about the pivotal role of the kill in the pleasure of the hunt. They argue, "To the sportsman the death of the game is not what interests him; that is not his purpose. What interests him is everything that he had to do to achieve that death—that is, the hunt" (Ortega y Gasset 1985, 96–97). Hell (1994) notes that the sociology of leisure, which has given rise to much of the recent research on tourism, has clear limits for the analysis of sport hunting and cannot do justice to this endeavor's highly ritualized and personalized spilling of blood, which asserts personal virility, mastery of property, social prestige, and cultural identity.

Whatever one's position on the pleasure of the kill, a broader acceptance of the hunt as crucial to humanity's relationship with animals demands that the internal motivations and practical scruples of each individual hunter become the relevant ethical boundary to recognize. Many advocates of hunting claim to consider it carefully indeed, developing clear and thoughtful guidelines regarding the material circumstances of the hunt. For these hunters, implicit in the idea of "trophy" is that the game pursued is a wild, free-ranging animal also available as quarry to other hunters, and never subject to practices such as stimulating antler growth with mineral blocks, hormones, or other substances. Any other approach is beyond acceptable ethical practice and diminishes the value of all trophies (Posewitz 1994).

They even extend the notion of ethics to the motivations of a given hunt

(Ritchi 1995), contrasting externally versus internally motivated hunters to extend the notion of hunting ethics to the pursuit of trophies:

> If you hunt these animals because they represent the survivors of many hunts, and you respect that achievement, then you have selected a high personal standard. If, on the other hand, you pursue a trophy to establish that you, as an individual hunter, are superior to other hunters, then you have done it to enhance your personal status, and that crosses the ethical line. (Posewitz 1994, 97)

Elephants are among the most important prestige trophies to be found. They are, along with a very few other species, arguably the most "externally motivated" trophy one can possibly obtain. The body is far too large to transport for a full-mount trophy that would rely on extensive taxidermy and require enormous display spaces. Rather, standard practice is to mount the head, featuring the tusks. In any collection, such trophies are almost always awarded a most prominent place.

This is true not only for Europeans on safari hunts but for Africans who have long used elephant tusks in architecture testifying to the political and spiritual power of particular leaders. For example, the classic ethnographic work *The Nuer* (Evans-Pritchard 1940) features photographs of the remarkable structures used by southern Sudanese prophets. These enormous mounds were each surrounded by rings of upright elephant tusks. Many of these structures, and indeed the men who controlled them, were targeted for destruction by British colonial forces. The British recognized their potentially revolutionary influence on fiercely autonomous peoples who were not necessarily prepared to submit to colonial rule. In fact, the long history of colonial suppression of indigenous African uses of elephants may explain why contemporary debates about legalizing elephant hunting are suffused with the politics of postcolonial political autonomy.

Consider recent debates about downlisting the status of elephants, to a less protected category under the Convention on International Trade in Endangered Species of Wild Fauna and Flora (see also Duffy, Chapter 22 in this volume). The official U.S. position under former President Clinton opposed this change. But Representative Richard Pombo (R-CA) traveled to Zimbabwe to make it clear to the official U.S. delegation, to the animal rights advocates, and to other countries, that "not everyone in the U.S. government agreed with this approach . . . nations should develop and control their own wildlife conservation programs" (Marlenee 1997). These divided opinions on the downlisting of elephants should raise questions about the homogeneity of African opinions, as well. A member of the non-official U.S. delegation writes of the successful vote to downlist elephants:

A hunter's trophy room in his California home.
Reproduced by permission from Safari Press, ed., Great Hunters, vol. 1:
Their Trophy Rooms and Collections *(Long Beach, CA: Safari Press, 1997)*

Immediately following the triumphant vote, elated Africans rose
in what seemed to be a spontaneous emotional act and began sing-
ing the southern African liberation hymn. It was a jubilant and
tremendously moving experience to watch as southern African
delegates celebrated what they said was a victory in winning respect
for their needs, their sovereignty and many years of hard work.
(National Animal Interest Alliance 2003)

In Africa, as in North America, it is worth asking whether governing elites truly represent public opinion and whether those most intimately involved with elephants are well-represented at such international negotiations. What is clear is that without the histories of domination through hunting practice, and domination's broader role in forging colonial control over African territories, there could not have been this particular nationalistic political impulse for concerted opposition to protection measures. Any ethical frameworks must be able to respond concretely to such crucial political challenges, as well as attending to the intimacy implicit in hunting's legacies of layered domination: humans over animals, and through that display of power, over other humans as well.

Intimacy

> All ethics so far evolved rest upon a single premise: that the individual is a member of a community of interdependent parts . . . a land ethic changes the role of *Homo sapiens* from conqueror of the land-community to plain member and citizen of it. It implies respect for his fellow-members, and also respect for the community as such.
>
> (Leopold 1949, 239–240)

Real intimacies with elephants reside in the details of daily life with them— a researcher's day in and day out walks to the same clearing; a safari guide's dogged tramping through urine-filled ditches in swampy terrain; a keeper's time spent cleaning parts of a zoo enclosure; a trainer's daily ritual for bathing a giant pachyderm. They are often at some remove from the pursuit of power, authority, and celebrity that most influences elephants' fates. Yet there are lessons from such intimacies that need to appear more, and appeal more, to key decision makers, many of whom are likely merely to "drop in" to such intimacies briefly for a tour or a trophy—if that.

Such decision makers are increasingly elevated, through the workings of social hierarchy and capital accumulation, far beyond the status of Leopold's "plain member and citizen" of society, let alone of nature. Those who wield the most power to appropriate nature hardly need bother, save perhaps to tip a few coins, with those who daily negotiate their relationships to the elements of nature. The latter individuals may or may not seek at times to dominate or appropriate elephants in the symbolically powerful ways previously described; but they balance such acts with others, inasmuch as they are the bathers and feeders, the trackers and trainers, the persistent ecologists counting elephants in clearings, or the passionate farmers angrily chasing them from trampled crops.

Not many people have this common experience of sharing some significant slice of their life with elephants. The salience of intimacy as a concept for elephant protection lies in its suggestion that those who fall into such a category might do well to consider the circumstances that unite them, rather than the convictions that divide them. In the current political climate, this would entail some acknowledgment of the ironies and contradictions inherent in articulated philosophies for and against hunting. Some hunting advocates truly believe the "anti-hunting movement reflects . . . increased distance from the environment, diminished awareness of how we interact with it, and denial of basic biological processes" (Nelson 1992, 31). Clearly, they have not been in the presence of those conservationists and animal advocates who have repeatedly risked—or even sustained—bodily injury in the course of their work with large animals, or who may have held a frothy-mouthed animal through death throes, trying to ascertain what viral agent could be causing an oncoming epidemic.

Conversely, many animal welfare advocates appear unable to acknowledge the profound ambivalence and power at the heart of hunting as an intimate confrontation with animality:

> Every good hunter is uneasy in the depths of his conscience when faced with the death he is about to inflict. . . . The generally problematic, equivocal nature of man's relationship with animals shines through the uneasiness. . . . Before and beyond all science, humanity sees itself as something emerging from animality, but it cannot be sure of having transcended that state completely. (Ortega y Gasset 1985, 88)

The framework of intimacy helps us acknowledge that the grounds on which we interact with elephants may shift, as our abilities to apprehend different dimensions of human and other life forms improve. It enables us to recognize and to respect historical relationships with elephants even as new circumstances of intimacy emerge that might displace or erase them. Research into cognition and memory, for example, may eventually shift conventional wisdom about what constitutes "personhood" (see Varner, Chapter 3 in this volume); recent work in science studies also takes up such questions (Haraway 2003).

Those who are most intimate with elephants historically may not have access to new academic developments and should not be penalized for that. Nor should they be excluded from dialogue with those who are more connected to new knowledge frontiers. An ethical intimacy framework recognizes that a researcher who has spent long hours studying the memory of elephants and a skilled tracker for elephant hunting both should have some say with respect to how policy and extractive practices are managed. In fact, each would have much to discuss with the other about the knowledge derived from their respective forms of intimacy with these creatures. A member of the South African secret police force in the early 1990s, however dependent on ivory revenues for political funding (Ellis 1994), would be out of place in such discussions because of his lack of intimacy with elephants. For him, ivory was a crucial but replaceable commodity and his reliance on it was not predicated on any real sharing of his life with elephants, but rather on the work of trackers and hunters and traders.

A practical ethic of intimacy for elephants calls for us to recognize and value historical and potential human communities of concern for, knowledge about, and interdependence with elephants. It also calls for members of such communities to consider the claims and common interests of others who have experienced or sincerely sought intimacy with elephants. I suggest that certain new forms of engagement with elephants are cause for hope that we may respond to this call. These include long-term research, tourism "off the beaten path" and out of motor vehicles, and perhaps forms of hunting that depart from the slaughter of elephants as mere replaceable commodities (as in the case of our hypothetical South African secret police officer), or for social display of wealth and power through trophies. Such new forms of touring and hunting could occur in historic elephant habitat among remaining wild populations, or in new contexts within which elephants gather, such as sanctuaries or very large ranches. They might even be virtual. The hunting might be motivated either by subsistence and ecological need, or, in the case of trophy hunting, (Posewitz 1994) by the accurate and respectful recognition of individual animals' life histories of maturity and successive triumphs over hunters. For those forms to take concrete shape, however, they would have to rely on other communities of intimate knowledge about elephants, as populations, and as individuals.

The social obstacles to an effective ethic of intimacy are enormous and a long time in the making. In large part, this is due to the profoundly gendered and racialized social differences within and among human societies at regional and international scales, reaching from the Middle Ages to contemporary tensions, such as postcolonialism, antifederalism, and increasing social and economic stratification. Until we are better able to confront and reconfigure the seductive cultural and interpersonal dynamics of these human divisions, the ethics of domination and appropriation will continue their twin reign. Their potential contributions to conservation processes at local levels notwithstanding, at a broader scale they will facilitate increasing human consumption of elephants as commodities within concentric local, regional, and global markets. This will likely mean the end of elephants living as complex groups in their natural habitats. It may even mean dramatic reduction of their multiple roles in human lives: as modes of transport, treasured companions, awe-inspiring prey, and remarkable competitors for ever more precious space, all on a planet where elephants are increasingly unable to live beyond the reach of human intimacies.

Acknowledgments

I am deeply indebted to the residents and staff of the Dzanga Sangha Reserve for their work with me and the tourists, guides, hunters, and researchers who suffered my questions and research presence. I am grateful for the support of McGill University's School of Environment and Department of Anthropology for research time and resources committed to the conception of this chapter. I am grateful for the support of the University of Michigan and the Harvard Academy of International and Area Studies for research time and resources committed toward its completion. Kate Christen, Nadine Naber, Damani Partridge, Harriet Ritvo, and Elizabeth Roberts all offered insightful comments on this manuscript. For information, corrections, and permission to reprint photos, I thank Ludo Wurfbain at Safari Press. Thanks to Andrea Steves, Jesse Worker, and Jason Yeo for their assistance with finalizing this manuscript.

Notes

1. It is useful to distinguish between descriptive ethics and normative ethics, particularly given that the two can overlap rather insidiously despite an author's best efforts at self-awareness on such matters. My work as an anthropologist can contribute to "environmental ethics" as thus defined: "the systematic and critical study of judgments and attitudes which (consciously or unconsciously) guide human beings in the way they behave toward nature" (Stenmark 2002).

2. This watershed between northern and more Mediterranean European hunting and forest use has foundations that emerge in the Middle Ages and can be seen in the very different legislative statutes for hunting across the two regions. In Italy, hunting was subordinated entirely to pastoralism and agriculture, in part because forests were scarcer and more feared (Chastel 1990). Further north, in the Rhine region, only the wolf could be freely exterminated, and the Badois (or Alsatian peasantry) were much more constrained in their hunting practices.

3. "Serions-nous incapables de faire simplement autant que les Anglais?" *Le Tourisme et la question Hoteliere en Afrique Equatoriale Française* (p. 5). Centre des Archives d'Outre Mer Series: Agence France Outre Mer Carton: 360 d. hotelier.

4. Poulenc. Bayanga. Hardin Interview and field notes from August 5, 1995, conducted with French Foreign Legion and Parachute troops deployed in the French bas Bouar for regional surveillance and security operations.

References

Anonymous. 1942. *Tourisme cynégétique en Oubangui-Chari.* Aix en Provence, Centre des Archives d'Outre Mer, Files from the Agence France Outre Mer, carton 547, d. 120.

Bandy, J. 1996. Managing the other of nature: Sustainability, spectacle, and global regimes of capital in ecotourism. *Public Culture* 8: 539–566.

Barnes, R. F. W. 2002 The bushmeat boom and bust in West and Central Africa. *Oryx* 36: 236–242.

Beinart, W. 1989. Empire, hunting and ecological change in southern and central Africa. *Past and Present* 128: 162–187.

Bird Rose, D. 1999. Indigenous ecologies and an ethic of connection. In N. Low (ed.), *Global ethics and environment* (pp. 175–187). New York: Routledge.

Cameron, Kenneth. 1990. *Into Africa: The story of the East African Safari.* London: Constable.

Causey, A. 1989. On the morality of hunting. *Environmental Ethics* 11: 327–343.

Chastel, A. (ed.). 1990. *Le chateau, la chasse et la forêt.* Luçon, France: Pollina.

Coquery-Vidrovitch, C. 1972. *Le Congo au temps des grandes compagnies concessionnaires, 1898–1930.* Paris: Mouton & Co.

Dyer, A. 1996. *Men for all seasons: The hunters and pioneers.* Agoura, CA: Trophy Room Books.

Ellis, S. J. 1994. Of elephants and men: Politics and nature conservation in South Africa. *Journal of Southern African Studies* 20: 53–69.

Evans-Pritchard, E. E. 1940. *The Nuer.* Oxford: Oxford University Press.

Fay, J. M. 1995. Pilot tourist safari hunting program in Northern Congo DRAFT. Ministere d'Agriculture, Elevage, Eaux, Forets, et Peches, Congo Forest Conservation Project (WCS), GTZ, SCI, Congo Safaris.

Giles-Vernick, T. 2002. *Cutting the vines of the past: Environmental histories of the Central African rainforest.* Charlottesville: University of Virginia Press.

Haraway, D. 1989. *Primate visions: Gender, race, and nature in the world of modern science.* New York: Routledge.

Haraway, D. 2003. *The companion species manifesto: Dogs, people, and significant otherness.* Chicago: Prickly Paradigm Press.

Hardin, R. 2000. Translating the forest: Tourism, trophy hunting, and the transformation of forest use in southwestern Central African Republic (CAR). PhD diss., Yale University.

Hardin, R. 2001. Concessionary politics in the Congo basin: History and culture in forest use practices. World Resources Institute and Central Africa Regional Program for the Environment, Working Papers Series, Institutions and Governance Program. Washington, DC: http://pubs.wri.org/pubs_description.cfm?PubID=3853.

Harms, R. 1981. *River of wealth, river of sorrow: The Central Zaire Basin in the era of the slave and ivory trade, 1500–1891.* New Haven, CT: Yale University Press.

Hell, B. 1994. *Le sang noir: Chasse et mythes du sauvage en Europe.* Mayenne, France: Flammarion.

Henson, P. 2002. Invading Arcadia: Women scientists in the field in Latin America, 1900–1950. *The Americas* 58: 577–600.

Jonas, H. 1990. *Le principe responsabilité: Une éthique pour la civilisation technologique.* Paris: Les Editions du Cerf. (Originally published 1972.)

Kohak, E. 2000. *The green halo: A bird's eye view of ecological ethics.* La Salle, IL: Open Court Publishing Co.

Leopold, A. 1949. *A Sand County almanac and sketches here and there.* New York: Oxford University Press.

Lynn, W. S. 2002. *Canis Lupus Cosmopolis:* Wolves in a cosmopolitan worldview. *Worldviews* 6: 300–327.

Mackenzie, J. M. 1987. Chivalry, Social Darwinism and ritualized killing: The hunting ethos in Central Africa up to 1914. In G. A. Anderson (ed.), *Conservation in Africa: Peoples, policies and practices* (pp. 41–62). Cambridge: Cambridge University Press.

Marlenee, R. 1997. Fox-Miller defeat a victory for hunters. *Safari Times* 9: 1, 3.

Naess, A. 1995. The deep ecological movement, some philosophical aspects. In G. Sessions (ed.), *Deep ecology for the 21st Century* (pp. 64–84). Boston: Shambhala.

Nash, D. 1996. *The anthropology of tourism.* Oxford: Pergamon.

National Animal Interest Alliance. 2003. Lions and tigers and elephants oh my! www.naiaonline.org/body/articles/archives/97cites1.htm.

Nelson, R. 1992. *Stalking the sacred game: Perspectives from Native American hunting traditions.* Bozeman, MT: Governor's Symposium on North America's Hunting Heritage.

Neumann, R. P. 1996. Dukes, earls, and ersatz Edens: Aristocratic nature preservationists in colonial Africa. *Environment and Planning: Society and Space* 14: 79–98.

Ortega y Gasset, J. 1985. *Meditations on hunting.* New York: Charles Scribner's Sons. (Originally published 1942.)

Ortner, S. 1974. Is female to male as nature is to culture? In M. Rosaldo and L. Lamphere (eds.), *Woman, culture, and society* (pp. 67–88). Stanford, CT: Stanford University Press.

Plender, J. 2001. Walking with animals, learning new tricks: Defenders of animal rights are mastering the arts of "econo-terrorism" and beating the capitalist Goliaths. *Financial Times.* March 17.

Posewitz, J. 1994. *Beyond fair chase: The ethic and tradition of hunting.* Helena, MT: Falcon Press.

Raffles, Hugh. 2002. *In Amazonia: A natural history.* Princeton, NJ: Princeton University Press.

Ritchie, L. 1995. *Hunting: Ethical and moral considerations.* www.geocities.com/Athens/2921/LRE001.html.

Rolston, H., III. 1994. Environmental ethics: Values and duties to the natural world. In L. Gruen and D. Jamieson (eds.), *Reflecting on nature: Readings in environmental philosophy* (pp. 65–84). New York: Oxford University Press.

Roulet, P. A. 2004. Chasseur blanc, cœur noir? La chasse sportive en Afrique centrale: Une analyse de son rôle dans la conservation de la faune sauvage et le développement rural au travers des programmes de gestion de la chasse communautaire. PhD diss., Universite d'Orleans.

Sawyer, S., and Agrawal, A. 1997. Environmental orientalisms. *Cultural Critique* 45: 71–108.

Scully, M. 2002. *Dominion: The power of man, the suffering of animals, and the call to mercy.* New York: St Martin's Press.

Singer, P. 1999. Ethics across the species boundary. In N. Low (ed.), *Global ethics and environment* (pp. 146–157). New York: Routledge.

Spurr, D. 1993. *The rhetoric of empire.* Durham, NC: Duke University Press.

Stenmark, M. 2002. *Environmental ethics and policy-making.* Burlington, VT: Ashgate.

Thom, M. 1990. Tribes within nations: The ancient Germans and the history of modern France. In H. K. Bhabha. (ed.), *Nation and narration* (pp. 23–43). New York: Routledge.

Turner, V. 1969. *The ritual process.* Chicago: Aldine.

Urry, J. 1995. *Consuming places.* New York: Routledge.

Warren, K. 1999. Care-sensitive ethics and situated universalism. In N. Low (ed.), *Global ethics and environment* (pp. 131–145). New York: Routledge.

Wilkie, D., and Carpenter, J. 1998. The potential role of safari hunting as a source of revenue for protected areas in the Congo Basin. *Oryx* 33: 333–339.

22 THE ETHICS OF GLOBAL ENFORCEMENT
ZIMBABWE AND THE POLITICS OF THE IVORY TRADE
ROSALEEN DUFFY

The transboundary nature of environmental resources has made them a key focus for global forms of management. In effect this means that resources held within national boundaries, such as wildlife, forests, and rivers are subject to global attempts to define and enforce appropriate environmental management to conserve them. The global management and use of wildlife in particular presents peculiar ethical concerns. In this chapter, I examine the debates raised by the global ivory trade and the key international regime for regulating it, the Convention on International Trade in Endangered Species of Wild Fauna and Flora (CITES). I investigate the competing environmental philosophies of preservation and sustainable utilization as the path to appropriate elephant conservation, and how they are deployed by various interest groups involved in the debates over whether the ivory trade should be banned.

Briefly, the overall policy direction of western states, South Asian states, East African states, and many environmental nongovernmental organizations (NGOs) has been informed by preservationist ideas that mean a total ban on the ivory trade is the only way to ensure the survival of elephants as a species. In contrast, local management practices of southern African elephant range states, especially Zimbabwe but also Botswana, Namibia, Zambia, and South Africa, have been determined by ideas of sustainable use, which allows for a limited ivory trade; this is intended to satisfy competing demands for economic development and environmental conservation in the developing world. As might be expected, these vastly different philosophical positions lead to complex ethical debates about who has the right to use elephants, how they can be used, and for what purpose. I also analyze the ways that these competing ideas of good conservation practice are created and embedded in international environmental institutions. I then analyze the preservationist approach to elephant management and examine the broad policy commitment to sustainable utilization by southern Africa, and Zimbabwe in particular, as the only ethically acceptable form of elephant management in the developing world. Finally, I investigate the global politics of the ivory ban and the problems associated with its enforcement at the local level.

Global Governance

First, I discuss the concepts and practices of the vast array of organizations, activities, and actors that fall under the term *global governance*. How and why do different interest groups claim knowledge that is universally valid and applicable? We can examine how these claims to expert knowledge can determine what constitutes good practice in conservation and how that informs policy decisions.

Global conventions have been established in response to transboundary problems where domestic legislation has proved inadequate, and they are especially significant in the area of environmental management. They are critical examples of global governance, which differs from "government" in that the location of power is diffused away from a single center with authority and control over a specific territory, activity, or group. For example, the Commission on Global Governance suggested that governance includes formal institutions and regimes empowered to enforce compliance, as well as the informal arrangements that people or institutions have agreed on or perceive to be in their interest (1995, 4). Global governance highlights a shift in the location of authority in the political, economic, and social realms. In the context of increasing globalization, concepts of global governance indicate a shift away from the state-centric view of global politics (Hewson and Sinclair 1999, 5–11). Because national borders do not bound the environment, global environmental management has proven to be central to the creation and extension of global governance. One form of global governance can be found in international agreements that are intended to govern a variety of local, national, regional, and global activities.

Global environmental regulation through international institutions, conventions, and legal frameworks has also raised questions about the contested nature and status of such norms. In particular, global conventions often rely on ideas of neutral and uncontested science that can be used to draw up universally applicable forms of environmental management. These are then used to justify and legitimate highly political global interventions at the local level. This goes further than suggesting neutral scientific information is used differently by various political actors such as elephant range states or environmental NGOs. Rather, I suggest that science and what is presented as scientific information is politicized. In effect, science produces the information to fit with the values that constitute the starting point for scientists themselves. (For further discussion on this, see the debate on indigenous technical knowledge, including Leach and Mearns [1996]; also see Litfin [1994, 51]; Neumann [2000]; and Sivaramakrishnan [1999]). Within global environmental debates, various interest groups have presented themselves as knowledge brokers, that is, as organizations that are able to frame and interpret complex scientific arguments for consumption by the public or policy makers. The ability of knowledge brokers, such as environmental NGOs, consultants, scientists, and policy makers, to frame and interpret scientific knowledge is a substantial source of political power. Such knowledge brokers are influential under conditions of scientific uncertainty that characterize environmental problems. These knowledge brokers are often referred to as "epistemic communities," or networks of persons and organizations deemed to provide expert information and advice on a specific issue, to inform, to guide, and even to determine policy outcomes at the local, national, or global levels (see Wynne 1992; Litfin 1994). It is in the spaces between competing scientific discourses that resistant actors have an opportunity to challenge global environmental norms, and argue that their local scientific knowledge takes account of micro-level issues and concerns.

Global Governance and CITES: Preservation versus Utilization

I now turn to considering the ways that broader debates about global governance are played out within CITES. In particular, I examine the competing elephant management philosophies of preservation and utilization and their mobilization in the debates surrounding politics and ethics of the ivory trade. The debate about the politicization of science is critical to an understanding of the competing discourses on the ivory trade ban. In the context of CITES, the idea of neutral scientific management can translate into the threat of or use of a global level prohibition (a trade ban) as a standard rule to regulate the behavior of all parties regardless of local specificities. To understand why such global level regulation is so controversial and politically contested, it is important to examine ways that the global legislative framework has been informed by a broader set of conservation ideologies, which are in turn presented as politically neutral environmental science. In examining the politics of global rule making, or governance, it is clear that such governance is often at odds with local contexts and norms that inform or even determine everyday activities and behaviors of people and policy makers within states that are signatories to CITES.

The ivory ban was instituted in 1989, and has been strongly resisted through campaigns to reopen a legal ivory trade. The disputes over the global trade ban on ivory sales has pitted ivory-producing states in southern Africa and East Asian trading states against a more preservationist international environmental movement and allied East African, South Asian, European, and North American states. These are broad characterizations. The interest groups and actors involved in each "side" of the ivory trade debate run a spectrum of views, and while they may agree on whether the trade should be banned, the reasons for doing so are not necessarily the same. For example, those in favor of a total ban on the ivory trade can take that stance because they are morally and ethically opposed to any such use of elephants, while others have no such moral standpoints. Instead they support a ban on the ivory trade for practical reasons because, in their view, there are no adequate controls in place to ensure a properly regulated legal trade. Now we will explore the differing conservation philosophies of preservation and utilization to illuminate why the disputes about the ivory trade have proved to be so controversial.

The preservationist stance is informed by the precautionary principle. The precautionary principle requires parties to demonstrate that trade is nondetrimental to the survival of the species being traded. If there is any doubt, states are asked to cease all trade until nondetriment findings are proven (Environmental Investigation Agency 1994, 5; Reeve 2002, 27–60). This has led to claims that long-term survival of elephant species can only be secured through a complete ban on the ivory trade. Preservationist interest groups have consistently argued that elephant range states can use elephants to generate funds for communities, private-sector operators, and governments through tourism, especially photographic tourism and ecotourism. In essence, "nonconsumptive use" of elephants means elephants can be used but only in ways that mean the elephant is not killed. According to this conservation philosophy they cannot be sport-hunted for trophies or used for meat, hides, or ivory products. In general, this strategy requires a commitment to protected areas in the traditional sense: national parks and wildlife reserves that are separated from areas used by human populations for agriculture, livestock production, and so on.

A more radical argument, put forward by some animal rights groups and others in a broad preservationist camp, is that elephants should be conserved for their intrinsic value and not because of their use value to humans. Conservation then means the creation of protectionist national parks, and any use at all, even for tourism purposes, is ruled out (for further discussion of "fortress conservation" see Brockington [2002] and Hutton and Dickson [2000]). Some opponents of Zimbabwe's conservation philosophy argue that it is essentially based on a hierarchical model of nature and accords human beings a primary position in that hierarchy, which gives humans dominion over wildlife. This is in conflict with ecological beliefs that environment is a living system of which humans are only one part. Opponents of utilization argue from the standpoint that biodiversity is crucial. Value is calculated in terms of the role each component part (including humans) plays in the ecosystem, rather than calculating it purely in terms of an economic or use value. In accordance with this theory all parts of an ecosystem have intrinsic value, and so all parts have a right to be preserved. Activist and journalist George Monbiot in particular argues that biodiversity matters "because it matters" and not because it has any specific utility, such as providing a cure for cancer or a new crop for human consumption.[1]

Zimbabwe's claim to be committed to using wildlife for human benefit and human welfare was forcefully deployed during debates about the status of the ivory ban during the 1990s; however, it is fundamentally at odds with theories that place humans in a web of nature on an equal footing with wildlife and other environmental resources. In addition, this is related to the belief that wilderness areas are priceless and that with utilization Zimbabwe is wrongly attempting to place a price on them. In effect, critics argue that wildlife has value beyond its economic worth since environmental, aesthetic, and cultural values are just as important. These ideas are extended to include the belief that human interference, such as the management of the environment, upsets the natural balance. Opponents of the utilization approach argue that humans are not given stewardship over nature and do not have a right to interfere with it.

In contrast, the sustainable utilization approach draws on the concept of a safe minimum standard of offtake of resources, and this assumes that the natural resource base will not be depleted faster than it is replaced. Notions of utilization of wildlife carry a strong political and moral message that defines animals as a resource to be used for human benefit. Critics of preservationist approaches suggest that what it means in practice (however unintentional) is that organizations and international treaties tend to treat wildlife as though it existed in a social and political vacuum, and that they employ notions of environmental science that imply that conservation does not affect people because it is purely about saving animals and habitats. Zimbabwe's viewpoint is diametrically opposed to preservationist or animal rights notions of preserving ecosystems and species for their own intrinsic value (see Duffy 2000, 9–21).

Researcher Simon Metcalfe argues that as governments in developing countries have difficulties in managing protected areas without the cooperation of local communities, wildlife policy should address questions of economic efficiency, environmental integrity, and equity and social justice (Metcalfe 1992a, 10). In line with this, the Zimbabwe Parks Department committed itself to a policy philosophy that elephant management is only likely to be successful if elephants can be used profitably and the primary benefits accrue to people with wildlife on their land.[2] Wildlife utilization is deemed to be the most effective means of establishing social and economic forces favorable toward conservation (Ministry of Natural Resources 1990, 9; also see Hulme and Murphree 2001; Metcalfe 1992, 4).

Critics of the preservationist approach argue that it effectively means that the poorest groups in the developing world (especially the rural poor) are expected to forgo the economic opportunities associated with the ivory trade but still live with all the costs of being in close proximity to elephants. For example, elephants in the communal lands of Zimbabwe have raided crops, which constitute the basic food supply of poor families. Those in favor of a utilizationist approach suggest it ensures that elephants will be conserved if they contribute to development and to meeting basic human needs through use of ivory trade, tourism, and sport hunting revenues for community projects. In particular, Zimbabwe's Campfire was one of the first community conservation projects in Africa, and its successes in improving wildlife management and community development meant that the model was picked up, modified, and applied across a number of other African states, including Zambia, Namibia, Kenya, and Mozambique (Hutton, Adams, and Murombedzi 2005).[3] Campfire was intended to alleviate poverty and stress on the environment through use of wildlife. Tolerance of wildlife, and especially of elephants in rural areas, is increased by giving them an economic value through selling sport hunting concessions, photographic safaris, production of meats, hides, and other wildlife products, either for local consumption or for sale.[4] These competing philosophical and ethical standpoints are used to define the status of elephants in relation to human welfare and define appropriate uses and conservation methodologies. These philosophies have played a vital role in outlining the terms of the global debate on the politics and ethics of the ban on the ivory trade. The wildlife trade is a global business that encompasses legal and illegal trade in animals and plants. Trade Records Analysis of Flora and Fauna in International Commerce (TRAFFIC) estimates that the value of the wildlife trade is second only to the drug trade among illegally traded goods; however, accurate figures are impossible to obtain because much of the wildlife trade is illegal and unrecorded. In 2002, Interpol estimated the value of the trade in plants, animals and their derivatives stood at around US$14 billion per annum, with the illegal trade in wildlife products accounting for approximately US$3.5 billion of that total.[5]

The function of CITES is to regulate this global trade. The basic regulatory tools and principles of CITES are set out in a system of appendices, where an Appendix I listing constitutes a trade ban, Appendix II allows controlled and monitored trading, and an Appendix III listing means the species is subject to regulation and requires the cooperation of other CITES signatories to ensure that trade does not lead to overexploitation. CITES does have provisions for species to be transferred between the appendices: for instance, to include, to delete, or to transfer a species requires a two-thirds majority vote of the parties at the biennial conferences. In addition, the convention has provisions for a split listing of a species. This allows certain local populations of a single species to be listed on a different Appendix from the rest of that species' populations (Hutton and Dickson 2000; Wijnstekers 2001, 393–421; Reeve 2002, 27–60). These appendices are the centerpiece of CITES' claims to be engaged in scientifically determined, politically neutral environmental management. Although most signatories to CITES would accept that the debates held at the biennial "Conference of the Parties" (COP) meetings are highly politicized, anti-ivory trade states and interest groups have maintained that CITES Appendix listings are scientifically determined (and therefore politically neutral).

The status of scientific knowledge within global institutions has been enhanced in the case of CITES by its peculiar relationship with international environmental NGOs. These NGOs have proved keen to appeal to notions of pure and uncorrupted science emanating from the nongovernmental sector. For example, their advocacy role has played a key part in defining and determining what gets counted as appropriate and acceptable elephant management (O'Brien et al. 2000, 109–123; Keck and Sikkink 1998). Preservationist NGOs, such as the International Fund for Animal Welfare (IFAW) and Environmental Investigation Agency, have played a central role in arguing for the ivory trade ban as a moral issue and the only ethically acceptable way of saving elephants. For them utilization through the ivory trade is not ethically justifiable, nor is it a practical answer to elephant conservation (see Duffy 2000, 113–140). More broadly, NGOs have retained a unique and powerful position in CITES; for example, they are allowed to take part in the biennial discussions but are not allowed to vote on policy decisions.[6]

For pro-ivory trade interest groups, the CITES appendices are an excellent example of the ways that global environmental institutions and regulations elevate scientific knowledge and the opinions of epistemic (or expert) communities to justify, from the global level to the local level, highly political environmental interventions, such as trade bans. For them it demonstrates that all populations of a single species, especially elephants, are treated equally under a particular Appendix listing, regardless of differences in its local, national, or regional status. The system of appendices has proved the most contentious issue for southern Africa because pro-utilization policy makers in government institutions, local NGOs, and community organizations argue that the basic philosophy that underlies most of the operation of CITES is the precautionary principle and is therefore not a kind of neutral or value free science. Consequently, those in favor of a pro-utilization approach (not incidentally, using their own supporting scientific information) suggest that calls for the ivory ban indicate a philosophically and politically inspired form of science, which is used to define highly controversial global policy decisions that also then become binding at the local level.

The Global Politics of the Ivory Ban

The global ban on the ivory trade has operated as a one-size-fits-all policy instrument. The ivory trade ban has been consistently resisted by southern African nations and in turn their lobbying has been challenged by various actors who support the international regimes and view global governance as beneficial. For pro-ivory trade interest groups, one of the problems with the debates over the ivory ban in the 1980s was that the African elephant was globally defined as a single population that covered most of the continent and was lumped together with Asian elephants that faced very different management challenges. In the post-ban period, Zimbabwe (in particular) has been keen to point out that different populations of African elephants require different management policies.

During the 1990s, Zimbabwe strongly resisted attempts to extend and consolidate global environmental governance over wildlife in Africa, in conjunction with its allies in Japan and Taiwan and the wider southern African region. In 1997, the tenth CITES conference, held in the heated atmosphere of Harare, Zimbabwe, proved to be the arena where all the main interest groups converged to begin the serious debates over reopening the ivory trade. The first proposal, to downlist the African elephant to Appendix II, was the most ambitious and politically contentious, but it was agreed among protrade interest groups that if the downlisting proposal failed, the second choice was to allow Namibia, Botswana, and Zimbabwe only to trade with Japan (as the largest ivory consumer) as long as the four states involved could prove that they had adequate controls to prevent illegal ivory trading. This was the first time that southern Africa had a real chance of overturning the ivory ban since 1989.[7]

Those in favor of reopening the ivory trade argued that the ban had not shut down markets for ivory as effectively as its supporters suggested. Zimbabwean delegates argued that since there was a continuing demand for ivory across the world, Zimbabwe had the right to use its resources to take advantage of such markets (see Dublin and Jachmann 1991, 62–63). They also suggested that moral or ethical standpoints aside, CITES, NGOs, and other interest groups had to accept that since the illicit ivory trade had never stopped completely, even by 1989, those states with stable and well-protected elephant populations should be allowed to make money from a trade in products that was going on regardless of global legislative framework banning it.

A major political bargaining chip at CITES conferences was that during the 1990s Zimbabwe held one of the largest and most secure elephant populations on the continent. According to CITES figures, in 1998 Zimbabwe claimed to have around 67,000 elephants, one of the largest populations in southern Africa.[8] The delegates at CITES meetings have always been aware of the importance of keeping such important elephant range states within the boundaries of the convention. The pro-ivory trade alliance also formed the Southern African Centre for Wildlife Management to lobby for a qualified lifting of the ivory ban (Department of National Parks and Wildlife Management 1992; Southern African Centre for Ivory Marketing 1994;). Consequently, during the Conference of the Parties to

CITES, it was inferred that if CITES members did not take account of southern African interests, then Zimbabwe would break the international trade ban. The then-minister of Environment and Tourism, Chen Chimutengwende, suggested that if CITES did not act in favor of Zimbabwe's interests then it would have made itself irrelevant.[9] That Zimbabwe has such a large elephant population meant none of the parties at the CITES COP could ignore their threats. It was a risky position for Zimbabwe to take because the country faced a possible withdrawal of aid not only for conservation but for other sectors as well and would have been treated as a pariah state for breaking the convention. Most parties to the convention have adhered (at least in formal if not in practical terms) to the stipulations of the convention, even when it runs counter to their domestic interests because signatories tend to fear the consequences of "going it alone." The ways that local communities have claimed rights to use wildlife have also been used to resist the global imposition of preservationist strategies, symbolized by the ivory ban. For example, the presence of Zimbabwean rural community groups at the conference was used to reemphasize that ivory trading was as much about rural development as about revenue generation. Then director of the Campfire Association, Taparendava Maveneke, stated that holding the 1997 CITES conference in Harare was important because it was the first time that rural communities had a chance to come face to face with key policy makers in the field of international wildlife trading.[10]

The ivory issue has also become part of the wider debate about debt and aid. Given that elephant range states are indebted countries and often in need of financial support for their conservation programs, numerous external actors have used offers of aid to influence policy making in range states or at the very least to support programs that intersect with their own conservation philosophy. The pro-ivory trade states and interest groups have linked elephant management to the debt and aid question. In line with this, southern African states put forward a proposal for a debt-for-ivory buyout. Zimbabwe's position has always been that the ivory stockpiles (produced through culling, natural death, problem animal control, and seizures of illegal ivory) constitute an unacceptable waste of a natural and renewable resource. It was suggested that the World Bank Global Environmental Facility could provide funding to environmental organizations to buy ivory stockpiles from African governments.[11] The newly purchased stockpiles could then be burnt to prevent them from entering a legal or illegal international trade. Those in favor of disposing of stockpiles in this way argued that the ivory would then have economic value to the producer states without allowing that ivory onto the international market where it might increase poaching and the illegal trade in ivory. In this way, burning stockpiles would neatly satisfy the demands of those advocating a utilizationist approach and a preservationist stance. However, very few offers of aid for ivory have been made, and those that have, have been very low. In 2002 the Humane Society of the United States offered to buy South Africa's stockpile of ivory and burn it. It offered US$250,000 for the 30–ton stockpile; the Humane Society argued that although the amount offered was low, South Africa would benefit from a significant increase in aid and donations to encourage "nonconsumptive use" options for elephants, such as photographic tourism and funding for protected areas. The South African National Parks Department described the offer as a joke, and pointed out that the ivory would be worth US$5 million on the open market.[12] In addition, CITES documents on ivory stockpiles have pointed to the failure of donors to fund elephant conservation plans and indicate that African elephant ranges states have also been unable to demonstrate they have adequate controls over the stocks.[13]

There were accusations at CITES in 1997 that environmental NGOs and preservationist western states were frustrating developing countries by imposing a new form of imperialism through global environmental governance. Such governance, critics argued, implied that African states, organizations, and communities should not be in control of environmental decision making but that global institutions should be. This dispute essentially grew from competing definitions of wildlife, as a national resource to be used for domestic benefits, versus the view of western states and global environmental NGOs that elephants were a world resource to be protected at all costs for the global good. This again raised ethical questions about who had the right to decide how wildlife should be conserved: should it be local communities eager to enter the international ivory trade, international NGOs keen to maintain a total trade ban, or global institutions invested with legislative powers that supersede national legislation of member states?

Many anti-ivory trade interest groups also used the debt and aid question to try to persuade pro-trading states into withdrawing their proposal for a partial lifting of the ivory ban, by suggesting further aid would be made available if they agreed to and implemented a global ivory ban. Some observers at the CITES conference believed that members of the pro-ban alliance had paid some African countries to vote against southern Africa, thereby exploiting the historical split between east and southern Africa on the ivory trade. It does not matter whether these rumors were true or verifiable; however, some delegates believed these rumors, which is important and interesting. Delegates' perceptions often influence or even determine their behavior, arguments, and positions at CITES meetings. The local press reported that at the time of the conference pro-ivory-ban interest groups, including animal rights lobbyists and many western states opposed to the trade, had threatened to withdraw development aid from states that supported pro-ivory-trading states, including Zimbabwe, Namibia, and

Botswana. It was also suggested that western delegates at CITES had stated they would only vote if there was a common African policy on the ivory trade.[14] In this way, the debt relief question was captured by animal rights NGOs and other anti-trade interest groups to try to influence policy making by elephant range states.

Finally, challenges posed to CITES regulations by the arguments put forward by southern Africa were successful, and at the 1997 meeting, Zimbabwe, Namibia, and Botswana had their elephant populations downlisted to Appendix II (Milliken 2002, 2). This decision came with the proviso that trading was only to be allowed in 1999 if adequate measures were introduced to ensure that trading was sustainable and free from illegally hunted ivory.[15] The decision meant that in 1999 Namibia, Zimbabwe, and Botswana reopened a restricted trade with Japan. The 1997 decisions paved the way for further relaxation of the ivory ban at the CITES conference in Chile in November 2002 that allowed Zimbabwe, Namibia, Botswana, and South Africa to begin planning sales of ivory stockpiles for 2004.[16] However, a one-time ivory sale to Japan was only to be allowed in mid-2004 if range states could prove there were sufficient controls in place to ensure that no illegally hunted ivory entered the system. Although South Africa, Namibia, and Botswana seemed likely to fulfill these criteria, Zimbabwe did not. Nevertheless, the decision was immediately met with the concern that this would provide a signal to poachers to restart the levels of commercial poaching witnessed in east Africa in the 1980s.[17]

Critics of the sustainable use arguments presented by southern African states have pointed out that local decisions can have global ramifications; for example, once ivory is in the global trading system it is very difficult to determine if it originated from a legal or illegal source. Following on from these arguments about the inability to control the illicit ivory trade, IFAW openly opposed Zimbabwe's position on the ivory trade. IFAW was part of the Species Survival Network, a coalition of fifty NGOs that included high-profile global NGOs such as Greenpeace. Species Survival Network was formed in 1992 and dedicated to strict enforcement of CITES regulations. IFAW and its allies argued that Zimbabwe and other southern African states would not be capable of preventing illegal exports if CITES allowed a partial or total lifting of the ivory ban.[18] The Species Survival Network took an approach that drew on the precautionary principle, arguing that a lack of information supporting the case for a renewed legal ivory trade meant that it would be premature to allow southern Africa to trade again and that any such trade would be harmful to wild-elephant populations in Africa and Asia.[19]

Those opposed to reopening the ivory trade pointed out that any discussion about reopening a legal ivory trade signaled poachers and traders to restart or increase their illicit activities.[20] For example, Malawian wildlife authorities indicated that they were worried that the sale of a 30–ton stock-pile by Botswana, Namibia, and South Africa, planned for 2004, would lead to an increase in poaching that was already evident within the country; they argued that their elephant populations were not as stable and not as well protected from poachers as those in southern Africa and therefore would be easy targets.[21] Furthermore, debate about reopening a legal ivory trade based on the relatively stable and well-protected populations of southern Africa sparked fears that such a trade would affect less stable and less se-cure elephant populations in Asia (for further discussion, see Martin and Stiles 2002; Milliken 2002, 3–5; Nash 1997). It is estimated that Asia has only 50,000 elephants left in the wild, and yet ivory pieces were found on sale across the region in 2002 though only Thailand had imported African ivory on sale.[22] In addition, anti-ivory-trade interests groups have argued that the ongoing and growing trade in illegal ivory proves that range states and consumer states would not have the capacities to ensure that illegally produced ivory did not enter the legally approved ivory trade. IFAW pointed to the discovery in 2003 of a consignment of sixty-five tusks at Bangkok airport, thirty-three tusks in Kenya, and 6 tons of ivory in Singapore, as ev-idence that the decision to allow a one-time sale of southern African ivory stockpiles had led directly to an increase in illegal hunting and trading of ivory.[23]

In addition, TRAFFIC indicated that Nigeria, Senegal and Ivory Coast were heavily involved in the illegal ivory trade in 2002–2003; one of the ma-jor sources of the illegal ivory was Democratic Republic of Congo,[24] where civil war broke out in 1997, and which, despite numerous peace agree-ments, remains turbulent and the site of a number of active rebel groups and of numerous interventions by neighboring states, including Uganda, Rwanda, Zimbabwe, and Angola. Uganda in particular was highlighted as a significant area for trafficking illegally hunted ivory from the Democratic Republic of Congo to Asia.[25] Clearly the conditions of warfare and conse-quent lack of control over wildlife mean that it presents a significant prob-lem for external forms of governance, such as CITES stipulations to prevent illegally produced ivory from entering international trade.

Wildlife conservation policy is not immune to local or national changes and turbulence. Proponents of the ban and even some of those in favor of sustainable use have accepted that any discussion of Zimbabwean elephant policy has to also take account of the fast-changing political environment in the country. The increasing use of violence by the government against op-position party members and their supporters has raised significant questions about the commitment of the Zimbabwean state to its own country's de-velopment. This has had an effect on environmental politics in Zimbabwe, with even the staunchest supporters of the utilization approach now ques-tioning whether the current state apparatus has the will or capacity to govern a legal ivory trade.[26] Critics of the Zimbabwe Parks Department's capacity to manage a legal ivory trade have pointed out that since the department had been systematically starved of funds by the central government, it needed every source of revenue available, including the sale of ivory, and that its lack of management capability made it unlikely that the money would be used effectively for elephant conservation.[27] Indeed, there were reports that amid the growing political chaos in Zimbabwe, the government had in fact illegally sold 8 tons of ivory to China to pay for Kalashnikov rifles.[28] Opponents of the ivory trade have argued that it is impossible to control the trade because of the levels of official complicity in poaching and trafficking elephant products. Elements within the Kenya Wildlife Service, the Zimbabwe Parks Department, and other wildlife authorities, government officials, and militaries across Africa have been accused of complicities in the illegal wildlife trade (see Ellis 1994; Duffy 1999).

In addition, from 2000, agricultural land in Zimbabwe was not the only target for invasion by the "war veterans." Protected areas, private game ranches, and privately owned conservancies were also caught up in the turbulent political conflict.[29] The Save Valley Conservancy was one of the worst affected. Its wildlife fences were cut and war veterans started fires to catch wildlife and set snares to trap game. An elephant refuge, this was also an important site for conservation of the rare black rhinoceros, *Diceros bicornis*. Many of these animals were left unguarded and exposed to poachers once the war veterans invaded the conservancy.[30] Given the current political situation in Zimbabwe, many were surprised, even greatly concerned that the CITES 2007 COP authorized a one-off sale of ivory stockpiles by that country (as well as by neighboring Botswana, South Africa, and Namibia). This recent decision underlines the importance of a discussion of the Zimbabwean standpoint, a discussion which also illuminates some broader debates about utilizationist versus preservationist conservation philosophies, also particularly relevant in the cases of Botswana, South Africa, and Namibia.

The case of the global trade ban on ivory represents the complex concerns that surround the creation and enforcement of global norms. Whereas environmental management requires global cooperation because of the transboundary nature of environmental problems, a focus on global-level concerns and ideas of good conservation practice means that other national or local ideas and practices are defined as defiant and resistant to what preservationist NGOs and interest groups associated with CITES present globally as "best practice." The (now changing) commitment to elephant conservation through the ivory ban in CITES is an example of how global norms can face significant challenges from various interest groups.

In particular, southern Africa's policy commitment to sustainable use as the only ethically acceptable form of elephant management in a developing country is legitimated as good conservation practice that satisfies a broader commitment to meeting the basic needs of human populations. The justification is based on the idea that conservation has to compete in social, economic, cultural, and environmental terms with other forms of development. As such, the southern African states have argued that wildlife constitutes a national resource, and because local communities are expected to live with all the costs of being in close proximity to wildlife, states and local communities have the right to use them for their own benefit. This stands in contravention to definitions of good conservation priorities under global environmental governance that view good conservation practice as the preserve of the global community, essentially that elephants constitute an important global resource and that their management cannot be left to national governments and local communities. There is no easy way out of this complex ethical quandary.

Notes

1. George Monbiot, "Natural Aesthetes," *Guardian* (UK), January 13, 2004.

2. Interview with Willie Nduku, Director, DNPWLM (Department of National Parks and Wildlife Management), July 7, 1995, Harare; interview with Rowan Martin, Assistant Director, Research, DNPWLM, May 29, 1995, Harare.

3. Interview with Keith Madders, Director Zimbabwe Trust (UK), October 12, 1994, Epsom, Surrey; interview with Brian Child, head of the Campfire Co-ordination Unit, DNPWLM, May 16, 1995, Harare; interview with Rob Cunliffe, Environmental Consultant, Harare, March 2, 2000; also see Hulme and Murphree (2001).

4. Interview with Professor Marshall Murphree, CASS (Centre for Applied Social Studies), University of Zimbabwe, July 18, 1995, Harare; interview with David Cumming, Director, WWF-Zimbabwe, February 22, 1995, Harare; interview with Stephen Kasere, Deputy Director, Campfire Association, February 14, 1995, Harare.

5. www.wwf.org.uk/wildlifetrade/law (accessed May 3, 2002); also see www.traffic.org; www.cites .org; www.wcmc.org.uk/CITES (accessed May 3, 2002) for further discussion.

6. www.cites.org (accessed May 3, 2002); also see WWF, undated; Princen and Finger (1994, 135–138).

7. See www.traffic.org; www.cites.org; www.wcmc .org.uk/CITES for further discussion (all accessed May 3, 2002).

8. Prop. 12.10, consideration of proposals for amendment of Appendices I and IIa, found at www.cites.org/eng/cop/11/prop/23.pdf (accessed May 3, 2004).

9. "CITES or No CITES, We'll Go It Alone," *Herald* (Zimbabwe), June 16, 1997; "Zimbabwe Backs Off Threat to Break Ivory Ban," *Pretoria News* (South Africa), June 17, 1997; "Harare Resists the Ivory Sale Ban," *Guardian* (UK), June 17, 1997; and "Ivory Trade States to Try again After Near Miss," *Guardian* (UK), June 18, 1997.

10. "Maveneke Rejects Debt Relief on Ivory Trade," *Herald* (Zimbabwe) June 13, 1997.

11. Interview with John Gripper, Sebakwe Black Rhino Trust, October 14, 1994, Ascott-Under-Wychwood, UK; "CITES Proposal to Buy Some of Africa's Ivory Stocks," *Pretoria News*, June 17, 1997; and "Maveneke Rejects Debt Relief on Ivory," *Herald* (Zimbabwe), June 13, 1997.

12. "Money to Burn," *Mail and Guardian* (South Africa), November 1, 2002.

13. CITES (2002), *Elephants: Conditions for the Disposal of Ivory Stocks and Generating Resources for Conservation in African Elephant Range States* www.cites.org/eng/decis/valid12/10–02.shtml (accessed February 13, 2004).

14. African States Divided on the Ivory Issue, *Herald* (Zimbabwe), June 12, 1997; and "Maveneke Rejects Debt Relief on Ivory," *Herald* (Zimbabwe), June 13, 1997.

15. Milliken, T. (1999), *African elephants and the 11th Meeting of the Conference of the Parties to CITES.* www.traffic.org/briefings/elephants-11th meeting.html (accessed May 3, 2002).

16. www.cites.org (accessed December 3, 2002); and TRAFFIC Recommendations to CITES at COP12, Chile 2002, www.traffic.org/cop12/proposal3_14.html#pro6 (accessed May 13, 2004).

17. Species Survival Network (2002), *CITES 2002:*

African Elephant, www.speciessurvivalnetwork.org (accessed May 13, 2004); "Wildlife Officials Brace for 2004 Ivory Sale," *Mail and Guardian* (South Africa), November 12, 2003; and "All Clear for Ivory Trade," *Mail and Guardian*, June 27, 1997; "Ivory Vote Sparks New Fears for Elephants," *Guardian* (UK), November 13, 2002.

18. "Reopening of Trade Slated as Danger to the Big Three," *Star* (South Africa), June 11, 1997; *Star*, "Massive Illegal Wildlife Trade Revealed," June 13, 1997; and "Downlisting of Jumbo: IFAW Still Opposed," *Herald* (Zimbabwe), June 11, 1997.

19. Species Survival Network (2002), *CITES 2002: African Elephant*, www.speciessurvivalnetwork.org (accessed February 13, 2004); Milliken (2002, 6–8); also see "Illegal Ivory Deal Buys Kalashnikovs for Mugabe," *Sunday Times* (UK) July 9, 2000; also see "Wildlife Officials Brace for 2004 Ivory Sale," *Mail and Guardian* (South Africa), November 12, 2003; and "Massive Thai Ivory Seizure Signals Danger for World's Elephants," *Mail and Guardian*, August 1, 2003.

20. "Slaughter of Elephants Starts Again; and Species Survival Network, 2002," *Observer* (UK), April 28, 2002; *CITES 2002: African Elephant*, www.speciessurvivalnetwork.org (accessed February 13, 2004).

21. "Wildlife Officials Brace for 2004 Ivory Sale," *Mail and Guardian* (South Africa), November 12, 2003; also see "Notorious Ivory Smuggler Beats Malawi Trap," *Guardian* (UK), December 27, 2002.

22. "Tusk Force: Illegal Asian Ivory Trade is Flourishing," *Guardian* (UK), February 27, 2002; "Massive Thai Ivory Seizure Signals Danger for World's Elephants" *Mail and Guardian* (South Africa), August 11, 2003; see also Milliken (2002, 1).

23. "Massive Thai Ivory Seizure Signals Danger for World's Elephants," *Mail and Guardian* (South Africa), August 11, 2003; "Tusks of 600 Elephants Make up Record Haul of Smuggled Ivory," *Guardian* (UK), July 13, 2002.

24. "West Africa Fuels Illegal Ivory Trade," *Guardian* (UK), December 15, 2003; "Slaughter of Elephants Starts Again," *Observer* (UK) April 28, 2002.

25. "Ivory Trafficking Booms in East Africa," *Mail and Guardian* (South Africa), August 29, 2003; and see "Ethiopia Returns Seized Jumbo Tusks to Kenya," *Mail and Guardian*, October 10, 2003.

26. Anonymous interviewee; also see "SA Cashes in on Zim Confusion," *Mail and Guardian* (South Africa), May 2, 2003; and Species Survival Network (2002), *CITES 2002: African Elephant*, www.speciessurvivalnetwork.org (accessed February 13, 2004).

27. "Zim's Illicit Ivory Trade Exposed," *Mail and Guardian* (South Africa), December 6, 1996; "Zim Parks in Trouble," *Mail and Guardian*, June 31, 1997; and Species Survival Network (2002), *CITES 2002: African Elephant*, www.speciessurvivalnetwork.org (accessed 13.02.04); and Environmental Investigation Agency, *November 2002: A Second One-Off Sale Agreed*, www.eia-international.org/cgi/news/news.cgi?a=140&t=template.htm (accessed February 13, 2004).

28. "Illegal Ivory Deal Buys Kalashnikovs for Mugabe," *Sunday Times* (UK), July 9, 2000; also see "Wildlife Officials Brace for 2004 Ivory Sale," *Mail and Guardian* (South Africa), November 12, 2003; and "Massive Thai Ivory Seizure Signals Danger for World's Elephants," *Mail and Guardian*, August 11, 2003.

29. "Ecological Crisis Looms in Zim," *Mail and Guardian* (South Africa) August 25, 2000; also see "No End to the Land-Grab in Zimbabwe; Zimbabwe Conservation Task Force, Zimbabwe's Tragedy," *Mail and Guardian*, April 12, 2002. Zimbabwe Conservation Task Force, Zimbabwe's Tragedy, www.zctf.mweb.co.zw/page6.html (accessed January 14, 2004).

30. "Ecological Crisis Looms in Zim," *Mail and Guardian* (South Africa) August 25, 2000; "Invasions Threaten Peace Park," *Mail and Guardian*, October 26, 2001; "A Devastating Year for Zimbabwe"; *Mail and Guardian*, December 22, 2000; "No End to the Land-Grab in Zimbabwe," *Mail and Guardian*, April 12, 2002; ; "Zim Land Reform Decimates Game," *Mail and Guardian*, August 16, 2002; "Zim Wildlife Pillage Continues," *Mail and Guardian*, August 13, 2003; Interview with Clive Stockil, Chair of the Save Valley Conservancy, May 13, 1996, Chiredzi.

References

Brockington, D. 2002. *Fortress conservation: The preservation of the Mkomazi Game Reserve, Tanzania*. Oxford: James Currey.

Commission on Global Governance. 1995. *Our global neighbourhood*. Oxford: Oxford University Press.

Department of National Parks and Wildlife Management. 1991. *Protected species of animals and plants in Zimbabwe*. Harare: Department of National Parks and Wildlife Management.

Department of National Parks and Wildlife Management. 1992. *Research plan: DNPWLM, Zimbabwe*. Harare: Branch of Aquatic Ecology and Branch of Terrestrial Ecology, Research Division, Department of National Parks and Wildlife Management.

Dublin, H. T., and Jachmann, H. 1991. *The impact of the ivory ban on illegal hunting of elephants in six range states in Africa.* WWF International Research Report. WWF Project No. 4578. Gland, Switzerland: World Wide Fund for Nature.

Duffy, R. 1999. The role and limitations of state coercion: Anti-poaching policies in Zimbabwe. *Journal of Contemporary African Studies* 17: 97–121.

Duffy, R. 2000. *Killing for conservation: Wildlife policy in Zimbabwe.* Oxford: James Currey.

Ellis, S. J. 1994. Of elephants and men: Politics and nature conservation in South Africa. *Journal of Southern African Studies* 20: 53–69.

Environmental Investigation Agency. 1992. *Under fire: Elephants in the front line.* London: Environmental Investigation Agency.

Environmental Investigation Agency. 1994. *CITES: Enforcement not extinction.* London: Environmental Investigation Agency.

Hewson, M., and Sinclair, T. J. 1999. The emergence of global governance theory. In M. Hewson and T. J. Sinclair (eds.), *Approaches to global governance theory* (pp. 3–22). Albany: State University of New York Press.

Hulme, D., and Murphree, M. (eds.). 2001. *African wildlife and livelihoods: The promise and performance of community conservation.* Oxford: James Currey.

Hutton, J., Adams, W. A., and Murombedzi, J. C. 2005. Back to the barriers? Changing narratives in biodiversity conservation. *Forum for Development Studies* 2: 341–370.

Hutton, J., and Dickson, B. (eds.). 2000. *Endangered species, threatened convention: The past, present and future of CITES.* London: Earthscan.

Keck, M. E., and Sikkink, K. 1998. *Activists beyond borders: Advocacy networks in international politics.* Ithaca, NY: Cornell University Press.

Leach, M., and Mearns, R. (eds.). 1996. *The lie of the land: Challenging received wisdom on the African environment.* Oxford: James Currey.

Litfin, K. 1994. *Ozone discourses: Science and politics in global environmental cooperation.* New York: Columbia University Press.

Martin, E., and Stiles, D. 2002. *The South and South East Asian ivory markets.* London: Save the Elephants.

Metcalfe, S. 1992. *Community natural resource management: How non-governmental organisations can support co-management conservation and development strategies between government and the public.* Harare: Zimbabwe Trust.

Milliken, T. 2002. *The world's unregulated domestic ivory markets.* Cambridge: TRAFFIC International.

Ministry of Natural Resources. 1990. *National conservation strategy.* Harare: Ministry of Natural Resources.

Nash, S. V. (ed.). 1997. *Still in business: The ivory trade in Asia seven years after the CITES ban.* Cambridge, UK: TRAFFIC International.

Neumann, R. P. 2000. Primitive ideas: Protected buffer zones and the politics of land in Africa. In V. Broch-Due and R. A. Schroeder (eds.), *Producing nature and poverty in Africa* (pp. 220–242). Uppsala: Nordiska Afrikainstitutet.

O'Brien, R., Goetz, A. M., Scholte, J. A., and Williams, M. 2000. *Contesting global governance: Multilateral economic institutions and global social movements.* Cambridge: Cambridge University Press.

Princen, T., and Finger, M. 1994. *Environmental NGOs in world politics: Linking the global and the local.* London: Routledge.

Reeve, R. 2002. *Policing the international trade in endangered species: The CITES treaty and compliance.* London: Royal Institute of International Affairs/Earthscan.

Sivaramakrishnan, K. 1999. *Modern forests: Statemaking and environmental change in colonial Eastern India.* Oxford: Oxford University Press.

Southern African Centre for Ivory Marketing. 1994. The SACIM treaty: Its aims and objectives: A position paper presented by Botswana, Malawi, Namibia, and Zimbabwe. Paper presented at the conference on the African Elephant in the Context of CITES, in Kasane, Botswana, September 19–23, 1994 (UK Department of Environment), Part B.

Wjinstekers, W. 2001. *The evolution of CITES.* 6th edition. Geneva: CITES Secretariat.

WWF. Undated. CITES: What is it? Godalming, Surrey, UK: WWF-UK.

Wynne, B. 1992. Uncertainty in environmental learning: Reconceiving science and policy in the preventive paradigm. *Global Environmental Change* 2: 111–127.

Zimbabwe Trust. 1992. *Wildlife: Relic of the past or resource of the future? The realities of Zimbabwe's wildlife policymaking and management.* Harare: Zimbabwe Trust.

CONTRIBUTORS

Lori Alward, PhD
Tacoma Community House
Tacoma, Washington, USA

Joseph C. E. Barber, PhD
Disney's Animal Kingdom
Orlando, Florida, USA

Janine L. Brown, PhD
Smithsonian's National
Zoological Park
Conservation and Research
Center
Front Royal, Virginia, USA

Jacob V. Cheeran, DVM
Project Elephant
Government of India
Thrissur, India

Catherine A. Christen, PhD
Smithsonian's National
Zoological Park Conserva-
tion and Research Center
Front Royal, Virginia, USA

Rosaleen Duffy, PhD
Centre for International
Politics
University of Manchester
Manchester, United Kingdom

Yudha Fahrimal, DVM
Universitas Syiah Kuala
Darussalam, Banda Aceh
Sumatra, Indonesia

Richard Fayrer-Hosken,
BVSc, PhD
Department of Large Animal
Medicine
College of Veterinary
Medicine
University of Georgia
Athens, Georgia, USA

Marie Galloway
Smithsonian's National
Zoological Park
Washington, DC, USA

Marion E. Garaï, PhD
Space for Elephants
Foundation
KwaZulu Natal, South Africa

Jane Garrison
Los Angeles, California, USA

Hank Hammatt
Elephant Care International
Hohenwald, Tennessee, USA

David Hancocks
Melbourne, Australia

Rebecca Hardin, PhD
Department of Anthropol-
ogy and School of Natural
Resources and Environment
University of Michigan
Ann Arbor, Michigan, USA
Harvard Academy for Inter-
national and Area Studies
Cambridge, Massachusetts,
USA

Michael Hutchins, PhD
The Wildlife Society
Bethesda, MD, USA
Graduate Program in
Conservation Biology and
Sustainable Development
University of Maryland
College Park
Maryland, USA
Center for Conservation and
Behavior
Georgia Institute of
Technology
Atlanta, Georgia, USA

Mike Keele
Oregon Zoo
Portland, Oregon, USA

Winnie Kiiru, MSc
Born Free Foundation
Nairobi, Kenya

Michael D. Kreger, PhD
U.S. Fish and Wildlife Service
Arlington, Virginia, USA

Fred Kurt, PhD
Research Institute of
Wildlife Ecology
University of Veterinary
Medicine
Vienna, Austria

Dhriti K. Lahiri Choudhury,
PhD
Department of English
Rabindra Bharati University
Calcutta, India

John Lehnhardt
Disney's Animal Kingdom
Orlando, Florida, USA

Khyne U Mar, PhD, FRCVS
Elephant Family-UK
London, United Kingdom

Jill D. Mellen, PhD
Disney's Animal Kingdom
Orlando, Florida, USA

Susan K. Mikota, DVM
Elephant Care International
Hohenwald, Tennessee, USA

Gary W. Miller
Disney's Animal Kingdom
Orlando, Florida, USA

Cynthia J. Moss, ScD
Amboseli Elephant Research
Project
Amboseli Trust for Elephants
Nairobi, Kenya

Joyce H. Poole, PhD
Amboseli Elephant Research
Project
Amboseli Trust for Elephants
Nairobi, Kenya

Greg D. Rossel
Motorola Corporation
Fort Worth, Texas, USA

Nigel Rothfels, PhD
College of Letters and Science
University of Wisconsin–
Milwaukee
Milwaukee, Wisconsin, USA

Dennis Schmitt, DVM, PhD
Ringling Bros. and Barnum
& Bailey
Department of Veterinary
Services
Polk City, Florida, USA
Department of Agriculture
Missouri State University
Springfield, Missouri, USA

John Seidensticker, PhD
Smithsonian's National
Zoological Park
Washington, DC, USA

Lalith Seneviratne
Colombo, Sri Lanka

Brandie Smith, MS
Department of Conservation
and Science
Association of Zoos and
Aquariums
Silver Spring, Maryland,
USA

Raman Sukumar, PhD
Centre for Ecological
Sciences
Indian Institute of Science
Bangalore, India

Gary Varner, PhD
Philosophy Department
Texas A&M University
College Station, Texas, USA

Christen Wemmer, PhD
Smithsonian's National
Zoological Park
Conservation and Research
Center
Front Royal, Virginia, USA
California Academy of
Sciences
San Francisco, California,
USA

Ian Whyte, PhD
South African National Parks
Kruger National Park, South
Africa

Nadja Wielebnowski, PhD
Chicago Zoological Society
Brookfield Zoo
Brookfield, Illinois, USA

INDEX

Page numbers in italics indicate photographs.

471